Mangrove Ecosystems of Asia

I. Faridah-Hanum • A. Latiff
Khalid Rehman Hakeem • Munir Ozturk
Editors

Mangrove Ecosystems of Asia

Status, Challenges and Management Strategies

Springer

Editors
I. Faridah-Hanum
Universiti Putra Malaysia
Selangor
Malaysia

A. Latiff
Fakulti Sains and Teknologi
Universiti Kebangsaan Malaysia
Banji
Malaysia

Khalid Rehman Hakeem
Universiti Putra Malaysia
Selangor
Malaysia

Munir Ozturk
Universiti Putra Malaysia
Selangor
Malaysia
and
Ege University
Izmer
Turkey

ISBN 978-1-4614-8581-0 ISBN 978-1-4614-8582-7 (eBook)
DOI 10.1007/978-1-4614-8582-7
Springer New York Heidelberg Dordrecht London

Library of Congress Control Number: 2013950763

Printed on acid-free paper

Springer is part of Springer Science+Business Media (www.springer.com)

Foreword

The present book is a welcome addition to the existing information available on these unique halophytes called mangroves. According to scientists they are botanical amphibians of seashore occupying a zone of desiccating heat, choking mud and salt levels lethal for ordinary plants. Thought to have originated in Southeast Asia, mangroves are found around the world but mostly within 30 degrees of the equator. A few have however adapted to temperate climates and one has even been reported from New Zealand, an account of which appears in this volume as well. Mangroves have ultra-filtration system to keep salts out and elaborate roots for respiration in water logged substrates. Many mangroves can be recognized by their prop roots that make them appear to be standing on stilts above water. Mangroves may have the highest net productivity of carbon of any natural ecosystem.

Mangroves serve as a buffer between land and sea and hence are protection against erosion and may even reduce the hazards of tsunamis and cyclones. They harbor a variety of life forms like fish, amphibians, reptiles, and birds are a good source of timber, fuel, fodder, tannins, honey, etc. Despite their importance, mangroves are constantly under threat worldwide. They are being sacrificed for civic facilities like house building, roads and hotels and are often additionally subjected to pollution and disruption of their sensitive water and salt balance. This book has highlighted some of these problems and suggested ways and means to deal with them. The main focus is on Indonesia, Malaysia, Philippines, Thailand, Timor-Leste, Iran, China, New Zealand and Bangladesh where large populations of mangroves are declining and the degraded condition of what remains, needs attention. The book contains 20 chapters covering topics like present status and distribution of mangroves, threats to the ecosystem due to deforestation and over exploitation of resources, effect of climate change and related adaptations, management options and challenges to reaching the goal. The authors and the editors of this book have done a commendable job in covering the diverse aspects of mangrove ecology and restoration.

M. Ajmal Khan
Qatar Shell Professorial Chair for Sustainable Development
Department of International Affairs
College of Arts and Sciences
Qatar University, 2713
Doha, Qatar

Preface

Over the past decades, we have witnessed an impressive socio-economic development in the Asia-Pacific region particularly to address poverty eradication and to provide a better livelihood for the rural population including coastal communities. For thousands of years, the coastal, estuarine and riverine communities in the Asia-Pacific region have been depended on the coastal resources including those of the mangrove swamp ecosystem. It has been estimated that the total mangrove area of the world was 137,760 km^2 in 2000, and the region has the largest mangrove covering areas in the world. The region has been endowed with the most mangrove swamp forests and ecosystems in the world and well over 54 % of mangroves are still present in the region. The region has also experienced recent economic growth and expects continuing development in the coming century. In all the Asia-Pacific countries the mangrove ecosystems have been subjected to various forms of natural and man-made threats.

In the last decade or so land-use changes and conversion have occurred throughout the region. As much of the lowland and hill forests have been subject to exploitation via logging to feed wood-based industries and to build new urban infrastructure, the mangrove swamp forests also were converted to agriculture, aquaculture, resort facilities and other infrastructures. From the earlier and fundamental ecological and biogeographical approaches to studying mangrove ecology, biologists have adopted more sophisticated field and laboratory techniques to understand the ecology of the mangroves. Today the approaches have moved to GIS and molecular biology to discern mangrove evolution and adaptation. Lately the use of morphology and anatomy combined with molecular studies was employed to construct mangrove phylogenies to understand the complex ecosystem with respect to the spatial distribution of the species in the region.

It is with great pleasure that we present this book entitled *Mangrove Ecosystems of Asia: Status, Challenges and Management Strategies* to address the extent, status, present strategies and future challenges in managing and conserving the ecosystem. This volume was inspired by the peril to which mangrove ecosystems have been exposed and addresses the issue of mangrove destruction in a scientific manner. The present volume brings together a series of active researchers and thinkers in mangrove ecology and biology from several countries (India, Bangladesh, China,

Japan, the Philippines, Timor-Leste, Indonesia, Australia, New Zealand, Thailand and Malaysia) to achieve a new synthesis of mangrove ecological issues. The focus was not simply to present the past results of research and surveys on mangrove ecology; the authors were challenged to bring new ideas on conservation strategies for the future management of the constantly depleting rich and valuable resources of the Asia-Pacific region.

The first part of the book reviews mangrove ecology and covers the current status, challenges and management in countries such as Timor-Leste, New Zealand, Bangladesh, the Philippines, Thailand, Iran, Indonesia and Malaysia. The second part reviews some specific issues and challenges such as economic sustainability, the relationship between mangrove deforestation, recent developments in GIS and remote sensing application and economic development, organic carbon storage and turn-over in the mangrove ecosystem, the effects of climate change on mangrove communities and options for managing the adaptation. The third part reviews some management strategies for sustainable exploitation of aquatic resources, economic sustainability of halophytes, and research development for sustainable management.

In the final analysis, further discussion and research regarding the mangroves of the Asia-Pacific region is suggested.

This is our opportunity to thank the authors who have given their time unselfishly to meet the deadlines for each chapter. We hope that the readers especially the students will benefit from reading this overview of mangroves as much as we have benefited from reading, evaluating and editing the chapters.

I. Faridah-Hanum
A. Latiff
Khalid Rehman Hakeem
Munir Ozturk

Contents

Contributors

Daniel M. Alongi Australian Institute of Marine Science, PMB 3 Townsville MC, 4810 Queensland, Australia

Mohti Azian Forestry and Environment Division, Forest Research Institute Malaysia (FRIM), 52109 Kepong, Selangor Darul Ehsan, Malaysia

Paul Bierman-Lytle Yale University USGBC LEED Technical Advisor, Pangaeon Global Denver, Colorado, 80237 Denver, United States of America

Sourav Chakma Forest Department, Ban Bhaban, Boyra, Khulna, Bangladesh

Joanna C. Ellison School of Geography and Environmental Studies, University of Tasmania, Locked Bag 1376, Launceston, Tasmania 7250, Australia

Kristine B. Garcia World Agroforestry Centre-Philippines, International Rice Research Institute, 2nd Fl., Khush Hall Bldg., Los Baños 4031, Philippines

Dixon T. Gevaña College of Forestry and Natural Resources, University of the Philippines Los Baños, Kanluran Rd, Los Baños 4031, Philippines

Rebecca V. Gladstone-Gallagher Department of Biological Sciences, University of Waikato, Private Bag 3105, Hamilton, New Zealand

Brij Gopal Centre for InlandWaters in South Asia, 41-B Shiv Shakti Nagar, Jagatpura Road, Rajasthan, 302017 Jaipur, India

Omar Hamdan Forestry and Environment Division, Forest Research Institute Malaysia (FRIM), 52109 Kepong, Selangor Darul Ehsan, Malaysia

I. Faridah-Hanum Faculty of Forestry, Universiti Putra Malaysia, 43400 Serdang, Selangor, Malaysia

M. Enamul Hoq Bangladesh Fisheries Research Institute, Mymensingh, Mymensingh 2201, Bangladesh

Saiful Islam Forest Department, Ban Bhaban, Boyra, Khulna, Bangladesh

Bhavanath Jha Discipline of Marine Biotechnology and Ecology, CSIR- Central Salt and Marine Chemicals Research Institute (CSIR-CSMCRI), Gijubhai Badheka Marg, 364002 Bhavnagar, Gujarat, India

Pastor L. Malabrigo, Jr. College of Forestry and Natural Resources, University of the Philippines Los Baños, Kanluran Rd, Los Baños 4031, Philippines

W. A. Wan Juliana Faculty of Science and Technology, Universiti Kebangsaan Malaysia, 43600 Banji, Selangor, Malaysia

Cecep Kusmana Department of Silviculture, Faculty of Forestry, Bogor Agricultural University, Bogor, Indonesia

A. Latiff Faculty of Science and Technology, Universiti Kebangsaan Malaysia, 43600 Banji, Selangor, Malaysia

Hongxiao Liu South China Botanical Garden, Chinese Academy of Sciences, 510650 Guangzhou, P. R China

Graduate University of Chinese Academy of Sciences, 100039 Beijing, P. R China

Kexin Liu State Key Laboratory of Nuclear Physics and Technology, Peking University, 100871 Beijing, P. R China

Carolyn J. Lundquist National Institute of Water & Atmospheric Research Ltd., PO Box 11115, Hamilton, New Zealand

Institute of Marine Science, University of Auckland, PO Box 349,Warkworth, New Zealand

Donald J. Morrisey National Institute of Water & Atmospheric Research Ltd., PO Box 893, Nelson, New Zealand

Toyokazu Naito Department of Human and Cultural Studies, Kyoto Gakuen University, 621-8555 Kameoka, Kyoto, Japan

Asish Kumar Parida Discipline of Marine Biotechnology and Ecology, CSIR-Central Salt and Marine Chemicals Research Institute (CSIR-CSMCRI), Gijubhai Badheka Marg, 364002 Bhavnagar, Gujarat, India

Ismail Parlan Forestry and Environment Division, Forest Research Institute Malaysia (FRIM), 52109 Kepong, Selangor Darul Ehsan, Malaysia

Nathsuda Pumijumnong Faculty of Environment and Resource Studies, Mahidol University, Salaya Phutthamonthon, Nakhon Pathom 73170, Thailand

Mizanur Rahman Forest Department, Ban Bhaban, Boyra, Khulna, Bangladesh

Muhammad Nawaz Rajpar Faculty of Forestry, Universiti Putra Malaysia, 43400 UPM Serdang, Selangor Darul Ehsan, Malaysia

Saeed Rashvand Research Center of Natural Resources and Agriculture of Qazvin Province, Qazvin, I.R. Iran

M. S. Razali Faculty of Science and Technology, Universiti Malaysia Terengganu, Kuala Terengganu, Terengganu, Malaysia

Hai Ren South China Botanical Garden, Chinese Academy of Sciences, 510650 Guangzhou, P. R China

Seyed Mousa Sadeghi Faculty of Forestry, University Putra Malaysia, 43400 Serdang, Selangor, Malaysia

Department of Natural resources research, Research Center of Natural Resources and Agriculture of Bushehr Province, Varzesh Avenue, Bushehr, I.R. Iran

Chengde Shen Key Laboratory of Isotope Geochronology and Geochemistry, Guangzhou Institute of Geochemistry, Chinese Academy of Sciences, 510640 Guangzhou, P. R China

Mohd Nazip Suratman Faculty of Applied Sciences and Centre for Biodiversity and Sustainable Development, Universiti Teknologi MARA, 40450 Shah Alam, Malaysia

Andrew Swales National Institute of Water & Atmospheric Research Ltd., PO Box 11115, Hamilton, New Zealand

Vivekanand Tiwari Discipline of Marine Biotechnology and Ecology, CSIR-Central Salt and Marine Chemicals Research Institute (CSIR-CSMCRI), Gijubhai Badheka Marg, 364002 Bhavnagar, Gujarat, India

Suphakarn Traesupap Coastal Development Centre, Faculty of Fisheries, Kasetsart University, 10900 Bangkok, Thailand

Lianlian Yuan South China Botanical Garden, Chinese Academy of Sciences, 510650 Guangzhou, P. R China

Mohamed Zakaria Faculty of Forestry, Universiti Putra Malaysia, 43400 UPM Serdang, Selangor Darul Ehsan, Malaysia

Jinping Zhang South China Botanical Garden, Chinese Academy of Sciences, 510650 Guangzhou, P. R China

Graduate University of Chinese Academy of Sciences, 100039 Beijing, P. R China

About the Editors

Prof. Dr. I. Faridah-Hanum received her Ph.D. in Botany from the University of Reading, England in 1989. She is currently a Professor of Forest Botany at the Faculty of Forestry, Universiti Putra Malaysia (UPM) and Dean, Faculty of Forestry (UPM). She has held the post of Head of Department of Forest Production, Faculty of Forestry, UPM twice in 1999 and 2009, and Deputy Director of UPM Research Management Centre in 2010.Amongst others, she is also the Fellow Academy of Science Malaysia, Executive Member of Asia Pacific Association of Forestry Research Institutes (APAFRI) and Foreign Specialist to Forest and Forest Products Research Institute (FFPRI) at Tsukuba, Japan. Presently, she is the chief editor of the 75-year old journal *The Malaysian Forester* and also editor of the Japanese journal TROPICS and *Pakistan Journal of Botany*. Prof. Faridah-Hanum has many years of experience in conducting research in the Malaysian forests. She also started the series of scientific expeditions in 1999 for Malaysia, led and participated in more than 35 expeditions botanising the Malaysian forests besides editing numerous chapters of the expedition findings. To date, Prof. Faridah has received a total of 20 research grants from both local and international organizations such as International Foundation of Science (Sweden), UNDP, Tsukuba Forest and Forest Products Research Institute (FFPRI), and the Centre of International Forestry Research (CIFOR). She has written, edited and published more than 250 papers, abstracts and reports in journals, books, proceedings and other popular publications. She has seven copyrights for the books authored in her area of specialization, and co-edited an important book on auxiliary plants in forestry and agriculture.

Prof. Emeritus Dato' Dr. A. Latiff joined Universiti Kebangsaan Malaysia (UKM) in 1979 as a lecturer. He was then appointed as Head of Botany Department in 1980–1984, later promoted to Associate Prof. (1983) and Professor of Botany (1991). Prof. Latiff was appointed as Dean of the Faculty of Science and Technology in 2002. Prof. Latiff is the winner of Malaysia-Toray Award in Science and Technology (1995) and the prestigious Langkawi Award (2004). His main research areas are plant taxonomy and biodiversity. Prof. Latiff has published more than 450 papers, abstracts and reports in journals, books, proceedings and other popular publications. He has so far edited more than 60 books and proceedings. Prof. Latiff is the Chairman of WWF

(Malaysia), Chairman of Pulau Banding Foundation (Malaysia), Fellow of Linnean Society of London, Fellow Academy of Science Malaysia and Trustee of Malaysian Nature Society.

Dr. Khalid Rehman Hakeem is the Fellow Researcher at Universiti Putra Malaysia (UPM), Serdang, Selangor, Malaysia. He has completed his Double Masters (Environmental Botany and Ecology) and Ph.D. (Botany) from Jamia Hamdard, New Delhi, India in 2006 and 2011 respectively. Dr. Hakeem has more than Eight years of teaching and research experience in plant eco-physiology, biotechnology, as well as environmental sciences. Recipient of several fellowships at both national and international levels, Dr. Hakeem has so far edited and authored more than seven books with international publishers. He also has to his credit more than 25 research publications in peer reviewed journals and 15 book chapters. Dr. Hakeem is currently engaged in studying the plant processes at ecophysiological as well as proteomic levels.

Prof. Munir Ozturk is a Consultant Fellow at Faculty of Forestry, Universiti Putra Malaysia. He has completed his Masters in Botany in 1964 Jammu & Kashmir University, India, Ph.D. and D.Sc. in 1970, 1975 in the field of Eco-Physiology from Ege University, Izmir, Turkey. Dr. Ozturk was appointed as full Professor in 1980 at the Ege University and served as Chairman of Botany from 1983–1986, and served as Director of Centre for Environmental Studies between 1990–1998. His main research areas are eco-physiology, phytoremediation, sabkha ecosystems and medicinal-aromatic plant diversity. He has published more than 400 papers in national and international journals, more than 50 book chapters and has edited more than 10 books with Springer and Cambridge Scholars. He is a fellow of the Islamic World Academy of Sciences.

Mangrove Ecosystem of Malaysia: Status, Challenges and Management Strategies

A. Latiff and I. Faridah-Hanum

Contents

A. Latiff (✉)
Faculty of Science and Technology, Universiti Kebangsaan Malaysia,
43600 Banji, Selangor, Malaysia
e-mail: latiff@ukm.my

I. Faridah-Hanum
Faculty of Forestry, Universiti Putra Malaysia, 43400 Serdang, Selangor, Malaysia

I. Faridah-Hanum et al. (eds.), *Mangrove Ecosystems of Asia*,
DOI 10.1007/978-1-4614-8582-7_1,

Abstract Unlike other types of forests which are more spatially widespread and can be found on varied soil types, the mangrove swamp forests are restricted to sheltered coasts, islands, lagoons, estuaries and rivers on muddy substrates. As an ecosystem it is an important habitat for diverse wildlife, including fishes, shells and microbes and a number of specialized plant forms. It is also of great socio-economic importance as a hydrological regulator, playing an important role in flood mitigation, buffering against saline intrusion and waves. It is also an important source for fuelwood, timber resources and provides a variety of produce used by local inhabitants. Despite these values, mangrove swamp forests are rapidly being cleared, degraded and transformed to other land-uses, especially for agriculture, aquaculture, resettlement, industrial and ecotourism infrastructures. In view of the recognized values, it is urgent that more suitable areas are protected for not only the biodiversity conservation purposes but also as a special and unique forest type. In Malaysia, a working plan for the Matang mangrove forest reserve, Perak (fifth revision) provides a comprehensive overview of the management and conservation of the mangrove ecosystem in Malaysia, which could also be a model for other mangrove areas in other states for their protection and management. In the long term, systematic and holistic planning represent the best means of achieving sustainable mangrove swamp forest management by incorporating conservation principles and forestry objectives.

1 Introduction

In Malaysia, the mangrove ecosystem has been an important resource for the coastal, estuarine and to a certain extent, the riverine communities. These communities have either been living within the mangrove swamp forests or at the fringes of inland mangroves and have been depending on it for their livelihood and socio-economic well-being(Lugo and Snedaker 1974). The mangrove swamp's aquatic resources have been important sources of fishes, shells and other gastropods for the fishermen, and the forests have been providing fuelwood, poles and other building materials for the local communities. In particular, all parts of *nipah* palms (*Nypa fruticans*) have been providing products to those communities all over the country (Latiff 2009). According to Chan (1987), such social-economic forestries and fisheries have been coexisting harmoniously for generations and have minimal impact on the ecosystem. However, with the recent resurgence of interests in aquaculture and agriculture, many areas of the inland mangrove forests had paved ways for these economic activities with some concomitant loss to the mangrove biodiversity, in general.

As the Forestry Departments realized the importance of mangrove resources to the communities, they began to take steps in forest protection and conservation by allocating some mangrove swamp forests as forest reserves, such as Virgin Forest Reserves, and effectively managed them based on good forestry management

Table 1 Extent of mangrove forest reserves (in ha) in Malaysia (Anon. 2003)

State	Total area	Note
Johor	21,180	Much of the mangrove forests both in east, south and west coasts of Johor are experiencing disturbances both from the natural hazards and socio-economic development and exploitation, especially in the Sungai Pulai area
Kedah	8,355	It is now reported that only 8,034 ha is left, mostly on the mainland at Sungai Merbuk; those of Langkawi Island experienced some forms of disturbances and exploitation. However, the mangroves of Kilim-Kisap area have been exploited for ecotourism
Perlis	Not available	There are patches of mangrove forest in the area of Kuala Perlis; possibly they are in stateland and not allocated as forest reserve
Negeri Sembilan	204	In 1994, it was reported that only 879 ha were left, but those in Sungai Linggi and Port Dickson areas were very disturbed. Much of the forests have been lost to socio-economic development
Pahang	3,916	The mangroves in the state are well-preserved, especially in the Sungai Kuantan and Sungai Rompin areas
Perak	40,683	Those in the Matang Forest Reserves, which now covers about 40,466 ha, are very well managed, but those in Bagan Datok areas are very disturbed as they are not allocated as forest reserves
Pulau Pinang	870	Very little is left but well-managed both on the mainland and island
Selangor	19,503	Recently, it has been reported that only 15,090 ha are left and those in Kuala Selangor and the islands off Port Kelang are well-managed
Kelantan	Not available	There are patches of mangrove forests in estuaries and their river banks in the state lands at the river mouth of Sungai Kelantan
Terengganu	1,822	There are patches of mangrove forests in estuaries and their river banks, and those in the area of Sungai Kemaman are well-managed
Melaka	80	There are patches of mangrove forests in estuaries and their river banks
Sub-total	97,517	In 1994, it was reported that 105,537 ha were present but after about 10 years only 97,517 were left
Sabah	340,689	Most of the mangrove forests are well-protected; only a small degree has been exploited. Those at Sunai Sugut and Sungai Kinabatangan are very rich in species diversity. There are also mangroves in sheltered river mouths and lagoons of some off-shore islands
Sarawak	126,400	Most of the mangrove forests are well-protected in the estuaries and banks of Sungai Sarawak, Sungai Rejang
Grand total	564,606	At the time of this writing the authors believe the total area of the mangrove forests has been somewhat diminished both by natural and human-induced activities

practices. The total extent of the mangrove forests reserves has been estimated to be about 580,000 ha of which 77.8 % is considered productive (Chan 1987). However, in Peninsular Malaysia about 105,537 ha is categorized under Permanent Forest Reserves and about 90 % of these occur on the more sheltered west coast while only 4 % occur on the more exposed east coast of Peninsular Malaysia (Table 1).

2 Biological Characteristics

2.1 Vegetation and Flora

The mangrove vegetation is evergreen and simple in physiognomic structure of two to three storeys varying from 5–25 m in height, depending on age and localities (MacNae 1968) and Snedaker (1978) (Tomlinson 1986; Ball 1988; Clough 1992; (Chapman 1975); Duke 1992; Smith 1992). The emergent layer usually consists of very few tall trees, and the canopy is comparatively even and closed, except where there are gaps, either natural or man-made. The understorey layer is poorly defined and merging with the ground layer which is devoid of growth except saplings. The principal mangrove species are characterized by special roots such as stilt roots and pneumatophores and also by their viviparous propagule habit. Several authors such as Wyatt-Smith (1960), Liew (1980), and Chai (1982) reported about 31 plant species which are exclusively found in the mangrove swamp forests, while a total of 51 species are non-exclusive or associates. However, Japar Sidik (1994) reported that Malaysia has 38 exclusive, 57 non-exclusive and nine associate mangrove species. The preliminary assessment of flowering plants of Matang Forest Reserves was made by Shamsul et al. (2005) and those of other areas were given by Norhayati et al. (2005) and Wan Juliana et al. (2010). Accounts of lower plants are very scarce, those of ferns and mosses for Matang Forest Reserves, for example, were reported by Jaman and Maideen (2005) and Damanhuri et al. (2005), respectively.

The vegetation of the mangrove forest is simple in structure thus the floristic composition is also low compared to other forest types. However, as an ecosystem the mangrove swamp forest is rich in flora and fauna. The perception among scientists is that the mangrove flora, structure, above ground biomass and net productivity on the west coast differ very much when compared with that of the east coast of Peninsular Malaysia, or even with those of Sabah and Sarawak. However, some studies showed that it is otherwise and some specific differences are observed. It has been found that the mangrove flora of Peninsular Malaysia's east coast mangroves is poorer in distribution and zonation is not obvious. This is probably because the east coast mangroves are exposed to larger waves of the South China Sea as compared to those in the west coast which are sheltered and confined within the Straits of Malacca. Early in the last century, Watson (1928) classified the mangroves in Peninsular Malaysia into five vegetation types based on species composition and dominance. They are *Avicennia-Sonneratia* type, *Bruguiera cylindrica* type, *Bruguiera parviflora* type, *Rhizophora* type and *Bruguiera gymnorhiza* type. A better and more comprehensive classification is given by Chai (1982) which included the inland mangroves. However, classification of the mangrove types of Terengganu is very commendable as it used aerial photography in addition to ground-truthing (Mohd. Lokman and Sulong 1990; Sulong and Ismail 1990). If recent techniques are employed to reclassify the mangrove types in Malaysia, it might put Watson's (1928) classification in a better perspective. This is important as in many localities the dominance of certain mangrove species has changed over the years due to habitat degradation, loss of species and exploitation.

There have been numerous studies on the flora of the mangrove forests, though the number of species is small compared to other ecosystems. The most recent are those of Wan Juliana et al. (2010) who illustrated and described the full mangrove flora of Langkawi Archipelago, Nilus et al. (2010) who illustrated the mangrove flora of Sabah and Azmil et al. (2012) who produced the checklist for the mangroves of Pulau Pangkor, Perak. Phang et al. (2005; 2007) discussed the enormous diversity of seaweed including that of mangroves. In Langkawi Islands and Sabah and probably elsewhere, *Caulerpa* spp., which in the former locality are found in the Sungai Kilim and Sungai Kisap, are edible as salad.

2.2 Fauna

Berry (1972) broadly categorized the animal communities in the mangroves into two components, namely, the aquatic and the terrestrial. The former consists of fishes, crabs, snails, worms and the bivalves, whilst the latter consists of insects (including the fireflies), birds (including the migratory species), lizards and monkeys. The mangroves of Pulau Langkawi, Matang, Port Kelang and Kukup are known for fishery, and those of Kampung Kuantan, Kampung Belimbing (Selangor), Sungai Kerian (Penang) and Matang (Perak) harbour fireflies which attracted ecotourists. The mangroves of Kuala Merbuk (Kedah), Kukup (Johor) and Kuala Gula (Perak) are equally known for supporting migratory birds.

The general fauna of the Matang mangroves was reported by Sasekumar (2005), but the specific avifauna was given by Noramly (2005), mammals by Shahrul Anuar et al. (2005) and fishes by Chong (2005) and Amiruddin et al. (2005). There are reports on fireflies (Zaidi et al. 2005), mudskippers (Faridah et al. 2005), zooplankton (Ooi et al. 2005) and others. It has been argued many times that the mangrove fauna excluding the fishes and shells have been underestimated in their value and significance to the overall mangrove biodiversity. Both the terrestrial and aquatic birds are indeed very important in the ecology and well-being of the mangrove habitats because these birds are predators of fishes and other invertebrates. The herpetofauna are poor except for a few species of mangrove vipers and lizards. However, the insect diversity, particularly the butterflies and macroinvertebrates, are quite high (Zaidi and Azman 2005; Zaidi et al. 2005).

2.3 Microorganisms

There are very few studies on the microorganisms in the mangroves except that of Kuthubutheen (1984) and Alias et al. (1995) who reported the phylloplane fungi on a few species of mangroves. The microscopic fungi and lichens are quite well represented in the mangrove forests but not studied and reported, as far as the authors know. The authors were also informed of other studies, especially the degradation of mangrove leaf litter by microbes which is important in nutrient recycling in the habitat. One of the most noticeable macroscopic fungi in the mangroves is the species of *Ganoderma,* which in recent years have been claimed to be medicinal. The checklist of mangrove and marine fungi was prepared by Siti Aisyah Alias (2007)who reported a total of more than 234 species identified and an additional 68 species unidentified. The ascomycetes were the largest group discovered followed by deutromycetes and basidiomycetes.

3 Exploitation of Mangrove Forests

3.1 Timber Extraction

Mangrove forests are quite an important source of timber and non-timber products but insignificant compared to the timbers of the lowland and hill dipterocarp forests which produce the heavy and hard wood timbers. The timbers of mangroves are harvested and converted to charcoal and pole production for mainly domestic market (Amir 2005). Among the non-timber products are a few medicinal plants, aquatic vertebrates and invertebrates, vinegar and *nipah* attaps. In the past the mangrove forests were exploited for the above products, amongst others, but today the aquatic invertebrates and fishes are among the productive resources. Aquaculture such as for fishes and prawns have proven more economical. However, in the last decade the mangrove services have been further exploited, especially for nature recreation and ecotourism activities. Some mangrove areas in Langkawi Island, Matang, Kuala Selangor, Lumut, and Kuantan have attracted entrepreneurs to start both recreational and ecotourism products. Awang Noor (2005) and Amir (2005) have summarized both the goods and services of mangrove forests, including the environmental values. The former discussed extensively the economic value of mangrove forests.

3.2 Mangrove Stocking

Ashari et al. (2005) discussed the management of stocks at Matang since 1904, which employed different rotations from period to period. Presently, Matang practices a 30-year rotation period but Johor practices a 20-year rotation period. The gross volume and basal area either increased or decreased due to the different regimes of thinning and rotation. Some studies had suggested the rotation period may be reduced to a 22-year period for optimal productivity. Juliana and Nizam (2005) had also described the mangrove structure and above-ground biomass as indicators of mangrove stocking and carbon sesquestration. In general, mangrove stocking is adequate for the supply of poles and wood for the local markets, either for construction and charcoal factories. In particular, the mangroves of Matang, Perak are managed like a commercial plantation on a 30-year rotation to supply produce for the local economic demands.

3.3 Mangrove Forest Conversion

3.3.1 Agriculture

In the state of Selangor, for example, the extent of mangrove forests in 1975 was 39,695 ha and by the year 1999 only 15,090 ha were left, a reduction of about 62 %, and much have been lost to land conversion to oil palm cultivation and aquaculture activities. The areas affected were the forest reserves of Kuala Sepang, Banjar Selatan, Teluk Gadong, and Jugra, among others (Haliza et al. 2005). Cultivation of oil palm in Malaysia has been the most profitable of all the agricultural crops and a vast amount of lands, including those of inland mangroves, have been lost to oil palm plantation. Pulau Carey, Selangor had substantial mangroves in the past but now the whole island is cultivated with oil palm. The authors believe similar trends also occur in other states, particularly Perak, Kedah, Negeri Sembilan, Sabah and Johor. In terms of productivity and economic benefits oil palm plantations are many times more economical than mangrove forests. In the next decade it is perceived that much of the mangrove swamp forests in Sarawak will be converted to oil palm plantations too as the state tries to eradicate poverty among the rural communities and empower them as settlers depending on the plantation.

3.3.2 Aquaculture

Fish and prawn cultures have been proven to be economically more profitable, especially for oversea markets. Similarly, many estuaries and rivers have been the sites for fish cultures, especially in the Kilim-Kisap area of Langkawi, Matang and Kuala Selangor. The prawn cultures have been developed somewhat inland but still in the mangrove forests. In Selangor, for example, aquacultures have significantly depleted mangrove areas in Kuala Bernam and Jugra (Haliza et al. 2005). Ong (1982) had already warned about the proliferation and expansion of aquaculture industries in Malaysia as the demand for fishes and other aquaculture produce are on the increase due to the popularity of seafoods. The conversion of mangrove areas to aquaculture farms not only was prominent in Selangor but also in almost all other states, notably in Langkawi, Perak, Johor, Sarawak and Sabah.

3.3.3 Resettlement

As the inland mangrove forest areas are converted to other land-use, especially the oil palm plantation, a small area in Selangor, about 412 ha of the mangrove forest, also gave way for resettlement of indigenous communities as in Kuala Sepang (Haliza et al. 2005). Similarly, in Langkawi some areas were converted to fish landing ports, and the mangroves of the Malut area were cleared, developed and later abandoned. However, this conversion to settlement area is insignificant.

4 Impacts on Mangroves

4.1 Pollution

Both the pollution in the estuaries and rivers has its sources in the inland industrial and agricultural activities and other land-use patterns upstream. Historically we have judged the quality of water in the mangrove areas by its foul smell and dark colour as being affected by pollutants. In Matang in particular, sediments from the developing town of Taiping flowed into the mangrove areas, making the muddy sediments sandy. Activities such as aquaculture, cockle harvesting, navigation and river settlement also contribute to river and estuarine pollution. Mohd. Kamil et al. (2005) have shown that the water quality of the Matang mangroves has deteriorated in the past few years, to cite an example. Almost all rivers in Malaysia which originate from the hinterland carrying loads of pollutants will pass through belts of mangroves on both sides of the rivers and estuaries. The authors wonder what would be the short-term and long-term effects of these pollutants, especially the grease and heavy metals, on the biodiversity of the mangrove areas downstream. There are not many studies on this aspect to discern.

5 Mangroves of Malaysia

5.1 Langkawi Islands

5.1.1 Floristic Composition

Pulau Langkawi has an exceptional natural settings and beautiful landscapes that attract both naturalists, scientists and tourists alike. The mangrove forests of the Kilim-Kisap areas in particular are testimony to the above statement. In addition, the mangrove forests there are found on the shallow limestone substratum, making them one of the most outstanding features in Peninsular Malaysia, and possibly in the world. The mangrove ecosystem is both dynamic and fragile and is very sensitive to both natural stochastic events and human activities. Though they provide many essential services such as storm protection, erosion control, waste-water clean-up, and forest products, they are consistently subjected to conversion to other land-use purposes of greater economic returns. Wan Juliana et al. (2010) described and illustrated a total of 76 species of mangroves trees (45 %), shrubs (25 %), ferns (9 %), climbers (9 %), herbs (7 %) and bryophytes (5 %) in 58 genera and 35 families. Out of the total 76 species, 32 are exclusive, 33 are non-exclusive and 11 are associate mangrove species. Comparatively, the mangrove forests in the Langkawi islands have a high diversity of mangrove plants in Peninsular Malaysia.

In 1980, the total mangrove area of the Langkawi Islands was 3657.67 ha and, about 11 years later, the extent of the mangroves area was reduced by 11.85 % to

3270 ha. Some of the mangrove areas had been earmarked for aquaculture ponds, chalets, navy facilities and other uses. Norhayati and Latiff (2001) had estimated the density of mangroves in a 1 ha plot as being 849 per ha, and the stands belonging to nine species and four families. The most dominant species is *Rhizophora apiculata* with an important value of 50.2 and tree density of 557 /ha, while the total above ground biomass was estimated at 115.07 t/ha.

5.1.2 Threats and Conservation

It is estimated that in 1988 there was a total of 4,165.29 ha of mangroves, in the year 1993 there were 3,902.85 and by the year 1999 there was a total of 3,764.97 ha. This means that between 1988 and 1999, a total of 400.32 ha or 36.39 ha per year were lost to other land-uses. From another study, in a period of five years (1988–1993) 6.3 % of the total mangrove areas were deforested, and in the next interval (1993–1999), a further 3.53 % were deforested. These activities coincided with the fact that Pulau Langkawi was declared as a free-trade zone in 1985.

As stated by Norhayati and Latiff (2001), the estimated above-ground biomass of mangroves in Pulau Langkawi 115.07 t/ha. Using this figure it could be estimated that the total amount of biomass lost in the last 11 years (1988–1999) was 46,064.82 tonnes. From 1988 to 1993 a total of 129.69 ha were lost, with only 26 % to agriculture (33.09 ha), and between 1933 and 1999, a total of 128.54 ha were lost, 23.94 ha to agriculture and 15.75 ha to aquaculture. The threats and management of the mangroves of Langkawi in particular has been discussed by Latiff (2012).

5.2 Mangroves of Selangor

5.2.1 Floristic Composition and Biomass

Soepadmo and Pandi Mat Zain (1989) surveyed the mangroves of Sementa, Selangor where 32 species of plants were found. The dominant species were *Avicennia alba* and *Sonneratia alba* in the *Avicennia* zone. In the mixed *Rhizophora* zone, the dominant species were *Rhizophora mucronata* and *R. apiculata*, and in the *Bruguiera* zone, *Bruguiera cylindrica* and *B. parvifolia*. The total number of stems differed from zone to zone, ranging from 4189 /ha in the *Avicennia* zone to 13,290 /ha in the *Bruguiera* zone, and the above-ground biomass ranged from 124.53 t/ha in the former zone and 150.78 t/ha in the latter zone. That of Kuala Selangor, to a certain extent, has been conserved with the establishment of a Nature Park.

5.2.2 Threats and Conservation

Nik Mohd. Shah et al. (2005) provided an excellent description of the management and conservation of mangroves in Selangor. In the year 2003 a total of 14,897 ha existed in the state, which fall under the categories of production forest, soil conservation forest, wildlife conservation forest and Virgin Jungle Reserves. Since 1920 most of the mangrove forests were allocated as Permanent Forest Reserves, and the first working plan was prepared in 1922 and the last one was for 2006–2015. The case of the Kuala Selangor mangroves illustrates the various threats faced and how management strategised their *in situ* conservation. Some pristine patches were developed as a nature park, those along the Sungai Selangor at Kampung Kuantan and Kampung Belimbing were developed for recreation and ecotourism as fireflies occur there, and rehabilitation and restoration were conducted where the mangroves were depleted by natural causes. However, with the construction of the Selangor Dam, some effects on the population of *Sonneratia caseolaris* along the Sungai Selangor have been observed. The threats and management of mangroves in Selangor has been discussed by Haliza et al. (2005).

5.3 Mangroves of Johor

Johor has a total of 20,533 ha of mangrove forests which are mostly found in Sungai Pulau Forest Reserve, Sungai Johor Forest Reserve, Sungai Santi Forest Reserve and Sungai Lebam Forest Reserve, and the first working plan for the state was developed in 1941. The latest integrated management plan (2000–2009) was developed primarily to conserve and manage the forests through sustainable regime to ensure that they contribute to the state and national economy and environmental stability. The threats to the present mangrove forests come from various sources. For example, large scale development projects for infrastructure, urban development, industries and harbours in and around Bandar Nusajaya would definitely affect the existing environment of the mangroves. In addition, the proposed petrochemical plant and the Iskandar Corridor development would also pose possible threats (Che Hashim et al. 2005; Maimon et al. 2008).

5.4 Mangroves of the East Coast of Peninsular Malaysia

5.4.1 Floristic Composition and Biomass

Soepadmo and Pandi Mat Zain (1989) surveyed the mangroves of Kuala Kemaman and Kg. Pantai Tinggi, Kemaman, Terengganu where only 24 species of plants are recorded. The dominant species were *Rhizophora apiculata* and *Bruguiera gymnorrhiza*. The total number of stems at Kuala Kemaman Forest Reserve was 5,340 /ha and the above-ground biomass was 199.13 t/ha, whereas those of Kampung Pantai Tinggi were 3,281 /ha and 163.10 t/ha, respectively. Mohd. Lokman and Sulong (2001) described the vegetation and flora of the mangroves of Terengganu.

There are other surveys and floristic studies on the mangroves of the other east-coast states of Kelantan, Pahang and Johor. However, those of Kelantan are situated on the statelands, and hence not protected. Furthermore, they occur in small patches at the river mouths and stand structure and composition is rather poor. There were some studies but neither published nor reported for reference. Those in Pahang are comparatively richer, especially in the Kuantan and Rompin districts.

5.4.2 Threats and Conservation

The mangroves in the east coast states of Peninsular Malaysia, particularly those in Terengganu and Pahang are not-well sheltered by lagoons and rivers, unlike those in the west coast states. Hence they are not as diverse and widely distributed. However, they are also threatened by similar factors such as strong waves, especially during the monsoon, small-scale agriculture, aquaculture, resettlements and construction of infrastructures, especially those in Setiu, Dungun and Kemaman, Terengganu (Gong et al. 1984). For the state of Terengganu in the last five years five compartments of Kuala Kemaman Forest Reserves were gazetted as Virgin Jungle Reserves. This exercise augurs very well for mangrove conservation in Peninsular Malaysia.

5.5 Mangroves of Sabah

Tangah (2005) stated that Sabah has about 316,024 ha of mangrove forests in the forest reserves and about an additional 25,000 ha are outside the reserves. Much of them are still pristine and not exploited for commercial purposes. A review of the past and current status of the mangrove forest management was conducted by Kugan (2003) who revealed that the state had embarked on production of chipwood and bark from mangrove trees on a commercial scale in the early 1970s. However, the insignificant contribution to the state's revenue and the damaging extraction method employed prompted the state government to discontinue it in 2001. The challenges that the state government had embarked were to store the timber stocks, to arrest the competing land-use, to diversify resource utilization, to maintain a healthy mangrove ecosystem and increase efforts in conservation (Liew 1980). Fatimah et al. (2012) illustrated the case of mangroves in the Kota Marudu area where the communities were engaged in both the conservation efforts and resource exploitation to eradicate poverty in the area.

5.5.1 Threats and Conservation

As stated earlier much of the mangroves in Sabah are still intact in their natural state. Several years ago the state government decided to exploit for rayon and only recently the project had been terminated. The authors do not foresee pertinent threats to the

Sabah mangroves as the demand for their exploitation is not significant. However, when the resources of the lowland dipterocarp forests of Sabah diminish there is the possibility the timber resources of the mangroves will be tapped.

5.6 *Mangroves of Sarawak*

According to the national figures, Sarawak has about 126,400 ha of mangroves. However, according to Marajan (2005), based on satellite imagery of the 740 km long coastline, some 142,693 ha are covered with mangrove forests. This illustrates very well that up-to-date techniques such as aerial photography and satellite imagery could enhance the inventory of resources. However, only about 48 % is under permanent forest reserves, the remainder are within the stateland. Similar to Sabah, Sarawak also went for chipwood and charcoal production for export and the annual production had been substantial. The poles and other non-wood products were for domestic use. The management plans were written in the 1950s and the main objectives were to satisfy the domestic demand for poles, firewood and charcoal and to export the surplus.

5.6.1 Threats and Conservation

The authors foresee pertinent threats to the Sarawak mangroves will occur in the next decade as the demand for their land conversion and exploitation of the rich resources are becoming more apparent (Ashton and McIntosh 2001). However, like Sabah when the resources of the lowland dipterocarp forests and peat swamp forests of Sarawak diminish, there is the possibility the timber resources of the mangroves will be tapped too.

6 Management and Conservation

The Departments of Forestry in Malaysia, as custodians and managers of the mangrove forests, are all committed to conservation of biodiversity which emphasize both the protection and sustainable utilization of the resources. In Malaysia, the basis and concept that underlines the practice of sustainable forestry is to set aside adequate natural forest lands, including mangrove forests, as Permanent Forest Estates (PFE) that are strategically located throughout the country. There are two types of PFEs, namely, the production and protection forests. While the cutting cycle for hill mixed dipterocarp forest is 25 years, that for peat swamp forest is 45 years and for the mangroves it is between 20 and 30 years and is kept unamended. The management and conservation of mangrove forests are discussed under the following sub-headings.

6.1 Sustainable Forest Management

All states in Malaysia are committed to forest conservation including the mangrove forests except those without substantial areas such as Kelantan, Melaka, Perlis and Pulau Pinang. While the rate of exploitation is higher in the states of Peninsular Malaysia, Sabah and Sarawak are beginning to demonstrate the value of both conservation and utilization of mangroves for their states' revenue in the near future. These commitments are illustrated by Shaharuddin et al. (2005), Marajan (2005) and Tangah (2005). It is presumed in the next decade that both the states of Sabah and Sarawak will embark on mangrove exploitation to give added value to their mangrove forests, in addition to more serious efforts in conservation. Sustainable mangrove forestry in Malaysia may prove as the way forward in ensuring a balance between exploitation and conservation by all states. The need for more studies and research to develop more products has also been discussed (Ibrahim and Husin 2005; Ong 2005; Latiff 2005).

6.2 Minimising Impacts and Promoting Wise Use of Resources

The keys to conservation are to protect the mangrove resources *in situ* and when the need to utilize the resources for economic purposes arises steps must be taken to minimize the impacts to the ecosystem. As the mangrove ecosystem is very fragile any form of disturbances no matter how small could possibly create long-lasting impacts. As stated by Kugan (2003) the harvesting of mangroves for woodchips in Sabah had critically damaged the mangrove vegetation so that the production was stopped by the state government after about 30 years of exploitation. The Environmental Impact Assessment regulation in Malaysia is already in place for land conversion and other prescribed activities.

6.3 Enhancing Biodiversity Management

The existence of a unique ecosystem diversity, rich species diversity, flora and fauna is well documented in the mangrove ecosystem. The mangroves of Malaysia are rich both in terms of flora and fauna (Aldrie and Latiff 2008). Some of the species have been exploited and utilized while some hold potential for the future economic benefits of the communities concerned. Once again the key is mangrove forest conservation to ensure the conservation of species and subsequent use of their genetic diversity. There is an urgent need though to strengthen both the institutional and research capacity to address this important issue. The Department of Forestry in all states is committed to conserve mangroves in their respective states, hence the capacity for research must be further strengthened.

6.4 Strengthening Mangrove Virgin Jungle Reserve

The Virgin Jungle Reserves (VJR) within the permanent forest reserves are established for the purpose of stock holdings of important habitats and species of forestry important for future silviculture, education and research. Strengthening the present VJRs would certainly ensure mangrove forest conservation in the country. The state of Terengganu in particular should be commended for establishing five new VJRs in the Kuala Kemaman Forest Reserves and the state of Perak for well-managed mangroves at Matang.

6.5 Enhancing Public Awareness

The public is the ultimate benefactor of mangrove conservation; hence to enhance public awareness on the importance of this ecosystem is the most important assurance for future generations. A step has been taken by the Malaysian Nature Society in establishing a mangrove Nature Study Centre supported by a private company in Terengganu. School children and university students are taken to the centre to do nature studies on the ecology of the mangrove flora and fauna. The local community in Setiu, Terengganu, with the assistance from WWF Malaysia and the Universiti Malaysia Terengganu, has also shown similar commitment by embarking on mangrove replanting. In fact, the 2004 tsunami had created the impetus in creating public awareness on the importance of the mangroves. Many states had embarked on mangrove tree planting in the last five years, some with great success and some with minor failures.

7 Functions of Mangroves

As sinks for waste-water borne pollutants
It has been shown that mangrove soils and roots could trap and immobilize heavy metals and nutrients from waste water originating from the hinterland. Hence it is believed that mangroves could function as a purifier of pollutants (Conley et al. 1991; Ambus and Lowrance 1991). This function has been taken for granted such that many inland factories and industries pollute the upstreams and pollutants flow downstream through the mangroves to the sea.

As a sediment removal system
As water flows slower in streams and rivers of mangrove areas than that of non-mangrove rivers, sediments tend to settle down to the bottom and that which flows outwards towards the sea is sediment-free (Wolanski 1995). As observed in Matang, Perak much of the muddy substratum had become sandy and this affected the cockle production which ultimately brought adverse implications to the cockle farming.

Coastal erosion prevention
The strong roots and buttress systems of the mangrove plants form a natural buffer between the land and sea. They also tend to break strong wind and wave actions. This had been proven in 2004 when a tsunami struck the coasts of the Langkawi Islands, Kedah and Perak. If not for this buffer effect much more damage to the estuaries and rivers would have occurred. In addition, mangroves also contribute to land-building through accretion (Othman 1994).

Recreational areas
Today mangrove areas are capable of generating some economic returns through boating, bird watching, jungle trekking, and other recreational activities. Kampung Kuantan and Kampung Belimbing in Kuala Selangor are known to attract eco-tourists as fireflies synchronizing light emitting become the attraction at night. In Lumut, Perak some recreational facilities have been constructed and developed to attract local visitors and tourists.

Education
As mangrove forests contain salt-tolerant plants and animals, they could play an important role in educating the public, especially school children on ecology. An excellent example is the Kuala Selangor Nature Park that has conducted many education programmes for the school children and the public at large by the Malaysian Nature Society. The area is about 95 ha and a total of 157 species of birds and 13 species of plants are present in the park.

8 Development of a Management Plan

Mangrove swamp forests are always under serious threats of various forms notably from conversion to other land-uses especially aquaculture and agriculture. Razani Ujang (1982) stated that between 1955 and 1980, a total of 10,500 ha of mangrove swamp forests have been converted and Selangor alone had lost about 7,500 ha or about 30 % of the total mangrove areas in the state. The state of Kedah including Pulau Langkawi is no exception. About 1,500 ha of the Sungai Merbok mangrove area had been converted to rice fields but those areas remained idle because of the acid sulphate soils that don't favour successful rice farming.

The problem lies in the difficulty in recognizing the indirect and direct benefits of the mangrove swamp forests. Since the mangrove ecosystem is an interphase between terrestrial and marine environments, there exists competition for various economic interests. Major industries in sectors such as forestry, fisheries and agriculture could claim mangroves as their administrative domain, and the policy that is best for one is detrimental for another. This is observed as happening in many states such as Perak and Selangor. Hence, trade-offs between alternative development and resource use must be examined more carefully and comprehensively. Current economic analysis can assist to identify the problem of using the cost-benefit approach to solve problems associated with a decision on coastal resource use, such as mangroves.

8.1 Matang Mangrove Forest

The Matang mangrove forest is taken here as a model for sustainable management because not only is it always claimed to be the best managed mangrove forest in Malaysia and probably also in the world but also it has a long history of management, as a first working plan was drawn in 1904 (Gan 1995). This is supported by Muda et al. (2005) who detailed the management system by introducing zoning, rotation, yield estimation and regulation as well as sound silvicultural practices (Hossain 2004).

9 Research and Development

Hamdan et al. (2012) outlined the various aspects of research and development of mangroves in Peninsular Malaysia with a focus on the rationale for rehabilitation and restoration of the health of mangroves. The policy and the role of various legislation in the country is quite clear but yet as shown by Haliza et al. (2005) in the case of the mangroves in Selangor, there are many conflicts on the ground with respect to implementation of legislations (e.g. National Forest Act 1984) and the various master plans at the local government level. Just like any other forest types, mangrove forests are also subjected to sustainable forest management and the states that adopted this are Perak, Johor, Selangor and Kedah where mangrove forests are very extensive. The objectives of sustainable management are to produce fuelwood, charcoal and poles, to protect the riverine and coastline ecosystems and to practice conservation (Ong 2003).

10 Mangrove Ecotourism

The initiative taken in Langkawi island and Kota Marudu, Sabah by the various authorities and stake-holders in promoting sustainable mangrove ecotourism is very commendable and it should be a model for other protected areas. Visitors and tourists were taken by boats to not only observe the beauty of mangrove vegetation, flora and fauna but also the activities of the local communities in small-scale exploitation of mangrove resources. These activities are both educational for the visitors and tourists and profit making for the local communities.

11 Challenges

Among the present and future challenges are:

a) To conserve adequate areas of riverine and coastal zones covering all forest types for the appropriate species. Of particular significance is the conservation of the *nipah* areas which are mostly outside the forest reserves. The species has been acclaimed as one of the important multi-purpose ones but conservation is not in sight, though exploitation has been minimal. There have been surveys and discussion on the possibility of converting the *nipah* sugar to biofuel.

b) To protect the coastal and estuarine ecosystems. The disaster of the tsunami of 2004 has probably taught us some important lessons of what the mangroves could do to protect the estuarine areas in particular. Likewise there are also many coastal areas which have been eroded by sea.
c) To introduce up-to-date and workable management regimes. The states of Perak, Johor, Sabah and Sarawak have updated their management plans to suit possible change in policy of exploitation and management.
d) To handle the land conversion issues. Both the local and state governments must adhere to the existing laws and regulations to ensure that land conversion issues are addressed in an appropriate manner in the future.

12 Management Strategies

The management strategies employed amongst others are:

a) To maintain and propagate the most productive forest subtypes, e.g. *Rhizophora* forest. Surveys as conducted by the Forestry Department Terengganu (Mohd. Lokman and Sulong (2001) are excellent examples to recognize the mangrove types and subtypes by zones. This classification would help the various state governments to manage their resources efficiently.
b) To encourage the propagation of other forest types and subtypes. Where the existing mangrove types have suffered damage efforts should be taken to undertake rehabilitation and restoration of mangrove belts.
c) To introduce high quality mangrove species, e.g. X*ylocarpus* species. In the past no efforts have been taken to improve the quality of mangrove species either by genetic selection or propagation.
d) To conserve all riverine and coastal mangroves. As stated above all inland mangroves in statelands, especially the *nipah* belt, should be conserved.
e) To create and maintain adequate mangrove wildlife. Where evidence is shown that there has been depletion or loss of certain animal species efforts should be taken to enrich the population.

To achieve the above strategies, all parties especially the Forestry Departments of all states, non-government organisations, schools, universities, research institutes and other stake-holders must agree to prioritise conservation and maintain sustainable timber production through management zoning, felling rotation, and best silvicultural practices. Understanding the ecology and biogeography of the mangroves are the key to successful management (Hamilton and Snedaker 1984).

Acknowledgements The authors would like to express their thanks to the various colleagues who have helped in reading and commenting on this manuscript, in particular Prof. Munir Ozturk who has been instrumental in initiating it.

References

Aldrie AA, Latiff A (2006) Hutan Simpan Merbok, Kedah: Warisan Kepelbagaian Hutan Paya Bakau. Jabatan Perhutanan Semenanjung Malaysia

Alias SA, Kuthubutheen AJ, Jones EBG (1995) Frequency of occurrence of fungi on wood in Malaysian mangroves. Hydrobiologia 295:97–106

Ambus P, Lowrance R (1991) Comparison of denitrification in two riparian soils. Soil Sc Soc Am J 55:994–998

Amir BHI (2005) Matang's charcoal and related industries. In: Shaharuddin MI, Muda A, Ujang R, Budin KA, Lim KL, Rosli S, Jalil MS, Latiff A(eds) Sustainable management of Matang mangroves 100 years and beyond. Forest biodiversity series, vol 4, Forestry Department Peninsular Malaysia, Kuala Lumpur, pp 520–531

Amiruddin A, Yahya Y, Siti Azizah MN (2005) Fish and fauna of Matang mangroves. In: Shaharuddin MI, Muda A, Ujang R, Budin KA, Lim KL, Rosli S, Jalil MS, Latiff A (eds) Sustainable management of Matang mangroves 100 years and beyond. Forest biodiversity series, vol 4, Forestry Department Peninsular Malaysia, Kuala Lumpur, pp 441–444

Anonymous (2003) Status Pengurusan Hutan Paya Laut Malaysia. Unpublished report of the Forestry Department Peninsular Malaysia

Ashari M, Awang Noor AG, Nobily F, Alek RJ (2005) Stand structure and species composition of mangrove forests in Kuala Trong range, Matang, Perak. In: Shaharuddin MI, Muda A, Ujang R, Budin KA, Lim KL, Rosli S, Jalil MS, Latiff A (eds) Sustainable management of Matang mangroves 100 years and beyond. Forest biodiversity series, vol 4, Forestry Department Peninsular Malaysia, Kuala Lumpur, pp 265–277

Ashton EC, Macintosh DJ (2001) Preliminary assessment of the plant diversity and community ecology of the Semantan mangrove forest, Sarawak, Malaysia. For Ecol Manag 66:111–129

Awang Noor AG (2005) Socio-economic benefits of mangroves with special reference to Matang mangrove forest. In: Shaharuddin MI, Muda A, Ujang R, Budin KA, Lim KL, Rosli S, Jalil MS, Latiff A (eds) Sustainable management of Matang mangroves 100 years and beyond. Forest biodiversity series, vol 4, Forestry Department Peninsular Malaysia, Kuala Lumpur, pp 464–494

Azmil AR, Wan Juliana WA, Razali MS, Shahir S, Al-Hadi M, Firdaus NA (2012) Flora bakau Pulau Pangkor. In: Abd Rahman Hj Abd Rahim, Masran Md Salleh, Mohd Nasir Abu Hassan, Muhamad Abdullah, A Latiff (eds) Hutan Pulau Pangkor Perak: Pengurusan Hutan Persekitaran Fizikal dan Kepelbagaian Biologi Siri Kepelbagaian Biologi Hutan, vol 19, pp 19–128

Ball MC (1988) Ecophysiology of mangroves. Trees 2:129–142

Berry AJ (1972) The natural history of west Malaysian mangrove fauna. Mal Nat J 25:136–162

Chai PK (1982) Ecological studies of mangrove forests in Sarawak. PhD Thesis, University of Malaya, Kuala Lumpur (unpublished)

Chan HT (1987) Country report on mangroves in Malaysia. In: Umali RM, Zamora PM, Gotera RR, Jara RS, Camacho, RS, Vannuchi M (eds) Mangroves of Asia and the Pacific: status and management. Technical report of the UNDP/UNESCO research and training pilot programme on mangrove ecosystems in Asia and the Pacific. JMC Press, pp 131–150

Chapman VJ (1975) Mangrove vegetation. In: Cramer J CVJ (eds) (1977) Ecosystems of the world: wet ecosystems vol 1. Elsevier, New York

Che Hashim H, Shahsiah AK, Che Ros A (2005) Management and conservation of mangroves: Johor experience. In: Shaharuddin MI, Muda A, Ujang R, Budin KA, Lim KL, Rosli S, Jalil MS, Latiff A(eds) Sustainable management of Matang mangroves 100 years and beyond. Forest biodiversity series, vol 4, Forestry Department Peninsular Malaysia, Kuala Lumpur, pp 69–80

Chong VC (2005) Fifteen years of mangrove fisheries research in Matang: What have we learnt? In: Shaharuddin MI, Muda A, Ujang R, Budin KA, Lim KL, Rosli S, Jalil MS, Latiff A (eds) Sustainable management of Matang mangroves 100 Years and beyond. Forest biodiversity series, vol 4, Forestry Department Peninsular Malaysia, Kuala Lumpur, pp 413–431

Clough BF (1992) Primary productivity and growth of mangrove forests. In: Robertson AI, Alongi DM (eds) Coastal and estuarine studies, American Geophysical Union, Washington, DC

Conley LM, Dick RI, Lion LW (1991) As assessment of the root zone method of waste water treatment. Res J Wat Pollut Cont Fed 63(3):239–247

Damanhuri A, Maideen H, Ong LP, Latiff QA, Lee GE (2005) Mosses of Matang mangrove forest reserve. In: Shaharuddin MI, Muda A, Ujang R, Budin KA, Lim KL, Rosli S, Jalil MS, Latiff A (eds) Sustainable management of Matang mangroves 100 years and beyond. Forest biodiversity series vol 4, Forestry Department Peninsular Malaysia, Kuala Lumpur, pp 249–253

Duke NC (1992) Mangrove floristics and biogeography. In: Robertson AI, Alongi DM (eds) Coastal and estuarine studies, American Geophysical Union, Washington, DC

Fatimah MY, F-H Ibrahim, Awang NAG (eds) (2012) Colours of Kota Marudu, Sabah, Malaysia. Universiti Putra Malaysia Press, Serdang

Faridah Y, Shukor MN, Mazlan AG (2005) Foraging behaviour and food selection of giant mudskippers, *Periophthalmodon schlosseri* Pallas 1770 at Kuala Gula, Matang mangrove reserve. In: Shaharuddin MI, Muda A, Ujang R, Budin KA, Lim KL, Rosli S, Jalil MS, Latiff A(eds) Sustainable management of Matang mangroves 100 years and beyond. Forest biodiversity series, vol 4, Forestry Department Peninsular Malaysia, Kuala Lumpur, pp 304–311

Gan BK (1995) A working plan for the Matang mangrove forest reserve, Perak. State forestry department of Perak

Gong WK, Ong JE, Leong YK (1984) The mangrove resources of the east coast of peninsula Malaysia— botanical aspects. In: Chua TE, Joseph KC (eds) Coastal resources of east coast peninsular Malaysia. An assessment in relation to potential oil spill. Penerbit Universiti Sains Malaysia, [Minden] Pulau, Pinang, pp 91–109

Haliza AR, Sarah A, Mokhtar M, Latiff A (2005) Perundangan yang berkaitan ekosistem hutan paya bakau di Selangor. In: Shaharuddin MI, Muda A, Ujang R, Budin KA, Lim KL, Rosli S, Jalil MS, Latiff A(eds) Sustainable management of Matang mangroves 100 years and beyond. Forest biodiversity series, vol 4, Forestry Department Peninsular Malaysia, Kuala Lumpur, pp 127–138

Hamdan O, Khali AH, Shamsudin I, Raja BRS (eds) (2012) Status of mangroves in Peninsular Malaysia. Ministry of Natural Resources and Environment Malaysia, Forest research Institute Malaysia and Department of Forestry Peninsular Malaysia

Hamilton LS, Snedaker SC (eds) (1984) Handbook for mangrove area management. UNEP and Environmental Policy Institute, East-West Center, Honolulu, Hawaii

Ibrahim S, Husin MN (2005) Future research and development of mangroves in Malaysia. In: Shaharuddin MI, Muda A, Ujang R, Budin KA, Lim KL, Rosli S, Jalil MS, Latiff A (eds) Sustainable management of Matang mangroves 100 years and beyond. Forest biodiversity series, vol 4, Forestry Department Peninsular Malaysia, Kuala Lumpur, pp 153–161

Jaman R, Maideen H (2005) Fern and fern-allies diversity of Matang mangrove forest. In:Shaharuddin MI, Muda A, Ujang R, Budin KA, Lim KL, Rosli S, Jalil MS, Latiff A (eds) Sustainable management of Matang mangroves 100 years and beyond. Forest biodiversity series, vol 4, Forestry Department Peninsular Malaysia, Kuala Lumpur, pp 246–248

Japar Sidik B (1994) Mangrove plant resources in ASEAN region. In: Green EP, Short FT (eds) Proceedings third ASEAN-Australian symposium on living coastal resources. Chulalongkorn University, Bangkok, pp 123–138

Japar Sidik B, Zakaria MH, Arshad A (2005) Mangrove ecosystem and associated plant resources. In: Shaharuddin MI, Muda A, Ujang R, Budin KA, Lim KL, Rosli S, Jalil MS, Latiff A (eds) Sustainable management of Matang mangroves 100 years and beyond. Forest biodiversity series, vol 4, Forestry Department Peninsular Malaysia, Kuala Lumpur, pp 392–412

Juliana WA, Nizam MN (2005) Forest structure and above ground biomass of two mangrove communities in Matang. In: Shaharuddin MI, Muda A, Ujang R, Budin KA, Lim KL, Rosli S, Jalil MS, Latiff A (eds) Sustainable management of Matang mangroves 100 years and beyond. Forest biodiversity series, vol 4, Forestry Department Peninsular Malaysia, Kuala Lumpur, pp 236–245

Kugan F (2003) Conservation and biodiversity of mangrove in Sabah. Paper presented at the national workshop on mangrove forest— new insights on the management and silviculture of mangroves, Taiping, Perak. 15–17 Jul, Malaysia

Kuthubutheen AJ (1984) Leaf surface fungi associated with *Avicennia alba* and *Rhizophora mucronata* in Malaysia. In: E. Soepadmo et al. (eds) Proceedings of the Asian symposium on mangrove environment— research and management, University of Malaya, Kuala Lumpur, pp 153–171

Latiff A (2005) Conservation and management of mangroves in Peninsular Malaysia. In: Shaharuddin MI, Muda A, Ujang R, Budin KA, Lim KL, Rosli S, Jalil MS, Latiff A (eds) Sustainable management of Matang mangroves 100 years and beyond. Forest biodiversity series, vol 4, Forestry Department Peninsular Malaysia, Kuala Lumpur, pp 139–152

Latiff A (2009) Malaysian nipah: an appraisal of its scientific, economic potential and future. In: Abd. Rahman, AR et al (eds) Prosiding Seminar Pengurusan dan Pemuliharaan Hutan Nipah 2008. Jabatan Perhutanan Semenanjung Malaysia, Kerajaan Negeri Pulau Pinang & Kementerian Sumber Asli & Alam Sekitar, pp 49–58

Latiff A (2012) Conservation strategies for endangered mangrove swamp forests in Malaysia. Pakistan J Bot 44(special issue):27–36

Liew TC (1980) Mangrove forest of Sabah. Proceedings of workshop on mangroves and estuarine vegetation Malaysia. Universiti Pertanian Malaysia, pp 354–377

Lugo AE, Snedaker SC (1974) The ecosystem of mangrove: Ann Rev Ecol Syst 5:39–64

MacNae W (1968) A general account of the fauna and flora of mangrove swamp and forest in the Indo-West pacific region. Adv Mar Biol 6:73–270

Hossain M (2004) Biomass, litter production and selected nutrients in *Bruguiera parvifolia* (Roxb.) Wight & Arn. dominated mangrove forest ecosystem at Kuala Selangor. PhD Thesis, Universiti Putra Malaysia

Maimon A, Wan JWA, Norhayati A, Shukor MN (eds) (2008) Biodiversity of Sungai Pulau, Ramsar site, Johor. Universiti Kebangsaan Malaysia, Jabatan Aolam Sekitar Malaysia & UEM Land

Marajan E (2005) Management and conservation of mangroves: Sarawak experience. In: Shaharuddin MI, Muda A, Ujang R, Budin KA, Lim KL, Rosli S, Jalil MS, Latiff A (eds) Sustainable management of Matang mangroves 100 years and beyond. Forest biodiversity series, vol 4, Forestry Department Peninsular Malaysia, Kuala Lumpur, pp 58–68

Mohd. Lokman H, Sulong I (2001) Mangrove of Terengganu. A joint publication between Kolej Universiti Sains dan Teknologi Malaysia and Forestry Department Peninsular Malaysia

Yusoff MK, Haron AR, Khairuddin NA, Jahaya JN, Ismail MZ (2005) Water quality assessment of Matang mangrove forest. In: Shaharuddin MI, Ujang R, Budin KL, Rosli S, Jalil MS, Latiff A (eds) Sustainable management of Matang mangroves 100 years and beyong. Forest Biodiversity SZeries, vol. 4, Forestry Department Peninsular Malaysia, Kuala Lumpur, pp 194–203

Muda A, Ahmad Z, MA, Lim KL (2005) Sustainable management and conservation of the Matang mangroves. In: Shaharuddin MI, Muda A, Ujang R, Budin KA, Lim KL, Rosli S, Jalil MS, Latiff A (eds) Sustainable management of Matang mangroves 100 years and beyond. Forest biodiversity series, vol 4, Forestry Department Peninsular Malaysia, Kuala Lumpur, pp 39–52

National Forestry Act 1984 (Akta 313) Jabatan Perhutanan Semenanjung Malaysia. International Law Book Services. pp 19

Nik Mohd. Shah NM, Shah Rani AZ, Zakaria AH (2005) Management and conservation of mangroves: Selangor experience. In: Shaharuddin MI, Muda A, Ujang R, Budin KA, Lim KL, Rosli S, Jalil MS, Latiff A (eds) Sustainable management of Matang mangroves 100 years and beyond. Forest biodiversity series, vol 4, Forestry Department Peninsular Malaysia, Kuala Lumpur, pp 91–101

Nilus R, Chung AYC, Pereira JT, Sugau JB, Tangah J, Sabran S, Chong REY (2010) Mangroves of Sabah— an introduction to the flora and fauna. Sabah forestry department, Malaysia

Noramly G (2005) Bird diversity of Matang mangroves. In: Shaharuddin MI, Muda A, Ujang R, Budin KA, Lim KL, Rosli S, Jalil MS, Latiff A (eds) Sustainable management of Matang mangroves 100 years and beyond. Forest Biodiversity Series, vol 4, Forestry Department Peninsular Malaysia, Kuala Lumpur, pp 289–296

Norhayati A, Latiff A (2001) Biomass and species composition of a mangrove forest in Pulau Langkawi, Malaysia. Malays Appl Biol 30(1–2):75–80

Norhayati A, Nizam MS, Juliana WA, Shukor MN, Fera Fizani AF, Hafizah S, Sariah J, Jamaliah J, Intan Nadira AK, Latiff A (2005) Sustainable management of Matang mangroves 100 years and beyong. Forest Biodiversity SZeries, vol. 4, Forestry Department Peninsular Malaysia, Kuala Lumpur, pp 54–264. (eds: Shaharuddin MI, Ujang R, Budin KL, Rosli S, Jalil MS, Latiff A)

Ong JE (1982) Mangrove and aquaculture in Malaysia. Ambio 11(51):252–257

Ong JE (2003) Plants of the Merbok Mangrove, Kedah, Malaysia and urgent need for their conservation. Folia malaysiana 4:1–18

Ong JE (2005) The need for ecological consideration in the sustainable management of Matang mangroves. In: Shaharuddin MI, Muda A, Ujang R, Budin KA, Lim KL, Rosli S, Jalil MS, Latiff A (eds) Sustainable management of Matang mangroves 100 years and beyond. Forest biodiversity series, vol 4, Forestry Department Peninsular Malaysia, Kuala Lumpur, pp 162–167

Ooi AL, Chew LL, Chong VC, Ogawa Y (2005) Diel abundance of zooplankton particularly fish larvae in the Sangga Kecil estuary, Matang mangrove forest reserve. In: Shaharuddin MI, Muda A, Ujang R, Budin KA, Lim KL, Rosli S, Jalil MS, Latiff A (eds) Sustainable management of Matang mangroves 100 years and beyond. Forest biodiversity series, vol 4, Forestry Department Peninsular Malaysia, Kuala Lumpur, pp 445–453

Othman MA (1994) Value of mangroves in coastal protection. Hydrobiologia 285:277–282

Phang SM, Wong CL, Lim PE (2005) Seaweed diversity of Langkawi with emphasis on northeast Langkawi. Malay J Sci 24:77–94

Phang SM, Wong CL, Lim PE, Ooi JLS, Gan SY, Ismail M, Yeong HY, Mustafa EM (2007) Seaweed diversity in Malaysia. In: Chua LSL, Kirton LG and Saw LG (eds) Status of biological diversity in Malaysia and threats assessment of plant species in Malaysia. Forest Research Institute, Malaysia, pp 185–210

Razani Ujang (1982) The role of mangrove forests in coastal zone management. Paper presented at the workshop on ecological basis for rational resource utilization in the humid tropics of Southeast Asia. Universiti Pertanian Malaysia

Sasekumar A (2005) Fauna of Matang mangrove forest. In: Shaharuddin MI, Muda A, Ujang R, Budin KA, Lim KL, Rosli S, Jalil MS, Latiff A (eds) Sustainable management of Matang mangroves 100 years and beyond. Forest biodiversity series, vol 4, Forestry Department Peninsular Malaysia, Kuala Lumpur, pp 278–282

Shaharuddin MI, Nazir Khan NK, Baharom Y, Othman R (2005) Management of Matang mangroves: historical perspectives. In: Shaharuddin MI, Muda A, Ujang R, Budin KA, Lim KL, Rosli S, Jalil MS, Latiff A (eds) Sustainable management of Matang mangroves 100 years and beyond. Forest biodiversity series, vol 4, Forestry Department Peninsular Malaysia, Kuala Lumpur, pp 27–38

Shahrul Annuar MA, Nor Zalipah M, Nurul Ain E, Ibrahim J, Rayan DM, Ganesan M, Nazri A, Hussain YM (2005) Survey of mammals and other vertebrate fauna of Matang mangrove swamp forest, Perak. In: Shaharuddin MI, Muda A, Ujang R, Budin KA, Lim KL, Rosli S, Jalil MS, Latiff A (eds) Sustainable management of Matang mangroves 100 years and beyond. Forest biodiversity series, vol 4, Forestry Department Peninsular Malaysia, Kuala Lumpur, pp 320–324

Shamsul K, Manaf TA, Ibrahim AZ, F-H Ibrahim, Latiff A (2005) Preliminary assessment of the flowering plants diversity of Matang mangrove forest. In: Shaharuddin MI, Muda A, Ujang R, Budin KA, Lim KL, Rosli S, Jalil MS, Latiff A (eds) Sustainable management of Matang mangroves 100 years and beyond. Forest biodiversity series, vol 4, Forestry Department Peninsular Malaysia, Kuala Lumpur, pp 223–235

Siti Aisyah Alias (2007) A checklist of manglicolous marine fungi from Malaysia. In: Chua LSL, Kirton LG, Saw LG (eds) Status of biological diversity in Malaysia and threats assessment of plant species in Malaysia, Forest Research Institute, Malaysia, pp 181–185

Smith TJ (1992) Forest structure. Chapter 5 in Coastal and estuarine studies. Robertson AI and Alongi DM (eds) American Geophysical Union, Washington, DC

Snedaker SC (1978) Mangroves: Their value and perpetuation. Nat Res 14:6–13

Soepadmo E, Pandi M Zain (1989) Floristic composition, structure and potential net primary productivity of residual mangrove communities in Peninsular Malaysia. Paper presented at the National Workshop on the AACMS Project-Coastal living resources. Awana Golf & Country Club, Genting Highlands, Pahang, Malaysia (mimeograph)

Sulong I, Ismail H (1990) Classification of mangrove forest by using 1:40 000-scale aerial photograph. For Ecol Manag 33–34:583–592

Tangah J (2005) Management and conservation of mangroves: Sabah experience. In: Shaharuddin MI, Muda A, Ujang R, Budin KA, Lim KL, Rosli S, Jalil MS, Latiff A (eds) Sustainable management of Matang mangroves 100 years and beyond. Forest biodiversity series, vol 4, Forestry Department Peninsular Malaysia, Kuala Lumpur, pp 53–57

Tomlinson PB (1986) The botany of mangroves. Cambridge tropical biology series, Cambridge, pp 413

Wan Juliana WA, Damanhuri A, Razali MS, Latiff A (2010) Mangrove flora of Langkawi. Pusat Penyelidikan Langkawi, LESTARI & LADA

Watson JG (1928) Mangrove forests of the Malay Peninsula. Malayan forest records. Fraser & Neave, Malaysia, 6:275

Wolanski E (1995) Transport and sediment in mangrove swamp. Hydrobiologia 295:31–42

Wyatt-Smith J (1960) Field key to the trees of mangrove forests in Malaya. Malay Forester 23:126–132

Zaidi MI, Azman S (2005) Butterfly fauna (Lepidoptera: Rhopalocera) of Langgun Island and Kisap area, Langkawi. In: Mohd. Shafie Leman, Ibrahim Komoo, Latiff A, Hood Salleh (eds) Kilim basin: culture and natural heritage of Langkawi, pp 279–288

Zaidi MI, Azman S, Wong CY (2005). Synchronous fireflies on *berembang* trees along Sungai Sepetang, Taiping, Perak. In: Shaharuddin MI, Muda A, Ujang R, Budin KA, Lim KL, Rosli S, Jalil MS, Latiff A (eds) Sustainable management of Matang mangroves 100 years and beyond. Forest biodiversity series, vol 4, Forestry Department Peninsular Malaysia, Kuala Lumpur, pp 283–288

Zaidi CC, Lihan T, Mohamad CR, Mazlan AG (2005) Macroinvertebrate fauna of Kilim mangroves, Langkawi. In Mohd. Shafie Leman, Ibrahim Komoo, Latiff A, Hood Salleh (eds) Kilim basin: culture and natural heritage of Langkawi, pp 269–278

Distribution and Rarity of Rhizophoraceae in Peninsular Malaysia

W. A. Wan Juliana, M. S. Razali and A. Latiff

Contents

Abstract This paper attempts to review the distributions and rarity of Rhizophoraceae in Peninsular Malaysia. Results presented were from plot studies (11 sites), random field surveys (3 sites) and previously published reports. The primary data were from four sites in Johor (Belungkor Forest Reserve, Pulai Forest Reserve, Santi Forest Reserve and Tanjung Piai), four in Langkawi (Ayer Hangat, Sungai Kilim, Kisap Forest Reserve and Selat Tuba), two in Matang Forest Reserve (Compartment 49 and VJR), one in Selangor (West Port, Klang), one in Terengganu (Kuala Kemaman Forest Reserve), and one each in Melaka (Tanjung Tuan) and

W. A. Wan Juliana (✉) · A. Latiff
Faculty of Science and Technology, Universiti Kebangsaan Malaysia,
43600 Banji, Selangor, Malaysia
e-mail: ayie@ukm.my

M. S. Razali
Faculty of Science and Technology, Universiti Malaysia Terengganu,
Kuala Terengganu, Terengganu, Malaysia

I. Faridah-Hanum et al. (eds.), *Mangrove Ecosystems of Asia*,

DOI 10.1007/978-1-4614-8582-7_2,

Negeri Sembilan (Port Dickson), respectively. Secondary data were obtained from reports of studies at Sungai Merbok Forest Reserve in Kedah, Matang mangrove forest in Perak, Sementa mangrove forest in Selangor, Kuala Sedili Forest Reserve, Johor, Terengganu mangrove forest, Terengganu and also a general survey throughout Peninsular Malaysia. All trees at 5 cm and above diameter were recorded in plot studies. A total of seven species from the family Rhizophoraceae were sampled at the study plots, the number of which ranged from two to six. The most common and highly abundant species in the study sites was *Rhizophora apiculata* with an estimated total of 1,184 trees (51.2 % of the total). Other abundant species were *R. mucronata* (25.5 %) and *Ceriops tagal* (12.2 %). The less abundant and restricted species in the study sites were *Bruguiera cylindrica* (3.80 %), *B. gymnorrhiza* (2.94 %), *B. parviflora* (4.06 %) and *B. sexangula* (0.30 %). *Rhizophora stylosa*, *Ceriops decandra*, *Kandelia kandel* and *Bruguiera hainesii* were only found from random surveys and not in plot studies and are considered rare species. The *Rhizophora x lamarckii* and *R. x annamalayana* are considered very rare and endangered and were only found in Selat Kuah, Langkawi and Pulai Forest Reserve and Merbok Forest Reserve, respectively.

1 Introduction

Mangrove trees are a major component of a mangrove ecosystem. According to Duke (1992), mangrove has been defined as a "community of trees, shrubs, palms or ground ferns, generally exceeding more than half a meter in height, and which normally grows above mean sea level in the intertidal zones of marine coastal environments, or estuarine margins." There are three components of a mangrove habitat: plants, aquatic animals and terrestrial animals.

2 Mangrove Forest in Malaysia

According to Japar Sidik (1994), a mangrove forest developed best in Malaysia where the highest number of species occur and is favoured by a humid tropical climate and high rainfall, which are usually accompanied by silt-laden rivers forming suitable mudflats. These mangrove forests are also found to develop further inland, up to where the tidal influence of the sea can be felt in the rivers or streams. The Malaysian mangrove is the third largest mangrove forest in the Asia-Pacific region after Indonesia and Thailand. It could be found mainly in the states of Perak, Kedah and Johor. Smaller mangrove areas are found in Kelantan, Terengganu and Pahang. In Sarawak, mangroves are found along the coastlines and estuaries of the Sarawak River, the Rajang Delta and the Trusan River and in Sabah they are found in the eastern and northern coastal areas of the state (Table 1).

In the case of Peninsular Malaysia, mangrove forests are well developed in the west due to relatively sheltered coasts. The seas of the west coast are calmer due to

Table 1 Occurrences of mangrove by state in Peninsular Malaysia, Sabah and Sarawak. (Source: Shaharuddin et al. 2005)

State	State land mangrove (ha)	Permanent reserve forest (ha)	Total area (ha)
Kedah	150	8,257	8,407
Pulau Pinang	494	376	870
Perak	122	41,302	41,424
Selangor	4,606	14,897	19,503
Negeri Sembilan	–	204	204
Melaka	–	80	80
Johor	3,348	17,832	21,180
Terengganu	692	1,130	1,822
Sub-total	9,412	84,078	93,490
Sarawak	93,200	33,200	126,400
Sabah	23,266	317,423	340,689
Total	125,878	434,701	560,579

the protection accorded by Sumatera and bordered by the Strait of Malacca that has a limited wind fetch, whereas the east coast is exposed to the South China Sea that has larger and more energetic waves (Aldrie 2002). Mangrove forest developments are inhibited by strong currents and wave action, especially due to the monsoon season (Gong et al. 1984; Mohd. Lokman and Yaakob 1995). About 96 % of these forest reserves are located on the west coast while only 4 % are located on the east coast. There are small patches of mangrove area on the east coast of Peninsular Malaysia and they are confined to river mouths in the states of Pahang, Terengganu and Kelantan. However most of these mangrove reserves are situated in Pahang, with 11 locations, compared to three locations in Terengganu, and no mangrove forest is set aside as a reserve in Kelantan (Japar Sidik 1994).

In Peninsular Malaysia, the total extent of mangrove forest reserve in 2005 was estimated to be about 102,541 ha. When compared to the total forest area of about 4,639,981 ha and total land area of about 13,167,245 ha, mangrove forests amount to only 2.2 % and 0.79 %, respectively (JPNT 2005). The largest mangrove forest in Peninsular Malaysia is the Matang Mangrove (covering about 41,000 ha), which has been managed for charcoal, firewood and poles by the Forestry Department since 1904. Studies on forest composition are very important as a part of the present environmental impact assessment and, more importantly, for management of natural resources, especially for monitoring changes in ecosystem quality.

3 Research on the Mangrove Forest in Malaysia

Published works on estuary mangrove swamps on the east coast of Peninsular Malaysia are scanty. On the other hand, in other states on the west coast of Peninsular Malaysia, mangrove areas have been well studied. The first study on floristic composition, structure and potential net primary production of mangrove forest in Kuala Kemaman was done by Soepadmo and Pandi Mad Zain (1989). This study surveyed the mangrove of Kuala Kemaman and Kampung Pantai Tinggi, Kemaman where only 24 species of plants are recorded. The dominant species were *Rhizophora apiculata* and *Bruguiera gymnorrhiza*.

Norhayati (1995), who studied the biomass and species composition in 1 ha stand of mangrove forest in Kisap Forest Reserve, Langkawi, recorded a total of 849 trees comprising nine mangrove tree species from four families. The dominant species is *Rhizophora apiculata* (65.5 %) followed by *Xylocarpus granatum* (10.6 %) and *Bruguiera parviflora* (9.8 %). Sulong and Ismail (1990) identified the species groups of mangrove forest from Kemaman to Kuantan and they recognized three mangrove forest types, namely, the *Rhizophora* type, *Avicennia/Sonneratia* type, and the mixed-mangrove type. An area of 2,214 ha is covered by mangrove forest, of which 2 % is the *Avicennia/Sonneratia* type, 24 % the *Rhizophora* type and 74 % the mixed-mangrove type. *Avicennia/Sonneratia* is found to have the highest stand density with 13,348 trees/ha, followed by *Rhizophora* with 6,697 and mixed-mangrove forestwith 1,997.

The study conducted by Hafizah Seman (2004) in the Kisap Forest Reserve, Langkawi in the 0.25 ha plot area found six species from four families, namely, *Rhizophora mucronata*, *R. apiculata*, *Xylocarpus* moluccensis, *Ceriops tagal*, *Lumnitzera littorea* and *Avicennia marina*. The most dominant species was *R. mucronata* (45.2 %), followed by *R. apiculata* (30.4 %) and *C. tagal* (14.2 %). The other study in Langkawi was conducted by Fera Fizani (2004) in the Ayer Hangat Forest Reserve, whereby 230 mangrove trees in the 0.25 ha plot area were sampled comprising seven different species from five families, namely, Rhizophoraceae, Meliaceae, Avicenniaceae, Combretaceae and Sonneratiaceae. The result showed that the dominant species is *Rhizophora mucronata* (58.7 %), followed by *Sonneratia alba* (14.9 %) and *Rhizophora apiculata* (13.4 %).

Research on the conservation value of mangrove has been carried out by Ashton and Macintosh (2001) at Semantan mangrove forest, Sarawak. They found that the uniqueness in the Semantan mangrove is related to the large strand of mature *X. granatum* that dominates the forest. Research about forest composition and biomass estimation of the mangrove at west Port, Klang, Selangor by Norhayati et al. (2007) determined that the total number of individual trees recorded in all ten plots was 222 from ten species of three families, namely, Rhizophoraceae, Avicenniaceae and Meliaceae. *Rhizophora apiculata* was the most dominant tree species (34.7 %) followed by *Ceriops tagal* with 58 trees forming 26.1 % of all trees. This study also found 26 trees of *Bruguiera cylindrica*, 11.7 % *R. mucronata*, 8.1 % *B. gymnorrhiza* and 0.9 % of *B. parviflora*.

Another study in Belungkor Reserve Forest, Johor conducted by Intan et al. (2003) sampled 196 trees in 0.1 ha. It included eight mangrove species from three families, namely, Rhizophoraceae, Euphorbiaceae and Meliaceae. From the total individuals, 168 trees are in the family Rhizophoraceae. *Rhizophora apiculata* was the dominant species, which covers 30.1 % (59 trees), followed by *R. mucronata* with 53 trees (27 %) and *Ceriops tagal* with 45 trees (23 %) (Table 2).

Table 2 Location of sampling plots in Peninsular Malaysia (areas in ha)

Location	Size	Plot size	Date	Sources
Sungai Pulai, Johor	7,600	0.1	12-16/07/2002	Jamaliah et al. (2003)
Sungai Belungkor, Johor	1,600	0.1	12-15/07/2002	Intan et al. (2003)
Sungai Santi, Johor	3,028	0.1	13-14/07/2002	Sariah (2003)
Matang: VJR C49	110	0.06	20-25/10/2002	Juliana and Nizam (2005)
Ayer Hangat, Langkawi	555	1.0	21-27/12/2003	Fera Fizani (2004)
Sungai Kisap, Langkawi	1,464	0.25	21-27/12/2003	Hafizah Seman (2004)
West Port, Klang	10,817	0.4	10/2003	Norhayati et al. (2007)
Port Dickson & Tanjung Tuan	60.7	0.06	25-27/08/2005	Juliana et al. (2007)
Kuala Kemaman F.R	816	1.0	2007	Ida Suzilawate (2007)

4 Study Sites & Methods

4.1 Southern Part of Peninsular Malaysia

The southern part of Peninsular Malaysia is located at the confluence of the South China Sea and the Strait of Malacca tidal regimes, leading to complex and strong tidal processes. The largest river discharging into the Johor Strait at the eastern side is Sungai Johor, while Sungai Pulai is the largest river on the western side. Sungai Santi is located at the southeast of Sungai Johor, near the southeastern tip of the peninsula and part of Kota Tinggi District, Johor.

4.2 Sungai Pulai Forest Reserve

The Sungai Pulai Forest Reserve is the largest forest reserve in South Johor (7,600 ha), stretching from Jeram Batu in the north to Tanjung Piai in the southwest and Tanjung Pelepas in the southeast. This reserve is also managed by the Johor State Forestry Department for sustainable forestry production, especially to supply wood for the charcoal industry.

4.3 Western Part of Peninsular Malaysia

At Matang, many big rivers discharge their water and effluents into the Malacca Strait, including Sungai Sangga Besar for the compartment 49 Virgin Jungle Reserve (VJR).

4.4 Matang Forest Reserve, Compartment 49

The VJR of compartment 49 covers a total area of 110 ha, which include the protective forest (85 ha) and dryland forest (25 ha). A plot of 30m × 20m was established at compartment 49.

4.5 Northern Part (Langkawi) of Peninsular Malaysia

In Langkawi, the major rivers that discharge into the Pulau Peluru Strait are Sungai Kisap and Sungai Kilim.

4.6 Ayer Hangat Forest Reserve

Gua Cherita Forest Reserve is located at Tanjung Rhu, Langkawi, and there are eight compartments in this reserve, which are located at the northeastern part of the main island of Langkawi. This study was conducted in compartments 3, 4, and 5. This reserve is managed by the Forestry Department of Kedah State. The sampling areas consisted of 50m × 10m plots, located randomly.

4.7 Sungai Kisap Forest Reserve

The Kisap Forest Reserve is located on the northeastern coast of Pulau Langkawi. Its border extends for about 27.5 km covering an area of 1,464 ha from a total of 3,270 ha or 45 %. The Kisap FR contains the largest mangrove area on the island. There are 17 compartments, six of which are under mangroves. These compartments are numbered 4, 5, 6, 7, 8 and 9. Two plots were set up in compartment 16 and one each in compartments 14, 15, and 17. The sampling plots consisted of five 50m × 10m plots, located randomly.

4.8 Mangrove in West Port, Klang

Plot establishment and field surveys on tree species composition and other measurements were conducted in October 2003. The first site was located at Pulau Che Mat Zin in compartment 35 of Che Mat Zin Forest Reserve, while the second site was at Pulau Indah, partly in compartments 10 and 3 of Pulau Indah Forest Reserve. Pulau Che Mat Zin is located between Pulau Selat Kering, Pulau Kelang and Pulau Indah. This island comprised mainly mangrove forest with some intertidal mudflats to the west and east. Pulau Indah is located in the innermost of the Klang Islands. The sampling plots consisted of five plots of 10m × 10m arranged along a line-transect at each site.

4.9 Mangrove Forest in Port Dickson, Negeri Sembilan and Tanjung Tuan, Melaka

Mukim Pasir Panjang is located 16 km from Port Dickson and 130 km from Kuala Lumpur. Tanjung Tuan is bordered with Port Dickson and located 16 km from this town. The total area of Tanjung Tuan is 60.7 ha. Plot studies for species composition and biomass was 0.06 ha with six quadrats of 10m × 10m at Sungai Menyala. All trees in the study plot at 5 cm and above diameter were recorded.

Secondary data were also included from Sungai Merbok Reserve Forest, Kedah, Matang Mangrove Forest, Perak, Sepang kecil, Selangor, Kuala Sedili Reserve Forest and Mangrove Flora of Tanjung Piai, Johor and also Terengganu Mangrove Forest.

4.10 Sampling Methods

Measurements of DBH involved marking a tree ≥ 5 cm DBH, at its point of measurement (p.o.m) 1.3 m above ground level or 20 cm above its buttress. Measurements were made by using fiberglass diameter tape. Standard procedures suggested by Lugo and Snedaker (1974) when measuring tree diameter were followed. When a stem forks below breast height, each branch was measured as a separate stem. When a stem forks at breast height or slightly above, the diameter was measured at breast height or just below the swelling caused by the fork. For stems with swellings, branches or abnormalities at the p.o.m., the diameters were measured slightly above the irregularity where it stopped affecting the normal form. Field survey on tree species composition and other measurements were also conducted at the studied sites.

5 Results and Discussion

The information of mangrove species in this study was collected from Japar Sidik (1994) who recorded 11 species of Rhizophoraceae present in Malaysia including: *Bruguiera cylindrica, B. gymnorrhiza B. hainesii B. parviflora B. sexangula Ceriops tagal C. decandra, Kandelia candel Rhizophora apiculata R. mucronata* and *R. stylosa*. The number of individuals of all species in the mangrove study site was shown in Appendix 1.

A total of seven species of mangrove plants was recorded at the study sites and the distribution of Rhizophoraceae was shown (Appendix 2). The number of species at the 11 sites ranged from as low as two to the highest of six species. *Rhizophora apiculata* was the most common species, and was mostly present at all sites (Appendix 2). There was an estimated total of 1,184 (51.2 %). The other abundant species were *R. mucronata* (25.5 %), which appeared at 11 sites, and *Ceriops tagal* (12.2 %), which also appeared at ten sites. Five species were classified as restricted because only 0.3 % – 4.0 % were present at the study site. They were *B. cylindrica, B. parviflora*

B. gymnorrhiza, and *B. sexangula*. Another species like *B. hainesii C. decandra*, *R. stylosa* and *K. candel* were classified as restricted and rare species at 12 study sites around Peninsular Malaysia.

The mangrove species composition of the west coast and east coast is different. For example, from the Sementa mangrove forest, Soepadmo and Pandi Mat Zain (1989) reported a total of 32 species found there, of which 18 are considered as principal mangrove species. On the east coast, especially in Terengganu mangroves, Mohd Lokman and Sulong (2001) listed a total of 55 species, with 29 exclusive mangrove species and a further 26 species as being non-exclusive.

Hafizah Seman (2004) and Fera Fizani (2004) had recorded a similar number of species—six species and seven species, respectively—and only three species from the family Rhizophoraceae. Another study by Norhayati (1995) recorded nine species in the area of the Kisap Forest Reserve. The species composition of these three studies was also similar. The scenario could be explained by the location of these three study sites as they were carried out in the west coast. In Matang Mangrove forest, Wan Juliana et al. (2005) established a seaward plot and an inland plot. This study only recorded four species and three families of mangrove species. The species composition is much lower compared to the other studies because the area of the study was a pure stand of *R. apiculata*. Studies at Merbok Forest Reserve, Kedah had also recorded a large number of species. This is because the sampling method used in this study was transects. It can be concluded that the study methods influence the species composition because the mangrove areas have zonation patterns in species composition.

The species *R. apiculata* and *R. mucronata* are widely distributed throughout mangrove forest areas in Peninsular Malaysia. Both species are of economic importance in forestry and fisheries industries. There is little information about the other species, *R. stylosa* or locally named as *Akik jalar* or *Bakau pasir*. This species does not grow extensively in all mangrove areas in Peninsular Malaysia and can only be found in very restricted locations, though it has a wide distribution in the Indo-Pacific region. It stretches from the Queensland coast to as far as Taiwan (Ding Hou 1960). It has been reported that in the peninsula, *R. stylosa* was only found in Pulau Langkawi, Melaka and Johor (Kochummen 1989). Other than these areas, our survey carried out in 2001 and 2002 showed that the species is also found in Sungai Kurung Tengar, Perlis, Bagan Lalang mangrove, Sepang, Selangor, Pulau Besar, Melaka, Pulau Burung, which is small rocky island off the coast of Port Dickson, two sites at Sungai Mawar, Endau, Johor, and Pulau Sibu and Pulau Tinggi, both islands of which are off the coast of Mersing, Johor (Nasir & Safiah 2007). *R. stylosa* grows best in hard sandy soil substrate or even on rocky islands. Its occurrence in muddy areas has rarely been reported. The species is not found in Matang mangrove areas which have soft muddy alluvial soil. Soil samples were collected from Bagan Lalang mangroves where *R. stylosa* was present. The soil was analyzed and results showed that this species grows best in areas with higher contents of sandy materials compared with other soil components such as silt and clay.

A total of 26 species in 12 families were found and 25 of them are exclusive mangrove species, that is, 66 % of all exclusive mangrove species found in Malaysia

(Japar Sidik 1994). All five members of *Bruguiera* were found in the Sungai Santi. One rare species, namely, *B. hainesii*, probably is a new record to Johor (Chan 1999). Previously in Peninsular Malaysia, *B. hainesii* was only found in the Matang forest reserve (Gan 1995). Other than *B. hainesii B. sexangula* and *C. decandra* are also rare species.

As a comparison, a study by Intan et al. (2003) at Hutan Simpan Belungkur, reported a total of 16 mangrove species from nine families. Johor Forestry Department (1999) recorded a total of 26 exclusive mangrove species were found from four separate studies in Johor. One of the exclusive species from the family Rhizophoraceae is *Kandelia candel*. Absence of this species from this survey cannot be ignored. A total of 35 species and a hybrid from 26 genera and 19 families of mangrove flora were collected during the survey on Tanjung Piai and Sungai Pulai, Johor. The Tanjung Piai mangrove species is represented 35 % taxa of Malaysia's mangrove species (Japar Sidik 1994) and 31 % of the world's mangrove plant. Out of the total mangrove flora recorded, 23 species were classified as exclusive species. A total of 15 mangrove species and a hybrid found in this survey were not recorded by Jamaliah et al. (2003) in their mangrove species survey in Pulai Reserve Forest, Johor. The survey has added a new record where a sterile hybrid between *R. stylosa* and *R. apiculata,* i.e. *R. x lamarckii,* was discovered. The *R. stylosa* is considered rare and endangered and the *R. x lamarckii* is considered very rare and endangered.

A report about the Biodiversity Audit and Conservation Plan for the Mangroves of Johor, which was collected from research on the mangrove at Johor (Sungai Sedili Kecil and Sungai Sedili Besar), revealed that along the eastern banks of the mouth of Sungai Sedili Kecil is a fringing belt of the tidal mangrove. The most common mangrove from the family Rhizophoraceae is *R. apiculata*. The other tree species include *B. cylindrica B. gymnorrhiza* and *R. mucronata* . Mangrove fringing rocky and sandy shores are encountered at Teluk Merbok, situated between the estuary of Sungai Sedili Kecil and Tanjung Sedili Kecil, which are located between the rocky promontories of the bay area along stretches of sandy beach with pockets of mudflats. Mangrove trees found include *B. cylindrica B. gymnorrhiza, R. apiculata* and *R. mucronata* . Sungai Sedili Besar is unique in that it contains mangroves (a restricted habitat on the east coast) and close association of mangroves with riverine (a severely threatened habitat in Peninsular Malaysia) and coastal forests upstream of the current Kuala Sedili Forest Reserve.

Kandelia candel occurs sporadically along banks of tidal rivers on the east coast but it is very rare on the west coast. During the survey it was found in Muar as well as in Sungai Sedili Besar. During the survey it was found in Sungai Sedili Besar that *B. sexangula* is the only species that sometimes forms stilt roots. It occupies the more inward parts of the mangrove and is less common than the similar *B. gymnorrhiza*. It has an ornamental potential. It was noted in Sungai Santi but also found in Sungai Pulai and at Muar, Johor.

Studies at Merbok Forest Reserve, Kedah had recorded a large number of species. Ong et al. (1980; 2003) made a comprehensively study at Merbok Reserve Forest and stated that forest is one of the mangrove forests with the highest density of plants. These forests have not less than 30 true mangrove species. Ong (2003) recorded 53 species from 25 families and Rhizophoraceae is the most abundantly distributed. The species of Rhizophoraceae presented were *B. cylindrica B. gymnorrhiza, B. parviflora B. sexangula C. tagal R. apiculata R. mucronata* and *R. x annamalayana.*

An inventory on mangrove flora was conducted around Pulau Tuba, Pulau Dayang Bunting, Pulau Ular and Pulau Singa Besar which represent Selat Kuah (Razali Salam et al. 2005). Random walk and boat surveys were employed to document all mangrove species found. In this survey, there are 15 species and a hybrid which was not recorded by Norhayati and Latiff (2001) in their survey of mangrove species in Kisap Forest Reserve. The study has also added three new records for the Kilim-Kisap area (Wan Juliana et al. 2005), namely, *Xylocarpus mekongensis, R. stylosa* and a sterile hybrid between *R. x lamarckii.* Selat Kuah is a suitable habitat for *R. apiculata* and *R. stylosa.* The mangrove forest in Selat Kuah is considered as having a high density of mangrove plants, followed by the Pulai Forest Reserve and Merbok Reserve Forest.

From the preliminary assessment of the flowering plant diversity of Matang mangrove forest (Shamsul et al. 2005.), seven species of Rhizophoraceae were present, namely, *B. cylindrica B. gymnorrhiza B. parviflora B. sexangula R. apiculata R. mucronata* and *C. tagal*. All the species are abundant at the Sungai Derhaka Besar, Jebong Forest Reserve. Importantly, *R. apiculata* dominated the area of study.

Through the observation in the Terengganu Mangrove Forest, there are 29 exclusive and 26 non-exclusive mangrove species in Terengganu. In contrast, Malaysia has 38 exclusive and 57 non-exclusive mangrove species. Thus, more than 50 % of the mangrove species of Malaysia are available in Terengganu. For Rhizophoraceae, *B. cylindrica B. gymnorrhiza, B. parviflora B. sexangula C. decandra, K. candel, R. apiculata* and *R. mucronata* were recorded.

6 Conclusion

From this study, Rhizophoraceae were distributed all over Peninsular Malaysia and *R. apiculata* was the dominant species in all areas followed by *R. mucronata. R. apiculata* dominated the mangroves on the east and west coasts of Peninsular Malaysia. *B. sexangula R. stylosa C. decandra, K. candel,* and *B. hainesii* were restricted species in Peninsular Malaysia and hybrids, i.e *R. x lamarckii* and *R. x annamalayana,* were the rare and endangered species. In conclusion, all Rhizophoraceae are present in Perlis, Kedah, Perak, Selangor, Negeri Sembilan, Melaka, Johor, Terengganu and Kelantan.

Acknowledgements The authors would like to acknowledge the Forestry Department Peninsular Malaysia, Kedah Forestry Department for their cooperation and support. Thanks to the School of Environmental and Natural Resources Sciences, Faculty of Science and Technology, Universiti Kebangsaan Malaysia and also the Biology Department, Faculty of Science and Technology, Universiti Malaysia Terengganu.

Appendix 1 Number of individuals for each species at the study sites for family Rhizophoraceae

Location	Species	No. of individuals	Source
Belungkur F.R., Johor	*R. apiculata*	59	Intan et al. (2003)
	R. mucronata	53	
	C. tagal	45	
	B. gymnorrhiza	5	
	B. cylindrica	6	
Sungai Pulai F.R.R, Johor	*B. cylindrica*	8	Jamaliah et al. (2003)
	C. tagal	62	
	R. apiculata	49	
	R. mucronata	68	
Sungai Santi F.R., Johor	*R. mucronata*	33	Sariah (2003)
	R. apiculaa	23	
	C. tagal	9	
	R. apiculata	15	
	R. mucronata	43	
Pulau LangkawiAyer Hangat F.R	*C. tagal*	15	Norhayati et al. (2005)
Kisap F. R	*R. apiculata*	10	
	R. mucronata	46	
	C. taga	29	
	R. apiculata	3	
	R. mucronata	64	
Kisap F.R.	*C. tagal*	2	Hafizah (2004)
	R. apiculata	114	
	R. mucronata	64	
Ayer Hangat F.R.	*C. tagal*	15	Fera Fizani (2004)
	R. apiculata	46	
	R. mucronata	134	
Pulau Langkawi M.F.	*R. apiculata*	557	Norhayati & Latiff (2001)
	R. mucronata	12	
	B. gymnorrhiza	26	
	B. parviflora	83	
	C. tagal	44	
Port Dickson, Negeri Sembilan & Tanjung Tuan, Melaka	*B. gymnorrhiza*	1	Gan (2006)
	C. tagal	4	
	R. apiculata	46	
Kuala Kemaman F.R. Terengganu	*B. cylindrica*	6	Ida Suzilawate (2007)
	B. gymnorrhiza	2	
	B. sexangula	7	
	R. apiculata	188	
	R. mucronata	20	
Matang: VJR C49	*R. apiculata*	62	Juliana & Nizam (2005)
	R. mucronata	20	
West Port, Klang	*R. apiculata*	77	Norhayati et al. 2007
	R. mucronata	25	
	B. cylindrica	26	
	B. gymnorrhiza	18	
	B. parviflora	2	
	C. tagal	58	
Total		2313	

Appendix 2 Distribution of Rhizophoraceae in Peninsular Malaysia

Location	Species											Source
	B. cylindrica	*B. gymnorrhiza*	*B. hainesii*	*B. parviflora*	*B. sexangula*	*C. tagal*	*C. decandra*	*K. candel*	*R. apiculata*	*R. mucronata*	*R. stylosa*	
Ayer Hangat F.R.	X	X		X		X			X	X		Norhayati et al. (2005), Fera Fizani (2004).
Kisap F.R.		X		X		X			X	X		Norhayati et al. (2005), Norhayati & Latiff (2001), Hafizah (2004).
Merbok F.R.	X	X		X	X	X	X		X	X		Aldrie & Latiff (2006)
Matang F.R.	X	X	X	X	X	X			X	X		Juliana & Nizam (2006), Shamsul et al. (2005)
West Port, Klang	X	X		X		X			X	X		Norhayati et al. (2007) –
Port Dickson & Tanjung.Tuan, Melaka		X				X			X			Gan (2006)
Belungkur F.R.	X	X				X			X	X		Intan et al. (2002)
Sungai Pulai F.R.	X				X	X			X	X	X	Jamaliah et al. (2003)
Tanjung Piai	X	X		X		X	X			X	X	Mangrove flora of Tanjung Piai, Johor
Santi F.R.	X	X	X		X	X			X	X	X	Sariah (2003)

References

Aldrie AA (2002) Pengurusan kearah pemuliharaan spesies bakau: Kajian kes di Hutan Simpan Merbok. MSc. Thesis, Universiti Kebangsaan Malaysia (unpublished)

Aldrie AA, Latiff A (2006) Hutan Simpan Merbok, Kedah: Warisan Kepelbagaian Hutan Paya Bakau. Jabatan Perhutanan Semenanjung Malaysia

Ashton EC, Macintosh DJ (2001) Preliminary assessment of the plant diversity and community ecology of the Semantan Mangrove Forest, Sarawak, Malaysia. Forest Ecol Manag 66:111–129

Chan HT (1999) Biodiversity survey of the mangroves of Johor: flora and vegetation. Preparation of an integrated management plan for sustainable use of Johor mangrove forest. Johor Forestry Department/ Danced/ Darudec

Ding H (1960) A review of the genus *Rhizophora* with special reference to the Pacific species. Blumea 10(2):625–634

Duke NC (1992) Mangrove floristics and biogeography. In Robertson AI, Alongi DM (eds) Tropical mangrove ecosystem, pp 63–100

Fera Fizani AF (2004) Komposisi spesies dan anggaran biojisom hutan paya bakau di hutan simpan Ayer Hangat, Pulau Langkawi. MSc. Thesis, Universiti Kebangsaan Malaysia (unpublished)

Gan BK (1995) A working plan for the Matang mangrove forest reserve, Perak. State Forestry Department of Perak

Gan SH (2006) Flora dan biojisim pokok di Hutan Paya Bakau Mukim Pasir Panjang, Port Diskson dan Tanjung Tuan, Melaka. BSc. Thesis. Universiti Kebangsaan Malaysia (unpublished)

Gong WK, Ong JE, Leong YK (1984) The mangrove resources of the east coast of peninsula Malaysia – botanical aspects. In. Chua TE, Joseph KC (eds) Coastal resources of east coast peninsular Malaysia. An assessment in relation to potential oil spill, pp 91–109

Hafizah S (2004) Biojisim dan komposisi spesies Hutan Paya Bakau di Hutan Simpan Kisap, Pulau Langkawi. MSc. Thesis. Universiti Kebangsaan Malaysia (unpublished)

Ida Suzilawate I (2007) Species composition, biomass and conservation value of a mangrove swamp forest, Kuala Kemaman forest reserve, Terengganu. MSc. Thesis, Universiti Kebangsaan Malaysia (unpublished)

Intan Nordira AK, Norhayati A, Shukor MN, Wan Juliana WA (2003) Komposisi spesies dan struktur hutan paya bakau Belungkor, Johor Proc Sem Keb. Pengurusan Persekitaran, pp 62–68

JPJ (1999) Jabatan Perhutan Johor

JPNT (2005) Laporan Tahunan Jabatan Perhutanan Negeri Terengganu

Jamaliah J, Norhayati A, Nizam MS, Shukor MN, Juliana WA (2003) Komposisi dan biojisim spesies pokok paya bakau Hutan Simpan Pulai, Johor Proc Sem Keb. Pengurusan Persekitaran, pp 62–68

Japar Sidik B (1994) Mangrove plants resources in the asian region. In: Wilkinson CR, Sudara S, Ming CL (eds) Proceedings third Asian-Australia symposium on living coastal resources, vol 1. Status reviews, pp 123–138. Chulalongkorn University, Bangkok

Juliana WA, Norhayati A, Nizam MS, Latiff A, Ahmad Farid AF, Mohd. Razali S (2005) Mangrove flora of Sungai Kilim, Pulau Langkawi. In: Mohd Shafeea Leman, Ibrahim Komoo, A Latiff, Hood Salleh (eds) Culture and natural heritage of Langkawi, Kilim Basin, LESTARI, UKM Selangor pp 289–298

Kochumen KM (1989) Rhizophoraceae. In: Ng FSP (eds) Tree flora of Malaya, vol 4. Longman Malaysia, Kuala Lumpur, pp 302–323

Lugo AE, Snedaker SC (1974) The ecosystem of mangrove. Anal Rev Ecol Syst 5:39–64

Micael J (1999) Biodiversity audit and conservation plan for the mangrove of Johor. Project document no.6, Forestry Department, Peninsular Malaysia/Johor State, pp 112–125

Mohd Lokman H, Sulong I (2001) Mangrove of Terengganu. A joint publication between Kolej Universiti Sains dan Teknologi Malaysia and Forestry Department Peninsular Malaysia

Mohd Lokman H, Yaakob R (1995) Beach erosion variability during a northeast monsoon: The Kuala Setiu coastline, Terengganu, Malaysia. Pertanika J Sci Tech 3(2):337–348

Nasir MH, Safiah Yusmah MY (2007) Distribution of *Rhizophora stylosa* in peninsular Malaysia. J Trop Forest Sci 19(1):57–60

Norhayati A, Nizam MS, Hafizah S, Fera Fizani AF, Latiff A (2005) Community structure and biomass estimation of trees at two mangrove forests in Pulau Langkawi. In: Mohd Shafeea Leman, Ibrahim Komoo, Abdul Latiff Mohamad, Hood Salleh (eds) Warisan Budaya dan Sumber Asli Langkawi, Lembangan Kilim, LESTARI, UKM Selangor pp 243–249

Norhayati A, Nizam MS, Juliana WA, Shukor MN, Fera Fizani AF, Hafizah S, Sariah H, Jamaliah J, Intan Nordira AK, Latiff A (2005) Comparision of mangrove tree species composition at selected location in Peninsular Malaysia. In: Shaharuddin, Azahar M, Razani U, Kamaruzaman AB, Lim KL, Suhaili R, Jalil MS, Latiff A (eds) Sustainable management of Matang mangrove: 100 years and beyond, vol 4. Forestry Department Peninsular Malaysia, pp 235–263

Norhayati A (1995) Biomass and species composition of mangrove forest Pulau Langkawi. MSc. Thesis, Universiti Kebangsaan Malaysia (unpublished)

Norhayati A, Latiff A (2001) Biomass and species composition of mangrove forest in Pulau Langkawi, Malaysia. Malays Appl Biol 30(1 & 2):75–80

Ong JE (2003) Plants of the Merbok mangrove, Kedah, Malaysia and urgent need for their conservation. Folia malaysiana 4:1–18

Ong JE, Gong WK, Wong CH (1980) Ecological survey of the Sungai Merbok estuarine mangrove ecosystem. Universiti Sains Malaysia

Razali SM, Wan Juliana WA, Aldrie AA, Latiff A (2005) Penemuan semula *Rhizophora stylosa* Griff. dan penemuan baru *Rhizophora x lamarckii* Montr. di Selat Kuah, Pulau Langkawi. In: Norhayati A, Kamal Roslan M, Shukor MN, Mustaffa O, Latiff A (eds) Straits of Kuah. Culture and nature heritage of Langkawi. Institute for Environment & Development (LESTARI), UKM, Forestry Department of Peninsular Malaysia & Langkawi Authority Development pp 323–326

Soepadmo E, Pandi Mat Z (1989) Floristic composition, structure and potential net primary productivity of residual mangrove communities in Peninsular Malaysia. Paper presented at the National Workshop on the AACMS Project-Coastal living resources, Awana Golf and Country Club, Genting Highland, Pahang, Malaysia

Shamsul K, Tajuddin AM, Ibrahim AZ, F-H Ibrahim, Latiff A (2005) Preliminary assessments of the flowering plant diversity of Matang mangrove forest. In: Shaharuddin, Azahar M, Razani U, Kamaruzaman AB, Lim KL, Suhaili R, Jalil MS, Latiff A. (eds) Sustainable management of Matang Mangrove: 100 years and beyond, vol 4, Forestry Department Peninsular Malaysia, pp 223–234

Shaharuddin MI, Azahar M, Razani U, Kamaruzaman AB, Lim KL, Rosli S, Jalil Md. S, Latiff A (2005) Sustainable management of Matang mangroves: 100 years and beyond. Forestry Department Peninsular Malaysia, Kuala Lumpur

Sulong I, Ismail H (1990) Classification of mangrove forest by using 1:40,000-scale aerial photograph. Forest Ecol Manag 33–34:583–592

Wan Juliana WA, Norhayati A, Nizam MS, Aldrie AA, Latiff A (2007) Mangrove flora of Selat Kuah, Langkawi. In: Norhayati A, Kamal Roslan M, Shukor MN, Mustaffa O, Latiff A (eds) Selat Kuah: Warisan Budaya dan Sumber Asli Langkawi. Institute for Environment & Development (LESTARI), UKM, Forestry Department of Peninsular Malaysia & Langkawi Authority Development, pp 326

Wan Juliana WA, Norhayati A, Nizam MS, Latiff A, Ahmad Farid AF, Razali MS (2005) Mangrove flora of Sungai Kilim, Pulau Langkawi. In: Mohd Shafea Leman, Ibrahim Komoo, Latiff A, Hood Salleh (eds) Warisan Budaya dan Sumber Asli Langkawi. Lembangan Kilim. LESTARI, UKM, Selangor, pp 289–298

Distribution and Current Status of Mangrove Forests in Indonesia

Cecep Kusmana

Contents

Abstract Indonesia is an archipelagic country of more than 17,504 islands with the length of coastline estimated at 95,181 km, which bears mangroves from several meters to several kilometers. They grow extensively along the inner facing coastlines of most of the large islands and estuarine. They consist of various community types, either mixed or pure stands, mainly distributed in the five big islands (Jawa, Sumatra, Kalimantan, Sulawesi, Papua). In 2009, the Agency of Survey Coordination and National Mapping (Republic of Indonesia) of Indonesia reported the existing mangrove forest area in Indonesia of about 3,244,018 ha; however, at 2007 the Directorate General of Land Rehabilitation and Social Forestry, Ministry of Forestry (Ditjen RLPS MoF) of Indonesia reported about 7,758,411 ha of mangrove area (including an existing vegetated mangrove area). It was further reported that of those mangroves 30.7 % were in good condition, 27.4 % moderately destroyed and 41.9 % heavily destroyed. There are at least five ministries responsible for mangrove resource allocation and management in Indonesia, in which the Ministry of Forestry has the major authority. Nowadays, two Bureaus of Mangrove Forest Management, the National Mangrove Working Group and the Local (Provincial and Regency/City) Mangrove Working Group, as well as the Presidential Decree (PerPres) No. 73/2012

C. Kusmana (✉)
Department of Silviculture, Faculty of Forestry,
Bogor Agricultural University, Bogor, Indonesia
e-mail: ckusmana@ymail.com

I. Faridah-Hanum et al. (eds.), *Mangrove Ecosystems of Asia*,
DOI 10.1007/978-1-4614-8582-7_3, © Springer Science+Business Media New York 2014

regarding National Strategy of Mangrove Management have been setup to strengthen the sustainable mangrove forest management. Currently the Indonesian Government leases a 85,000-ha mangrove forest in Bintuni, Papua and 28,280 ha in Batu Ampar, West Kalimantan to three forest concessioner companies to be harvested using seed tree method silvicultural systems. To enhance the conservation focus as stated on the Presidential Decree (Kepres) No. 32/1990, the width of the mangrove green belt in any coastal area should be set up about 130 × annual average of the difference between the highest and lowest tides. In Indonesia some mangrove forests have been destroyed by various causes, mainly conversion to other uses. In order to recover the destroyed mangroves, the Indonesian Government (c.q. Ministry of Forestry and Ministry of Marine and Fishery) collaborated with stakeholders (domestic and international) and executed rehabilitation as well as restoration of those destroyed mangroves, either in or outside state forest area.

1 Introduction

Indonesia, with its wide range of natural habitats, rich plant and animal resources and high numbers of island endemic species, is recognized as a major world center for biodiversity (Department of Forestry 1997). Although only covering 1.3 % of the world's total area, it is home to 10 % of the flowering plants, 12 % of the mammals, 16 % of the reptiles and amphibians, 17 % of the birds and 35 % of the fishes of the world. The great expanse of Indonesia's territorial waters and richness of the Indo-West Pacific seas further add to the country's biodiversity. It supports a rich variety of coastal and marine habitats including the world's largest area of mangroves, and extensive reef ecosystems that are among the world's richest in species of corals, fishes and other reef organisms.

The biological resources of the mangrove ecosystem, which are believed to be highly productive, are not only able to provide various valuable forest products, but also maintain estuarine water quality as a habitat for many commercially important species of fish and prawns. For tropical countries, the mangrove is one of the important natural resources for the development sector in order to enhance human welfare through resource exploitation and environmental stability. Therefore, an adequate balance must be sought between the environmental benefits of the marginal mangroves and the productive role of these ecosystems on a sustained management basis (FAO 1982). As such, the mangrove forests should be managed to obtain the main objectives of mangrove forest management, i.e. to minimize the destruction or conversion of the mangrove forests, to utilize the mangrove resources on a sustained-yield basis, to preserve the unique flora and fauna, to establish a mangrove protection forest and recreational forest, and to avoid or minimize environmental degradation (Soerianegara, unpublished report).

Indonesia, as an archipelagic country comprising more than 17,504 islands (28 big islands and 17,475 small islands), has an extremely long coast line. The overall length of the Indonesian coast is estimated to be 95,181 km (DKP DKI Jakarta 2009) with a varied climate and physical environment. A substantial proportion of this coastal area bears mangroves of various extents, from several meters to several kilometers.

The mangrove resources in Indonesia involve the flora, fauna and land resources which are needed for supporting many kinds of human needs. In Indonesia, the mangroves developed well along the inner facing coastlines of most of the large islands and estuaries. They are composed of trees, shrubs, herbs and grasses, epiphytes and parasites (Kusmana 1993). Those various kinds of mangrove flora have been supporting the daily life needs for local people living in the surrounding mangroves.

For centuries the Indonesian people have traditionally utilized mangroves, mainly for firewood, charcoal, tannin, dyes, food and beverages, medicine, pole and timber. At an early stage of commercialization, fishing and charcoal making are generally the basic economic activities in the mangrove areas. However, in the following period a large scale of commercial mangrove exploitation in Indonesia began with production of logs, charcoal and chip-woods. At the same time, the increasing population growth and economic development in this country resulted in the destruction, even disappearance, of many mangroves through conversion of them to fishponds, industrial estates, transportation and recreation infrastructure, resettlement, tin mining, agricultural activities, and other uses.

The multiple role of the mangroves as a renewable resource in the coastal area in relation to serving valuable forest products and environmental services for the coastal population is well recognized in Indonesia, so that degraded mangroves must be rehabilitated and mangrove plantations should be established in some intertidal areas to enrich land productivity as well as environmental quality of the ecosystem.

2 Mangrove Area and Distribution

Mangrove forests in Indonesia which grow at the coastal areas are belonging to 257 regencies/cities. According to the latest information, the mangrove vegetated area in Indonesia is estimated at 3.2 million hectares (Agency of Survey Coordination and National Mapping, Republic of Indonesia 2009). On the other hand, the Ministry of Forestry (2007) reported that the potential area to be planted by mangroves (including the mangrove vegetated area) is estimated at 7.8 million hectares (30.7 % in good condition, 27.4 % moderately destroyed, 41.9 % heavily destroyed) as shown on Table 1. They are more developed on the five big islands, i.e. Java, Sumatra, Kalimantan, Sulawesi and Papua.

It is reported that while large portions of the mangrove forests have been commercially exploited, the mangrove areas as land resources have been converted to other uses (agriculture, fishery, urbanization, mining and salt ponds) which often raised conflict of interest among users. In some places, over-exploitation and the reclaiming of mangrove areas may result in a degradation and disappearance of mangroves. Consequently, the management and utilization planning program involving mangrove resources must seek a balance between the economic and ecological viewpoints. To achieve this, the current status of the mangrove resource management and utilization should be known in order to identify the kind of important resources, resource users and the problems involving mangroves. As a result the planning program to solve the problems involving mangrove resources could be determined wisely.

Table 1 Mangrove vegetated area and potential area to be planted by mangrove (including mangrove vegetated area) in Indonesia. (Source: Center for Marine Natural Resources, Agency of Survey Coordination and National Mapping, Republic of Indonesia (2009))

No.	Province	Area of Mangroves (ha)	
		Agency of Survey Coordination and National Mapping, Republic of Indonesia 2009	RLPS-MoF 2007
1	Nanggroe Aceh Darussalam	22,950.321	422,703.000
2	North Sumatera	50,369.793	364,581.150
3	Bengkulu	2,321.870	0.000
4	Jambi	12,528.323	52,566.880
5	Riau	206,292.642	261,285.327
6	Kepulauan Riau	54,681.915	178,417.549
7	West Sumatera	3,002.689	61,534.000
8	Bangka Belitung	64,567.396	273,692.820
9	South Sumatera	149,707.431	1,693,112.110
10	Lampung	10,533.676	866,149.000
11	DKI Jakarta	500.675	259.930
12	Banten	2,936.188	1,180.484
13	West Java	7,932.953	13,883.195
14	Central Java	4,857.939	50,690.000
15	East Java	18,253.871	272,230.300
16	D.I. Yogyakarta	0	0
17	Bali	1,925.046	2,215.500
18	West Nusa Tenggara	11,921.179	18,356.880
19	East Nusa Tenggara	20,678.450	40,640.850
20	West Kalimantan	149,344.189	342,600.120
21	Central Kalimantan	68,132.451	30,497.710
22	South Kalimantan	56,552.064	116,824.000
23	East Kalimantan	364,254.989	883,379.000
24	North Sulawesi	7,348.676	32,384.490
25	Gorontalo	12,315.465	32,934.620
26	Central Sulawesi	67,320.130	29,621.560
27	South Sulawesi	12,821.497	28,978.300
28	South East Sulawesi	44,030.338	74,348.820
29	West Sulawesi	3,182.201	3,000.000
30	North Maluku	39,659.729	43,887.000
31	Maluku	139,090.920	128,035.000
32	Papua and West Papua	1,634,003.454	1,438,421.000
Total		*3,244,018.460*	*7,758,410.595*

3 Mangrove Flora

Soemodihardjo et al. (1993) reported that there are about 157 species of flora growing in mangroves in Indonesia consisting of 52 species of trees, 21 species of shrubs, 13 species of liana, seven species of palms, 14 species of grasses, eight species of herbs, three species of parasites, 36 species of epiphytes and three species of ferns (Table 2).

Table 2 Mangrove flora in Indonesia (Soemodihardjo et al. 1993)

Fern		*Herb*	
Pteridaceae	*Acrostichum aureum*	Acanthaceae	*Acanthus ebracteatus*
	Acrostichum speciosum		*A. ilicifolius*
Blechnaceae	*Stenochlaena palustris*		*A. volubilis*
		Aizoaceae	*Sesuvium portulacastrum*
Ephyphite		Asteraceae	*Pluchea indica*
Adianthaceae	*Vittaria* sp.	Chenopodiaceae	*Tectocornia australica*
Aspleniaceae	*Asplenium nidus*	Araceae	*Colocasia esculenta*
Davalliaceae	*Davallia* sp.		*Cryptocorina ciliata*
	Humata parvula		
		Herb	
Cycads		Cyperaceae	*Cyperus compactus*
Cycadaceae	*Cycas rumphii*		*C. compressus*
Polypodiaceae	*Cyclophorus cinnamoneous*		*C. javanicus*
	Drymoglessum heterophyllum		*C. malacensis*
	Drynaria sp.		*Fimbristylis ferruginea*
	D. rigidula		*Scirpus grossus*
	D. sparsisora		*Thoracostachyum sumtranum*
	Nephrolepis acutifolia	Poaceae	*Chloris gayana*
	Phymatodes scolopendria	(Gramineae)	*Cynodon dactylon*
	Ph. Sinuosa		*Dyplachne fusca*
	Platicerium coronarium		*Paspalum scrobiculatum*
Schizaeaceae	*Lygodium laxum*		*P. vaginatum*
			Phragmites karka
Ephyphite			*Sporobolus virginicus*
Asclepiadaceae	*Dischidia benghalensis*		
	D. rafflessia	*Pandan*	
	D. mommularia	Pandanaceae	*Pandanus tectorus*
	Hoya sp.	*Palma*	
Orchidaceae	*Aerides odorata*	Palmae	*Calamus erinaceus*
	Anota violaceae	(Araceae)	*Licuala* sp.
	Bulbophyllum xylocarpi		*Livistonia saribus*
	Cymbidium sp.		*Nypa fruticans*
	Dendrobium aloifolium		*Oncosperma tigillarium*
	D. callybotrys		*Phoenix paludosa*
	D. pachyphyllum		
	D. prostratum	Liana	
	D. rhyzophoreti	Asclepiadaceae	*Cynanchum carnosium*
	D. subulatum		*Finlaysonia obovata*
	D. teretifolium		*Gymmanthera paludosa*
	Oberonia laeta		*Sarcolobus banksii*
	O. rhizophoreti	Asteraceae	*Wedelia biflora*
Malastomalaceae	*Prachycentria constrica*	Leguminosae	
	Plethiandra sessifolia	Caesalpinioideae	*Caesalpinia bonduc*
Rubiaceae	*Hydnophytum formicarum*		*C. crista*
	Myrmecodia sp.	Papilionoideae	*Aganope heptaphylla*
			Dalbergia candenatensis
Parasite			*D. menoides*
Loranthaceae	*Amyema grafis*		*Derris trifoliata*
	A. mackayense	Rhanaceae	*Smythea lancaeta*
	Viscum ovalifolum	Verbenaceae	*Clerodendron inerme*

Table 2 (continued)

Anacardiaceae	*Gluta velutina*		*E. indica*
Apocynaceae	*Voacanga grandiflora*	Flocourtiaceae	*Scolopia macrophylla*
Bataceae	*Batis agillicola*	Guttiferae	*Calophyllum inophyllum*
Chenopodiaceae	*Halosarcia indica*	Lecythideceae	*Barringtonia asiatica*
Euphorbiaceae	*Glochidion littorale*		*B. racemosa*
Gooddeniaceae	*Scaevola sericea*		
Leguminosae		Leguminosae	
Papilinoideae	*Desmodium embellatum*	Caesalpinioideae	*Cynometra iripa*
Lythraceae	*Aegiceras corniculatum*		*C. ramiflora*
	A. floridum	Mimosaceae	*Pithecelobium umbellatum*
	Ardisia elliptica		*Serianthes spp.*
Myrtaceae	*Osbornia octodonta*		*Pongamia pinnata*
Plumbaginaceae	*Aegialitis annundata*	Malvaceae	*Hibiscus granatum*
Rubiaceae	*Ixora timorensis*		*Thespesia populnea*
	Scyphiphora hydrophyllaceae	Meliaceae	*Xylocarpus granatum*
Rutaceae	*Paramygna angulata*		*X. mekongensis*
Sapindaceae	*Allophyllus cobbe*		*X. moluccensis*
Tiliaceae	*Brownlowia argentata*	Moraceae	*Ficus microcarpa*
	B. tersa	Myristaceae	*Myristica hollrungii*
Verbanaceae	*Prenma obtusifolia*	Rhizophoraceae	*Bruguiera cylindrica*
Tree			*B. exaristata*
Apocynaceae	*Cerbera manghas*		*B. gymnorrhiza*
	C. odollam		*B. hainessi*
Avicennaceae	*Avicennia alba*		*B. parviflora*
	A. eucalyptifolia		*B. sexangula*
	A. marina		*Ceriops decandra*
	A. officinalis		
Bignoniaceae	*Dolichandrone spathaceae*		*C. tagal*
Bombaceae	*Camptostemon philipinense*		*Kandelia candel*
	C. schultzii		*Rhizophora apiculata*
Celastraceae	*Cassine viburnifolia*		*R. mucronata*
Combretaceae	*Lumnitzera littorea*		*R. stylosa*
	L. racemosa	Sapotaceae	*Pouteria obovata*
	Terminalia catappa	Sonneratiaceae	*Sonneratia alba*
Ebenaceae	*Diospyros littorea*		*S. caseolaris*
			S. ovata
		Sterculiaceae	*Heritiera littoralis*
			H. globosa

Furthermore, Kusmana (1993) reported that there are approximately 202 mangrove plant species comprising of 89 species of tree, five species of palm, 19 species of liana, 44 species of soil herbs, 44 species of epiphyte, and one species of fern. Out of the total 202 species, 43 species are true mangroves and the rest are associate mangrove. About 166 are found in Java, 157 in Sumatra, 150 in Kalimantan, 142 in Papua, 135 in Sulawesi, 133 in Maluku and 120 in Lesser Sunda Islands. The distribution of mangroves within the main islands of Indonesia can be seen in Table 3.

Table 3 Distribution of some major and minor mangrove species in the main islands of Indonesia. (Source: Kusmana et al. (1993))

No.	Species	Island						
		Java	Bali&LSI[a]	Sumatra	Kalimantan	Sulawesi	Maluku	Papua
1	*Aegiceras corniculatum*	+	+	+	+	+	+	+
2	*Aegiceras floridum*		+			+	+	+
3	*Avicennia alba*	+	+	+	+	+	+	+
4	*Avicennia lanata*				+	+		
5	*Avicennia marina*	+	+	+	+	+	+	+
6	*Avicennia officinalis*	+	+	+	+	+	+	+
7	*Bruguiera cylindrica*	+	+	+	+	+	+	+
8	*Bruguiera gymnorrhiza*	+	+	+	+	+	+	+
9	*Bruguiera parviflora*	+	+	+	+	+	+	+
10	*Bruguiera sexangula*	+	+	+		+	+	+
11	*Ceriops decandra*	+	+	+	+	+	+	+
12	*Ceriops tagal*	+	+	+	+	+	+	+
13	*Dolichandrone spathacea*	+					+	
14	*Excoecaria agallocha*	+	+	+	+	+	+	+
15	*Heritiera littoralis*	+	+	+	+	+	+	+
16	*Kandelia candel*			+	+			
17	*Lumnitzera littorea*	+	+	+	+	+	+	+
18	*Lumnitzera racemosa*	+	+	+	+	+		+
19	*Nypa fruticans*	+	+	+	+	+	+	+
20	*Osbornea octodonta*	+	+			+	+	+
21	*Phoenix paludosa*			+				
22	*Pemphis acidula*	+	+					+
23	*Rhizophora apiculata*	+	+	+	+	+	+	+
24	*Rhizophora lamarckii*		+	+			+	+
25	*Rhizophora mucronata*	+	+	+	+	+	+	+
26	*Rhizophora stylosa*	+	+	+	+	+	+	+
27	*Scyphiphora hydrophyllacea*	+	+	+	+	+	+	+
28	*Sonneratia alba*	+	+	+	+	+	+	+
29	*Sonneratia caseolaris*	+	+	+	+	+	+	+
30	*Sonneratia ovata*	+		+	+	+	+	+
31	*Xylocarpus granatum*	+	+	+	+	+	+	+
32	*Xylocarpus moluccensis*	+	+	+	+	+	+	+
33	*Xylocarpus rumphii*	+	+				+	+

Note: + present

[a]Lesser Sunda Islands (LSI)

4 Mangrove Fauna

According to Soemodihardjo et al. (1993), Kartawinata and Waluyo (1977), Darnaedi and Budiman (1984) and Mustafa et al. (1979), there are about 118 species of marine fauna associated with mangroves in Indonesia, consisting of 48 species of Gastropoda, nine species of Bivalvia and 61 species of Crustacea (Table 4).

In 1984, the Ecology Team of Faculty of Fishery of IPB reported 45 species of fishes live in mangroves of Segara Anakan-Central Java. They are dominated by *Mugil* sp., *Sillago* sp., *Johnius* sp., *Trachiphalus* sp., *Cynoglossus* sp., *Setipine* sp. and *Leiognathus* sp. The common fish species of commercial interest in Indonesia are mullets (*Mugil* sp.), milkfish (*Chanos chanos*), tilapia (*Chichlidae* spp.), snappers (*Lutjanidae* spp.) and sea bass (*Lates calcarifer*). The most common fish is perhaps the mudskippers (*Perioptthalmus* spp.), which is endemic to the mangroves.

In Indonesia, terrestrial mangrove fauna consists of 16 species of mammals, 49 species of reptiles, six species of amphibian and 76 species of birds (LPP Mangrove 2000) as shown in Table 5.

5 Mangrove Habitat

Based on tree dominant species, the mangrove community in Indonesia can be viewed as association (mix stand) and consociation (pure stand). There are five consociations commonly found in Indonesian mangrove, namely, *Avicennia, Rhizophora, Sonneratia*, *Bruguiera* and *Nypa* consociations. Regarding mix stand, association between *Bruguiera* spp. and *Rhizophora* spp. are frequently found mainly landward. In general, because of a large variety of local habitat, mangrove communities in Indonesia differ among islands.

Based on Sukardjo et al. (1984), the mangrove community in Indonesia consists of:

1. Shrub community
   This mangrove community is formed by mangrove tree pioneer species growing at the coastal line or new delta. Tree species are dominated by *Avicennia marina*, *A. alba*, and *Sonneratia caseolaris*. The seedlings of *Ceriop tagal* grow in this community in the transition area between high and low tide. Sometimes some non-mangrove species, i.e. *Phragmites karka*, *Pandanus* spp., and *Glochidion littorale* grow in this community.
2. Young mangrove community
   This community has one layer of forest canopy formed mainly by the species of *Rhizophora* spp. In unsuitable habitat for *Rhizophora, Avicennia* and *Sonneratia* were grown. Upon further development, there will be mixed stand between *Rhizophora* and the other mangroves such as *Bruguiera* and *Xylocarpus* as well as *Excoecaria agallocha* far landward.

Table 4 Mangrove marine fauna in Indonesia

Gastropoda		Amphibolidae	*S. fragilis* (Lamarck)
Potamididae	*Terebralia palustris* (Linnaeus)	Cerithidae	*Cerithium morum* Lamarck
	T. sulcata (Born)		*C. patulum*
	Telescopium telescopium Linnaeus		*Clypeomorus granosum*
	T. mauritsi Butot	Melangenidae	*Melangena galeodes* Lamarck
	Cerithidea djadjarensis (Martin)	Trochidae	*Monodonta labio* (Linnaeus)
	C. alata (Philippi)	Assimineidae	*Syncera breviculata* (Pfeiffer)
	C. obtusa (Lamarck)		*S. javana* (Thielf)
	C. quadrata Sowerby		*S. nitida* (Pease)
	C. weyersi Datzenberg		*S. woodmasoniana* (Nevill)
	C. cingulata (Gmelin)	Stenothyridae	*Stenothyra glabrata* (A. adams)
Ellobiidae	*Cassidula aurisfelis* Bruguire	Muricidae	*Chicoreus adustus*
	C. lutescens Butot		*Drupa margariticola*
	C. mustelina Deshayes	Nassariidae	*Nassa olivacea*
	C. triparietalis (Martens)		*Alectrion taenia*
	C. sulculosa (Musson)	*Bivalvia*	
	Auriculastra subula (Quoy et Gaimard)	Corbiculidae	*Polymesoda coaxans* Gmelin
	A. elongate		*P. expansa* (Mousson)
	Ellobium aurisjudae Linnaeus	Verenidae	*Gafrarium tumidum* Roding
	E. aurismidae (Linnaeus)	Anomiidae	*Enigmonia aenigmatica* (Chemnitz)
	E. polita	Ostreidae	*Crassostrea cucullata* Born
	E. tornatelliforme (Petit)	Chamidae	*Chama fragum*
	Phytia plicata (Ferussac)	Mytilidae	*Brachyodontes bilocularis*
	P. trigona (Troschel)	Spondylidae	*Spondylus hystrix*
	P. pantherina	Arcidae	*Anadara artiquata* Linnaeus
	Melampus singaporensis (Pfeiffer)	*Crustacea*	
	M. pulchellus Petit	Grapsidae	*Sarmatium incidum*
	M. semisulcatus Mousson		*S. crassum*
Littorinidae	*Littorina scabra* (Linnaeus)		*M. crassipes*
	L. carinifera (Menke)		*Sesarma taeniolata* White
	L. intermedia Philippi		*S. meinerti* De Man
	L. melanostoma Gray		*S. edwarsii*
	L. undulata Gray		*S. bataviana* De Man
Neritidae	*Nerita planospira* Anto		*S. moeschi*
	N. Albicilla Linnaeus		*S. cumolpe* De Man
	N. chameleon		*S. smithi* H. Milne-Edwards
	Neritina violaceae (Gmelin)		*S. bocourti* A. Milne-Edwards
	N. turrita (Gmelin)		*S. fasciata* Lancherter
	N. bicanaliculata		*S. palawensis*
	N. zigzag Lamarck		*S. videns* De Hans
	N. variegata Lesson		*S. onychophora* De Man
	N. auriculata Lamarck		*S. rousseauxi* H. Milne-Edwards
	Clithon corona (Linnaeus)		*S. erythrodeactylum* Hess
	C. ovalaensis		*S. longipes* (Krauss)

Table 4 (continued)

Thiaridae	*Melanoides riqueti* (Grateloup)		*Metapograpsus latifrons* (White)
	M. tuberculata (Muller)		*Uca vocans* Linnaeus
Amphibolidae	*Salinator burmana* (Blanford)		*U. lactea* (De Haan)
Ocypodidae	*U. signatus* (Hess)	Ocypodidae	*O. arenaria* De Man
	U. consobrinus (De Man)		*O. cardimana*
	U. anulipes (H. Milne-Edwards)		*Ilyoplax delsmani* De Man
	U. dussumieri (H. Milne-Edwards)		*Tylodiplax indian*
	U triangularis A. Milne-Edwards	Portunidae	*Scylla serrata* (Forskal)
	U. marionis	Gegarcinidae	*Cardisoma carnifex* (Herbst)
	U. coartasus	Thalassinidae	*Thalassina anomala* Herbst
	U. rosea	Alpheidae	*Alpheus crassimanus* Heller
	Macrophtalmus convexus Stimpson		*A. bisincisus* De Man
	M. telescopicus Owen	Paguridae	*Caenobita cavipes* Stimpson
	M. tridentatum	Balanidae	*Balanus* spp.
	M. definitus Adam et White		*Clibanarius* spp.
	Ocypoda ceratophthalamus (Phallas)		

3. Old mangrove community
   This community is often called mangrove climax dominated by big trees of *Rhizophora* and *Bruguiera*. Commonly, some species such as *R. mucronata* and *R. apiculata* dominate soft mud soils, *R. stylosa* dominate sandy soils, and *Bruguiera* spp. dominate firm mud habitat. At gaps or opening areas, some ground cover species are grown, such as *Acrostichum aureum*, *Derris* spp., and *Acanthus illicifolius*.
4. Nypa community
   In this community, *Nypa fruticans* grow extensively formed pure stand and sometimes grow mixed sporadically with other trees species (*Lumnitzera* spp., *E. agallocha*, *Heritiera littoralis*, *Instia bijuga*, *Kandelia candel*, and *Cerbera manghas*).

Other mangrove community types have been found in several regions in Indonesia as shown on Table 6.

6 Management of Mangroves Ecosystem in Indonesia

According to Soemodihardjo and Soerianegara (1989), in Indonesia there are at least five ministries that are directly or indirectly involved in determining the mangrove resource allocation and management. They are the Ministry of Forestry, the Ministry of Marine and Fishery, the Ministry of Home Affairs, National Land Bureau (BPN), and the Ministry of Life Environment. However, the Ministry of Forestry has the major authority to manage the mangrove resources. Of the other three ministries,

Table 5 Species of terrestrial mangrove fauna in Indonesia

No.	Items	Species	Common Name
	Aves		
1		*Alcedo caerulescens* (L)	Small blue Kingfisher
2		*Halcyon cyanoventris* (L)	Javan Kingfisher
3		*Todirhampus chloris* (L)	White Collared Kingfisher
4		*Todirhampus sanctus* (L)	Sacred Kingfisher
5		*Pelargopsis capensis* (L)	Stork-billed Kingfisher
6		*Alcedo meninting* (L)	Blue-eared Kingfisher
7		*Anas gibberifrons*	Grey Teal
8		*Anhinga melanogaster* (L)	Oriental Darter
9		*Collocalia fuciphaga*	Edible-nest Swiftlet
10		*Collocalia esculenta*	White bellied Swiftlet
11		*Apus affinis*	House Swift
12		*Apus pacificus*	Fork-tailed Swift
13		*Ardea cinerea*	White bellied Swiftlet
14		*Ardea purpurea*	Purple Heron
15		*Egretta garzetta*	Little Egret
16		*Egretta intermedia* (L)	Plumed Egret
17		*Nycticorax nycticorax*	Black-crowned Night Heron
18		*Ardeola speciosa*	Javan Pond Heron
19		*Butorides striatus*	Little Heron
20		*Bubulcus ibis*	Cattle Egret
21		*Artamus leucorynchus*	White-breasted Wood Swallow
22		*Lalage nigra*	Pied Triller
23		*Caprimulgus affinis*	Savannah Nightjar
24		*Aegithina tiphia*	Common Iora
25		*Mycteria cinerea*	Milky Stork
26		*Streptopelia chinensis*	Spotted Dove
27		*Treron vernans*	Pink-necked Pigeon
28		*Macropygia emiliana*	Red Cuckoo Dove
29		*Geopelia sriata*	Peaceful Dove
30		*Crypsirina temia*	Racket-tailed Treepie
31		*Cacomantis merulinus*	Plaintive Cuckoo
32		*Centropus nigrorufus*	Sunda Coucal
33		*Centropus bengalensis*	Lasser Caucal
34		*Dicaeum trochileum*	Scarlet-headed Flowwerpecker
35		*Hirundo tahitica*	Pasific Swallow
36		*Hirundo rustica*	BarnSwallow
37		*Lanius schach*	Long-tailed Shrike
38		*Motacilla flava*	Yellow Wagtail
39		*Rhipidura javanica* (L)	Pied Fantail
40		*Cyornis rufigastra*	Mangrove Blue Flycather
41		*Muscicapa sibirica*	Asian Brown Flycather
42		*Nectarinia jugularis* (L)	Olive-backed Sunbird
43		*Nectarinia calcostheta* (L)	Copper-throated Sunbird
44		*Anthreptes malacensis* (L)	Brown-throated Sunbird
45		*Anthreptes singalensis* (L)	Ruby-cheeked Sunbird
46		*Oriolus chinensis*	Black-naped Oriole
47		*Parus major*	Great Tit
48		*Phalacrocorax sulcirostris*	Litle Black Commorant

Table 5 (continued)

No.	Items	Species	Common Name
49		*Phalacrocorax niger*	Litle Cormorant
50		*Picoides macei*	Fulvous-breasted Woodpecker
51		*Picoides maluccensis*	Brown-capped Woodpecker
52		*Lonchura punctulata*	Scaly-breasted Munia
53		*Lonchura leucogastroides*	Javan Munia
54		*Paser montanus*	Eurasian Tree Sparrow
55		*Psittacula alexandri*	Moustached Parakeet
56		*Loriculus galgulus*	Blue-crowned Hanging-Parrot
57		*Cacatua alba*	White Cacatoo
58		*Pycnonotus aurigaster*	Sooty-headed Bulbul
59		*Pycnonotus goiavier*	Yellow-vented Bulbul
60		*Amaurornis phoenicurus*	White-breasted Waterhen
61		*Porphyrio porphyria*	Purple Swamphen
62		*Calidris ferruginea*	Curwel Sandpiper
63		*Tringa hypoleucos*	Common Sandpiper
64		*Prinia familiaris*	Bar-winged Prinia
65		*Prinia polychroa*	Brown Prinia
66		*Orthotomus sepium*	Olive-backed Tailorbird
67		*Orthotomus ruficeps*	Ashy Tailorbird
68		*Orthotomus sutorius*	Common Tailorbird
69		*Gerygone sulphurea*	Golden-bellied Gerygone
70		*Acrocephaus stentoreus*	Clamourus Reed-warbler
71		*Sterna nilotica*	Gull-billed tern
72		*Sterna bergii*	Great Crested-Tern
73		*Acridotheres javanicus*	Javan Myna
74		*Zoothera interpres*	Chesnut-capped Thrush
75		*Zosterops chloris*	Lemon-bellied White-eye
76		*Zosterops palpebrosus*	Oriental White-eye
	Mammal		
	Carnivora		
1		*Vulpes bengalensis*	Bengal fox
2		*Canis aureus*	Jackal
3		*Lutra perspicillata*	Smooth otter
4		*Amblonyx cinerea*	Otter
5		*Herpestes edwardsi*	Mongoose
6		*H. javanica*	Java mongoose
7		*Paradoxurus hemaphroditus*	Palm civet
8		*Viverra zibetha*	Large Indian civet
9		*Panthera tigris*	Sumatra tiger
10		*Felis viverrina*	Fishing cat
11		*F. bengalensis*	Leopard cat
12		*F. haus*	Jungle cat
	Artiodactyla		
1		*Sus scropa*	Wild boar
2		*Muntiacus muncak*	Barking deer
3		*Axis axis*	Spotted deer
4		*Tragulus javanicus*	Mouse deer
	Amfibi		
1		*Bufo melanostictus*	Toad
2		*Rhacophorus maculatus*	Tree frog

Table 5 (continued)

No.	Items	Species	Common Name
3		*Rana cyanophlyctia*	Frog
4		*R. limnocharis*	
5		*R. tigrina*	
6		*Microhyla ornata*	
	Reptilia Crocodilia		
1		*Crocodilus siamensis*	
2		*C. nova guinea*	Freshwater New Guinea crocodile
3		*C. porosus*	Estuarine crocodile
4		*Tomistoma schlegeli*	False gavial
	Squamata		
1		*Hemidactylus flaviviridis*	Wall gecko
2		*Eublepharis fasciolatus*	Leopard gecko
3		*Gecko gecko*	Tokay
4		*Mabuya multifasciata*	Common skink
5		*Calotes versicolor*	Lizard
6		*Chamaeleon zeylanicus*	Indian chameleon
7		*Varanus sp.*	Bengal monitor
8		*V. salvator*	Yellow monitor
9		*V. flavescens*	Ruddy sub-nosed monitor
10		*Naja naja*	Cobra
11		*Typhlops porractus*	Blind snake
12		*T. acutus*	Blind snake
13		*Ahaetula ahaetulla*	Whip snake
14		*A. cyanochloris*	
15		*Python reticulatus*	Python
16		*Natrix stolata*	Keel back
17		*Enhydris enhydris*	
18		*Fordonia leucobalia*	
19		*Bungards lividus*	Krait
20		*Acrochordus granulatus*	Wart snake
21		*Hydrophis obscurus*	
22		*H. nigrocinctus*	
23		*Microcephalophis cantoris*	Sea snake
24		*Enhydrina achistoss*	Beaked deep sea snake
25		*Cerberas thynchops*	
26		*Ptyas mucosus*	Rat snake
27		*Spalerosophis diadema*	
28		*Vivera russeli*	Russell's viper
29		*Pligodon arnensis*	Kukri snake
30		*Oligodon dorsalis*	
31		*Dryophis mycterigans*	Tree snake
32		*Lycondon aulicus*	Common wolf snake
33		*Eryx conicus*	Russel's wolf snake
34		*Psammophis condouarus*	
	Testudinate		
1		*Pelochelys bironi*	Coast shell-turtle
2		*Morenia petersi*	Bengal terrapin
3		*Batagur baska*	River terrapin
4		*Lepidochelys olivaca*	Ridley turtle
5		*Chelonia mydas*	Green turtle

Table 5 (continued)

No.	Items	Species	Common Name
6		*Tryonix hurun*	Peacock soft-shell turtle
7		*T. gageticus*	Ganges soft-shell turtle
8		*Lissemys punctata*	Indian flap-shell turtle
9		*Kachuga tecta*	India roofted turtle
10		*K. smiti*	
11		*K. kachuga*	

the Ministry of Marine and Fishery has the foremost concern with the mangrove resources for the well-known important contribution of the mangrove to the coastal fishery. The authority of the Ministry of Home Affairs and BPN is concerned with the agrarian or land use aspects and the Ministry of Life Environment with the well-being of the environment as a whole.

Years ago the Indonesian government initiated setting up institutions to strengthen the mangrove forest management, such as follows:

1. Mangrove Forest Management Bureau (*Balai Pengelolaan Hutan Mangrove* or BPHM)
   This bureau was established based on the Decree of the Minister of Forestry No. P.04/Menhut-II/2007, 6 Februari 2007 consisting of:
   a) BPHM Region I located at Denpasar Bali having the mandate for managing mangrove in Java, Bali, Madura, Nusa Tenggara, Sulawesi, Maluku and Papua.
   b) BPHM Region II located at Medan, with the mandate for managing mangrove at Sumatra and Kalimantan
2. National Mangrove Working Group (*Kelompok Kerja Mangrove Nasional* or KKMN) and Local Mangrove Working Group (*Kelompok Kerja Mangrove Daerah* or KKMD)
   KKMN is a working group from which the members come from inter-sector/institution/NGO. Nowadays, 23 KKMD at the province level and 16 KKMD at the regency/city level have been established to strengthen the capacity building for mangrove forest management. Fortunately, in 2012 the President of the Republic of Indonesia enacted the PerPres No. 73/2012 regarding the National Strategy of Mangrove Ecosystem Management to realize the Sustainable Mangrove Ecosystem Management and to improve the welfare of the local community-based mangrove resources.

Mangrove resource management in Indonesia is involved with the management of the mangrove forest exploitation, mangrove resource protection and mangrove rehabilitation (mangrove afforestation or reforestation).

Management of the mangrove forest exploitation in Indonesia is controlled by two major kinds of regulations. The first controls the silvicultural practices in the mangrove harvesting and the second controls the leasing arrangements for allocating the mangrove forest concessions.

Table 6 Mangrove community type in Indonesia

No.	Location	Community type	Species richness	Source
A.	*Java Island*			
1	Cilacap	*Aegiceras corniculatus—Ficus retusa* *Avicennia marina—Sonneratia alba* *Rhizophora mucronata—Bruguiera cylindrica*	14	Marsono (1989)
2	Ujung Karawang	*Avicennia marina—Avicennia corniculatus*	9	Djaja et al. (1984)
3	Indramayu	*Avicennia marina—Avicennia alba*	9	Sukardjo (1980)
4	Pulau Rambut	*Rhizophora mucronata—Rhizophora stylosa* *Rhizophora mucronata* *Schyphyphora hydrophyllacea—Lumnitzera racemosa*	13	Kartawinata and Waluyo (1977)
5	Pulau Dua	*Rhizophora stylosa—Rhizophora apiculata*	12	Buadi (1979)
6	Baluran	*Rhizophora stylosa—Rhizophora apiculata*	16	Indiarto et al. (1987)
7	Grajagan	*Rhizophora apiculata—Avicennia spp.*	14	Sukardjo, unpublished report
8	Muara Angke	*Avicennia alba—Avicennia marina* *Avicennia marina—Rhizophora mucronata*	11	Kusmana (1983)
B.	*Outside Java Island*			
1	Kangean isles	*Rhizophora stylosa* *Rhizophora apiculata* *Ceriops tagal*	12	Soemodihardjo, unpublished report
2	Tanjung Apar (East Kalimantan)	*Rhizophora apiculata—Avicennia alba* *Avicennia officinalis—Avicennia alba* *Ceriops tagal—Rhizophora apiculata*	13	Sukardjo, unpublished report
3	Tanjung Kasam (Riau)	*Xylocarpus granatus—Lumnitzera racemosa* *Rhizophora apiculata—Xylocarpus granatus*	12	Sukardjo, unpublished report
4	Way Sekampung (Lampung)	*Avicennia spp.* *Hibiscus tiliaceus—Pongamia pinnata*	14	Sukardjo (1979)
5	Banyuasin (South Sumatra)	*Rhizophora apiculata* *Bruguiera gymnorrhiza—Rhizophora apiculata*	9	Yamada and Sukardjo (1980)
6	Tanjung Bungin (South Sumatra)	*Rhizophora apiculata—Nypa fruticans* *Nypa fruticans—Rhizophora apiculata*	9	Sukardjo et al. (1984)

Table 6 (continued)

No.	Location	Community type	Species richness	Source
7	Talidendang Besar (Riau)	*Bruguiera parviflora* *Bruguiera sexangula* *Bruguiera sexangula—Nypa fruticans*	8	Kusmana and Watanabe (1992)
8	Sungai Gaung dan Mandah (Riau)	*Rhizophora apiculata—Rhizophora mucronata* *Bruguiera parviflora—Bruguiera sexangula* *Aegiceras corniculatus—Nypa fruticans*	7	Al Rasjid (1984)
9	Central Sulawesi			
	Ranu	Rhizophora apiculata—Ceriops tagal	3	Darnaedi and Budiman (1984)
	Lapangga	Rhizophora apiculata—Ceriops tagal	8	
	Matube	Rhizophora mucronata	3	
	Morowali	Rhizophora apiculata	5	
10	Halmahera (Maluku)	*Sonneratia alba* *Bruguiera gymnorrhiza—Xylocarpus granatus* *Rhizophora apiculata—Bruguiera gymnorrhiza* *Nypa fruticans—Rhizophora stylosa*	14	Komiyama et al. (1988)
11	Bone–bone (South Sulawesi)	*Sonneratia alba—Rhizophora apiculata* *Rhizophora mucronata* *Nypa fruticans—Rhizophora stylosa*	20	Ahmad (1989)
12	Simpang Ulim (Aceh)	*Rhizophora apiculata—Bruguiera gymnorrhiza*	8	Al Rasjid (1983)

7 Silvicultural Practices

For the first time, Kantor Besar Dinas Kesehatan Rakyat, through regulation no. 669/c, dated January 7, 1933 advocated a law to regulate the mangrove harvesting. Based upon this regulation, it was prohibited to cut mangroves within three kilometers from a village in order to control the mosquito populations. Later, a regulation incorporating the silvicultural guidelines was enacted through regulation no. 13062/465/BIR, dated July 1, 1938 in order to control the development of the mangrove forest in Cilacap, Central Java. According to this regulation, the forest should be divided into three management areas such as follows:

1. Mangrove production forest, where *Rhizophora* formed the main species. In this area the clear cutting would be practiced leaving 60–100 seed trees (mother trees) with a minimum diameter of 20 cm per ha to facilitate the regeneration of the clear-cut areas

2. Mangrove considered unsuitable for production
3. Protection forest areas along the coast and river bank where Avicennia and the other mangroves formed the dominant vegetation.

Unfortunately, the application of this regulation to the other mangrove forests in Indonesia was interrupted by World War II and the other mangroves formed the dominant vegetation.

The research and experimentation continued after World War II, however, the standardized mangrove management regulation in Indonesia was not put into the official law until 1978. In order to evaluate the effect of the application of the 1938 regulation on the regeneration of different mangrove species, Versteegh (1952) did research on the methods of regeneration of the various commercial species which had largely been ignored in Indonesia. Based upon his experimental results obtained in a mangrove forest of Bengkalis, Riau, he recommended that the clearcutting system was only suitable for areas frequently flooded by tides and an artificial as well as a natural regeneration of commercial species must be made. He introduced the working plan through an Area Method with a 30-year cutting cycle and leaving 64 seed trees/ha having a circumference of 45 cm distributed in a regular spacing throughout an over-logged area to manage a mangrove forest in Bengkalis. According to this method, the mangrove forest was divided into sub-blocks of 120 ha each where 4 ha (1/30 of sub-block) should be felled every year. But, Vesteegh's recommendations appeared not to have had much impact until the late 1970s. Instead a follow-up study of the Cilacap mangrove forest led to the adoption of the 1938 regulation, and the Standard Clear-Cutting System as a silvicultural practice which was recommended by the Forest Research Institute in 1956 was the main thrust of mangrove management in Indonesia until 1978.

In 1972, a Modified Clear-Cutting System, which is also called Stripwise-Selective-Felling System, was recommended by the Forest Planning and Production Division of the Directorate General of Forestry with the suggestions as follows (Wiroatmodjo and Judi 1979):

1. No logging activity is allowed within 50 m of the coastal limit of a mangrove or within 10 m along a river bank
2. Logging is allowed in 50 m wide strips at right angles to the coast line, while 20 m wide strips have to be left between the harvested areas to provide seeds for the natural regeneration
3. Only trees with a DBH (diameter at breast-height) of 7 cm or more can be cut in the production strips
4. If the natural regeneration in a large area is inadequate, enrichment planting with 2 × 3 m spacing must be carried out
5. Logs should be removed by rafting, boats and artificial canals
6. A rotation is set for 20 years.

This system was implemented by the mangrove forest concessionaries; however, it has never been written into the official law.

From the ecological viewpoint, this silvicultural system may cause the fish, shrimp and other marine organisms to accumulate in certain areas, i.e. in unharvested strip areas, so that predators (birds, snake, etc.) may prey them easily (Kusmana 1991). Consequently, this silvicultural system may cause the decrease of fish and shrimp production which could be taken by the fishermen. To improve the management system of the mangrove forest, the Government of Indonesia (c.q. Directorate General of Forestry) introduced the new silvicultural system which is called Seed-Tree Method through a Decree No. 60/Kpts/Dj/I/1978. The silent points of this system are as follows:

1. Felling rotation is set for 30 years, where an annual working plan is divided into about 100-ha felling blocks and each felling block itself must be divided into about 10–50 ha compartments depending on the forest condition. The felling rotation can be modified by concessionaires based on the habitat condition, ecological reasons and forest management objectives after getting an agreement from the Directorate General of Forestry.
2. Before felling, the trees in the compartments must be inventorized using a systematic strip sampling with a strip width of 10 m and distance between strips about 200 m. The inventory of the concession must be carried out by the concessionaires. Based on the results of this inventory, the Directorate General of Forestry will determine whether the forest is suitable for felling or thinning, and determine the limit of the annual allowable cut.
3. Trees to be cut must have a diameter of at least 10 cm at 20 cm above the highest prop-roots or buttress. Only axes, machettes and mechanical saws are used for felling the trees.
4. Cutting can only proceed in those areas where 40 seed trees of commercial species with a minimum diameter of 20 cm and spaced at 17 m from each other per hectare can be left for seed and seedling production. Clearcutting is permissible if about 2,500 seedlings/ha, which are distributed with a distance of 2 m or less from each other over the whole area, are available. Only species of Rhizophora, Bruguiera and Ceriops may be counted as seed trees. Also in order to improve the tree growth, thinning should be undertaken at a period of 15–20 years after the first felling, if more than 1,100 trees/ha in this secondary forest are available.
5. Logs must be transported by raft, boat or wooden carriage through the rivers, artificial canals, or railroads where the distance between canals and railroads must not be less than 200 m and the slash must be removed from the felling areas.
6. The hoarding log area is limited to about 0.1 ha in every 10 ha felling area.
7. Regeneration studies must be carried out to determine the effectiveness of the cutting and regeneration cycle.
8. The protective green belt is determined about 50 m along the coast line and 10 m along the river bank, waterways and main roads.

8 Leasing Arrangement of the Mangrove Forest Exploitation

The issuance of the leasing permit to exploit a mangrove forest is clarified in two categories depending on the extent of the mangrove area to be leased. Prior to 1970, the provincial government had the authority to issue all the permits, regardless of the extent of the mangrove area to be leased. However, in 1970 the Government of Indonesia (c.q Directorate General of Forestry) based upon *Undang-undang Pokok Kehutanan* (Basic Law of Forestry) No. 5, 1967 enacted *Peraturan Pemerintah* No. 21, 1970 which altered the leasing process. According to this regulation, the Minister of Agriculture, acting on behalf of the central government, had the authority to issue the licence for leasing a mangrove forest greater than 100 ha for a 30-year lease period. But, from 1983 to 2002 the permission for leasing the forests has been enacted by the Minister of Forestry. This regulation also permitted the provincial government to grant a two-year lease for a mangrove area of equal to or less than 100 ha. The shift of the major responsibility from the provincial to the central government for leasing a mangrove area greater than 100 ha was aimed at stimulating and facilitating foreign investment in the mangrove resources. Starting from 2003, the leasing of mangrove forest exploitation was only enacted by the central government (c.q. Ministry of Forestry). Now, there are three mangrove forest concession companies in Indonesia, i.e. PT. Bintuni Utama Murni Wood Industry in Papua (±85,000 ha), PT. BIOS (±10,100 ha) and PT. Kandelia Alam (±18,180 ha) in West Kalimantan.

9 Mangrove Resources Protection

Mangrove resource protection entails the designation of a proportion of an undisturbed mangrove area for natural conservation and a green belt (buffer zone) along the coast or river bank.

The mangrove forests in Pulau Rambut and Pulau Dua (West Java) were designated as wildlife reserves for bird sanctuaries. While there are five Biosphere Reserves in Indonesia, there currently is no Biosphere Reserve specifically dedicated to the mangrove. Nevertheless, Tanjung Puting (Kalimantan) and Bali Barat (Bali) National Parks include substantial areas of mangrove.

Because of the important function of mangroves in the coastal ecosystem, in the 1990s the government of Indonesia (c.q. Directorate General of Forest Protection and Nature Conservation) proposed a number of areas bearing mangroves as nature reserves. Among them, the mangrove areas at Muara Gembong, Muara Cimanuk, Muara Sedari and Muara Kamal (north coast of West Java) were nominated as protected areas because they serve as feeding grounds for the birds residing in Pulau Rambut (north of Jakarta). Recently, there have been at least 17 mangrove-bearing wildlife protection areas allocated in Indonesia (Table 7).

In Indonesia, due to the lack of a scientific database, the width of the mangrove green belt was determined arbitrarily. For example, in 1975 the Directorate General of Fishery, through Instruction No. H.I/4/2/1975, dated November 22, 1975 obliged

Table 7 Mangrove-bearing wildlife protection areas in Indonesia

No.	Location	Total area (ha)	The main protected wildlife
1	Berbak, Sumatra	8,500	*Crocodilus* spp.
2	Kuala Langka, Sumatra	1,000	*Crocodilus* spp.
3	Kuala Jambuaye, Sumatra	3,000	*Crocodilus* spp.
4	Muara Angke, Jawa	15	*Egretta* spp. *Haleyon* spp. *Anhinga* spp.
5	Muara Cimanuk, Jawa	7,100	*Ibis* spp.
6	Muara Mauk, Jawa	1,000	*Bubulens ibis*
7	Pulau Sepanjang, Madura	2,430	*Ibis cinereus* *Haleyon* spp. *Ciconia epsicopus*
8	Teluk Kelumpang, Kalimantan	13,750	*Nasalis larvatus*
9	Pamuka, Kalimantan	10,000	*Nasalis larvatus*
10	Muara Kendawangan, Kalimantan	150,000	*Nasalis larvatus*
11	Tanjung Puting, Kalimantan	11,000	*Nasalis larvatus* *Anhinga* sp. *Ibis cinerus*
12	Muara Kahayan, Kalimantan	150,000	*Nasalis larvatus*
13	Teluk Adeng dan Teluk Apar, Kalimantan	128,000	*Crocodilus* spp.
14	Gunung Lorentz, Papua		*Crocodilus* spp. *Haleyon* sp. *Ciconia episcopus*
15	Pulau Dolok, Papua	105,000	*Crocodilus* spp.
16	Bali Barat, Bali		Jalak Bali
17	Ujung Kulon, Jawa		Badak

a mangrove green belt of 400 m wide along the river bank. Because of this contrasting condition, the Minister of Forestry and the Minister of Agriculture issued a joint decree (SKB Menteri Pertanian dan Menteri Kehutanan No. KB 550/246/Kpts/4/1984 dan No. 082/Kpts-II/1984, 30 April 1984) involving the width of a mangrove green belt of 200 m wide.

Through Surat Edaran No. 507/IV-BPHH/1990, the Ministry of Forestry (c.q. Directorate General of Forest Utilization) suggested that the width of the green belt should be set at about 200 m along the coast line and 50 m along the river bank. Recently, according to the ecological studies related to organic matter production of the mangrove forest and the productivity of the fish and shrimps, Soerianegara et al. (1986) suggested that the width of the green belt should be set at 130 × the largest tidal range. The result of this study was stated on the Presidential Decree (Keppres) No. 32/1990 (article 27) that the width of the mangrove green belt is about 130 × the annual average of the difference between the highest and lowest tides.

10 Mangrove Forest Rehabilitation

Along the north coast of Java in which many land-hungry people live, the mangroves are being degraded and the problems involving land tenure of the mangrove areas have raised a conflict of interest among users. Although the mangrove reforestation or afforestation of newly formed land in the prograding coast is often hindered by human encroachment, since the 1960s Perum Perhutani (State Forest Cooperation) has eagerly rehabilitated the mangrove areas in this region. Soemodihardjo and Soerianegara (1989) reported that on the north coast of Java, before the land reaches an elevation above the sea surface at low tide, the land-hungry people would already lay a claim of ownership or at least of land use rights for the new land by sticking wooden posts onto the sea floor to mark the border line. Thus, newly formed land will directly be converted to brackish water fish ponds. In order to find out the best way to save the existing mangrove forest without ignoring the needs of the land-hungry people who live in the surrounding areas of mangroves, Perum Perhutani advocated a *tambak tumpangsari* which is also called *hutan tambak* or *tambak empang parit*. *Tambak tumpang sari* (forest-canal fish pond system) is made up of many smaller units in which each unit consists of a canal of 2 to 5 m wide and 1 m deep enclosing a rehabilitated mangrove stand in the middle. The proportion between the canal fish pond and the forest may vary, for example, the proportion of the fish pond to the forest is 20–80 % in Cikeong (Ujung Karawang) and 40–60 % in Cilacap (Kusmana et al. 1989). But, the optimal proportion is 54 % fish pond and 46 % forest (Zuna 1998). The species raised in the *tambak* are usually bandeng (*Chanos chanos*), mujair (*Tilapia mosambica*), udang windu (*Panaeus monodon*) and udang putih (*Panaeus merguiensis*). Widiarti and Effendi (1989) reported that a tambak-farmer in Blanakan and Cangkring villages (northern part of West Java) has an income of about Rp. 101,420 to Rp. 166,780 in a month, through cultivating the species of the above-mentioned fish.

Several years ago, the Government of Indonesia (c.q. Ministry of Forestry) had the high commitment to execute mangrove rehabilitation through the programs of National Action of Land and Forest Rehabilitation (*Gerakan Nasional Rehabilitasi Hutan dan Lahan* or *GN-RHL*) and routine rehabilitation activities. Besides, significant efforts to plant mangrove is also shown by the Ministry of Marine and Fishery and many stakeholders, mainly Stated Owned Corporation (*BUMN*) and private companies through Corporate Social Responsibility (CSR) programs as well as various levels of action from the community.

It was reported that the Ministry of Forestry has rehabilitated mangrove areas amounting to 37,539 ha until the year 2008. In the period of 2010 to 2014, the Ministry of Forestry planned to do mangrove rehabilitation at about 10,000 ha/year through the Mangrove People Nursery (*Kebun Bibit Rakyat* or *KBR*) program. In 2013, the target of the mangrove rehabilitation project will be raised up to 15,000 ha through the programs of Land Forest Rehabilitation, People Nursery and Social Aid. Beside the government, many international donor institutions set up joint work to execute mangrove rehabilitation in Indonesia, some of them include:

Fig. 1 View of planted Rhizophora seedlings using *guludan* technique in the surrounding area of Sedyatmo Highway, North Jakarta

a. Asian Development Bank or ADB (mangrove management project at Sulawesi, 1997)
b. International Tropical Timber Organization or ITTO (proposal of mangrove forest management)
c. UNDP-IUCN (program mangrove forest for the Future or MFF)
d. Yamamoto (Mangrove rehabilitation at Riau 500 ha; Jambi 20,000 ha; South Sumatra 20,000 ha; Bangka-Belitung 10,000 ha)

In Jakarta, Marine and Agriculture Services of DKI Jakarta Province joined with the others (Faculty of Forestry IPB, Jasa Marga, Bank Mandiri, Pertamina, Perusahaan Gas Negara, United Tractor, PT. Garuda Indonesia, AEON, etc.) rehabilitated destroyed mangrove areas surrounding Sedyatmo highway using the *Guludan* Technique introduced by Kusmana at 2005 (Fig. 1). *Rhizophora* spp. seedlings were used for this mangrove rehabilitation project which totaled more than 150,000 seedlings.

References

Ahmad T (1989) Potentialities of mangrove forest related to coastal aquaculture: a case study in Bone-bone, Luwu, South Sulawesi. In: Soerianegara I, Zamora PM, Kartawinata K, Umaly RC, Tjitrosomo S, Sitompul DM, Rosalina U (eds) Symposium on mangrove management: its ecological and economic considerations. Biotrop Special Publication, Bogor 37:287–299

Al Rasjid H (1983) Pengaruh pembinaan sistem silvikultur pohon induk terhadap perkembangan hutan payau di Simpang Ulim, Aceh. Lap Puslithut 416:8–30

Al Rasjid H (1984) Penilaian permudaan alam hutan mangrove antara sungai Gaung dan Sungai Mandah, Riau. In: Soemodihardjo S, Soerianegara I, Sutisna M, Kartawinata K, Supardi N, Al Rasjid H (eds) Prosiding seminar II ekosistem mangrove. MAB-LIPI, Jakarta, pp 314–320

Agency of Survey Coordination and National Mapping, Republic of Indonesia (2009) Peta Mangroves Indonesia. Cibinong: Pusat Survey Sumber Daya Alam Laut Badan Koordinasi Survey dan Pemetaan Nasional

Buadi (1979) Hutan Bakau di Pulau Dua, Teluk Banten, Jawa Baratl. In: Soemodihardjo S, Nontji A, DJamali A (eds) Prosiding seminar ekosistem hutan mangrove. MAB-LIPI, Jakarta, pp 69–71

Darnaedi D, Budiman A (1984) Analisis vegetasi hutan mangrove Morowali, Sulawesi Tengah. In: Soemodihardjo S, Soerianegara I, Sutisna M, Kartawinata K, Supardi NN, Al Rasjid H (eds) Prosiding Seminar II Ekosistem Mangrove. MAB-LIPI, Jakarta, pp 314–320

Department of Forestry of Republic of Indonesia (1997) National strategy for mangrove management in Indonesia. Department of Forestry of Republic of Indonesia, Jakarta

Djaja B, Sudargo G, Indrasuseno B (1984) Hutan mangrove di Tanjung Karawang, Bekasi, Jawa Barat. In: Soemadihardjo S, Nontji A, Djamali A (eds) Prosiding seminar II ekosistem mangrove. MAB-LIPI, Jakarta, pp 156–161

DKP DKI Jakarta: Marine and Agricultural Services Jakarta DKI Province. Jumlah pulau kecil. http://www.ppk-kp3kdkp.go.id/index.php?option = com_content. Accessed 7 July 2009

FAO (1982) Management and utilization of mangroves in Asia and the Pacific. FAO Environmental Paper 3. FAO, Rome, pp 160

Indiarto Y, Prawiroatmodjo S, Mulyadi, Kabinawa IN (1987) Analisis vegetasi hutan mangrove Baluran Jawa Timur. In: Soerianegara I, Adisoemarto S, Soemodihardjo S, Hardjowigeno S, Sudomo M, Ongkosongo OSR (eds) Prosiding seminar III ekosistem mangrove. MAB-LIPI, Jakarta, pp 92–97

Kartawinata K, Waluyo EB (1977) A preliminary study of the mangrove forest on Pulau Rambut. Mar Res Indon 18:119–129

Komiyama A, Moriya H, Prawiroatmodjo S, Toma T, Ogino K (1988) Forest as ecosystem, its structure and function: 1. Floristic composition and stand structure. In: Ogino K, Chinara M (eds) Biological system of mangroves: a report of East Indonesian mangrove expedition 1986. Ehime University, Ehime, pp 85–96

Kusmana C (1983) Analisis vegetasi hutan mangrove Muara Angke Jakarta [Skripsi]. Fakultas Kehutanan IPB, Bogor

Kusmana C (1991) Silvicultural practices of mangrove forests in Southeast Asian countries. Technical notes, vol. III, no. 1, July 1991. Faculty of Forestry Bogor Agricultural University, Bogor

Kusmana C (1993) The current status of mangrove forest management in Indonesia. Faculty of Forestry, Bogor Agricultural University, Bogor (Unpublished)

Kusmana C, Watanabe H (1992) Production structure of main commercial tree species in a mangrove forest in East Sumatra Indonesia. Biotropia 5:1–9

Kusmana C, Yunia C, Setyobudiandi I, Soendjoto MA (1989) Training course on the management of mangrove forest: the fieldtrip report to a mangrove forest of Cikeong, Muara Angke-Kapuk, and Cilacap. SEAMEO-BIOTROP, Bogor

LPP Mangrove (2000) Draft strategi nasional pengelolaan ekosistem mangrove Indonesia. LPP Mangrove, Bogor

Marsono DJ (1989) Structure and composition of mangrove vegetation in Cilacap. In: Soerianegara I, Zamora PM, Kartawinata K, Umaly RC, Tjtrosomo S, SItompul DM, Rosalina U (eds) Symposium on mangrove management: Its ecological and economic considerations. Biotrop Special Publication, Bogor 37:73–114

Mustafa M, Nurkim SH, Sutika N, Sanusi H (1979) Penelitian komunitas lingkungan dan regenerasi serta pengembangan hutan mangrove di Sulawesi Selatan, Universitas Hasanuddin, Ujung Pandang

Soemodihardjo S, Soerianegara I (1989) The status of mangrove forests in Indonesia. In: Soerianegera I, Zamora PM, Kartawinata K, Umaly RC, Tjitrosomo S, Sitompul DM, Rosalina U (eds) Proceedings symposium on mangrove management: its ecological and economics considerations. SEAMEO–BIOTROP, Bogor, pp 73–114

Soerianegara I, Naamin N, Hardjowigeno S, Abdullah A, Soedomo M (1986) Prosiding diskusi panel dayaguna dan batas lebar jalur hijau hutan mangrove. MAB-LIPI, Jakarta

Soemodihardjo S, Wiroatmodjo P, Abdullah A, Tantra IGM, Soegiarto A (1993) In: Clough BF (ed) The economic and environmental values of mangrove forests and their present state of conservation in the South-East Asia/Pasific region, Hal. ISME-ITTO-JIAM, Jepang, pp 17–40

Sukardjo S (1979) Hutan payau di Kuala Sekampung, Lampung Selatan, Sumatera. In: Soemodihardjo S, Nontji A, Djamali A (eds) Prosiding seminar ekosistem hutan mangrove. LIPI, Jakarta, 59–68

Sukardjo S (1980) The mangrove on the new cimanuk delta. Presented at the LIPI-UNO join seminar on the Coastal Resources of the Cimanuk, West Java, Indonesia

Sukardjo S, Kartawinata K, Yamada I (1984) The mangrove forest in Bungin river, Banyuasin, South Sumatra. In: Soepadmo E, Rao N, MacIntosh DS (eds) Proceedings Asian symposium on mangrove environment: research and management. UNESCO, Paris, pp 121–141

Versteegh F (1952) Problems of silviculture and management of mangrove forests. Asia and Pacific Forestry Commission, 1–13 December 1952, Singapore

Widiarti A, Effendi R (1989) Socio-economic aspects of tambak forest in mangrove forest complex. In: Soerianegara I, Zamora PM, Kartawinata K, Umaly RC, Tjitrosomo S, Sitompul DM, Rosalina U (eds) Proceedings symposium on mangrove management: its ecological and economic considerations. SEAMEO-BIOTROP, Bogor, pp 275–279

Wiroatmodjo P, Judi DM (1979) Pengelolaan hutan payau di Indonesia. In: Soemodihardjo N, Djamali A (eds) Prosiding seminar ekosistem hutan mangrove. MAB–LIPI, Jakarta, pp 191–198

Yamada I, Sukardjo S (1980) Ecological study of mangrove and swamp forest in South Sumatra. In: Tsubouchi Y, Ilyas N, Takaya Y, Hanafiah AR (eds) South Sumatra, man and agriculture. The Center for Southeast Asian Studies, Kyoto University, Jepang, pp 1–34

Zuna HY (1998) Analisis Ekologi-Ekonomi SIstem Tambak Tumpangsari di RPH Poponcol, Desa Mayangan, Kabupaten Dati II Subang [Thesis]. Graduate School of Bogor Agricultural University, Bogor

Mangrove Forests in Thailand

Nathsuda Pumijumnong

Contents

Abstract Thailand's coastal zone is covered by rich mangrove forests that support a vital ecosystem. The mangrove system provides food for the local people and nutrients to the surrounding seas. Further, these forests protect the local environment by acting like a green wall that reduces coastal erosion and helps reduce effects of heavy waves and strong winds on the coast. Climate change will undoubtedly adversely affect this ecosystem. Rising sea levels will impact the chemical and physical properties of mangroves, resulting in harm of both plant and animal species. Mangrove forests

N. Pumijumnong (✉)
Faculty of Environment and Resource Studies, Mahidol University,
Salaya Phutthamonthon, Nakhon Pathom 73170, Thailand
e-mail: nathsuda@gmail.com

I. Faridah-Hanum et al. (eds.), *Mangrove Ecosystems of Asia*,
DOI 10.1007/978-1-4614-8582-7_4, © Springer Science+Business Media New York 2014

in Thailand have already suffered destruction by intensive aquaculture encroachment and urban area extension. The Thai government is implementing new strategies to protect, preserve, and reforest certain areas; however, urbanization continues to release detrimental heavy metal discharge into waterways. Continued accumulation of these metals into the sediment will result in a long-term effect that will not be easily mitigated. Mangrove trees are fast growing and can also serve as carbon sinks. The impact of their ability to mitigate greenhouse effects when faced with toxic metal discharge is unknown. Studies determining a mangrove forest's ability to act as a carbon sink, even under the negative impact of human activity, will be important to preserve this ecosystem.

1 Introduction

Mangrove forests are found on tropical and subtropical coasts, existing at the cross bridge between inland and sea. The mangrove ecosystem is the only known blackish water one, and trees receive organic materials from estuarine and oceanic ecosystems (Ong 1993; Kristensen et al. 2008).

Mangrove forests have an economic value on the order of 200,000–900,000 USD/ ha (McLeod and Salm 2006). Regardless, mangrove ecosystems are critical habitats in developing countries, playing a key role in human sustainability (Alongi 2002) for food, timber, fuel, and medicine (Saenger 2002). Mangrove forests have played a role in reducing strong waves and, specifically, minimizing damage from the 2004 Tsunami. These tidal forests are often important nursery grounds and breeding sites for birds, mammals, fish, crustaceans, shellfish, and reptiles. They serve as a renewable resource of wood, and are important sites for accumulation of sediment, nutrients, and contaminants (Twilley 1995; Twilley et al. 1996; Kathiresan and Bingham 2001; Ellison 2008).

In Thailand, mangrove trees grow on sheltered muddy shores and low-lying bogs of river and stream estuaries at levels between low and high tides. High-density areas include the Gulf of Thailand banks on the west coast of the east peninsula. Mature natural mangrove forests remain only along the west coast of the peninsula in the provinces of Ranong, Phang-Nga, and Trang. Mangroves along the Gulf of Thailand are young, especially in Petchaburi, Samut Prakarn, Samut Sakhon, and Samut Songkram provinces (Aksornkoae 1993).

Increasing urbanization, aquaculture, agriculture, and industry are major sources of mangrove deterioration, not only due to area loss, but also because of pollution. Pollutants negatively impact the environment. For example, heavy metals exhibit toxic effects on living things and at high concentrations they are indestructible (MacFarlane and Burchett 2002). Mangrove forests in Thailand link both marine and terrestrial ecosystems (Alongi et al. 2001). Unfortunately, they are also becoming a sink for heavy metals and other pollutants. Multifaceted knowledge of this important ecosystem is important for reforestation development, forest management promotion, and resource conservation.

2 Mangrove Natural Distribution

2.1 Climate Conditions of Mangrove Forests in Thailand

Climate conditions of mangroves in coastal areas of Thailand are such that the average annual rainfall is 1,556 mm, maximum monthly rainfall occurs in September (378.3 mm), while December experiences the smallest amount (4.6 mm). The average annual temperature is 27.6 °C, which is highest in April (29.9 °C) and lowest in January (25.5 °C). The average annual relative humidity is 76.1 %, with a high in October (81.4 %) and low in January (70.0 %). The climate type is a tropical savanna climate with little rainfall and severe drought during winter and summer (Aksornkoae 1993).

2.2 Mangrove Soil

It is formed by the accumulation of sediment derived from coastal or river bank erosion and eroded soil from higher areas that is transported via rivers and canals. Some soil originates from sedimentation of colloidal materials and particles. River and canal sediment is fine and muddy, while coastal sediment is sand. Degraded organic matter also makes up mangrove soil (Aksornkoae 1993). In Thailand, mangrove soil is unripe or entisol. Ninety percent is composed of mainly fine clay particles. Mottle has laterite, pH is neutral or weakly acidic, and the color is grey-dark reddish brown (Mangrove for the Future 2011).

2.3 Mangrove Area in Thailand

Mangrove forests are on muddy tidal flats at river mouths and along the southern and eastern coasts, specifically on the Gulf of Thailand and heavily concentrated on the Andaman Sea. The Chao Phraya Delta is also home to a considerable forest family. Mangroves form two-story forests. The upper layer grows to 20 m in height and is dominated by *Rhizophora apiculata*, *Rhizophora mucronata* (both locally named kongkang), *Heritiera littoralis* (ngon kai), and *Xylocarpus mekongensis* (syn *X. moluccensis*). Common species of the lower layer include *Bruguiera cylindrica* (thua khao), *Bruguiera parviflora* (thua dam), *Bruguiera sexangula* (prasak nu), *Ceriops decandra*, and *Ceriops tagal* (the latter both called prong). Prasak (*Bruguiera gymnorrhiza*) can emerge up to 40 m above the forest and is 2 m in girth. Toward land, where mud accumulates, dryer soil is overgrown with ferns and herbs, which comprise evergreen forests. The chak palm *Nypa fruticans* is common on creek edges (FAO 2005). Mangrove species relate to their location, influenced by chemical, physical, and saline properties of sediment, water drainage and current, sediment moisture, and flooding frequency (Aksornkoae 1999). Mangrove tree species typically grow along the river bank, and species are clearly separate. The pioneer species is Avicenniaceae (*Avicennia alba* and *Avicennia officinalis*) followed by Rhizophoraceae (*Rhizophora apiculata* and *Rhizophora mucronata*), and Sonneratiaceae (*Soneneratia caseolaris*). *Avicennia alba* and *Avicennia officinalis* grow in separate locations—*Avicennia alba* is found predominantly around the river mouth, where soil is silt and clay with a loam texture. *Avicennia officinalis* is often found along small canals, where sediment is sandy clay loam. *Avicennia officinalis* grows better in coarse-grained sediment compared to *Avicennia alba* (Office of Mangrove Conservation Department 2010).

In 1975, mangroves in Thailand covered approximately 2 million Rai (1 Rai = 0.16 ha), but by 1996, the number decreased to around 1 million Rai. 2004 saw an increase in mangrove area to 1.5 million Rai as a result of conservation and rehabilitation efforts. Current mangrove issues can be categorized into 3 types: (1) remaining mangrove areas in which inhabitants continually utilize the area, specifically seen along the Andaman coastline of Ranong, Krabi, Trang, Satun, and Phuket; (2) encroaching of mangrove areas by shrimp farmers, as seen in the eastern Thai gulf and in Chantaburi, Surat Thani, and Nakhon Si Thammarat provinces; and (3) selling mangrove land titles due to losses from earlier utilization. Additionally, inappropriate land utilization continues to cause coastal erosion, specifically seen in Samut Prakarn, Samut Sakhon, Samut Songkhram, Phetchaburi, and Chacheongsao provinces. In 2007, through remote sensing technology and interpretation of LANDSAT 5 satellite images, mangrove area in Thailand was estimated to cover 1,435,116 Rai, with the largest area in Phang Nga province (18.55 % of the total mangrove area in Thailand).

3 Current Status of Mangrove Area

The last 46 years (1961–2007) saw a dramatic decline in the Thailand mangrove area, primarily due to encroachment. The major cause of initial mangrove encroachment was overexploitation for charcoal. Later, shrimp farming was the main factor. Other encroachment causes include urban expansion, industrial expansion, pier and road construction, and using mangrove areas for agriculture.

Extraordinary wetland ecosystems in Thailand, covered in mangrove forests, are Ramsa sites, including Don Hoi Lot in Samut Songkhram Province. Mangroves are highly populated along the shoreline on the east side of the Mae Klong River. Chao Mai Marine National Park-Ta Libong Island Non-Hunting Area-Trang River Estuaries are located in Trat Province. The ecosystems are diversified, comprising of riverine, estuarine, and coastal wetland, and include mangroves and Nypa. Kaper Estuary-Lamson Marine National Park-Kraburi Estuary is located in Ranong Province. This area has been declared a UNESCO Biosphere Reserve. Krabi Estuary National Reserve Forest is located in Krabi Province. Mu Koh Ang Thong Marine National Park is located in Surat Thani Province and is comprised of 42 small islands, where young mangrove forests are distributed along the coast. Pang Nga Bay Marine

National Park is located in Pang-Nga Province and contains at least 28 species of mangroves (Giesen et al. 2006). These preservation areas continue to increase the total mangrove area in Thailand.

Mangroves are vital food resources and protection zones for people living in coastal areas. Therefore, mangroves should be preserved, and cooperation among stakeholders related to mangrove conservation, rehabilitation, and research, should be encouraged. Information from these actions may lead to effective balance and sustainable management in the future. Table 1 illustrates the mangrove distribution in Thailand (Department of Marine and Coastal Resources 2005).

4 Mangrove Reforestation: Success and Failure

Mangrove forest areas have been dramatically decreasing, but several reforestation projects are in place. Success is limited due to each tree species requiring its own specific soil properties and optimal slope, tide, and water quality. These factors should be considered before implementing reforestation initiatives. Additionally, current projects are limited by planting only *Rhizophora mucronata* or *Rhizophora apiculata* because seeds are easy to plant in nurseries before transplantation. Mangrove trees reproduce using species seeds. These seeds actually start growing while still attached to the parent tree so that they can root themselves quickly dispersed into water.

Panapitukkul et al. (1998) examined the growth of mangrove forests in Pak Phanang Bay of southern Thailand and found an increase of 53.12 m/year during a 30-year period, specifically in *Avicennia alba* followed by *Sonneratia caseolaris* and *Rhizophora apiculata*. They concluded that the natural mangrove can be a rapid process if sufficient propagules. However mangrove areas have been dramatically changed to shrimp farms. Vaiphasa et al. (2007) examined solid shrimp pond waste on mangrove growth and mortality in Pak Phanang, Thailand. Excess sediment discharged from nearby ponds reduced mangrove growth rate and increased mortality. *Avicennia marina*, *Excoecaria agallocha*, and *Lumnitzera racemosa* tolerate this stress better than *Bruguiera cylindrical*. Further research assessing the impact of sedimentation and chemicals from shrimp farms on mangrove forests is needed. Other factors that may affect the success of the mangrove reforestation such as Buajan and Pumijumnong (2012) reported that for *Rhizophora mucronata*, *Avicennia alba*, and *A. marina* of the inner gulf, cambial activity significantly correlated with sea water level but insignificantly with salinity and climate. For *A. marina* and *A. alba*, more than one cambium existed simultaneously. The efficiency of this mechanism can improve tree growth under appropriate environmental conditions.

5 Carbon Content in Mangrove Trees

Department of Marine and Coastal Resources (DMCR) (2008) examined the carbon content in plants of Chumphon province in southern Thailand. Average carbon content in stems, branches, and leaves was 47.98 %, 47.68 %, and 43.33 % dry weight,

Table 1 Mangrove distribution in Thailand (by region and province)

Province	Year 1961[a]	1975[a]	1979[a]	1986[a]	1989[a]	1991[a]	1993[a]	1996[a]	2000[a]	2004[a]	2007[b]
Central region	*418,063*	*228,125*	*195,200*	*6,349*	*3,725*	*2,538*	*33,519*	*34,067*	*75,335*	*69,374*	*50,103*
Samut Prakarn	78,856	3,750	6,500	644	–	–	1,950	1,868	1,999	9,249	9,663
Bangkok	11,925	–	–	–	–	–	1,250	1,236	–	2,627	2,676
Samut Sakorn	176,519	115,625	90,100	887	–	–	11,369	10,602	21,144	19,253	14,503
Samut Songkhram	68,338	51,250	47,800	306	–	–	5,775	7,156	15,351	15,957	13,759
Phetchaburi	74,300	55,000	48,700	3,606	2,100	2,100	12,925	12,936	35,919	19,166	6,338
Prachuab Khiri Khan	8,125	2,500	2,100	906	438	438	250	269	922	3,122	3,164
Eastern region	*342,781*	*306,250*	*275,90*	*174,879*	*129,430*	*69,277*	*81,548*	*79,112*	*142,131*	*165,205*	*139,831*
Trat	90,663	66,250	61,500	55,112	53,987	48,438	47,925	47,087	57,787	59,482	56,668
Chantaburi	176,181	163,125	150,400	90,668	54,350	16,644	25,450	24,332	62,360	78,580	62,151
Rayong	27,650	34,375	28,800	15,112	10,987	963	4,250	4,103	8,322	11,764	8,924
Chonburi	23,906	23,750	20,700	9,362	6,550	938	575	575	6,519	4,461	4,870
Chachoeng sao	24,381	18,750	14,500	4,625	3,556	2,294	3,348	3,015	7,143	10,918	7,218

Table 1 (continued)

Province	Year 1961[a]	1975[a]	1979[a]	1986[a]	1989[a]	1991[a]	1993[a]	1996[a]	2000[a]	2004[a]	2007[b]
East coast of southern region	*352,807*	*221,875*	*211,100*	*122,772*	*106,775*	*87,375*	*102,654*	*103,571*	*205,053*	*212,894*	*169,772*
Chumphon	66,450	46,250	43,300	22,662	14,156	11,363	20,584	19,699	50,024	45,292	33,351
Surat Thani	73,769	23,125	36,300	26,774	23,544	13,775	19,775	19,586	22,078	58,127	38,392
Nakhon Si Thammarat	135,106	96,875	80,200	55,224	53,256	50,156	49,975	52,601	61,718	59,876	65,990
Phatthalung	15,819	11,875	10,200	656	525	375	800	881	19,747	1,354	2,067
Songkla	37,994	36,875	32,400	6,031	4,300	1,431	3,425	3,897	29,153	21,805	7,182
Pattani	23,669	6,875	8,700	11,425	10,994	10,275	8,095	6,907	22,333	26,440	22,538
Narathiwat	–	–	–	–	–	–	–	–	–	–	252
West coast of southern region	*1,213,577*	*1,198,125*	*1,213,475*	*923,724*	*888,864*	*927,198*	*836,545*	*830,650*	*1,067,043*	*1,133,634*	*1,075,410*
Ranong	168,963	151,250	141,200	135,087	132,688	121,688	120,675	120,229	157,948	170,335	155,062
Phang Nga	274,869	319,375	304,475	227,625	222,663	209,438	191,976	190,265	248,113	262,737	266,168
Phuket	17,313	19,375	17,800	12,094	11,163	9,713	9,675	9,448	11,990	11,725	11,496
Krabi	249,331	206,250	198,500	189,450	185,269	199,469	178,292	176,709	218,727	219,338	222,457
Trang	249,488	212,500	205,400	164,225	156,500	192,806	152,050	150,597	209,375	223,677	207,191
Satun	253,613	289,375	346,100	195,243	181,581	194,084	193,877	183,402	220,890	245,822	213,036
Total	*2,327,228*	*1,954,375*	*1,895,675*	*1,227,724*	*1,128,794*	*1,086,388*	*1,054,266*	*1,047,400*	*1,489,562*	*1,581,107*	*1,435,116*

Note: No data
1 Rai = 0.16 hectare
[a] Department of Marine and Coastal Resources, Natural Environmental Status Volume 1 Mangroves (2005)
[b] LANDSAT-5 Satellite interpretation 2007

respectively. *Ceriop tagal*, *Lumnitzera littorea*, and *Ceriops decandra* exhibited highest average carbon content (49.28 %, 48.49 %, and 46.63 % dry weight, respectively). Conversion to carbon storage in the mangrove forest was 144.21, 92.05 and 55.45 t ha^{-1}, respectively compared to the mangrove forest in Ranong province, where the average carbon content in the stem, branches, and leaves was 47.84 %, 47.96 %, and 46.23 % dry weight, respectively. *Ceriop tagal* has high average carbon content in its parts (48.91 %, 48.72 %, and 48.16 % dry weight, respectively). Mangrove forest carbon storage was found to be 112.59, 78.46, and 17.29 t ha^{-1}, respectively. DMCR (2010) surveyed the mangrove forest of the Surat Thani province, where above-ground biomass was evaluated to be 17,884.15 kg/Rai and the average carbon sequestration was 44.67 %. Meepol (2009) calculated mangrove forest carbon content and sequestration on abandoned shrimp in Donsak, Surat Thani Province. The average carbon content in each part of 7–10-year-old *Rhizophora mucronata* Lamk. was 45.30 %, 46.47 %, 45.62 % and 45.01 % dry weight, respectively. Meepol (2010) determined carbon sequestration in the Ranong Biosphere Reserve by harvesting 121 trees of 11 species. Samples of stems, branches, leaves, and stilt roots were collected for carbon content analysis via a dry combustion method. Results showed that the mangrove tree density, sampling, and seeding was 1,905, 1,105, and 22,762 stem/ha, respectively. The average biomass was 119.76 ton/ha, equivalent to 57.85 ton carbon/ha. Carbon content varied among species with an average of 47.74 % dry weight. The estimated total amount of carbon stored in the Ranong Biosphere Reserve was 398,971 tons, equivalent to 1.46 million tons of carbon dioxide.

Sriladda and Puangchit (2009) explored carbon sequestration in mangrove plantations in Pak Phanang, Nakhon Si Thammarat Province. This study covered five age classes: 4, 10, 14, 20 and 25 years old. Results showed that biomass production increased with age, except in the 25-year-old group. The highest biomass production was 267.79 tons/ha for the 20-year-old plantation, equivalent to 121.72 tons carbon/ha or 446.73 tons carbon dioxide/ha. Carbon content was not significantly different among species and plantation age, with an average of 45 % dry weight. Carbon content was significantly different among different tree parts with high carbon content in the stems (roots, branches, and leaves). Leaf carbon content was significantly different in other parts. The total carbon amount stored in whole mangrove plantations at Pak Phanang was approximately 0.225 million tons. Danpradit (2012) determined the carbon content of each tree part through a combustion technique. Results revealed that *Sonneratia caseolaris* (47.37 % dry weight) accumulated the most carbon. Average carbon content in stems, branches, leaves, and roots was 45.88 %, 45.39 %, 45.07 %, and 43.47 % dry weight, respectively. Carbon content was different among species, with an average of 45.38 % dry weight. ANOVA determined that carbon content was not significantly different among species, but was among plant parts ($p < 0.05$). Meepol (2010) reported that the average carbon content was 47.72 % dry weight, a result which was higher than Danpradit (45.38 % dry weight). The Office of Mangrove Conservation, Department of Marine and Coastal Resources (2010) reported an average of ten mangrove species (44.67 % dry weight) lower than the Danpradit result.

6 Heavy Metals in Mangrove Sediment

Buajan and Pumijumnong (2010) investigated the distribution of heavy metals in both dry and wet mangrove sediment in the Samut Sakhon province of the inner Gulf of Thailand. Physical and chemical properties (e.g. soil texture, pH, CEC, OM) were analyzed via standard methods, and heavy metals were analyzed by atomic absorption spectroscopy. For heavy metals, ranges were expressed in mg/kg: Cd (0.035 to 0.070), Cu (7.90 to 21.91), Pb (11.91 to 25.74), and Zn (55.99 to 75.05). Heavy metal concentration was higher in the dry season than in the wet season. Heavy metal levels were Zn > Pb > Cu > Cd. Factors responsible for heavy metal accumulation are organic matter, cation exchange capacity, and sediment texture. The concentrations of these metals were found to be higher inland, and decreased with closer proximity to the sea. Lattanasuttipong (2001) examined heavy metal accumulation—specifically Cd, Cu, Pb, and Zn—in sediment and in *Avicenna alba* Bl from the lower Tha-Chin river of the inner gulf. Results showed old leaves were rich in these heavy metals, specifically in the root and bark. Lertprasert (2006) studied the accumulation and distribution of heavy metals (Cd, Cu, Fe, Pb and Zn) in water and sediment in Ipomoea aquatic Forsk and *Rhizophora apiculata* Blume in the Phi Lok canal system of Samut Songkhram province. Heavy metal concentrations in water and aquatic plants were determined. Metal content in sediment from highest to lowest amounts were Fe > Zn > Cu > Pb > Cd. The average heavy metal concentrations in shoots and roots of *Rhizophora apiculata* and *I. aquatic* were determined, from highest to lowest, as Fe > Zn > Cu > Pb > Cd. Danpradit (2012) investigated all ten transecting lines of mangrove sediment distribution along the coastal line of the Surat Thani province. The concentration of Cd was at the level of noise (<0.4 mg. kg^{-1}, dry weight); therefore, the results were omitted. On the Estuary of Tapee (T3) transect line, high concentrations of Cu (13.75 ± 0.39 mg. kg^{-1}, dry weight) accumulated, whereas the Koh Klong Chaiya (T5) transect line contained a lower amount (2.90 ± 0.52 mg. kg^{-1}, dry weight). Pb contamination was higher in the Klong Leelet (T1) transect line (25.99 ± 3.76 mg.kg^{-1}, dry weight) and lower for the Klong Chaiya (T4) transect line (8.51 ± 0.45 mg.kg^{-1}, dry weight). Zn concentration in the Klong Leelet (T1) transect line (46.43 ± 2.19 mg. kg^{-1}, dry weight) was high and low for the Right of Klong Chaya (T10) transect line (7.50 ± 0.60 mg. kg^{-1}, dry weight).

The results of Danpradit (2012) showed that the average concentration of heavy metals in the T1 and T3 transect line were higher than in the other transect line. Because both lines were located near the Tapees' estuary, into which a river with an urban path flows, industrial and agricultural waste most likely transports pollutants to these study areas. Mean heavy metal concentrations in sediment in decreasing order are: Zn > Pb > Cu > Cd. This pattern was similar to that found in the Samut Sakhon province (Buajan and Pumijumnong 2010), Panta mala Bay, Panama (Defew et al. 2005), and Mai Po, Hong Kong (Che 1999).

Comparing average concentrations of selected heavy metals in sediment with values from other countries showed that concentrations in Surat Thani province (Danpradit 2012) were lower than in both Panta mala Bay, Panama (Defew et al. 2005)

Table 2 Comparison of heavy metal concentration in sediment

Location	Concentration of heavy metal (mg. kg^{-1})				References
	Cadmium	Copper	Lead	Zinc	
Surat Thani province, Thailand	<0.4	2.90–13.75	8.15–25.99	7.50–46.43	Danpradit (2012)
Chumphon province, Thailand	0.002–0.037	2.26–51.19	0.24–12.36	0.53–40.85	Pumijumnong and Uppadit (2012)
Tha Chin Estuary, Thailand	0.035–0.070	7.90–21.91	11.91–25.74	55.99–75.05	Buajan and Pumijumnong (2010)
Peninsular, Malaysia	0.5–0.8	2.0–31.9	3.1–83.1	0.1–4.3	Nazli and Hashim (2010)
Panta mala Bay, Panama	<10	56.3	78.2	105	Defew et al. (2005)
Mai Po, Hong Kong	1.1–1.14	51–87	69–220	130–308	Che (1999)
World median	0.35	30	35	90	Bowen (1979)
Soil quality standard, Thailand	<37	–	<400	–	PCD (2006)

and Mai Po, Hong Kong (Che 1999). Cu and Zn concentrations in Danpradit's investigation were lower than the concentrations of Cu and Zn in that of Buajan and Pumijumnong (2010). Moreover, we found that Cd, Pb and Zn concentrations in this study were higher than those in Chumphon province (Pumijumnong and Uppadit 2012), and Cd, Cu, and Pb concentrations were lower than in sediment from Peninsular, Malaysia (Nazli and Hashim 2010). Overall, the concentrations were lower than the world median, with the exception of Cd. Concentrations of Cd were lower than the Soil Quality Standards of the Pollution Control Department (Pollution Control Department [PCD] 2006). Importantly, the concentrations of all heavy metals presented no hazard to mangrove species. This is compared to reports of MacFarlane and Burchett (2002) and Rahman et al. (2011) in Table 2.

7 Heavy Metal Concentrations in Plants

Danpradit (2012) collected wood from nine species from ten transect lines. Wood can provide a clear indication of heavy metal content in plants. Root directly absorbs heavy metals from water, resulting in overestimation in sediment. Additionally, leaves absorb heavy metals directly from air and discharge them via salt glands (Defew et al. 2005). Cd concentrations measured from stems ranged from 0.148 to 0.001 $\mu g.g^{-1}$, with the highest value found in *Sonneratia caseolaris* at T10. Cu accumulation was higher in *Sonneratia caseolaris* at T4 and ranged from 3.23 to 0.39 $\mu g.g^{-1}$. Pb concentration ranged from 0.83 to 0.10 $\mu g.g^{-1}$, with the highest value for *Sonneratia caseolaris* and *Avicennia alba* at T1. *Avicennia alba* at T1 exhibited a high Zn concentration ranging from 9.77 to 0.57 $\mu g.g^{-1}$. Table 3 shows a comparison of heavy metal concentration in plant parts.

Table 3 Comparison of heavy metal concentrations in parts of the plant

Location	Parts of a tree	Scientific name	Concentration (μg. g^{-1})				References
			Cadmium	Copper	Lead	Zinc	
Surat Thani, Thailand	Stem	*Avicennia alba* Bl.	0.036–0.006	1.31–0.71	0.83–0.39	9.77–4.58	Danpradit (2012)
		Avicennia officinalis Linn.	0.033–0.010	1.81–0.76	0.55–0.15	9.33–5.20	
		Bruguiera parviflora Wight & Arn.ex Griff.	0.091–0.007	2.90–1.14	0.37–0.17	4.46–1.47	
		Ceriops tagal C.B. Robinson.	0.004	0.94-	0.45	3.87	
		Kandelia candel Druce.	0.013	0.70	0.40	4.49	
		Rhizophora apiculata Bl.	0.057–0.001	1.36–0.39	0.41–0.12	4.45–0.57	
		Rhizophora mucronata Poir.	0.036–0.006	1.24–0.036	0.34–0.21	4.10–1.97	
		Soneneratia caseolaris Engler.	0.148–0.011	3.23–0.043	0.83–0.10	5.25–1.03	
		Xylocarpus granatum Koen.	0.038–0.022	1.37–1.09	0.35–0.24	2.28–1.66	
Gujarat, India	Leaves	*Avicennia marina* (Forsk.) Vierh	5.71	–	0.84	2.89	Nazli and Hashim (2010)
	Stem		1.83	–	0.38	1.49	
	Root		4.05	–	1.57	4.00	
Mai Po, Hong Kong	Root	*Acanthus ilicifolius* Linn.	0.5–0.05	65.4–25.2	60.0–20.0	324.8–134.1	Che (1999)
		Aegicerus corniculatum (L.) Blanco.	0.5–0.07	46.2–22.4	60.0–20.0	136.8–44.0	
		Kandelia candel Druce.	0.6–0.1	29.4–19.6	60.0–20.0	208.0–81.6	
Hainan Island, China	Propagule	*Acanthus ilicifolius* Linn.	0.041	2.1	0.028	60.0	Lian et al. (1999)
		Aegicers corniculatum (L.) Blanco.	0.024	7.7	0.018	27.0	
		Avicennia marina (Forsk.) Vierh.	0.019	6.3	0.023	20.0	
		Bruguiera gymnorrhiza (L.) Lamk.	0.014	7.8	0.021	5.7	
		Bruguiera sexangula Poir.	0.031	7.8	0.026	30.0	
		Ceriops tagal C.B. Robinson.	0.033	5.4	0.029	6.8	
		Kandelia candel Druce.	0.057	3.1	0.038	9.8	
		Rhizophora stylosa Griff.	0.047	2.4	0.036	5.7	
		Sonneratia caseolaris Engler.	0.031	6.8	0.027	36.0	
		Sonneratia ovate Back.	0.041	7.4	0.021	19.0	

Table 3 (continued)

Location	Parts of a tree	Scientific name	Concentration (µg. g^{-1})				References
			Cadmium	Copper	Lead	Zinc	
Bhitarkanika, India	Leaves	*Avicennia officinalis* Linn.	–	2.1–3.7	–	0.7–1.5	Sarangi et al. (2002)
		Bruguiera cylindrical (L.) Blume.	–	1.7–2.1	–	0.8–2.0	
		Ceriops decandra (Griffith) Ding Hou.	–	0.8–1.7	–	0.3–1.0	
		Rhizophora mucronata Poir.	–	1.5-1.9	–	0.7–1.1	
		Xylocarpus granatum Koen.	–	2.05–2.2	–	0.4–0.6	
Terengganu, Malaysia	Leaves	*Rhizophora apiculata* Bl.	–	2.73	1.3	–	Kamaruzzaman et al. (2009)
	Bark		–	3.94	1.38	–	
	Root		–	5.21	2.05	–	
Samut Song Khram, Thailand	Shoot	*Rhizophora apiculata* Bl.	0.0015–0.0023	3.56–6.44	0.34–0.47	8.88–12.06	Lertprasert (2006)
	Root		0.0052–0.0084	4.72–9.55	1.06–1.40	13.54–15.22	
Peninsular, Malaysia	Leaves	*Sonneratia caseolaris* Engler.	1	26.8	35.5	5.9	Nazli and Hashim (2010)
	Root		0.6	31.2	92.9	10.0	

8 Community Good Practices in Mangrove Forest Rehabilitation

Several implemented, successful practices within communities were selected as case studies to contribute to a productive learning process. These are regarded both nationally and internationally as good practices based on results produced for 10–20 years. The good practices are drawn from four locations in the Gulf of Thailand and Andaman Sea coasts: (1) Ban Prednai, Trat province; (2) Bang Khunsai, Phetburi province; and (3) Ban Bang Tip, Phang Nga province (Chotthong and Aksornkoae 2006); and (4) Kho Kham community, Samut sakhon province.

8.1 Ban Prednai Community

Ban Prednai, Huang Nam Kao sub-district, Muang district of Trat province, has a population of 600 people. It is situated on the Gulf of Thailand's eastern area within a coastal plain. Its mangrove forest of 2,000 ha is located on the southern end of the community where a number of fishermen are based. In 1982, a logging concession overlapped with conservation of forest area, and the community began to take care of the forests.

Logging activities were not consistent with concessionaire requirements. Dikes were constructed to prevent saline water from entering the area, deteriorating mangroves. In 1983, villagers rallied to attempt to remove concessionaries, which resulted in armed violence. A letter submitted to the central government office in 1986 outlining the ongoing conflict resulted in abrupt resignation of the provincial governor. Subsequently, local military units dismantled them so seawater could restore the marine lifecycle. In 1998, villagers established the "Ban Prednai Mangrove Forest Conservation and Development Group."

With the removal of concessionaires, the community turned its attention to local and outside mangrove forest encroachers by launching reforestation campaigns to implement workable alternatives. Regulations were issued regarding using mangrove forests. Patrols were undertaken to both enforce regulations and to prevent illegal logging. Consultations with nearby communities and local government organizations related to push nets and trawler's fishing operations along the coastline were conducted. To sustain mangrove crab communities, catching during egg-laying and immature periods were banned. Due to these local efforts, after a period of only a year, mangrove crab numbers increased dramatically even during the dry season. The mangrove forest conservation group currently acts as a coordinating body for promoting various activities in the community. Monthly meetings are called to facilitate regular sharing of experiences.

8.2 Bang Khunsai Community

This is a sub-district of 11 villages situated within the Ban Laem district of Phetchaburi province in the western area of the Gulf of Thailand. This area receives silt from various rivers, which is a rich food source for marine life. Wind and waves are calm, making it a major breeding area for marine life, especially cockles, for which Khunsai is the largest production source in Thailand. Local people still collect cockles using traditional methods.

In 1991, investors and fishermen entered the area using cockle-raking trawlers to amass large quantities of cockles in a short time. Regulations prohibited collection of cockles no greater than 6 mm. The collection area was divided into two: one for tradition cockle collectors and the other for collectors using modern technology. Three-kilometre areas along the coast were designated for conservation, but the total cooperation of state agencies was not achieved. Thus, villagers established the "Bang Khunsai Marine Resource Conservation Group" to arrest those who encroached on conservation areas. The group also fosters the conservation area as well as a 100-ha mangrove forest. Moreover, the group is working with a local government agency to reforest areas and organize a learning centre for the public. Higher levels of hope for sustainability have been achieved by core members of the mangrove forest conservation group being elected to the Bang Khunsai Tambon Administration Organization (TAO) council, which is equipped with the authority and funds to push attempts to conserve and protect local resources.

8.3 Ban Bang Tip Community

This village belongs to the Bang Wan Sub-district of the Kuraburi District within Phang Nga Province, and has a population of about 2,000 people. The community is located on the Andaman Sea Coast in southern Thailand. The eastern side is mountainous with rain and deciduous forests. The land slopes down toward the western seashore that is home to 6,000 Rai of mangrove forests that were previously used for charcoal production, resulting in deteriorated forest conditions. In some areas, only dead stumps, traces of fish dynamiting, and exploitation from push nets and trawlers remain.

Village officials encouraged village youth and others to train for natural resource rehabilitation and management through classes organized by state agencies, helping to initiate the "Bang Tip Resource Administration and Management Group" in 1996. After the logging concession in 1997 expired, this group discussed village rehabilitation efforts with state agencies. The village was initially given 100 ha, which they subsequently began to reforest once a month through involving local children and youth. Marine life also revived during this period. The community also maintains a patrol in the mangrove forest as protection from encroachers by building barricades of bamboo stems pitched in rows. Confidence from this success promoted villagers' interest to an extent to women's groups.

8.4 Kho Kham Community, Samut Sakhon province

Villagers in Khok Kham community are an example of finding the appropriate solution of coastal erosion's problem. They used a bamboo stick to break the wave

Fig. 1 Comparison of the coastal line of Kok Kham sub district, Samut Sakhon province (photos by author). **a** Before using the bamboo stick technique. **b** After using the bamboo stick technique

intensity, and increasing of sediment can be successful. The bamboo stick wall at Kho Kham district, Samut Sakhon province stems from the local wisdom initiated in 2005 which can reduce strong waves, minimize coastal erosion and also increase sediment behind the bamboo wall, covering an area of Pantai Norsasing sub district of about 800 Rai and Bang Yapheg sub district by 400 Rai. By using this new technique, the sediment at the backside of the bamboo stick is about 1.5 m high, where afterwards the local people replant the mangrove forest into the sediment. Currently this new style is being extended to nearby areas (Figure 1).

9 Thailand's Policy on Mangrove and Coastal Ecosystem Management

9.1 Policies and Plans for the Enhancement and Conservation of National Environmental Quality (2540–2559)

The formulation of policies and plans for the enhancement and conservation of national environmental quality is in compliance with the Enhancement and Conservation of National Environmental Quality Act, B.E. 2535. In article (1), the National Environment Board shall submit plans for the enhancement and conservation of environmental quality to the Cabinet for approval. The Cabinet Resolution on 26th November 1996 approved the submitted policy. The related goal for coastal resources was designation of mangrove conservation areas of no less than 1 million Rai (1 Rai = 0.16 ha), including the conservation and rehabilitation of all coastal resources. Reported data from the Department of Marine and Coastal Resources showed 1.1 million Rai of protected mangrove areas. Still, effective control and management of environmental impacts on quality of coastal areas should be practiced.

9.2 *Main Principles of the Enhancement and Conservation of National Environmental Quality Act, B.E. 2535*

The Enhancement and Conservation of the National Environmental Quality Act, B.E. 2535, has considerably changed the pattern of environment and natural resources management in Thailand. It is a foundation for pragmatic utilization control and protection and of natural resources. Moreover, action plans, including the authority of officers to supervise all developments, are specified, which are responsive and conformed to sustainable development principles.

The main principles of the Enhancement and Conservation of National Environmental Quality Act, B.E. 2535, include the appointment of a National Environmental Board to emphasize environmental protection. Some duties of the Board are:

1. To submit policies and plans for enhancing and conserving national environmental quality for the Cabinet's approval
2. To recommend regarding financial, fiscal, taxation, and investment promotion measures for the implementation of the policies and plans for B.E. 2535 to the Cabinet
3. To supervise the management and administration of the Environmental Fund

According to the environmental protection issue, the Act focuses on four main parts which are: (1) environmental quality standards, (2) environmental quality management planning, (3) conservation and environmentally protected areas, and (4) environmental impact assessment report.

9.3 *Global Warming Preparation under the Framework of National Economic and Social Development Plan*

The 11th National Economic and Social Development Plan (2012–2016) focuses on Thailand's global warming impact which is grouped into four dimensions:

Natural resources dimension: Trends for more frequent and more severe forest fire, coastal/land ecosystem loss, biodiversity depletion, disasters and more frequent and severe floods and droughts	**Physical dimension:** Changes in sea-level, temperature, precipitation fluctuation and more severe coastal erosion
Economic dimension: Decreasing agricultural products, livestock and fishery, increasing cost of production, more intense environmental related trade measures, tourism and infra-structure destruction	**Social dimension:** Increased migration and career opportunities, decreased health status, high-risk disease situation (increased number of vector reproduction due to appropriate temperature for epidemic disease)

The 11th Plan framework for global warming suggested that for Thailand's future opportunities and survival, she should reform her economic structure toward a greener economy. This scheme could be practically adapted through activities promoting protection and rehabilitation of degraded natural resources and environment, as well as benefits of conservation measures such as Reducing Emission from Deforestation and Degradation (REDD), Payment for Ecosystem Services (PES), biodiversity offsets practices, production sector adaptation toward green/low carbon, and renewable energy consumption.

10 Conclusion

Awareness for mangrove conservation and utilization in Thailand is increasing, and with stakeholder participation, mangrove management is more acceptable. Nevertheless the impact of waste water discharge and accumulation of heavy metals along the coastal area remain problematic. Mangrove trees adapt to sea level rises from climate change; this effect has been neglected in studies and is a significant topic for further research.

References

Aksornkoae S (1993) Ecology and management of mangroves. International Union for Conservation of Nature & Natural Resources, Bangkok

Aksornkoae S (1999) Mangrove: ecology and management. Kasetsart University, Bangkok

Alongi DM (2002) Present state and future of the world's mangrove forests. Env Conser 29(3): 331–349

Alongi DM, Wattayakorn G, Pfitzner J, Tirendi F, Zagorskis I, Brunskill GL, Davidson A, Clough BF (2001) Organic carbon accumulation and metabolic pathways in sediments of mangrove forest in Southern Thailand. Marine Geo 179:85–103

Bowen HJM (1979) The environmental chemistry of the elements. Academic Press, London, p 1158–1166

Buajan S, Pumijumnong N (2010) Distribution of heavy metals in mangrove sediment at the tha-chin estuary, samut sakhon province. Thailand. J Environ Res 32(2):61–77

Buajan S, Pumijumnong N (2012) Seasonal cambial activity of some mangrove trees in Inner Gulf of Thailand in dependence on climate. Songklanakarin J Sci Technol 34(3):337–344

Che RGO (1999) Concentration of 7 heavy metals in sediments and mangrove root samples from Mai Po, Hong Kong. Mar Poll Bull 19:269–279

Chotthong B, Aksornkoae S (2006) Sustainable community-based mangrove rehabilitation in Thailand. Thailand Environment Institute, Nonthaburi

Danpradit S (2012) Carbon and heavy metal accumulations in mangrove sediment and wood in Surat thani province, Thailand. M.Sc Thesis, Faculty of Graduate Studies, Mahidol University, Thailand

Defew LH, Mair JM, Guzman HM (2005) An assessment of metal contamination in mangrove sediments and leaves from Punta Mala Bay, Pacific Panama. Mar Poll Bull 50(5):547–552

Department of Marine and Coastal Resources (DMCR) (2005) Natural environmental status, vol 1, mangroves. DMCR, Bangkok

Department of Marine and Coastal Resource (DMCR) (2008) Final report of economic value and resource dependence in mangrove forest. DMCR, Bangkok

Department of Marine and Coastal Resource (DMCR) (2010) The change situation and distribution of mangrove forest. DMCR, Bangkok

Ellison JC (2008) Managing mangroves with benthic biodiversity in mind: moving beyond roving banditry. J Sea Res 59:2–15

FAO (2005) Global Forest Resources Assessment 2005 thematic study on mangroves, Thailand: DRAFT, AUGUST 2005. FAO Forestry Department, Rome

Giesen W, Walffraat S, Zieren M, Scholten L (2006) Mangrove guidebook for Southeast Asia. FAO and Wetland International, Dharmasarn Co. Ltd

Kamaruzzaman BY, Ong MC, Jalal KCA, Shahbudin S, Mohd Nor O (2009) Accumulate of lead and copper in Rhizophora apiculata from Setiu mangrove forest. Terengganu, Malaysia. J Environ Biol 30(5):821–824

Kathiresan K, Bingham BL (2001) Biology of mangrove and mangrove ecosystems. Adv Mar Biol 40:81–251

Kristensen E, Bouillon S, Dittmar T, Marchand C (2008) Organic carbon dynamics in mangrove ecosystem: A review. Aquatic Botany 89(2):201–219

Lattanasuttipong K (2001) Accumulation of heavy metals in sediment and Avicennia alba Bl. from the lower Tha-chin river. Chulalongkorn University, Bangkok

Lertprasert N (2006) Accumulation and distribution of heavy metals in water, sediment. *Ipomoea aquatic* Forsk. and *Rhizophora apiculata* Blume. In the Phi Lok Canal System, Samut Songkhram Province, Thailand. M.Sc Thesis, Faculty of Graduate Studies, Mahidol University

Lian Y, Xu J, Lin P, Meguro S, Kawachi S (1999) Five heavy metals in propagules of ten mangrove species of China. J Wood Sci 45(4):343–347

MacFarlane GR, Burchett MD (2002) Toxicity, growth and accumulation relationships of copper, lead and zinc in the grey mangrove *Avicennia marina* (Forsk.) Vierh. Mar Env Res 54:65–84

Mangrove For the Future (MFF) (2011) Thailand National Strategy and Actions Plan 2011–2013. Produced by MFF with the financial support of Norad and Sida, p 67

McLeod E, Salm RV (2006) Managing Mangroves for Resilience to Climate Change. IUCN, Gland, Switzerland

Meepol W (2009) Growth and litter production of 7–10 years old *rhizophora mucronata* lamk. Planted on abandoned shrimp farm area at donsak, surat thani province. J Forest Manag 3(6):35–55

Meepol W (2010) Carbon sequestration of mangrove forest at ranong biosphere reserve. J For Manag 4(7):29–44

Nazli MF, Hashim NR (2010) Heavy metal concentration in an important mangrove species, *Sonneratia caseolaris*, in Peninsular Malaysia. Env Asia 3:50–55

Office of Mangrove Conservation (2010) Biodiversity in mangrove forest: upper southern of Thailand. Agricultural co-operative Federation of Thailand, Bangkok

Ong JE (1993) Mangroves—a carbon source and sink. Chemosphere 27:1097–1107

Panapitukkul N, Duarte CM, Thampanya U, Kheowvongsri P, Srichai N, Geertz-Hansen O, Terrados J, Boromthanarath (1998) Mangrove colonization: mangrove progression over the growing pak phanang (SE Thailand) mud flat. Estuar Coast Shelf Sci 47:51–61

Pollution Control Department (PCD) (2006) Soil quality standard. PCD, Bangkok

Pumijumnong N, Uppadit B (2012) Accumulation of heavy metals in mangrove sediments of chumphon province, Thailand. J Environ Res 34(2):21–38

Rahman M, Chongling Y, Islam K (2011) Accumulation, distribution and toxicological effects induced by cadmium on the development of mangrove plant *Kandelia candel* (L.). Contem Probl Ecol 4(2):133–139

Sarangi RK, Kathiresan K, Subramnian AN (2002) Metal concentrations in five mangrove species of the Bhitarkanika, Orissa, east coast of India. Ind J Mar Sci 31(3):251–253

Saenger P (2002) Mangrove ecology, silviculture, and conservation. Kluwer Academic Publishers, Dordrecht

Sriladda C, Puangchit L (2009) Carbon sequestration of mangrove plantations at pak phanang, nakhon si thammarat province. In: The conference on mangrove ecology, 12–14 September, 2008, Cha-um, Petchburi, pp 379–389
Twilley RR (1995) Properties of mangrove ecosystems and their relationship to the energy signature of coastal environments. In: Hall CAS (ed) Maximum power. Colorado Press, Colorado, p 43–62
Twilley RR, Snedaker SC, Yàez-Arancibia A, Medina E (1996) Biodiversity and ecosystem processes in tropical estuaries: perspective of mangrove ecosystems. In: Mooney HA, Cushman JH, Medina E, Sala OE, Schulze E-D (eds) Functional roles of biodiversity: a global perspective. Wiley
Vaiphasa C, de Boer WF, Skidmore AK, Panitchart S, Vaiphasa T, Bamrongrugsa N, Santitamnont P (2007) Impact of solid shrimp pond waste materials on mangrove growth and mortality: a case study from Pak Phanang. Hydrobiologia. Thailand 591:47–57

Philippines' Mangrove Ecosystem: Status, Threats and Conservation

Kristine B. Garcia, Pastor L. Malabrigo, Jr. and Dixon T. Gevaña

Contents

Abstract The Philippines has very rich biodiversity in terms of number and percentage. It is regarded as one of 17 mega biodiversity countries due to its geographical isolation, diverse habitats and high rates of endemism. It ranks fifth globally in terms of the number of plant species and maintains 5 % of the world's flora. In mangroves alone, the country holds at least 50 % mangrove species of the world's approximately 65 species. However, due to anthropogenic activities as well as natural disturbances, the country continues to lose its rich biodiversity resources including mangroves. This chapter revisits the status of Philippines' mangroves, its current and future threats and analyzes the mechanisms on how various stakeholders put efforts to address those threats. We found out that while a number of successful conservation and restoration efforts have been made, there are still clear gaps on how different stakeholders can turn their commitments and initiatives into actions to conserve and rehabilitate Philippines' mangrove for human well-being and sustainable development.

K. B. Garcia (✉)
World Agroforestry Centre-Philippines, International Rice Research Institute,
2nd Fl., Khush Hall Bldg., Los Baños 4031, Philippines
e-mail: k.garcia2@cgiar.org

P. L. Malabrigo, Jr. · D. T. Gevaña
College of Forestry and Natural Resources, University of the Philippines Los Baños,
Kanluran Rd, Los Baños 4031, Philippines

I. Faridah-Hanum et al. (eds.), *Mangrove Ecosystems of Asia*,
DOI 10.1007/978-1-4614-8582-7_5,

1 Mangrove Distribution in the Philippines

As an archipelagic country made up of more than 7,000 islands, the Philippines has one of the longest coastlines in the world extending up to 36,289 km. It is located at 13 °00 °N, 122 °00 °E, along the tropical band where mangroves thrive. Hence, the diversity of mangroves is relatively high due to its geographical location. The country holds at least 50 % (Primavera et al. 2004) of the world's approximately 65 mangrove species (Kathiresan and Bingham 2001). It is also considered as one of the top 15 most mangrove-rich countries in the world according to Long and Giri (2011).

The Philippine Government adopts the Food and Agriculture Organization (FAO) definition of forest as "an area of more than 0.5 ha and tree crown cover (or equivalent stocking level) of more than 10 % which includes natural and plantation and production forests" (Lasco et al. 2012). Based on this definition, the Department of Environment and Natural Resources (DENR) estimates that 7.2 million ha comprise the forest ecosystem, which is approximately 24 % of the total land area as of 2003 (FMB 2007). Three percent of the remaining forest cover in the country is considered as mangrove forests. Generally, mangrove area is declared by the Philippine government under Presidential Decree (PD) 705 as forest land. Mangrove forest is defined as a type of forest on tidal mudflats along the sea coast extending along the streams where the water is brackish. Mature mangrove areas do not exceed 20,000 ha, of which approximately two-thirds are in Palawan. Consequently, around 80,000 ha of mangroves left in the country were declared as wilderness and forest reserves in 1981, including all the 40,000 ha of pristine mangroves in Palawan (Primavera 2002).

The Philippines used to be covered by 400,000–500,000 ha of mangroves in 1920 but it declined to around 120,000 ha in 1994 (Chapman 1976; Brown and Fischer 1918; Primavera 2000). The decline may be attributed to overexploitation by coastal dwellers, and conversion to agriculture, salt ponds, industry and settlements (Primavera 2000). Recent estimates suggest that the mangrove area has increased to 247,362 ha (FMB 2007); however, it still fell short by almost half of its original area. This loss resulted in a significant decrease in mangrove ecosystem services including fish production and carbon sequestration. Primavera (1997) demonstrated the correlation in comparable decline in Philippine mangrove areas and production from near-shore municipal fisheries that contrasts with the increase in brackish water pond area and aquaculture contribution to total fish production (Fig. 1).

According to the estimate of Long and Giri (2011), using remotely sensed satellite observations for the year 2000, 66 out of the 82 provinces in the country contain mangroves with a total covered area of 256,185 ha. The estimate of Long and Giri (2011) from 2000 is slightly higher than that of DENR's estimate in 2003 (Fig. 2). In the same paper, they estimated that 19 % (49,363 ha) of the Philippines' total mangrove area is located within existing protected area networks (International Union for Conservation of Nature (IUCN) protected areas categories, I–VI), with the greatest area of protected mangroves located on Palawan.

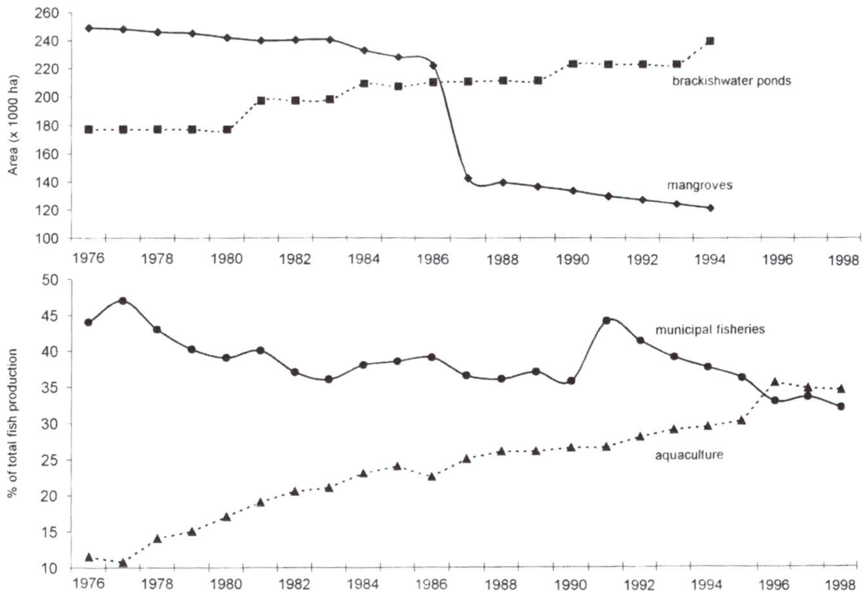

Fig. 1 Changes in mangrove and brackish water pond area (**a**) and contribution of municipal fisheries and aquaculture (**b**) to total fisheries production in the Philippines, 1976–1990. (Primavera 1997)

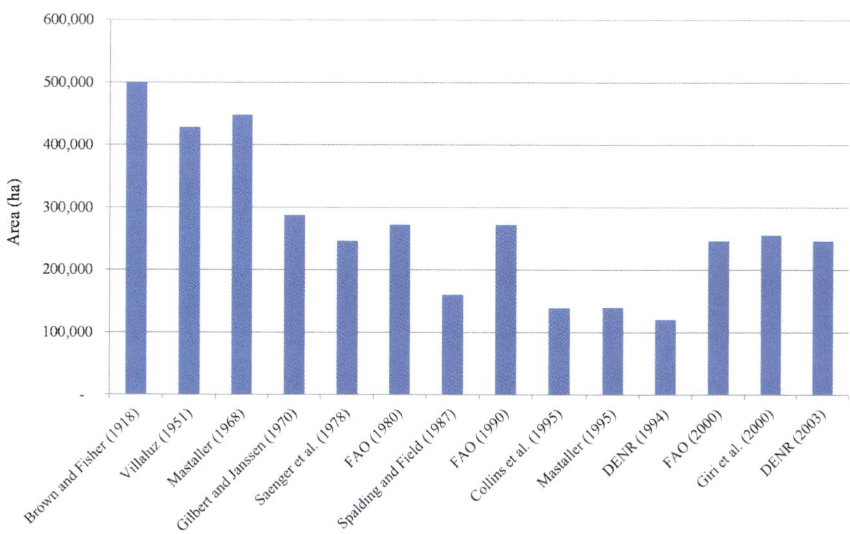

Fig. 2 Comparison of areal estimates of mangrove forest for the Philippines. Dates indicate year of estimate. (Long and Giri 2011)

Table 1 Number of subgeneric taxa in each true mangrove genera

Family	Genus	Number of species	Number of subspecies	Number of hybrids
ACANTHACEAE	*Acanthus*	2 (2)		
	Avicennia	8 (2)	4 (3)	
BIGNONIACEAE	*Dolichandrone*	1 (1)		
COMBRETACEAE	*Conocarpus*	1		
	Laguncularia	1		
	Lumnitzera	2 (2)		1
EUPHORBIACEAE	*Excoecaria*	3 (1)		
FABACEAE	*Cynometra*	1		
LYTHRACEAE	*Pemphis*	2 (1)		
	Sonneratia	6 (3)		2
MALVACEAE	*Camptostemon*	2 (1)		
	Heritiera	2		
MELIACEAE	*Aglaia*	1		
	Xylocarpus	2 (2)		
MYRSINACEAE	*Aegiceras*	2 (2)		
MYRTACEAE	*Osbornia*	1 (1)		
PELLICIERACEAE	*Pelliciera*	1		
PLUMBAGINACEAE	*Aegialitis*	2		
RHIZOPHORACEAE	*Bruguiera*	6 (4)		
	Ceriops	5 (2)		
	Kandelia	2 (1)		
	Rhizophora	6 (3)		4
RUBIACEAE	*Scyphiphora*	1 (1)		
Total 14 (11)	*23 (16)*	60 (29)	4 (3)	7 (0)

Note: Numbers inside the parenthesis are number of taxa present in the Philippines

2 Mangrove Taxonomy in the Philippines

Diversity of mangroves is extremely low as compared to that of tropical rainforests. However, mangroves can be considered one of the most taxonomically complex plant groups. Different mangrove species share a lot of common morphological characters that makes identification a very confusing task. Mangrove biologists generally classify mangrove plants into two: the true mangroves—species that are limited to the mangrove habitat; and the mangrove associates—mainly distributed in a terrestrial or aquatic habitat but also occur in the mangrove ecosystem (FAO 2007; Macintosh and Ashton 2002; Jayatissa et al. 2002; Duke et al. 1998). The true mangroves are further distinguished as major mangroves, which are tree species capable of forming dense pure stands, and minor mangroves, denoted by their inability to form a conspicuous element of the mangrove vegetation (Polidoro et al. 2010; Tomlinson 1986). There has been a long outstanding debate on these classifications, but during recent years, the increasing number of molecular works in mangroves has somehow developed a taxonomic consensus among mangrove biologists. Table 1 presents the most updated global mangroves' taxonomic list, which is a modification of the works of Kathiresan and Bingham (2001), to include the recent taxonomic revisions resulting

Table 2 Major and minor mangroves in the Philippines. (Primavera 2000)

I. ACANTHACEAE	*1. Acanthus ebracteatus*
	2. Acanthus ilicifolius
II. AVICENNIACEAE	*3. Avicennia alba*
	4. Avicennia officinalis
	5. Avicennia marina
	6. Avicennia rumphiana
III. BOMBACACEAE	*7. Camptostemon philippinensis*
	8. Captostemon schultzii
IV. COMBRETACEAE	*9. Lumnitzera littorea*
	10. Lumnitzera racemosa
	11. Lumnitzera rosea
V. EUPHORBIACEAE	*12. Excoecaria agallocha*
VI. LYTHRACEAE	*13. Pemphis acidula*
VII. MELIACEAE	*14. Xylocarpus granatum*
	15. Xylocarpus mekongensis
VIII. MYRSINACEAE	*16. Aegiceras corniculatum*
	17. Aegiceras floridum
IX. MYRTACEAE	*18. Osbornia octodonta*
X. PALMAE	*19. Nypa fruticans*
XI. PLUMBAGINACEAE	*20. Aegialitis annulata*
XII. RHIZOPHORACEAE	*21. Bruguiera cylindrica*
	22. Bruguiera exaristata
	23. Bruguiera hainesii
	24. Bruguiera gymnorrhiza
	25. Bruguiera parviflora
	26. Bruguiera sexangula
	27. Ceriops decandra
	28. Ceriops tagal
	29. Kandelia candel
	30. Rhizophora apiculata
	31. Rhizophora lamarckii
	32. Rhizophora mucronata
	33. Rhizophora stylosa
XIII. RUBIACEAE	*34. Scyphiphora hydrophyllacea*
XIV. SONNERATIACEAE	*35. Sonneratia alba*
	36. Sonneratia caseolaris
	37. Sonneratia gulngai
	38. Sonneratia lanceolata
	39. Sonneratia ovata

from various molecular works. Primavera (2000) also presented a list totaling some 35–44 major and minor mangrove species belonging to 14 families that can be found in the Philippines (Table 2).

3 Importance of Mangrove in the Philippines

Mangroves provide tremendous values and benefits to mankind and other marine organisms. They are a source of valuable plant products used as food, traditional herbal medicine and other wood and forest products. Mangrove forests serve as nesting grounds for hundreds of bird species, as well as nurseries, and are home to a wide variety of reptile, amphibian, mammals, fish, crabs, shrimps, mollusks and many other invertebrates (Nagelkerken et al. 2008). Being archipelagic in nature, a large part of the population of the Philippines depend on the mangroves for food, livelihood, and shelter derived from the mangrove ecosystem. In fact, more than half of the country's 1,500 towns and 42,000 villages depend on these marine habitats for food and other goods and services (Primavera 2000).

Recognizing the vulnerability of the country to storm surges and strong winds due to typhoons, planting of mangroves has been identified as one of the adaptation strategies to such climatic events. For instance, on the eastern coast of the Samar Island, the mangrove forest plays an important role in the protection of the coastline for coconut plantations. Mendoza and Alura (2001) noted that in areas without mangroves, the coconut trees were uprooted due to wave action during stormy weather. The event did not occur in coastal areas where a strip of mangroves was easily eroded compared to those with mangrove trees. In coastal areas directly exposed to the strong wave action of the Pacific Ocean, coastal erosion was reduced either by mangrove trees or cliffs. Mangroves also act synergistically with adjacent ecosystems such as seagrass and coral reef communities for coastal protection.

In the face of climate change, many of the regulating services of mangroves are actually becoming more necessary and valuable, especially their buffering capacity against storms and flooding. Mangroves can hold back the sea waves and reduce wave forces with their extensive and dense above-ground roots by an estimated 70–90 % on average (Macintosh 2010). Furthermore, in a study conducted by Harada et al. (2002) they demonstrated that mangroves are as effective as concrete seawall structures for reduction of tsunami-hit house damage behind the forest. Moreso, a six-year old mangrove forests of 1.5 km width reduce the sea waves by 20-fold, from 1 m high waves in the open sea to 0.05 m at the coast (Mazda et al. 1997).

Mangroves are also potential sources of livelihood for the community in the Philippines through the development of policies and programs that can help provide incentives to local people who are largely dependent on mangroves (Camacho et al. 2011). For instance, Camacho et al. (2011) wrote that Banacon Island in the Province of Bohol is perhaps one of the best when it comes to illustrating the carbon sink potential of mangroves in the Philippines. Banacon mangroves are in a vigorous condition and capable of storing vast amounts of carbon. They estimated that the 40-year-old plantation has the largest carbon density with 370.7 tons per ha, followed by the 15-year-old plantation with 208.5 tons per ha, 20-year-old plantation with 149.5 ton ha per ha, and lastly by natural stand with 145.6 tons per ha. They recommended that adopting incentive-based conservation programs such as payment for environmental services (PES) and Reducing Emissions from Deforestation and Forest Degradation projects (REDD) should also be explored in order to stimulate protection and enhance biodiversity, carbon stocks, water, aesthetics and local livelihoods.

Table 3 List of Mangrove Species Included in IUCN Red List. (after Polidoro et al. 2010)

Family	Species	Red List category
ACANTHACEAE	*Avicennia bicolor* Standley	VU
	Avicennia integra Duke	VU
	Avicennia rumphiana Hallier f.	VU
BIGNONIACEAE	*Tabebuia palustris* Hemsley	VU
FABACEAE	*Mora oleifera* (Hemsl.) Duke	VU
LYTHRACEAE	*Sonneratia griffithii* Kurz	CR
MALVACEAE	*Camptostemon philippinense* (Vidal) Becc.	EN
	Heritiera fomes Buch.-Ham.	EN
	Heritiera globosa Kostermans	EN
RHIZOPHORACEAE	*Bruguiera hainesii* C. G. Rogers	CR
PELLICIERACEAE	*Pelliciera rhizophorae* Triana and Planchon	VU

4 Conservation Efforts Addressing Threats

4.1 Current and Potential Threats to Mangrove Ecosystem Rehabilitation and Conservation

It is no doubt that mangrove forests are one of the world's most threatened tropical ecosystems. In fact, 11 true mangrove species (Table 3) qualified for the IUCN Red List categories of threat including two critically endangered, three endangered, and six vulnerable species (Polidoro et al. 2010). For these reasons, many tropical countries have considered the sustainable management of mangroves as major priorities in biodiversity conservation (Macintosh and Ashton 2002). In addition, several countries have already come up with their local mangroves Red List of threatened species. For instance, 12 species of true mangroves in India are considered to be 'critically endangered' and a total of 57 mangrove and mangrove-associated species are considered threatened (Kathiresan and Bingham 2001). Sri Lanka categorized eight mangrove species as locally threatened (Bambaradeniya et al. 2002). Ironically, in the Philippines, not a single mangrove species is included in the National Red List crafted by the Philippine Plant Conservation Committee and issued as a DENR Administrative Order (DAO) 2007–01.

Many reports have identified major causes of mangrove deforestation in the country including practices that pose potential threats to the diversity of mangrove species. While aquaculture development was identified as the most significant cause of mangrove degradation since the early years until present, there are also a number of serious threats including urbanization, conversion to agriculture, overharvesting for industrial uses such as timber and charcoal, and climate change, among others (Agaloos 1994; Alongi 2002; Primavera 2000; Boquiren et al. 2010).

4.1.1 Aquaculture Development

Aquaculture development, wherein ponds were built up into cultured ponds for production of shrimp, fish, and other aquatic resources, is known to be the leading cause of mangrove loss in the country. For instance, between 1968 and 1983, 237,000 ha of mangroves were lost for pond construction. This was almost half of the total national mangrove area (Fernandez 1978) at that time. Similarly, Agaloos (1994) and Primavera (2000) estimated that around half of the 279,000 ha of mangroves lost from 1951 to 1988 were developed into culture ponds (Agaloos 1994; Primavera 2000). Not only does aquaculture decrease the mangrove area, it also pollutes the mangrove ecosystem with effluents which in turn affect the services that a healthy ecosystem can provide. Shrimp aquaculture operates extensively normally for three to ten years after which the production decreases, and then abandonment occurs. Pollution and problems are often left behind (de la Torre and Barnhizer 2003). Once the operation is halted, aquaculture operators find another new location containing a healthy mangrove ecosystem and again deplete the resources (Ellison 2008). If this trend continues, mangrove areas in the country will be in serious threat. Although greater conservation and rehabilitation efforts have been in place (Samson and Rollon 2008), it is expected that the mangrove ecosystem in the country will continue to face degradation (Fortes 2004).

The municipal fishing sector comprises 68 % of the one million people engaged in the fishing industry in the Philippines, but it contributes only about 30 % of the total fish catch, while the 28 % engaged in aquaculture and only 4 % in commercial fishing contribute 60 % of the national fish catch (BFAR 1997). However, these figures do not reflect the negative impact of aquaculture to the mangrove ecosystem and to other marine ecosystems nearby such as the sea grass and coral reef.

4.1.2 Conversion to Agriculture

As opposed to aquaculture development, there were no significant accounts on mangrove area conversion to agriculture purposes in the country. However, it does not mean that this threat is far from beyond happening. Due to continued urbanization, some of the prime agricultural lands in the country are now being converted to settlements, hence the decrease in the available land for agriculture. As mangrove areas are rich in organic soils, they are prime locations for conversion into agricultural land, especially rice paddies and palm oil plantations to sustain the growing need for food. The possible greater threat from this happening is the drying and rapid and irreversible acidification of soils which can result in unusable land. In addition, as farmers often use fertilizers and chemicals, runoff containing these pollutants makes its way into water supplies. Despite their resilience, mangroves can tolerate only a limited amount of industrial and agricultural pollution without dying (American Museum of Natural History, n.d.).

4.1.3 Urbanization, Industry and Settlement

Extensive mangrove plantations found in Manila Bay in the early 1900s were subsequently replaced by fish ponds, settlements and port infrastructure (Brown and

Fischer 1920; Cabahug et al. 1986). In Bais Bay and Banacon Island, Philippines, cutting to make space for residential settlements has dramatically reduced the distribution on mangroves in the area (Walters 2003). The building of a causeway on Daco Island in 1950 and perimeter roads hastened further in-migration which caused local population to dramatically increase. The concentration of homes along the shore prompted mangrove cutting there. Backyard planting became widespread in the 1970s, but plantation expansion was later offset by the further cutting of mangroves from the landward side (Walters 2003).

Recently, the Philippine Reclamation Administration allowed the implementation of a 635-hectare reclamation project in Manila Bay beside the 175-hectare protected mangroves, lagoons and ponds known as the Las Pinas-Paranaque Critical Habitat and Ecotourism Area. While there has been a huge opposition to the project, the Court of Appeals approved the reclamation. Opposing parties which include politicians, socio-civic organizations and non-government organizations, proclaimed that reclamation is a passport for the destruction of Manila Bay and will allow imminent threats to livelihood and local fisheries (Punay 2013). It has also been raised that in pursuit of continued growth and economic development, the government failed to consider the ecological aspect of approved projects.

4.1.4 Cutting of Timber, Fuel and Charcoal

Due to an increase in the prices and access to commodities such as fuel and construction materials, people are forced to look for cheaper and alternative resources. Because of its physical characteristics, mangroves are often chosen as a primary option. Mangrove wood burns exceptionally hot and evenly and so has long been preferred as both a domestic cooking fuel and a fuel for commercial bakeries in the Philippines. In Bais Bay, for the past century people living along the coast have been relying heavily on cutting mangroves for domestic fuel and construction wood, especially for use as posts in fish weirs, called *Bunsod*, which are abundant in the shallow waters of North and South Bais Bay (Walters 2004).

4.2 Conservation and Rehabilitation Efforts: Failures and Future Directions

A number of efforts on mangrove conservation and rehabilitation have been completed in the country. Some were successful, some were not. Primavera and Esteban (2008) reviewed eight mangrove rehabilitation projects in the Philippines and found out that despite heavy funding in the hundreds of millions of dollars to rehabilitate thousands of hectares of mangroves over the last two decades, the long-term survival rates of mangroves are generally low at 10–20 %. Two of the main reasons cited are inappropriate species and sites because the ideal sites have been converted to brackish water fishponds. The favoured but unsuitable *Rhizophora* are planted in sandy substrates of exposed coastlines instead of the natural colonizers *Avicennia* and *Sonneratia*. Mangroves should be planted where fishponds are, not on seagrass beds and tidal flats where they never existed.

In addition, among the issues that were identified that impede success of mangrove rehabilitation and conservation efforts include lack of awareness, complexity of interactions between natural systems, social systems, and human values across temporal and spatial scales, weak and inadequate manpower, and lack of political will to enforce the laws (Primavera and Esteban 2008; Farley et al. 2009).

Among the reforestation projects that were implemented, community involvement is identified as they key factor for success (Alcala 1998; Primavera and Esteban 2008; Farley et al. 2009; Camacho et al. 2011). Involving the community is a more sustainable approach to reforestation and maintenance of existing resources because participatory approaches empower local communities to contribute more effectively to forest management (Contreras 2003). A popular success story involving a community that manages its natural resources is that of Banacon in Bohol Island. Recognizing the dire local needs for fuel wood and construction materials for building boats and houses due to mangrove scarcity after decades of continued exploitation, residents on Banacon have come to appreciate the benefits of owning their own mangrove plantations, and have so continued to plant vigorously even after the island was designated a protected area (Walters 2003; Camacho et al. 2011). Mangrove reforestation in Banacon is a community-initiated effort that started in 1957. Currently, mangrove plantations of Banacon are being managed by the local community with assistance from the DENR (Camacho et al. 2011).

Lasco et al. (2012) calculated the rate of change of each forest type based on official government data on forest cover as of 2003 and from the latest FAO Forest Resource Assessment (FRA) report for the Philippines (FAO 2010). They found that there has been a positive change for mangrove forests of about 0.008 % per year. This positive change in mangrove area may be attributed to some successes in mangrove reforestation and rehabilitation projects in the country. However, this positive change is still far from bringing back the mangrove area to its original extent, hence more is needed to be done.

In a recently concluded study by Calumpong and Cadiz (2012), it was recommended that to shore up its fish population and sustain its food supply, the Philippines must pursue a program to expand its mangrove forests from the current 140,000 hectares to approximately their 1920 level of 500,000 hectares. The researchers encouraged multi-species mangrove reforestation instead of dependence on monospecies stands of *Rhizophora* spp. or bakawan, which can be risky, since it is prone to pest attacks.

Another promising approach that is being developed in the country to encourage mangrove rehabilitation and conservation is the establishment of ecotourism in mangrove areas. The Philippines defined its ecotourism goal and described ecotourism as a form of sustainable tourism within a natural and cultural heritage area where community participation, protection, and management of natural resources, culture and indigenous knowledge and practices, environmental education and ethics as well as economic benefits are fostered and pursued for the enrichment of host communities and satisfaction of visitors (NESC 2002).

Currently, there are only a handful of mangrove-based ecotourism sites in the country. One is the Pagbilao Mangrove Experimental Forest in Quezon province which has the largest number of mangrove species of any stand in the Philippines (Bennagen and Cabahug 1992). The administration constructed a boardwalk wherein visitors can see clearly different mangrove and faunal species in the area. Another example is Banacon Island, which is the oldest of mangrove-based ecotourism sites. Likewise, Olango Island in Cebu province serves as recreational grounds for bird watching and observation of other wildlife. The development of ecotourism in mangrove areas provides cultural benefits. People from urban areas desire to experience the atmosphere of the mangrove ecosystem. The diverse mangrove plants and animals and their adaptations make the mangrove ecosystem an ideal ecological destination and field laboratories for biology and ecology students and researchers.

However, concerns are also being raised on ecotourism as it may also bring potential threats to the mangrove ecosystem. Among those are establishment of commercial areas, indirect costs of the damages to the services of the mangrove ecosystem, pollution, and waste. For instance, population density of tourists and frequency of visits for ecotourism activities might affect the natural vegetation and fauna in the mangrove areas. The noise and presence of people affects sensitive species of wildlife unlike tolerant species.

5 Conclusion and Future Perspectives

Mangroves are unique ecosystems which offer tremendous values and benefits. Philippine mangroves are very much diverse but facing tremendous threat. While previous major mangrove reforestation/rehabilitation in the Philippines is a big failure, there are also success stories that encourage continuing implementation of reforestation and conservation programs. We encourage those future programs to take into consideration the following recommendations:

a. Strengthening the information, education and communication program for the protection and conservation of mangrove areas.
b. Successful projects always start with proper awareness. There is a need for a more effective awareness campaign on the ecological and socio-economic importance of mangrove forests and other ecosystems. The government should implement new mangrove planting guidelines to enhance the survival rate of the mangrove species. The scientific community needs to provide the decision-makers with relevant information.
c. We should also continue to closely engage the local community in the management of resources as it is a more sustainable approach. They must be given technical assistance, training, education and diverse livelihood programs to enhance their capability. There is also a greater need for conservation that integrates research, training, advocacy and action including all sectors of the society at all levels.

d. If ecotourism is to be developed in a mangrove area, sustainable development and a holistic approach must be strengthened in the management, conservation, protection and utilization of the services provided by a mangrove ecosystem. The area must have management zonation with a strict protection zone and multiple use zones which includes ecotourism designated area. Each local government unit, DENR, people's organizations, private organizations and non-government organizations must cooperate in the management. The revenues gained from the collection of fees must be for the conservation and maintenance of the area.

References

Agaloos BD (1994) Re-afforestation of mangrove forests in the Republic of the Philippines. In: Proceedings of the international tropical timber organization workshop: development and dissemination of re-afforestation techniques of mangrove forests. Japan Association for Mangroves, Japan, pp 75–98

Alcala A (1998) Community-based coastal resource management in the Philippines: a case study. Ocean Coastal Manag 38:179–186

Alongi DM (2002) Present state and future of the world's mangrove forests. Env Conser 29(3):331–349

American Museum of Natural History (2013) Mangrove threats and solutions. http://www.amnh.org/explore/science-bulletins/bio/documentaries/mangroves-the-roots-of-the-sea/mangrove-threats-and-solutions. Accessed 8 May 2013

Bambaradeniya CNB, Ekanayake SP, Kekulandala LDCB, Fernando RHSS, Samarawickrama VAP, Priyadharshana TGM (2002) An assessment of the status of biodiversity in the Maduganga mangrove estuary. Occasional papers of IUCN Sri Lanka (no. 1), Nov. 2002

Bennagen E, Cabahug D (1992) Mangrove forest accounting: physical accounts (technical appendices). The Philippine Natural Resources Accounting Project (NRAP Phase 1), Manila

Boquiren R, Di Carlo G, Quibilan M (eds) (2010) Climate change vulnerability assessment of the Verde Island Passage, Philippines. Technical report. Conservation international, Virginia, pp 100

BFAR (Bureau of Fisheries and Aquatic Resources) (1997) 1996 Philippine profile, Department of Agriculture—Bureau of Fisheries and Aquatic Resources, Manila

Brown WH, Fischer FA (1918) Philippine mangrove swamps. Bulletin No. 17, Bureau of Forestry, Department of Agriculture and Natural Resources, Manila, Philippines

Brown WH, Fischer AF (1920) Philippine mangrove swamps. In: Brown WH, editor. Minor products of Philippine forests I. Bureau Forestry Bull 22:9–125

Cabahug DM Jr, Ambi FM, Nisperos SO, Truzan NC Jr (1986) Impact of community-based mangrove forestation to mangrove dependent families and to nearby coastal areas in Central Visayas: a case example. In: National Mangrove Committee (ed) Mangroves of Asia and the Pacific: status and management. Natural Resources Management Center, Ministry of Natural Resources, Quezon City, pp 441–466

Calumpong H, Pablina C (2012) Mangrove rehabilitation in Ticao Island, Masbate, Philippines. SEARCA Agriculture and Development Discussion Paper Series no. 2012–2014, p 41

Camacho L, Gevaña D, Carandang A, Camacho S, Combalicer E, Rebugio L, Youn Y (2011) Tree biomass and carbon stock of a community-managed mangrove forest in Bohol, Philippines. Forest Sci Tech 7(4):161–167

Chapman VJ (1976) Mangrove vegetation. J. Cramer, New York, p 477

Contreras A (ed) (2003) Creating space for local forest management in the Philippines. De La Salle University Press, Manila

de la Torre I, Barnhizer D (2003) The blues of a revolution: the damaging impacts of shrimp farming. Industrial Shrimp Action Network, Tacoma, WA

Duke NC, Ball MC, Ellison JC (1998) Factors influencing biodiversity and distributional gradients in mangroves. Global Eco Biogeo Letters 7:27–47

Ellison AM (2008) Managing mangroves with benthic biodiversity in mind: moving beyond roving banditry. J Sea Res 59:2–15

FAO (Food and Agriculture Organization) (2007) The world's mangroves 1980–2005: a thematic study in the framework of the global forest resources assessment 2005. Food and Agriculture Organization of the United Nations, Rome, 2007, pp 1–74. http://www.fao.org/docrep/010/a1427e/a1427e00.htm. Accessed 27 March 2013

FAO (Food Agriculture Organization) (2010) Global forest resources assessment 2010 key findings. United Nations Food and Agriculture Organization (FAO), Rome

Farley J, Batker D, Torre Idl, Hudspeth T (2009) Conserving mangrove ecosystems in the Philippines: transcending disciplinary and institutional borders. Env Manag 7:321–32

Fernandez PM (1978) Regulations and their impact on mangrove areas for aquaculture the Philippine experience. Workshop proceeding on the conversion of mangrove areas to aquaculture, Iloilo, pp 78–84

FMB (Forest Management Bureau) (2007) Philippine forestry statistics: forest cover within forest lands: 2003. Forest Management Bureau, Department of Environment and Natural Resources, Quezon City

Fortes M (2004) National report on seagrass in the South China Sea-Philippines. Reversing environmental degradation trends in the South China Sea and Gulf of Thailand. UNEP/GEF South China Sea Project

Harada K, Imamura F, Hiraishi T (2002) Experimental study on the effect in reducing tsunami by the coastal permeable structures. Final Proc. Int. Offshore Polar Eng. Conf. USA, 652–658

Kathiresan K, Bingham B (2001) Biology of mangroves and mangrove ecosystems. Adv Mar Biol 40:81–251

Jayatissa LP, Dahdouh-Guebas F, Koedam N (2002) A revision of the floral composition and distribution of mangroves in Sri Lanka. Bot J Linnaean Soc 138:29–43

Lasco RD, Veridiano RA, Habito M, Pulhin FB (2012) Reducing emissions from deforestation and forest degradation plus (REDD +) in the Philippines: will it make a difference in financing forest development? Mitig Adapt Strateg Glob Change. doi:10.1007/s11027-012-9411-5

Long J, Giri C (2011) Mapping the Philippines' mangrove forests using landsat imagery. Sensors 2011, 11:2972–2981. doi:10.3390/s110302972

Macintosh DJ (2010) Coastal Community Livelihoods: implication of intact ecosystem services. Paper presented at the 2010 Katoomba Meeting XVII held June 23-24, 2010 in Hanoi, Vietnam. http://www.ecosystemmarketplace.com/documents/acrobat/k17/Don%20Macintosh.pdf. Accessed 8 May 2013

Macintosh DJ, Ashton EC (2002) A review of mangrove biodiversity conservation and management. Centre for tropical ecosystems research, University of Aarhus, Denmark

Mazda Y, Magi M, Kogo M, Hong PN (1997) Mangrove as a coastal protection from waves in the Tong King delta, Vietnam. Mangroves Salt Marshes, 1. pp 127–135

Mendoza A, Alura D (2001) Mangrove structure on the eastern coast of Samar Island, Philippines (pp 423–425). In: DE Stott, RH Mohtar, GC Steinhard (eds) Sustaining the global farm. Selected papers from the 10th International Soil Conservation Organization Meeting held May 24–29, 1999 at Purdue University and the USDA-ARS National Soil Erosion Research Laboratory

Nagelkerken I, Blaber S, Bouillon S, Green P, Haywood M, Kirton LG, Meynecke JO, Pawlik J, Penrose HM, Sasekumar A, Somerfield PJ (2008) The habitat function of mangroves for terrestrial and marina fauna: a review. Aqu Bot 89(2):155–185

NESC (National Ecotourism Steering Committee) (2002) National ecotourism strategy. Department of Wildlife and Natural Services, Republic of the Philippines, pp 119

Polidoro BA, Carpenter KE, Collins L, Duke NC, Ellison AM, Ellison JC, Farnsworth EJ, Fernando ES, Kathiresan K, Koedam NE, Livingstone SR, Miyagi T, Moore GE, Ngoc Nam V, Ong JE, Primavera JH, Salmo III SG, Sanciangco JC, Sukardjo S, Wang Y, Hong Yong JW (2010) The

loss of species: mangrove extinction risk and geographic areas of global concern. PLoS ONE 5(4):e10095
Primavera JH (1997) Socioeconomic impacts of shrimp culture. Aquacult Res 28:815–827
Primavera JH (2000) Development and conservation of the Philippine mangroves: institutional issues. Ecol 35:91–106
Primavera JH (2002) Management and conservation of mangroves in the Philippines. In Joint UNU-Iwate-UNESCO International Conference. July 8–13, 2002. http://landbase.hq.unu.edu/Conference/Abstracts/Primavera.htm. Accessed 8 May 2013
Primavera JH, Sadaba RB, Lebata MJHL, Altamirano JP (2004) Handbook of mangroves in the Philippines—Panay. SEAFDEC Aquaculture Department, Iloilo
Primavera JH, Esteban JMA (2008) A review of mangrove rehabilitation in the Philippines: successes, failures and future prospects. Wetlands Ecol Manag 16(3):173–253
Punay E (2013). CA Junks Petition vs. Php 14 billion reclamation. http://www.philstar.com/metro/2013/05/04/938120/ca-junks-petition-vs-p14-b-reclamation-project. Accessed 8 May 2013
Samson MS, Rollon RN (2008) Growth performance of planted mangroves in the Philippines: revisiting forest management strategies. Ambio 37(4):234–240
Tomlinson PB (1986) The botany of mangroves. Cambridge University Press
Walters BB (2003) People and mangroves in the Philippines: fifty years of coastal environmental change. Env Conser 30:97–107
Walters BB (2004) Local management of mangrove forests: effective conservation or efficient resource exploitation? Human Ecol 32:177–195

Distribution, Characteristics and Economic Importance of Mangrove Forests in Iran

Saeed Rashvand and Seyed Mousa Sadeghi

Contents

S. M. Sadeghi (✉)
Faculty of Forestry, University Putra Malaysia, 43400 Serdang, Selangor, Malaysia
e-mail: smbooraki@gmail.com

Department of Natural resources research, Research Center of Natural Resources and Agriculture of Bushehr Province, Varzesh avenue, Bushehr, I.R. Iran

S. Rashvand
Research Center of Natural Resources and Agriculture of Qazvin Province, Qazvin, I.R. Iran

I. Faridah-Hanum et al. (eds.), *Mangrove Ecosystems of Asia*,
DOI 10.1007/978-1-4614-8582-7_6, © Springer Science+Business Media New York 2014

Abstract Mangrove forests of southern Iran are located in the earth's sub-tropical zone. The trees species composition of these forests is limited to only Hara (*Avicennia marina*) and Chandal (*Rhizophora mucronata*). There are not considerable differences in terms of vertical structure of mangrove forests; however, the tree density is reduced from east to west in southern Iran. The forests are located at estuaries which can receive fresh water in the rainy season. Those forests can play an eco-socio-environmental role in local society. In addition, the biodiversity conservation, carbon sequestration and other non-wood products are some aspects which can improve the importance of mangrove forests. The results provide some clear characteristics of mangrove habitat even though we considered some limiting and threatening factors of mangrove forests. According to those points we suggested a guideline for improvement of mangrove forests in southern Iran.

1 Introduction

Vegetation is considered one of the important available resources in ecosystems. As primary producers they play an important role in the life of other organisms' preservation, and development of plants have a special situation in Iran because desert makes up such a large part of it. It should be noted that mangrove forests have a valuable role in a marine chain ecosystem (Nybakken 1993). The coastal zone forms the basis element of marine life and supports plenty of marine living resources. This zone provides nursing or feeding places for many coastal and marine species and has more variation compared to other marine parts. So, not only the entire ecosystems should be noted by management, but also all fundamental processes must be preserved, especially in critical areas. Mangrove forests are samples and need considerable attention (Clark et al. 1992). Mangroves are woody plants that grow between land and sea in the tropical and sub-tropical regions. Mangrove plant communities thrive and survive in environments that are high salinity, frequent tides fill the environment, sometimes heavy storms occur, the average annual temperature is high and there are anaerobic conditions in the sites of the habitats. In such circumstances, most specific species have adapted. Mangroves construct a unique ecological environment that can host communities which have a rich variety of species (Kathiresan and Bingham 2001). The ecological importance of mangrove forests is more than it has ever been recognized. These forests are a specific landscape of tropical and semi-tropical coastal regions and have a considerable effect on their established environment. Their root systems result in stabilizing sediments and their communities reduce the wave energy. In contrast, habitat and shelter for many organisms are made. Mangrove enriches biomass more than any other plant communities (Rodtiguez and Feller 2004). The annual economic value of mangrove for coastal protection and providing ecological services is about 200,000–900,000 USD per hectare (Wells et al. 2006). The degradation rate of mangrove increases in some tropical areas due to human activities including the rapid expansion of shrimp farming for export (Pillay 2004). This has raised a considerable environmental concern among people and environmentalists as well as conservationists.

Despite the low level of these types of forests (12,585 ha) in Iran, it plays an important role in balancing the Persian Gulf and Oman Sea ecosystems. Moreover, ecotourism and photographers' attention to mangrove habitats, available wildlife and natural attractions are considered as economic potential in these areas. Today, forest reserves are vitally important in the world. Forestry and environmental organizations spend most on separation, management and running conservation programs, studying and conducting research. Mangrove forests in Iran are among protected and supported forests. A wide variety of species can be found along the different levels of latitude from north to south in Iran. One of communities and specific plant ecosystems of the country is mangrove forest, which is located in coastal regions of the Persian Gulf and Oman Sea. Despite conducting numerous studies on mangroves, generally decision-makers and planners have little understanding of these ecosystems. Due to this fact, the destruction and extinction continues. Thus enhancing the understanding of national and local decision-makers as well as planners and stating direct and indirect environmental and economic values are still a basic need to preserve this ecosystem (Kathiresan and Bingham 2001).

2 Mangrove Forests

Mangrove forests are coastal tidal formations in tropical and subtropical regions. The term mangrove refers to unique plants on the site, whereas mangrove forests are representative of all the communities that are formed by these plants. Mangrove plants are flowering and terrestrial halophile optional. Due to the inability to compete with other terrestrial species of plants, they entered into tidal areas and are accustomed to live in these conditions. Enduring the harsh conditions of life between sea and land, the mangrove forest has dominated in this area. Hence, a small number of plants are able to compete with them in the mangrove occupied zone (Danehkar 1995). These ecosystems are accounted among the richest and most fertile ecosystems in the world, and more than 80 % of total world fisheries are dependent on a rich ecosystem of mangroves, estuaries and coral reef habitat. The available plants in this ecosystem are a collection of halophile plants which are resistant to Salinity. They are comprised of various trees, shrubs, palms, epiphytes, bushes, grasses and ferns (Majnoonian 2001). Having 63 species of trees and shrubs in mangrove forests and an ample number of woody and non-woody plants, they have constructed a sustainable and self-regulated environment in tropical coasts (Danehkar 1995). Exclusive physiological processes allow mangrove forests to grow in a constantly changing environment. Mangroves should resist against drying effects of sun and wind, the imbalance caused by high salinity of seawater, lack of oxygen and soil saturation of the water. Vivipary, pneumatophore production, and mechanisms of desalinity enabled these plant communities to survive from the flooded salty environment (Danehkar 2001). They have adapted against high temperature and evaporation because of fleshy leaves, thick cutin, small leaf, small stomata and sinking of the stomata in the epidermis (Safiari 2001).

3 Distribution of Mangrove Forest Habitat in Iran

The Iranian mangrove forests are located at coastal seas of the Persian Gulf and Oman Sea in the range of 1,830 km from east to west in southern Iran. They are considered as the last distribution of mangrove forests in Southeast Asia. These forests are distributed in three provinces of Iran, namely, Hormozgan, Bushehr and Sistan va Balochestan. The climate of the area is sub-tropical and the annual precipitation is less than 200 mm. The rainfall is mostly in autumn and winter. But in the eastern part, monsoon rains occur. Mangrove forest in Iran is limited to two species, namely, *Avicennia marina* and *Rhizophora mucronata*. However, the majority of these mangrove forests are comprised of *Avicennia marina* (Avicenniaceae), which is established in all habitats of southern Iran. *Avicennia* is given in the honor of Abu Ali Sina who was a famous Persian scientist (Safiari 2001). Mangrove trees are seen with a big crown and umbrella canopy in which they are reflected as green communities on the water. Although many branches are produced from the trunk, there is no bole, thus identifying the main trunk from the branches is difficult. The colour of the tree stems is greenish grey in the young stage and it becomes light grey when it ages. The *Avicennia* leaves are leathery, lanceolate or ovate and opposite in arrangement. They are basically narrow and have a short stalk. The upper surface of the leaves is dark green and its lower surface is covered with tiny hairs. These leaves last for several years and are stable too. Mangrove shrubs bloom in June. Mangrove inflorescence is a cluster of small bright yellow blooms; this blossom is small with 1–2 ovate Brakteh (sepal) grouped in apical pistil. The petal is short and the upper surface of the calyx is glucose while the lower surface is velvety. The stigma in these types of flowers has almost no style. The mangrove fruit is almond shaped and small and *Avicennia* yield fruit in late July to mid-August. The fruit size is 1-3 × 1-1 cm. Mangrove fruit is unlike the seed of terrestrial trees because it buds on the mother tree and then is fallen into bog. As a result a seedling is reproduced and that is why it is called pseudo viviparous. A proper time to collect mangrove seeds is when the exocarp starts to wrinkle and wrinkle lines of fruit exocarp can be spotted as the exocarp is smooth prior to ripping. Regeneration under a mother tree canopy can be done in the second half of July to mid-August. In a typical and proper condition, 90 % of the seeds are germinated in the period of 10–15 days, and after 2–3 months they will be 30–40 cm. Their height will be 1–1.5 m after 2–3 years. The proper time of transferring the seedlings from nursery to planting area is from the second half of November. These seedlings are planted in holes with dimensions of 30 × 40 cm and 2 m away from each other. Experience of planting trees in Hormozgan has shown that seedlings which were planted in the last frontier tides had better quality of growth. In addition, the mortality rate of seedlings which were planted off-line of flood tide was 100 % because desalinity did not occur.

The "Chandal" species (*Rhizophora mucronata*) are from a family of Avicenniaceae. Less than 4 % of mangrove forests are from *Rhizophora,* which can be viewed at Jask port, Iran. Both *Avicennia* and *Rhizophora* are in shrub formation (Danehkar 1993). Because the tree has less tolerance to salinity than *Avicennia*, *Rhizophora* grows inside the tidal area. However, the *Rhizophora* can be seen out of the tidal border in some specific locations. This tree can grow on the soils with fine texture, particularly in heavy clay soil. It should be mentioned that the optimum rate of *Rhizophora* growth has been witnessed at salinities less than 50 ppt. This tree, with small diameter and a height between 2.5 and 4.5 m and its crown, is oval to elliptical shape and found in SiriK habitats in Hormozgan province, Iran. The trunks of trees are smooth and the skin has bumps and grooves and a longitudinal surface. The leaves are thick, simple, leathery, and the arrangement is opposite in which their track remains after deciduous time. The leaf margin is smooth and pointed. Likewise, the stipules fall swiftly. Flowers are grouped at the end of the branches and they have four petals, four sepals, and 8–12 anther. The ovary is semi-inferior and has two parts and two ovules in each segment. The fruit is conical, woody, unblooming and contains a seed. *Rhizophora* seed starts growing and produces a seedling with a long root (30 cm) on the mother tree. This young tree goes into mud after being removed from the tree and leaves begins to grow at the end of the growing point. Pneumatophore of these trees is originated from the stem and goes into the mud like an arc which is called "stilt". This tree is unlike *Avicennia* pneumatophore which exits from the soils like a column. The roots are sometimes 1 m long and play the role as a guardian. The roots help the trees to be established and breathe. The wood of this species is very hard and strong and shows tolerance against pests and diseases. They are also resistant to termites. According to studies, the average mass of *Rhizophora* wood is about 0.9 and it has a high heat value. Five tons of *Rhizophora* wood can produce energy equal to 2–3 tons of coal. The skin of this spices is rich in terms of tannin, which is used in leather production, medicines and dyeing (Sistani 1990).

The plants grow in the border between land and water among the margin of the estuary. The plants grow in larger scale or in a small gulf, which are far from the sea waves. Mangrove forests are established in the fen soil and they are exposed to the tide permanently. Shrubs and the forest floors can be observed when the tide is shifting. In this situation the mangrove forests are seen as a maze and networks or scattered islands. All or partial tree crown dips into the water at high tide. It can be concluded that the proper area for mangrove planation are the sites which dip into the water when there is a high raise. The mangrove sites are distributed in three provinces which are described as follows.

3.1 Sistan va Baloochestan Province (Erfani 2007)

Avicennia Marina is established in Gooatr Gulf in the estuary of BahooKalat River (671.53 ha).

3.2 Hormozgan Province (from East to West) (Danehkar 2006)

Avicennia Marina is distributed in the region of Jask in Gabrik, Jagin, in the entrance of Shahr-e-No, in Lash's coasts, yekdar and Sorgelam estuaries and also in the Kashi's river lagoons which covers an area of 643.9 ha.

- SiriK region: *Avicennia marina* occupy an area of 773 ha in Nakhle ziarat, Pachoor, Ziarat, Garenho, Ganari and Kortan estuaries.
- The region of Koolaghan: Jalabi, hassan Langi estuaries are in the vicinity of the Shoor river along with the *Avicennia* species as well as the region of Kalahi in Mashdar, Behine, Kargan, and Minab estuaries. They cover an area of 1513.3 ha.
- Khamir Port zone: The entrance of Mehran's River to the Persian Gulf and in north west estuaries of Qeshm Island in the Khur-e Khoran estuary, from sandy islands in front of Tabl and Laft village to Kooran in which *Avicennia marina* covers an area of 853.3 ha.

3.3 Bushehr Province

Nayband Gulf is comprised of two estuaries, namely, Asalooyeh and Basatin, which are covered with the *Avicennia marina* species; that habitat's area is 377 ha. A thorough explanation is provided for each in the following.

Fig. 1 A landscape of mangrove forest at Asalooyeh estuary. (Source: Google Earth 2012)

Fig. 2 Natural regeneration of *Avicennia marina* at Asalooyeh estuary

3.3.1 Asalooyeh Estuary

This estuary is located in the north of Nayband Gulf. Mangrove stands are distributed in the margin of the estuary. The area of these sites is 237 ha. One hundred eight hectare of that area consist of massive and small streams and sandy hills without any trees. The forest quality in this estuary is in good condition. Sixty percent of them are very dense stands and 30 % of them are dense stand and 10 % of the remaining population is scattered in low dense stands (Figs. 1 and 2; Rashvand 1997).

3.3.2 Basatin Estuary

This estuary is situated in the north east of Nayband Gulf. Mangrove forest stands are distributed in the form of scattered stands. The mangrove habitat area for this

Fig. 3 Location of mangrove forest at Basatin estuary. (Source: Google Earth 2012)

Fig. 4 Natural regeneration of *Avicennia marina* at Basatin estuary

estuary is 120 ha of which 40 ha form big and small streams. *Avicennia* stands in this site have lower density compared with the Asalooyeh estuary. Twenty-five percent of total forests are very dense, 45 % is dense, 20 % is scattered and 10 % of them have low density (Figs. 3 and 4; Rashvand 1997).

3.3.3 Dayyer Port

The site is located 2 km from the margin of the southern Dayyer city. The surface of this site is three hectares. The *Avicennia* stands have good quality of height and it can be classified as a dense stand. The forest at this site consists of dense stands (Figs. 5 and 6; Rashvand 1997).

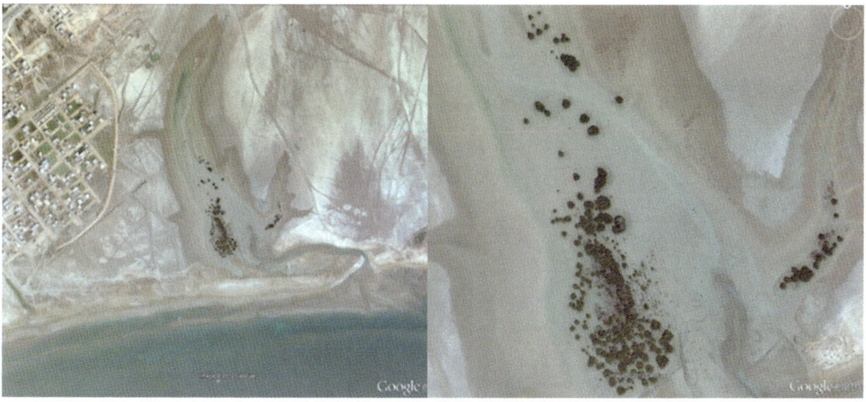

Fig. 5 A landscape of mangrove forest at the Dayyer site. (Source: Google Earth 2012)

Fig. 6 Mangrove forest at Dayyer habitat

3.3.4 Male-Gonze Site

This site is located 35 km west of Dayyer city. The surface of this area is 22 ha. The surface of the main stream and tributary is 10 ha. This stand has plenty of aging trees and it has differences from sites in terms of physiography. *Avicennia* stands in this site are directly in contact with the waves of the Persian Gulf. However, in other sites, sea waves move toward the land through the streams at high tide and affect the forest stands (Figs. 7 and 8; Rashvand 1997).

Fig. 7 *Avicennia marina* at Male-Gonze site

4 Species Composition of the Ecotone Zone of Bushehr Province

The halophile species occupy the ecotone zones which are located out of the tidal border. The formation of these plants consists of bushes and herbs. These plant communities are established after mangrove forests. The flora consists of 16 species, 15 genera, and eight families which belong to eight plant communities. According to the condition of the sites and their resistance against salinity, the description for those communities is presented as follows.

4.1 Mangrove Forest of Nayband

1. *Halocnemum strobilaceum* community from Chenopodiaceae
2. *Limonium gemelim* community from Plumbaginaceae

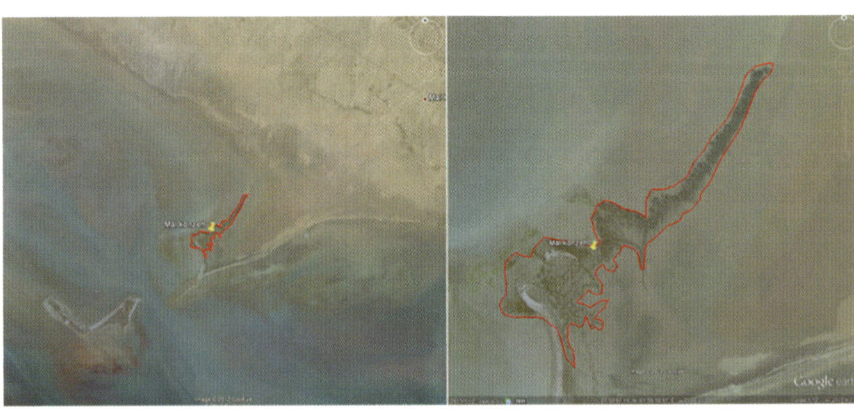

Fig. 8 A landscape of mangrove forest at Male-Gonze estuary. (Source: Google Earth 2012)

3. *Aeluropus lagopoides* community from Poaceae
4. *Puccinella distans* community from Poaceae
5. *Suaeda fruticosa* community from Chenopodiaceae

In the *Puccinella distans* community, there are *Cyperus arenrius* and *Psylliostachys spicata* species.

In addition, the *Tamarix* sp., *Cistanche tubulosa* and *Atriplex leucoclada* species are seen sporadically among the above-mentioned communities.

4.2 Mangrove Forest of Dayyer Port

Diversity and density of vegetation cover in this habitat from sea to land is limited to two communities as follows:

1. *Halocnemum strobilaceum* community from Chenopodiaceae
2. *Cressa cretica* community from Convolvulaceae

In the *Cressa cretica* community two "one-year-old" species from the Chenopodiaceae family, namely, *Salsola baryosma* and *Suaeda aegyptiaca* are distributed among plants.

4.3 Mangrove Forest of Male-Gonze Site

In this site, there are some communities in the vicinity of *Avicennia* forests in which their direction is from sea to land and they are established in the following order:

1. *Halocnemum strobilaceum* community from Chenopodiaceae
2. *Suaeda fruticosa* community from Chenopodiaceae
3. *Aeluropus lagopoides* community from Poaceae
4. *Hammada salicornicum* community from Chenopodiaceae
5. *Limonium gemelim* community from Plumbaginaceae
6. *Gymnocarpus decander* community from Caryophyllaceae

Scattered bushes from *Cistanche tubulosa* species (Orobancaceae) are in the above-said communities. Generally, the *Halocnemum strobilaceum* community is the closest strip of vegetation to the sea. The closer we get to the sea, its frequency increases, but the percentage of canopy cover is reduced and the density of the individuals decrease. This plant has entered into the *Avicennia* community in a scatter form and has composed a specific ecotone. This species is the only heliophile plant which can be established in the flooding condition. Next, the most frequent species are *Limonium gemelim, Aeluropus lagopoides* and *Puccinella distans* (Rashvand 1997).

Table 1 Chemical and physical results of soil analysis at Asalooyeh estuary

Soil factors	Depth of sampling (cm)	Direction of sampling from sea (1) to land (5)					
		1	2	3	4	5	6
Texture	0–25	L	C.L	Si.C.L	C	Si.C.L	Si.L
	25–50	Si.L	C.L	Si.C.L	C.L	Si.L	L
Clay (%)	0–25	33.6	31.6	31	26.8	30.2	4.2
	25–50	33.6	34.3	32.7	22.4	26.5	28.6
Fine silt (%)	0–25	5.1	5.1	4.7	5.5	4.9	4.1
	25–50	32.1	7.1	5.5	3.7	5	6.1
Coarse silt (%)	0–25	29	48.9	40.2	44.3	50.6	32.7
	25–50	17.3	42.2	39.8	51.3	50.4	22.4
Sand (%)	0–25	20.4	14.3	24.1	23.5	14.3	20.4
	25–50	39.8	17.9	20.4	23.5	17.9	42.9
EC (dS/m)	0–25	48.9	41.8	44	41.1	44.5	45.4
	25–50	26.2	44.2	39.2	40.7	44	40.7
pH	0–25	6.7	6.8	6.8	7.3	7.1	6.8
	25–50	7.6	6.8	6.7	7.4	7.1	7
Organic matters (%)	0–25	0	5.28	5.88	0	0.98	0
	25–50	0	0	0	0	0	0
Color	0–25 (Wet)	2.5Y 6/2	5Y 5/1	2.5Y 5/1	5Y 5/2	5Y 5/2	2.5Y 5/2
	0–25 (Dry)	2.5Y 6/3	5Y 6/1	2.5Y 6/1	5Y 6/2	5Y 6/2	2.5Y 7/2
	25–50 (Wet)	2.5Y 5/2	5Y 6/1	2.5 Y 4/2	2.5Y 5/2	2.5Y 5/2	2.5Y 7/3
	25–50 (Dry)	2.5Y 6/2	5Y 5/1	5Y 5/2	2.5Y 6/2	2.5Y 6/2	2.5Y 8/2

L loamy, *CL* clay-loamy, *SiCL* silty-clay-loamy, *C* clay, *SiL* silty-loamy, *EC* electrical conductivity

5 Soil Characters of Mangrove Forest Sites

5.1 *Avicennia Site in Nayband Gulf*

The soil surrounding the Nayband site is part of flood plain soil and some other halophile plants are growing in this area. Soil texture is so heavy and draining of lands is very poor. This zone includes two *Avicennia* communities, namely, the Asalooyeh estuary and the Basatin estuary.

5.1.1 Physical and Chemical Changing Processes of Soil in Asalooyeh Estuary

- Soil texture: Soil texture of 0–25 cm from sea to land is silty loam, loamy, clay loam, silty clay loam, clay, and silty clay loam, whereas the soil texture of the 25–50 cm layer is clay loam, clay loam silty loam silty loam, silty clay loam and loamy, respectively.
- It can be viewed that the type of soil texture is changing from moderate to heavy and then is changing to medium when we get further away from the sea (Table 1 and Fig. 9).
- Soil electrical conductivity (Ec): Soil Ec does not follow a specific process in the tidal area from sea to land. The minimum soil Ec is 39.2 dS/m and the maximum value of soil Ec is 48.9 dS/m (Table 1).

Fig. 9 Soil samples which had been collected from mangrove forest in Bushehr province

- Soil pH: The trend of pH flux does not follow a clear pattern. The maximum pH range is 7.6 at the entrance of steams and the minimum is 6.7 inside mangrove forests (Table 1).
- Soil organic matter[1]: The majority of soil organic matter was related to mangrove stand soil which was 5.88 %[1]. The percentage of organic matter reduced as we moved from sea toward land (Table 1).

5.1.2 Physical and Chemical Changing Processes of Soil in Basatin Estuary

- Soil texture: Soil texture of 0–25 cm from sea to land is silty-clay, silty-loam, clay, clay, clay and sandy while whereas the soil texture of the 25–50 cm layer is clay-loam, clay-loam, clay-loam, clay, sandy-clay loam and sandy-clay-loam. Far away from the sea, the type of soil texture changes from moderate to heavy and then changes to light soil upon reaching the land (Table 2).
- Soil Ec: Soil Ec does not follow a specific process in the tidal area from sea to land. The minimum soil Ec is 39.2dS/meter and the maximum value of soil Ec is 48.9 dS/m. The minimum soil Ec would be 27.7 dS/m and the maximum value is 55.9dS/m (Table 2).
- Soil pH: The trend of pH flux follows a clear pattern from sea to land. The maximum pH range is 8.4 and the minimum is 6.7. They both were related to outside of the mangrove forest and inside the *Avicennia* stands, respectively (Table 2).
- Organic matter: The majority of soil organic matter was related to mangrove stands soil which was 5.6 % (Table 2).

[1] The cold chemicals method had been undertaken to measure the chemical composition of organic matter in all samples as follows: the method of measurement of the consumed dichromate by standardized ammonium ferrous sulfate.

Table 2 Chemical and physical results of soil analysis at Basatin estuary

Soil factors	Depth of sampling (cm)	Direction of sampling from sea (1) to land (6)					
		1	2	3	4	5	6
Texture	0–25	S	C	C	C	S.L	C
	25–50	S	C	C.L	S.C.L	S.C.L	C.L
Clay (%)	0–25	8.2	44.9	46.9	40.4	19	54.7
	25–50	6	50	33.7	20.4	24.1	32.2
Fine silt (%)	0–25	2	3.1	6.9	5.5	1.8	8.2
	25–50	3.1	6.2	5.1	4.1	5.7	5.1
Coarse silt (%)	0–25	3.1	27.5	18.2	16.3	6.8	16.7
	25–50	3.1	21.4	21.9	24.5	7.3	24.5
Sand (%)	0–25	86.7	24.5	27.9	37.8	72.4	20.4
	25–50	87.8	22.4	39.8	51	62.9	38.2
EC (dS/m)	0–25	32.3	49.5	54.4	36.8	35.2	36.8
	25–50	27.7	36.7	40.1	46.2	30.6	44.5
pH	0–25	7.5	7.1	7.9	7	7.5	7.1
	25–50	7.5	6.9	6.9	7.4	7.8	7.5
Organic matters (%)	0–25	2.8	–	5.6	–	0.95	–
	25–50	1.01	4.44	2.41	2.11	1.48	0.58
Color	0–25 (wet)	2.5Y 5/2	5Y 5/1	2.5Y 5/2	2.5Y 5/3	2.5Y 5/2	10YR 5/3
	0–25 (dry)	5Y 6/1	5Y 6/2	2.5Y 6/3	2.5Y 6/3	2.5Y 6/3	10YR 6/4
	25–50 (wet)	5Y 5/2	5Y 5/2	5Y 5/2	5Y 5/1	5R 6/2	2.5Y 6/3
	25–50 (dry)	5Y 6/2	5Y 6/2	5Y 6/2	5Y 6/2	5Y 7/1	2.5Y 6/4

S sandy, *C* clay, *CL* clay-loamy, *SCL* sandy-clay-loamy

5.2 *The Mangrove Site of Bardestan Estuary (Dayyer)*

From the south, this region is connected to the Persian Gulf, from the north, it is surrounded by hilly lands, and from the east and west sides, it is linked with flood plains. This land is classified as flood plains with high salinity and heavy texture. The wheat farming takes place is some areas which are away from the sea. The slope range of this land is 1–5 % (Rashvand 1997).

5.2.1 Physical and Chemical Changing Processes of Soil in Bardestan Estuary

- Soil texture: Soil texture of 0–25 cm from sea to land consisted of sandy-loam, silty-loam, and silty-loam, loamy-loamy and loamy and at a depth of 25–50 cm it was comprised of loamy, sandy-clay loam, loamy, loamy, loamy and loamy, respectively. Generally, the soil type in this site is relatively light (Table 3 and Fig. 9).
- Soil Ec: The maximum soil Ec is 38.5 dS/m and the minimum is 30.9 dS/m (Table 3).
- Soil pH: Changes in soil pH can be seen from sea to land. The maximum soil pH was 8.1 and the minimum of 7.1 But the trend of change did not follow a systematic pattern (Table 3).

Table 3 Chemical and physical results of soil analysis at Dayyer estuary

Soil factors	Depth of sampling (cm)	Direction of sampling from sea (1) to land (6)					
		1	2	3	4	5	6
Texture	0–25	S.L	Si.L	Si.L	L	L	L
	25–50	L	S.C.L	L	L	L	L
Clay (%)	0–25	10.2	21.4	20.8	23.1	14.3	25.1
	25–50	17.9	20	22.4	26.5	18.8	14.3
Fine silt (%)	0–25	4.1	4.1	8.2	9.2	10.2	2
	25–50	10.2	6.1	8.2	9.2	8.2	6.1
Coarse silt (%)	0–25	12.2	34.7	30.6	27.6	24.5	69.4
	25–50	20.4	36.7	24.5	23.5	18.4	61.2
Sand (%)	0–25	57.1	24.5	18.4	14.3	8.2	20.4
	25–50	42.9	22.4	24.5	12.2	12.2	28.6
EC (dS/m)	0–25	44.1	55	43.1	32.8	29	51.2
	25–50	51.7	43.4	50.7	37.9	31	7.6
pH	0–25	7.4	6.8	7.7	7	7.4	7.6
	25–50	7.4	6.8	7.5	6.8	7.6	7.4
Organic matters (%)	0–25	2.02	5.34	–	–	1.62	–
	25–50	–	–	–	–	–	–
Color	0–25 (wet)	10YR 5/1	5Y 5/1	10YR 5/2	10YR6/3	2.5Y 6/3	2.5Y 7/4
	0–25 (dry)	10YR 6/1	5Y 5/2	10YR 7/3	10YR 7/2	2.5Y 7/3	2.5Y 8/2
	25–50 (wet)	5Y 5/1	2.5Y 5/2	2.5Y 5/1	2.5Y 5/3	10YR 5/3	5Y 4/1
	25–50 (dry)	5Y 7/1	2.5Y 6/2	2.5Y 6/1	2.5Y 7/3	10YR 6/3	5Y 4/1

L loamy, *SL* sandy-loamy, *SCL* sandy-clay-loamy, *SiL* silty-loamy

5.3 *The Avicennia Site in Male-Gonze Estuary*

On one side, this area is connected to the sea and on the other side it is surrounded with the flood plains of Chah-Bahman. The type of Male-Gonze lands is low land with high salinity and heavy texture. Due to heavy texture of the soil, draining is very poor and the plants of this area are highly resistant to salinity (Rashvand 1997).

5.3.1 Physical and Chemical Changing Processes of Soil in Male-Gonze Estuary

- Soil texture: Soil texture of 0–25 cm consisted of clay, clay, clay, sandy-clay, clay-loam and clay, but at a depth of 25–50 cm clay, clay, clay, loamy, clay-loam and silty-loam can be found. The type of soil in this site has changed to heavy soil from moderate. They can be spotted when moving toward land from sea. But the *Avicennia* site is located on the flood plains which have a heavy texture with high level of clay. This can be discovered only a few kilometers away from the sea (Table 4 and Fig. 9).

Table 4 Chemical and physical results of soil analysis at Male-Gonze estuary

Soil factors	Depth of sampling (cm)	Direction of sampling from sea (1) to land (6)					
		1	2	3	4	5	6
Texture	0–25	S.L	C.L	C	C	C	C
	25–50	L	C.L	C	C	C	C
Clay (%)	0–25	26.5	36.7	42.8	48.9	57.1	8.2
	25–50	26.5	34.7	42.8	55.1	61.2	4.1

L loamy, *SL* sandy-loamy, *SCL* sandy-clay-loamy, *SiL* silty-loamy

Table 5 The results of the water analysis of the *Avicennia* site in the Nayband zone

Number	pH	Ec (dS/m)
1	8.10	45.4
2	8.08	45.4
3	8.10	44.8
4	8.20	44.1
5	8.06	48.3
6	8.25	43.9
7	8.22	46.5
8	8.29	43.2
9	8.16	46.2
10	8.14	46.2
11	8.27	43.8
12	8.23	44.3
The control sample (seawater)	8.13	43.2

6 Ec and pH of Water in the *Avicennia* Sites of the Nayband Gulf in Bushehr Province

According to our methods, water samples along the transect of 2.5 km were taken in the Basatin estuary at regular intervals from beginning to end. However, sampling time decreased in order to increase the accuracy of field working. In addition, the sampling procedure had been undertaken in a tidal situation. The range of Ec was at least 43.2–48.3 dS/m and also the pH ranged from 8.08 to 8.29 within mangrove forest habitat in October 1999. Water pH in the *Avicennia* site was higher than the Hara habitat of seawater pH. The Ec of seawater outside of the mangrove site was 43.2 dS/m (Table 5; Rashvand 1997).

7 Role of Sea Tides on Soil at the *Avicennia* Site

The permanent sea tidal waves help in reducing soil salinity in the warm season so that the amount of soil salinity of the site reduces and conditions are optimized for *Avicennia* growth.

8 Specific Characteristics of the *Avicennia Marina*

8.1 Pneumatophore

This plant has one short main root which is similar to other mangrove plants that can cope with flooding conditions and bogged soils. The *Avicennia* root can usually penetrate into the soil from 60 to 80 cm. At that point, it reaches the hard pan of soil that cannot be dipped into. Hence, the roots grow in a horizontal level, initiating from the main root to the surrounding area. It can be scattered around two or three times larger than the size of tree crown. Pneumatophore grow in a vertical direction and they can come out of the soil. They provide the required oxygen for the plant. Pneumatophore root density depends on two main factors. First, it depends on the size of the plant. Next, it relies on the location of the plant. The longer the plant is in water, the more oxygen is needed. Therefore, the density of pneumatophore will be increased and the next issue is about the height of pneumatophore. As water depth increases, the pneumatophore raises. The maximum height of pneumatophore recorded at the Nayband Gulf site was 60 cm which was related to low height locations. The water drain is later than the other areas. This can particularly be seen in extraneous streams. These roots are kind of adaptations to the environmental circumstances. The density and size of pneumatophore around streams was bigger than inside the *Avicennia* stand. The pneumatophore height could reach to the minimum size in the silty and sandy soils. Their root system causes stabilization of the soil and their communities could reduce the tidal energy. So it makes a more suitable habitat and shelter for a number of other organisms. Mangroves have high biomass value compared to other communities (Rodriguez and Feller 2004).

8.2 Features of Established Mangrove Forest Sites

All of the mangrove sites are influenced by the tides which are located around the main and extraneous estuary. These sites have some constraints including daily flooding condition, lack of oxygen of the soil, low depth of soil, and high salinity of water. These factors limit the establishment and growing of plants in the mangrove area. Then, only mangroves can establish in this area and other species are not able to grow. Mangrove forests in Iran are composed of two species, including *Avicennia marina* and *Rhizophora mucronata*. They reduce wave energy and create stability in sediments by their root system, and also they provide a good shelter and habitat for other biodiversity and support the food network beside the beach. *Avicennia marina* has the unique characteristic of being highly resistant to salinity. The front line of defense in many mangrove plants is to prevent the entrance of salt into the plant. This action is done by semi permeability of root, selective absorption and filtering the mineral matters. Another way for adaptation is rapid excretion of salt that has entered into the plant cells. The leaves of many species of mangroves have special salt glands. In some species of the plants, salt is seen on their leaves. Hence, those characters are some adaptations of the plant in the environment.

8.3 The Characteristics of Avicennia Seed

The *Avicennia* species is viviparous or semi-viviparous. The seedlings of this species can grow on their mother tree and fall down in the water. They are located in the crust, floating on water, and in normal circumstances they live 3–4 days. In the case of establishment in the soil they can survive and grow.

8.4 Plant Rapid Reproduction

In the proper conditions such as flooding of the soil, permanent daily tidal and high level of organic matter of soil, the plant can reproduce rapidly (Safiari 2001).

8.5 Changes in the Anatomical Structure of Plant Leaves

Due to bearing special traits, some changes occur in the anatomical structure in the plants so that their leaves resemble other xerophyte plant leaves (Safiari 2001).

9 Effective Factors in Mangrove Distribution

Establishment of the mangrove forests depends on climate, land form in coastal areas, tidal range, and type of soil and availability of fresh water. Factors such as weather conditions, tides, salinity and soil texture and structure can limit the mangrove habitat range and distribution of mangrove forest as follows.

9.1 Temperature

Mangrove forests are distributed in warm climates and the minimum temperature of the mangrove habitat is 19 degrees Celsius (°C). Therefore, those areas which have an annual mean temperature below 19 °C cannot be classified as a mangrove forest habitat ecologically. These plants can not tolerate large fluctuations in temperature (up to 10 °C) even in their natural habitat and the establishment and growth are not acceptable. Generally, in areas where mangrove grows, the water is always warm.

9.2 *Quiet Places Like Estuaries and Deltas Are the Best Zones for Mangrove Forests*

Severe bank waves and strong tidal activity can limit mangrove improvement in coastal areas. Seedling mortality and soil erosion can be increased under intensely wavy conditions. Owing to some processes in the site soil, the best mangrove distribution is in the areas with the highest tides.

9.3 *Fine Texture Sediments*

Seedlings of these plants grow very well when soil has proper conditions such as fine texture (silty and clay) and organic matter. Volcanic soil makes the mangrove sites fertile (although quartz sediments and granite soil are poor soil). As a result, the soil of mangrove communities has fine and heavy texture along with organic matter and sulfide-rich material.

10 Goiter Gulf Mangrove Forest

The Goiter Gulf is located in the southeastern part of Iran in Sistan and Baloochestan (Goiter and Bahoo estuary) province. Monitoring of changes in relationship with the extent of mangrove forests in the Gulf of Goiter was performed by using aerial photographs in 1955 and 1964, satellite images of Landsat TM and ETM in 1998 and 2001 and satellite images IRS-LISSLLL in 2001. In general, the results showed that the changes of mangrove forest surface were positive from 1995 to 2001. In addition, the mangrove forest area had increased from 246.01 to 671.53 ha. Majority of the improvements was between 1964 and 1998. The higher growth potential of these forests in the Bahoo estuary rather than the Goiter estuary was observed. Similarly, this estuary in relation to change rate was more dynamic than the Goiter estuary in that period of time (Erfani 2007).

The mangrove forest in the Goiter Gulf is comprised of only one species, namely, Hara (*Avicennia marina*). Tree density per hectare in the Goiter Gulf varied between 5,900 and 200, while the mean value was 1,623 individuals ha^{-1}. The average height of trees and canopy was 2.65 and 1.81 m, respectively. The canopy diameter and average value of tree canopy cover was 2.63 m and 11.66 m^2, respectively. The maximum value of canopy cover of one individual was 35.934 m^2 at Bahoo estuary. Average canopy cover was 54.60 %. The maximum and minimum canopy cover was 100 % and 14.32 %, which were related to Goiter and Bahoo estuaries, respectively. The average density of pneumatophore and average height were 111.14/m^2 and 10.8 cm. The average number of dropped leaves in autumn and winter was 14.32/m^2 (Erfani 2007).

11 The Quantitative Characteristics of Mangrove Forests in Bushehr Province

11.1 *The Asalooyeh Estuary*

11.1.1 The Changing Trend of Mangrove Forest from Sea to Land

- **The height of trees:** Although the *Avicennia marina* is a tree, at this site all individuals were in shrub formation. The most frequent tree height was 1.5 m which showed a young stand. As we move away from sea to land, the height of Hara has a decreasing trend. The maximum measured height in this young stand was 2.7 m which was related to trees at the vicinity of the waterway margin (Fig. 2).
- **The diameter of canopy:** The most frequency for the beginning of Hara stand is in the vicinity of low tide. The changing canopy diameter from sea to land followed the tree height tendency and has a decreasing trend.
- **The distance of shrubs:** Soil quality (organic matter) and stand quality (the height of Hara) in habitat doesn't change; distance between the individuals does not change appreciably and the stand is not separable and zoning. This is true in the area of this habitat, and means changes in habitat quality are slow and distance of individuals change subtlety until the site is supposedly homogeneous.
- **The number of seedlings:** Most of Asalooyeh estuary is covered by high density stand. High density of this site caused lack of space and light, consequently, the shortage of seedlings (height < 0.8 m) is considered.
- **Pneumatophores:** The density of pneumatophores in the streamside is higher than other areas of the site.

11.2 *Basatin Estuary*

11.2.1 Changing Trend of Mangrove Forest from Sea to Land and the Heights of Shrubs

The maximum tree height was 3.5 m and the most frequently occurring height was 2.3 m. The forest is middle aged and there is a specific homogeneous stand height frequency. As we move away from sea to land, the height of individuals has a decreasing trend.

- **The diameter of canopy:** The maximum diameter of canopy was 4.75 m and the most frequent canopy size was 1.5 m. The diameter of canopy following from heights has almost a regular homogenity. The changing of canopy diameter from sea to land following the height of trees has a decreasing trend.

- **The stand density:** When soil quality and stand quality (the height of seedling) are fixed, there is no change in the distance of individuals. As the soil quality changed, the stand density decreased. In this estuary, the stand density is homogeneous at the beginning of stand from the sea. But the number of individuals decreased at the end of stand toward the land. Hence, in some parts of the site we can classify the stand based on tree density from sea to land into two zones: zone 1 which is highly dense and zone 2 which is semi-dense.
- **The regeneration:** The density of seedlings increases from sea to land. It is considered that the seedling number increased in the second zone in comparison with the first zone which is denser than the first zone.
- **Pneumatophores:** The number and the size of pneumatophore at streamside (at zone 2) are bigger than that value inside the stand which is located in zone 1.

11.3 Dayyer Site

11.3.1 Changing Trend of Mangrove Forest from Sea to Land

- **The height of trees:** Frequency in height categories hasn't any particular order. The most frequent height class is < 0.8 m followed by 1.8 m. The amount of individuals with average height is not a maximum value and the height frequency does not form a normal curve diagram. The absence of a normal curve of tree height frequency is due to severe harvesting of forest by domestic people to provide firewood. We did not observe any specific trend of change in terms of tree height from sea to land due to traditional logging operations. The forest stand is middle aged, although a few old individuals can be considered and the maximum measured height is 3.6 m.
- **Diameter of canopy:** The most frequent canopy diameter is related to individuals with canopy diameter less than one meter which follows the height frequency. The highest diameter of canopy is 5.5 m. The changing trend of the canopy diameter doesn't follow a normal trend from sea to land.
- **The distances of trees:** The stand of the beginning point of the forest from the sea is denser than that value near to land. As we move from sea to land the distances between individuals increased.
- **The number of seedlings:** The seedling density decreased as we moved from sea to land. The biggest value of regeneration was between old trees.

11.4 Male-Gonze Habitat

A Changing Trend of Mangrove Forest from Sea to Land

- **The tree height:** The maximum height of trees was 4.2 m and the most frequent height class was < 1 m. The regular order in height frequency was not seen and the changing factor of height was 67 %. The older individuals were dominant in this forest. The height of trees decreased from sea to land.

Table 6 Quantitative and qualitative characteristics of mangrove forest in Bushehr habitats

Site	Quantitative variables								Qualitative variables
	Crown surface (M2)	ATH (m)	Density ha^{-1}	Pneumatophore		Bole form		DS/m^2	
				Density/ m^2	Height (cm)	S	NS		Succulent
Nayband gulf									
Asalooyeh estuary	0.7	1.2	6,296	550	15		NS	8	√
Basatin estuary	2.9	1.8	4,657	238	17		NS	8	√
Dayyer	6.2	1.9	2,916	138	11		NS	3	√
Male-Gonze	9.4	2.3	556	300	12		NS	38	√
Total	19.2	7.2	14,425	1,226	55			57	
Average	4.8	1.8	3,606.25	306.5	13.75			14.25	

ATH average tree height, *S* single trunk, *NS* no single trunk, *DS* density of seedlings/m^2

- **The diameter of canopy:** The most frequent canopy diameter size class was < 1 m and the biggest canopy diameter size was 8.45 m. The canopy diameter distribution did not has a regular pattern. However, the canopy diameter size declined from sea to land. Decreasing intensity of canopy diameter and the height of trees from sea to land depends on the amount of slope of the forest floor. This site is under wavy sea conditions. In addition, the forest stand received direct waves from the sea. Hence, the largest canopy diameter and the highest tree height were related to trees which were located at the beginning of the site from the sea.
- **The distance of trees:** As we move from sea to land the distances between individuals increased. So, we can classify the stand into two main zones: (1) a dense zone which is near the sea, (2) a scattered zone which is located at the outside part of the site from the sea.
- **The number of seedlings:** This site had more seedlings in comparison with the other site (Table 6). This is due to the relatively high rate of seed production by mother trees which are distributed more than the other stands. Likewise, the greater distance of trees in this site compared to the rest, provides more space for seedlings.

As we can see in Table 6 the maximum height of pneumatophore is 55 cm and the maximum average height of trees is 2.3 cm.

12 The Effective Factors in Destroying Hara Habitats in Bushehr Province

The mangrove forest of Iran is under continuous anthropogenic disturbances due to disregard of the environmental considerations. Those sites which are located in Bushehr province are threatened by extra disturbances which are related to southern pars gas projects. For instance, in the Nayband Gulf the gas plan development, infrastructures and other service provider projects can strongly affect the mangrove forest. The Iran Department of Environment has reported: "the number of migratory

Table 7 Checklist of pests of mangrove forest in Bushehr Province

Order	Family	Economic importance
Coleoptera	Scarabaeidae	No
Lepidoptera	Nuctuidae	No
Coleoptera	Cicindelidae	Yes (predator)
Hemiptera	Pentatomidae	No
Diptera	Dolichopodidae	No
Diptera	Tephritidae	Yes
Diptera	Not identity	No
Coleoptera	Bruchidae	No
Hymenoptera	Cephidae	No
Diptera	Psychodidae	No
Coleoptera	Chrysomelidae	No
Coleoptera	Oedemeridae	No
Lepidoptera	Not identity	No

birds and other aquatic organisms in these habitats has reduced". They also emphasized: "Birds need tranquility and scattered in loud voices from environment and then far from their habitat. Construction of the hotel and any other buildings or areas within the national park or vicinity of it eliminate the ecosystem's balance of that region". Sulfur pollution, oil spills into the water, and tree felling could damage one of the most valuable mangrove forest reserves in Bushehr province. Another factor of destruction of these forests in recent decades is the Persian Gulf War which damaged much of the oil reserves. Consequently, the oil resources and other toxic material was distributed seaside. Those polluted waters affected the mangrove forests and other fauna and flora resources.

Grazing is another element which affected the mangrove forests of Iran. The residents of villages in Bushehr, Hormozgan and Sistan VA Baloochestan provinces cut the tree branches to provide food resources for their domestic animals (goats, cows and camels). In addition, biotic factors such as insects, fungi, virus and bacteria are other groups which impacted the mangrove forest in Iran (Table 7).

Harvesting of the river's fresh water, which provides the fresh water of mangrove sites, is a trend which has increased sharply. Consequently, due to a lack of fresh water for mangrove forests the water salinity could rise and affect the biotic and abiotic elements of the mangrove ecosystem. One to several species of mollusks can damage these forests.

12.1 Insects of the Mangrove Forest in Iran

12.1.1 Beetles Order (Colcoptera)

- **Scarabidae**
  Two genera from this family were collected from mangrove forests in Bushehr province at Bordekhoon and Male-Gonze regions that had not been identified yet.

All insects of this family have 8–10 joint tentacles. They have 5–7 ending joints which are sheet-shaped. Larvae are C-shaped and are named white worms. The diversity of this family is high but small numbers of them can damage the plant in the agricultural sector.

- **Chysomelidae**
  One genus of this family was collected from mangrove trees in the region of Male-Gonze that had not yet been identified. Members of this family are convex and spherical. The species has five joints but sound four joints and its tentacles have 11 joints. Larvae are Carabeiform and their thorax has three pairs of legs; its head is typically small and is hidden under the first chest.
- **Scolytidae**
  Damaged trees were remarkably observed in this area. The suspected insects were from Scolytidae. Some of the mangrove trees in the area of Male-Gonze had dried and were eliminated and the damage effect is very similar to the damage effects of this family, but the research crew could not find any individual of that family. These insects are very small and their sizes are 2–3 mm. Their head is small and hidden under the chest. They have pin-shaped antenna and compound eyes which are deeply depressed. The larvae are small and have a milky color and no feet. These larvae live between bark and wood of the tree trunk and feed the latex. Consequently, the tree will die.
  These insects are secondary pests and they will attack the weakened trees. They can make two types of corridors (mature insect corridors and larvae corridors). The larvae corridors cross the mothers' corridor. The larvae life cycle will be completed in this corridor and the mature insect will make a hole and come out of the trunk. So, the trunk and branches of those infected trees are seen as punctured.
- **Cicindelidae**
  This beetle is collected on the mangrove trees in the Male-Gonze region. The insect is a useful hunter that plays a role in controlling pests of mangrove trees.

12.1.2 Butterfly (Lepidoptera)

- **Noctuidae**
  The members of this family are collected from mangrove trees in Bushehr province but their genus and species are not identified yet. The members of this family contributed to produce the biggest family between Lepidoptera.

Generally their color is dark and opaque and the length of body is 0–25 mm. Their front wings are narrow and hind wings are broad. Many individuals have a round and a bean-shaped spot on their front wings. Their antennae are long and filamentous. The wing veins have a very specific situation, i.e., that the m2 vein in the front wing is closer to m3 than m1. The larvae have a smooth and dark body and have a pair of abdominal feet. The larvae feed the leaves and they overwinter as pupa. Many of them are pests which can damage the forest.

12.1.3 Hemiptera Order

- **Pentatomidae**
  Members of this family have been collected by light trap in the mangrove forests of the Male-Gonze region. Their body is shield-shaped with a 5-joint antenna. They often have a bad smell and they are called foul sun pests. These sun pests are pests of many plants and overwinter as a mature insect below the litter fall. They are mass spawning and have 2–3 generations in a year.

12.1.4 Homoptera Order

Fulgoroidea and Cicadoidae included many families which were collected from mangrove trees in Male-Gonze and the Asalooyeh zone. These pests feed on plants, after which the plants weaken.

12.1.5 Diptera Order

Many species of Diptera were collected. These small-sized flies are seen in different colors such as brown, blue, etc. In addition, the damage of the above-mentioned insects was observed; however, there was not documented research in relation to this aspect.

13 The Importance of Mangrove Forests

The mangrove ecosystems perform as a biodiversity conservation resource as shelter, a place of nurturing and nursing for many plants and animals. The main character of these ecosystems is biomass production in the tidal area of the sea, which is a big containment of food. Tropical mangrove swamps in comparison with other marshes have been disturbed by anthropogenic disturbances directly. The International Union for Conservation of Nature (IUCN) in the revision of the mangrove ecosystems has introduced 22 main usages of these ecosystems. Usually, logging is the most important economic use of these ecosystems. Due to improvement of commercial use of mangrove forests, in most areas they are managed as forest reserve to conform the forest regeneration. Unfortunately, malicious use of the area has been improved as a normal activity which elaborated the big concern in relationship with well-functioning of those environments (Pillay 2004). The overall production of Fiddler Crabs of mangrove forests in SiriK, Iran was 37.90 and 10.05 g dry weight/10 m^2 for males and females, respectively (Mokhtari et al. 2008).

Mangrove forests have a specific condition which can play a very good role in the fishery and fish farming industry. Some species of fishes spend a part of their life cycle in the mangrove forest. Among these, some oyster that attach to *Avicennia* roots and can grow up in these forests. The life of some snails and shrimp are dependent on mangrove forests. But today, due to anthropogenic disturbances (pollution and logging) in the mangrove forest, the fishing in this area has been reduced substantially. For instance, many fishes that spawn between the branches and roots of mangroves have lost their habitat now.

The pneumatophore structure can reduce the marine current's energy that is able to elaborate the sedimentary process and soil preservation. Deposition and decomposition of litter fall in the forest floor can develop the forest soil in terms of quality and quantity. In fact, the mangrove forests will stabilize land owing to increase in sedimentary phenomena, sediment protection and soil formation. Consequently, the risk of soil erosion will reduce, especially in high energy tidal conditions such as tropical storms.

14 Economical Uses of Mangrove Habitats in Southern Iran

In addition to wood products, many countries benefit from mangrove forest which own these resources (Danehkar 1993). Fishing ranked first among all types of harvesting of mangrove forest followed by logging and coal production which were each at 1.1 %. Unfortunately, the logging operations in most developing countries are ongoing. All species of mangroves are used as firewood by rural people. For instance the *Rhizophora* species have suitable quality for coal production (heavy wood, steady heat when burning and producing low smoke); consequently, they are more popular among the locals. According to those factors, in some areas, the main purpose of mangrove management is improvement of firewood and the charcoal industry (Sistani 1990). The mangrove trees are the primary producers of the chain food in their environment. They play an important role in maintaining the activity of plankton and phytoplankton. The plankton community can provide food for some marine animals, especially for shrimp. Tides bring the food materials from the sea into mangroves. In contrast, organic materials are removed from mangroves to the sea. Much of the mangrove forests in Iran are used by villagers who are around these habitats. Domestic animal grazing is one of the most common and illogical uses of this plant. Because wetland halophyte species are rich in terms of salt and iodine in their texture, they are highly effective in fattening. According to that, in some countries, they use the forage of mangrove halophytes in high ratio in animal feeding (Abbasi 1990). The domestic animal grazing (camel and cow) and harvesting of mangrove forest foliage are two chief methods of forage harvesting, which is at a high level in southern Iran. Fishing in the (large and small) estuary habitats ranked second as economic activity among all the economical uses of mangrove forest. In recent years, forests, rangelands and watershed organizations of Iran have improved the mangrove plantation in three provinces (Bushehr, Hormozgan and Sistan VA Baloochestan) of southern Iran to increase the socio-eco-environmental products of mangrove habitats. For example, the mangrove plantation around Qeshm Island has been established to develop those above-mentioned objectives.

In addition, mangrove forests are important due to bee breeding and honey production. The flower of mangrove trees is fragrant and nectar-filled. The abundance of wild bees in the flowering season around the Harra branches is considerable. The honey of wild bees in Qeshm Island, which make their hives among mangrove trees, is very delicious and aromatic. This bee produces a honey in the summer season which has the smell of Harra flowers nectar. The local people believe this honey has great therapeutic properties which result from the mild flavor and pleasant fragrance of Harra trees. That scent is observable in the early summer from the forest. The habitat of the mangrove forest including the sea view and the green islands can also provide non-wood products from the forest. These environmental products attract visitors and affect their mind strongly, which can be remembered for a long time. Further, the mangrove forest can be designated as a biodiversity conservation and ecotourism destination.

15 Characteristics of Mangrove Forest Formations

The most important factors which can influence the mangrove formation are: climate, land forms in coastal areas, tidal range and soil type. Hence, those factors can influence strongly the species richness of mangrove forest. Temperature, water salinity, smooth flow of the sea and soil type are the most important elements in distribution of mangrove forest on the southern coast of the Iran.

Community evolution of mangrove forests has been made gradually over millions of years and amazingly adapted with salinity and established into a tidal shores condition. Basically, these plants are not able to tolerate temperatures less than 5 °C. Moreover, salinity between 20 and 32 per thousand is suitable for their growth. The mangrove plant is one of the strongest of the flora that developed in the tidal area with minimum demands in terms of nutrient components. One of the characteristics of Hara trees is its special roots. The main roots are short and shallow. The vertical lateral and small and sponge roots grow from the main root and perform the pneumatophore, which averages 30 cm above ground level. As the tree grows, the network-shaped roots (pneumatophore) improve gradually. Thus, the optimum condition for mangrove tree establishment will be provided. The root network will reduce the salinity of habitat, and then the plant can absorb the nutrient materials from the saline environment. The mangrove forest bathes twice a day. At high tide the fish move into the forest and local fishermen know this point as well. They will fish at this time. The wet mangrove floor can provide a comfortable environment for fish-eating birds, especially for white Egret (*Egretta Alba*). In addition, some immigrant species from Caraderidae (Morgh baran/Abchilik) remain in the mangrove zone in winter. This environment is a comfortable habitat for immigrant birds. A considerable number of European-Asian species of birds (about 43 species of fish eaters) will migrate to Africa. One of the migration routes is in the south of Iran, in the Hormoz strait, which is between Qeshm and the Abou-mousa islands in the Persian Gulf. The protected mangrove tidal areas are a suitable environment for resting and feeding for fish-eating birds and other species of migratory birds.

16 The Environmental Importance of Mangrove Forests in Southern Iran

The primarily importance of mangrove forests is related to shoreline protection. In addition, wood products of mangrove forests are secondary positive points of those forests (Ewel et al. 1998). The mangrove create a good habitat for mollusks, crustaceans and fish that are important food sources for fish and wild birds. The relationship between recirculating ecological primary production (mangrove ecosystem) and production of valuable fish and shellfish in other tropical areas is well known. Also mangrove forests, due to favorable ecological conditions, are a safe habitat for migratory birds such as *Egretta alba* (great Egret), *Ardea goliath* (great Havasil), *Egretta gularis* (Beach Egret), *Areola grayii* (Indian Egret), *Platalea ieucorodia* (spoonbill), *Pluvialis squatarola* (gray Salim) and *Numenius tenuirostris* (Gilanshah) from tropical regions.

In addition, mangrove forests can provide food resources for domestic animals such as cow and camel, but they should be managed under sustainable management. Protection of coastal areas is a concern of conservationists as good managers. Developing the coastal area under sustainable management is complex and multi propose management. Mangrove forests can protect the coastal area as a natural protective wall. Furthermore, the mangrove ecosystems can play a multi-purpose role for biodiversity conservation in the tidal zone.

The key points for determining a suitable habitat for mangrove plantations in southern Iran are as follows:

1. The location should be in the tidal area.
2. It is necessary that the slope of the shoreline seaward is a gently till (slope $< 1\,\%$), so that the intensity of the waves hitting the beach will be minimal.
3. Soil texture should be clay.
4. The soil pollution results from waste oil and other pollution resources.

For instance, the residual materials from ship engines can affect the water and soil quality and consequently the forest ecosystem can be influenced. To protect the plantation area from those pollution resources, the plantation area should not be adjacent to the dock or mooring buoys near the bays of ships and boats. The mangrove tidal area is created as a suitable habitat for mollusks, crustaceans and fish as an important food source for fish and wild birds.

17 Limiting and Threatening Factors of Mangrove Forests in Southern Iran

From 1980 to 2001 around 35 % of mangrove forests had been lost all over the world, which exceeded other losses of two well-known threatened habitats: tropical rain forests and coral reefs (Valiela et al. 2001). In general, two main groups of factors can affect the mangrove forests of Iran as follows:

17.1 Natural Factors

1. Lack of nutrient components of the soil
2. Sedimentation in habitat can affect the establishment and growth of seedlings
3. Climate change, which can affect the drought regime. Consequently, the precipitation and evaporation will be changed and the salinity will increase and the mangrove forests will be limited
4. In the south-east of the country, the monsoon storms can destroy the trees and seedlings
5. Inappropriate morphological land

17.2 Anthropogenic Factors

Development of infrastructure, logging, grazing, fish farming, immigration to coastal cities, increasing of pollution in the environment and oil pollution affect the mangrove forest in the Persian Gulf and Oman Sea.

Unfortunately, road construction in Bushehr Province in the Nayband forests affected the mangrove forest greatly. The road disconnected the water circulation in the estuary. The shortage of water in the site could increase the tree damage. In the past, there have been several considerable mangrove forests in the south of Iran. But there is not a big area of mangrove forest in those zones. For instance, in the Shoor river estuary, Hassan Langi estuary, Gaz estuary, Hivi estuary, Birizak estuary, Rangi estuary, Gorazi estuary, Chalpy estuary and Toorkande river estuary and Gabrik and Kashi wetlands were instances of a considerable mangrove forest in the past but today they have vanished.

In humid areas some factors such as incorrect logging overharvesting of forests, shifting to agricultural land, fish-farm development and changes in drinking water usage are considered as major causes of damage.

In arid and semi-arid zones overgrazing by domestic animals, foliage harvesting, timber and non-timber logging of forest, road development and soil salinity are threatening factors.

In addition, in the other parts of the mangrove forest migration of rural people to the city, urbanization in coastal areas, development of industries at coastal area and pollution are important factors which have threatened the mangrove forests. Oil pollution in the sea is the most dangerous type of pollution in mangrove habitat. Invertebrates' mortality, falling tree leaves and death of seedlings are short-term effects of oil pollution. Mortality of mature trees is the long-term influence of pollution. The oil can enter into the cells and consequently change the osmotic pressure of the organism, and the plant will die. Another long-term impact of oil pollution is on the soil (Sistani 1990). A thick black layer of oil residue on the soil in the northern part of the Persian Gulf area was observed until 2005. This pollution was from the Persian Gulf War in 1991.

18 Conclusion and Future Prospects

The results show that the mangrove forests are distributed in the south of Iran on the soil with clay and sandy texture and a low slope (< 1 %) coastal area. Their habitats are not exposed to significant waves and are located on estuaries. The vertical structure of forest stands in three provinces (Sistan VA Baloochestan, Hormozgan and Bushehr) is same. A considerable difference was related to tree density and the width of forest stands from sea to land (Table 6). Due to the increase of salinity from east to west (Sistan VA Baloochestan to Bushehr) the tree density and width of forest was reduced (Table 6). Likewise, the temperature declined from east to west which could affect the forest density and forest distribution. For instance, after the Male-Gonze habitat, the natural stand of mangrove forest cannot be seen. Hence, plantation, biomass production and carbon sequestration are essential priorities which must be investigated.

The mangrove forest in southern Iran is under continuous threat of anthropogenic disturbances: oil pollution, logging, grazing, fish farming and development of infrastructures. Amongst all types of disturbances oil pollution is the most important threat which is increasing because of the development of oil harvesting operations and war in the Persian Gulf in recent decades. However, overseeing that forest under sustainable management according to the protection and improvement of the present forest level and rehabilitation should be the first priority of decision makers. Forest monitoring in permanent plots on a large scale and over the long run in the region is another priority of research. In addition, improvement of local people life (educational, economic, cultural and infrastructural improvement) can help the managers to protect and develop the forests as well.

Mother trees are the most important element in the regeneration procedure. Selection of elite trees, which could produce seeds and support the recruitment in the forest, might improve the forest reproduction. Under the sustainable forest management program, forest managers can determine and protect the mother trees for future seed production programs. Although the main source of water (sea) is available and the ecological situation is suitable, the forest rehabilitation programs need human support. For instance, the seedling production in the nursery should be another implementation which must be done by decision makers. Likewise, research in terms of seedling production and rehabilitation can be another essential priority.

Ecological factors may affect the establishment of seedlings and saplings and development of established forest. According to the ecological situation we can prioritize the habitat for plantation and development of mangrove forests in southern Iran. From east to west the latitude increased and the temperature decreased. Further, as the availability of ocean resources declined the water and soil salinity increased. Consequently, the mangrove forest density declined from east to west and the border of mangrove forests is located at Male-Gonze. Hence, we can prioritize the plantation zone based on provinces as follows: (1) Sistan VA Baloochestan, (2) Hormozgan and (3) Bushehr.

Climate change and global warming can influence the future of tropical forests (Wright 2010). Those main elements can change the environment which can promote new habitat in a micro habitat scale for new species. For example, invasive species can threaten the endemic species. Monitoring of change in the forest surface, flux in water level of the sea, trends of climate factors changing, sedimentation trend and anthropogenic disturbances are the necessary research program in a mangrove forest. Their results can improve the programs which could be undertaken by managers in the forestry sector. Likewise, forest protection, insects and diseases, regeneration, silviculture, plantation, biomass production, carbon offset, biodiversity conservation and beach protection are future research plans.

Acknowledgements The authors would like to thank the director of Research and Education Organization of Ministry of Jahad agriculture who provided a grant for conducting this research. Moreover, our appreciation goes to Mr. Seyed Ali Rezvani Kalajahi who offered support to finalize this manuscript.

References

Abbasi H (1990) Wet land forests in the world and Iran and their eco-social importance. Paper presented at the the conference on plantation in wet lands, Tehran, Iran

Clark JR, Garcia S, Caddy J (1992) Integrated management of coastal zones. FAO, Rome

Danehkar A (1993) Investigation on mangrove forest at Sirik area in Hormozgan Provience. Master thesis University of Tarbiat Modarres, Tehran, Iran

Danehkar A (1995) Biology and ecophysiology of mangrove trees: part 1. Iranian J Rang For 25(Athumn):24–29

Danehkar A (2001) Zoning and planing of management of Biospher of *Avecinnia*, Iran Department of Environment, Tehran. Iranian J Environ (36):25–33

Danehkar A (2006) Project of management and development of mangrove forests in the Hormozgan pr Province. Department of Natural Resources of Hormozgan province, Bandarabbas, I.R. Iran

Erfani M (2007) Investigation on structure and comparision of grey mangrove vegetation aera at Govater bay and Gur-e-bahu international wetland in Sistan va baluchestan province for selection of protected zones. Master thesis, University of Tehran, Tehran, Iran

Ewel KC, Twilley RR, Ong JE (1998) Different kinds of mangrove forests provide different goods and services. Glob Eco Biogeo let 83–94

Kathiresan K, Bingham BL (2001) Biology of mangroves and mangrove ecosystems. Advan Mar Biolo 40:81–251

Majnoonian HMP (2001) Coastal-Marin protected areas (functions and values). Iran Department of Environment, Tehran

Mokhtari M, Savari A, Rezai H, Kochanian P, Bitaab A (2008) Population ecology of fiddler crab, Uca lactea annulipes (Decapoda: Ocypodidae) in Sirik mangrove estuary, Iran. J Estua Coast Shel Scie 76(2):273–281

Nybakken JW (1993) Marine biology: an ecological approach. Harper Collins College Publishers, New York

Pillay T (2004) Aquaculture and the environment. Wiley-Blackwell, UK

Rashvand S (1997) Stratural investigation of mangrove forest of Bushehr province. Master, Gorgan University of Agricaltural sciences and Natural Resources, Gorgan, Iran

Rodriguez W, Feller IC (2004) Mangrove landscape characterization and change in Twin Cays, Belize using aerial photography and IKONOS satellite data. Atol Resear Bulle 509:1–22

Safiari S (2001) Mangrove forests, part 2, mangrove forests of Iran. Research institute of Rangelands and Forests, Iran

Sistani D (1990) An overview on mangrove forest of Iran and world. Iran Department of Environment, Tehran

Wells S, Ravilious C, Corcoran, E (2006) In the front line: shoreline protection and other ecosystem services from mangroves and coral reefs. UNEP-WCMC Cambridge, UK

Valiela I, Bowen JL, York JK (2001) Mangrove forests: one of the world's threatened major tropical environments. BioScience 51(10):807–815

Wright SJ (2010) The future of tropical forests. Annals New York Acad Scie 1195(1):1–27

Plant Diversity and Forest Structure of the Three Protected Areas (Wildlife Sanctuaries) of Bangladesh Sundarbans: Current Status and Management Strategies

Saiful Islam, Mizanur Rahman and Sourav Chakma

Contents

Abstract The protected areas of Sundarbans maintain a globally significant ecosystem and provide ideal habitats for aquatic resources (fish, shrimp), birds, and wildlife. The UNESCO World Heritage Committee declared the whole Sundarbans as its 798th heritage site including three wildlife sanctuaries in order to conserve all flora and fauna. The study area, covering an area of 139,700 ha, was surveyed in the year 2012–2013 by stratified systematic sampling. The sampling sites were previously stratified on the vegetation map depending on the available forest types. A total of 31 families with 63 species and 11,619 individuals were enumerated from the study site across all size classes of trees including non tree plants. Leguminosae was found to be a comparatively diverse family having ten species, and Euphorbiaceae was dominated by a large number of individuals. The plant species richness (herbs, shrubs, climbers, etc.) was predominantly confined to the streamside than the 'forest proper', and the mean of these two groups was statistically significant. Due to domination by a few species, the overall Shannon's value of the study area was only 2.19. Mean number of species significantly varied between the protected areas. Species accumulation

S. Islam (✉) · M. Rahman · S. Chakma
Forest Department, Ban Bhaban, Boyra, Khulna, Bangladesh
e-mail: sislam47@hotmail.com

I. Faridah-Hanum et al. (eds.), *Mangrove Ecosystems of Asia*,
DOI 10.1007/978-1-4614-8582-7_7,

curves did not follow regular fashion due to domination by few species. The cluster diagram has validated environmental factors to discriminate species composition in protected areas. All other stand parameters also varied. Results are compared with the few available studies. Management of the study area should incorporate research activities to determine successional change due to dynamic ecological process.

1 Introduction

Since ancient times, emperors and kings took initiatives for the protection of animals, fish and forests. The earliest instance of such deliberate establishment of what we call today the protected areas was in fact the way forward to the modern concept of conservation—the wise maintenance and utilization of earth's natural resources. The principle is to plan resource management on the basis of accurate inventory and to take protection measures to ensure that resources do not become exhausted. Protected areas when designed and managed appropriately are now recognized as offering major sustainable benefits to society. In other words, the establishments and management of protected areas is one of the most important ways of ensuring that the world's natural resources are conserved so that they can better meet the material and cultural needs of mankind now and in the future.

The World Conservation Strategy (WCS), prepared by four of the world's leading conservation agencies—the International Union for Conservation of Nature and Natural resources (IUCN), World Wild Life Fund (WWF), Food and Agriculture Organization of the United Nations (FAO), and the United Nations Environment Programme (UNEP), demonstrates how the conservation of living resources is essential for sustaining development by maintaining the essential ecological processes and life-support systems on which human survival and development depends. In general terms, the maintenance of species and ecosystems requires that the use of living resources within a healthy environment must be done on a sustainable basis. Among other more specific actions, it requires the establishment of networks of natural protected areas for the conservation of species and ecosystems in wild environments. Most tropical countries have established protected areas to ensure long-term maintenance of conservation of biological diversity.

The Bangladesh Wildlife (Preservation) (Amendment) Act of 1974 provides for the establishment of national parks, wildlife sanctuaries and game reserves. Protected areas include the Wildlife Sanctuary (WS) National Park and Game Reserve. Article 23 of the Bangladesh Wildlife (preservation) Order of 1973 has provisions for declaration of protected areas and also has regulations prohibiting activities in the protected areas. The first Sundarban East Wildlife Sanctuary (WS) was established in 1960 through the notification of Forest Act 1927. Later on, in 1977 the Sundarban South WS and Sundarban West WS were declared through notifications in the official gazette in accordance with the provisions of Article 23(1) of the Bangladesh Wildlife (preservation) Order of 1973.

The Bangladesh Wildlife Preservation Order 1973 (amended in 1974) defines a Sanctuary as "exclusively undisturbed ground primarily for the protection of wildlife inclusive of all natural resources such as vegetation, soil and water". While biodiversity conservation is the principle objective of the sanctuaries, non-consumptive use (education, tourism) may be allowed. Through the circulation of the Wildlife (Preservation) Act in 1974, the National Forest Policy in 1994, the Forest Act 1927 (amendment in 2000) and the latest Wildlife (Protection and Safety) Act 2012, the emphasis of forests management has gradually shifted from timber production to ecological requirements, conservation of biological diversity, meeting bona-fide subsistence consumption needs of local people, and climate change mitigation and adaptation functions. But today, the habitat of protected areas are changing with tidal surges, cyclonic storm, increased soil salinity and water salinity due to the effect of sea level rise and decreased freshwater flow from upstream.

The protected areas of Sundarbans carry a globally significant ecosystem and show ideal habitats for aquatic resources (fish, shrimp), birds, and wildlife. Environmental conservation of protected areas offers coastal protection by creating a living shelterbelt in southwest Bangladesh. Commercial cutting of trees had been banned since 1989 by the Government in Protected Areas to ensure additional protection for wildlife habitat and natural resources. However, the wildlife sanctuaries must be managed under a multi-objective conservation management system with due attention to capacity building of the personnel including logistic support. The UNESCO World Heritage Committee declared the whole Sundarbans as its 798th heritage site on the sixth of December 1997, including three wildlife sanctuaries with an area of 139,700 ha in order to conserve all flora and fauna of Sundarbans (Anon 1998).

Scientific evidence accumulates that many of the earth's ecosystems have become severely degraded and that restoration will take decades if not centuries. Enhancing ecosystem health represents purposeful objectives without which, the very foundations of our social and economic systems are undermined (Rapport 1998). Various approaches to the question of what constitutes ecosystem health have been examined (Rapport 1995, 1998). While most people visualize instinctively a healthy ecosystem as being pristine or at least appearing to be minimally altered by human action or natural cause, in fact, there is no universal conception of ecosystem health, thus there is considerable variation in the concept being described or defined (Callow 1992; De Leo and Levin 1997). However, in general, ecosystem health has been seen as the preferred state of ecosystems modified by human activity (e.g., farm land, urban environments, and managed forests), and in contrast, ecological integrity as an unimpaired condition in which ecosystems show little or no influence from human actions (Callow 1992; De Leo and Levin 1997). Natural ecosystems, by definition would continue to function in essentially the same way if humans were removed (Anderson 1991).

Biological diversity is certainly an important element in understanding the structure and function of ecosystems. In defining ecosystem health, therefore, scientific information is not only important but also even essential. Monitoring biodiversity is the first step in systematic conservation planning and management. Biodiversity conservation at landscape has emerged as a global priority (Meyer and Turner 1992). For example, the largest share of the European Commission investments in biodiversity is in support of protected areas, representing 35 % of the biodiversity contribution (Anon 2012). This includes projects to strengthen local capacities to maintain and value protected areas, to promote income generating activities in the protected areas and their buffer zones and to support scientific monitoring. The second most significant type of biodiversity related activity is support to the sustainable management of forests, representing 31 % of the total European Commission biodiversity finance (Anon 2012). This involves efforts to elaborate forest sustainable management plans, to address forest governance issues, to combat illegal logging and to design strategies to mitigate climate change through the prevention of deforestation.

While monitoring of the ecosystem is an important tool for the management to assess the impacts of prevailing management practices, on the other hand, assessment of the existing human resource management in terms of capacity building and infrastructure development in the protected areas is also equally important. In a complex ecosystem like the Sundarbans, susceptible to continuous river erosion and accretion, and also exhibiting impacts of climate change, monitoring is a real challenge for an organization. Such natural processes and disasters may be important contributors in vegetation dynamics of Sundarbans (Iftekhar and Saenger 2007). It has been observed by Revilla et al. (1998) that 3,026 ha of land were eroded by the rivers during the period of 1981–1997. Cyclones, storms, tidal surges and rapid siltation are some natural disasters in Sundarbans. As such, selection of appropriate sampling techniques that capture systematically all variations in the habitat is also imperative. There are important disadvantages of random sampling as compared to systematic sampling for which systematic sampling is widely practiced in tropical forest inventory (e.g. FAO 1994; Wood 1990; Sutherland 2000). The first systematic survey of the Sundarbans was carried out during 1926–1928 by Curtis (1933), which was followed by Forestal (1960), Chaffey and Sandom (1985), and Revilla et al. (1998). However, the results of these surveys were limited to timber volume and stocking statistics by species and compartments looking primarily at commercial tree species only. Studies on plant species diversity and forest structure of three protected areas of Bangladesh Sundarbans are fairly unknown.

The objective of this study was to evaluate the current status of the protected areas in terms of plant species composition, diversity and structure of the forest in order to recommend strategies for conservation and management.

2 Study Site and Methodology

2.1 Study Site

The Sundarbans mangrove is a salt-tolerant wetland forest ecosystem like the other coastal mangroves of Southeast Asia, and it is the largest contiguous patch of mangrove forest in the world (Chaudhuri and Naithani 1985) . It is located in the estuary of the river Ganges in the extreme southwestern corner of Bangladesh.

It is the last large wilderness area of significant natural beauty in Bangladesh and potential site for recreation. The forest covers an area of about 6,017 km^2 of which 62 % falls within the territory of Bangladesh, between the latitudes 21° 31′ and 22° 30′ N and between the longitudes 89° and 90° E (Fig. 1, location map), while the remaining area belongs to India. Of the total land area of Sundarbans, approximately 70 % are flat lands with occasional depressions, and the remaining 30 % comprises a complex network of streams and rivers varying considerably in width and depth (Fig. 1). The study site is composed of three protected areas-wildlife sanctuaries with an area of 139,700 ha and situated on the south of the Sundarbans Reserved Forest along the coast of the Bay of Bengal (Fig. 1). These three protected areas are the Sundarbans East Wildlife Sanctuary, Sundarbans South Wildlife Sanctuary and Sundarbans West Wildlife Sanctuary (Fig. 1).

The elevation of the study site is hardly 3 m above mean sea level (Siddiqi 2001). The soils are finely textured silty clay loam and the sub-soil is stratified with alternate layers of clay and sand but is compacted at greater depth (Choudhury 1968). The forest area of Sundarbans East Wildlife Sanctuary is fertile due to the fresh silt supply by rivers compared to the southern and western sanctuaries. On the western part of the study site, due to low silt deposition, the forest floor is compacted and does not support healthy tree growth. On the other hand, too much silt deposition in the eastern part of the forest causes a rise of the forest floor with irregular tidal inundation. The soil of the overall Sundarbans in general is alkaline with a pH range from 7.0 to 8.0 in most of the forest areas.

The climatic condition in and around the forest shows distinct seasonal variation with highest temperature occurring in April and May and the lowest in December and January. Mean annual maximum and minimum temperatures vary between 32° and 20 °C and ten-year average temperature was 26.0 °C (Canonizado and Hossain 1998). The region has high relative humidity, and average annual relative humidity ranges from 77 to 80 % with high humidity peaking at around 95 % in June–October and low (70 %) in the months of February and March (Anon 1998). The mean annual rainfall ranges from about 1,900 mm to about 2,500 mm. June, July, August and September are the wettest months and December, January and February the driest.

The water level inside the Sundarbans is highly dependent on the tidal oscillation and to a lesser extent on the quantity of freshwater flow from upstream. The change in spatial pattern of salinity inside the Sundarbans is related to the changes in the volume of fresh water flow from upstream rivers. In general, across the whole Sundarbans, salinity increases from east to west direction. On the basis of degree of salinity, the Sundarbans follow a definite pattern of ecological succession in the three distinct ecological zones: less saline zone, moderately saline zone and strong saline zone. In addition to the depth, duration and frequency of tidal inundation, the level of salinity greatly influences the distribution of species. In the dynamic process of ecological succession, one community is replaced by another over time, and in the process, newly accreted lands are first colonized by the different species of grasses and sedges, which are subsequently replaced by pioneer shrubs and trees in the soft soil. Changes in the vegetation community continue to occur till the final or climax stage is attained in a mature soil. Sundri (*Heritiera fomes*) is the characteristic

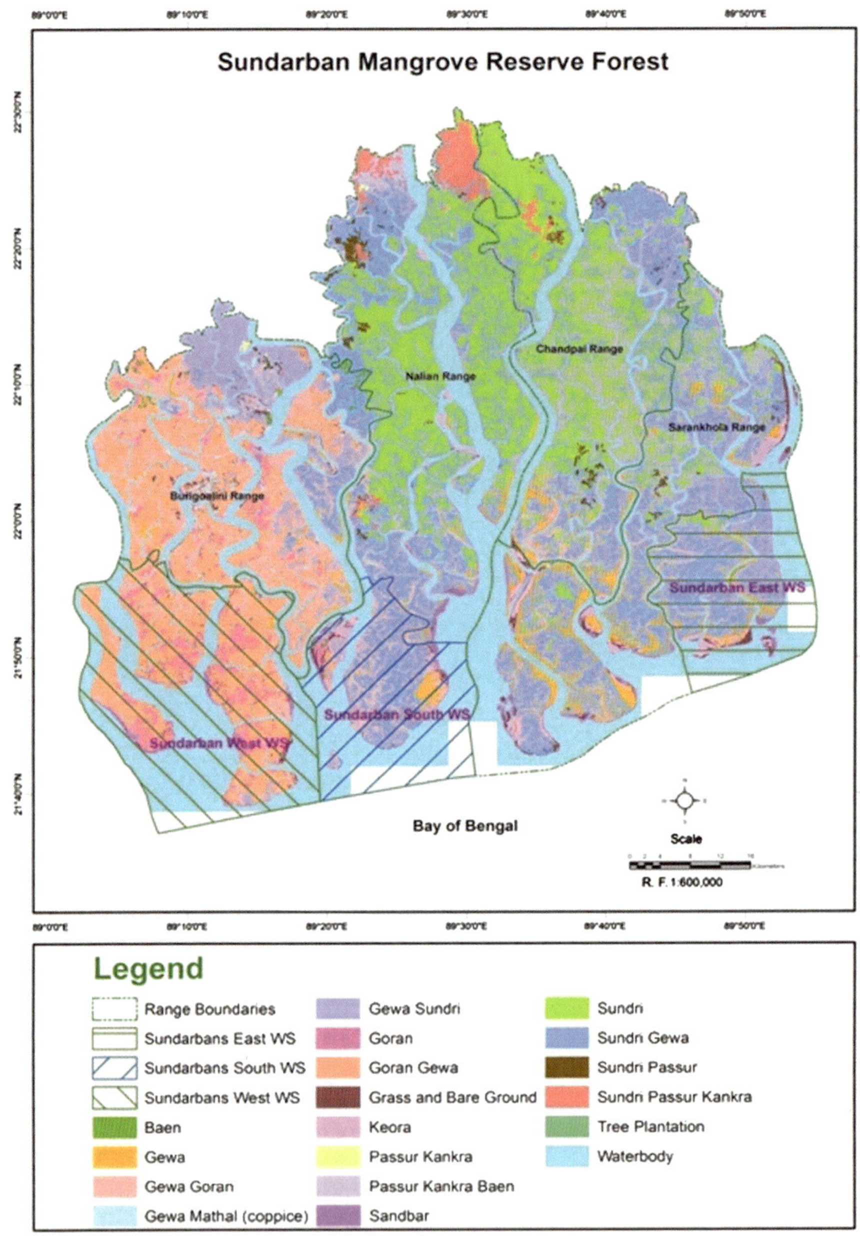

Fig. 1 Vegetation map of Sundarban Reserve Forest showing three Wildlife Sanctuaries (WS) situated along the coast of the Bay of Bengal. Major forest types by species (local name) are shown with different colors

Fig. 2 View of typical mangrove forest of the study area

dominant climax species of the less saline zone, Gewa (*Excoecaria agallocha*) of the moderately saline zone and Goran (*Ceriops decandra*) is a typical species of a strongly saline zone. A view of a typical mangrove forest of the study area is shown in Fig. 2.

2.2 Methodology

2.2.1 Transect Sampling and Plot Design

The study area, covering an area of 139,700 ha was surveyed in the year 2012–2013 by laying out of a series of line transects (Figs. 3 and 4) with positioning of plots on the line at a definite interval to capture maximum diversity within a site. The sampling sites were previously stratified on the vegetation map depending on the available forest types (see Fig. 1). Due to systematic coverage of the line transect in a sample site, all rare niches were included. Both on the vegetation map and in the ground two major habitat types such as streamside and 'forest proper' were

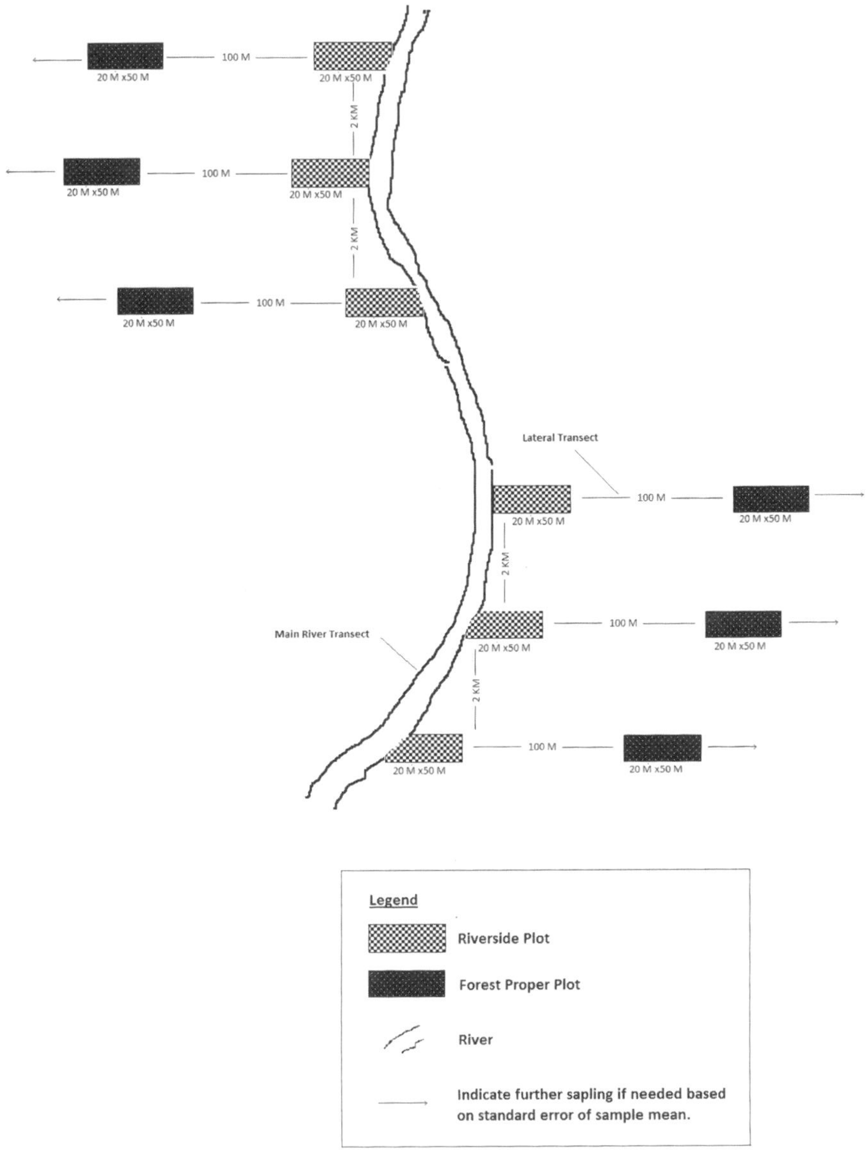

Fig. 3 Schematic diagram showing lay-out of main and lateral transects and positioning of plots in the sampling site having wide rivers (Design 1)

identified for sampling. River track was considered as the main transect line and the lateral transects were established at right angles to the main transect to sample forest proper plots, spaced systematically at a 100-m distance inside the forest. But these lateral transects also had been positioned alternately on either side of the main river

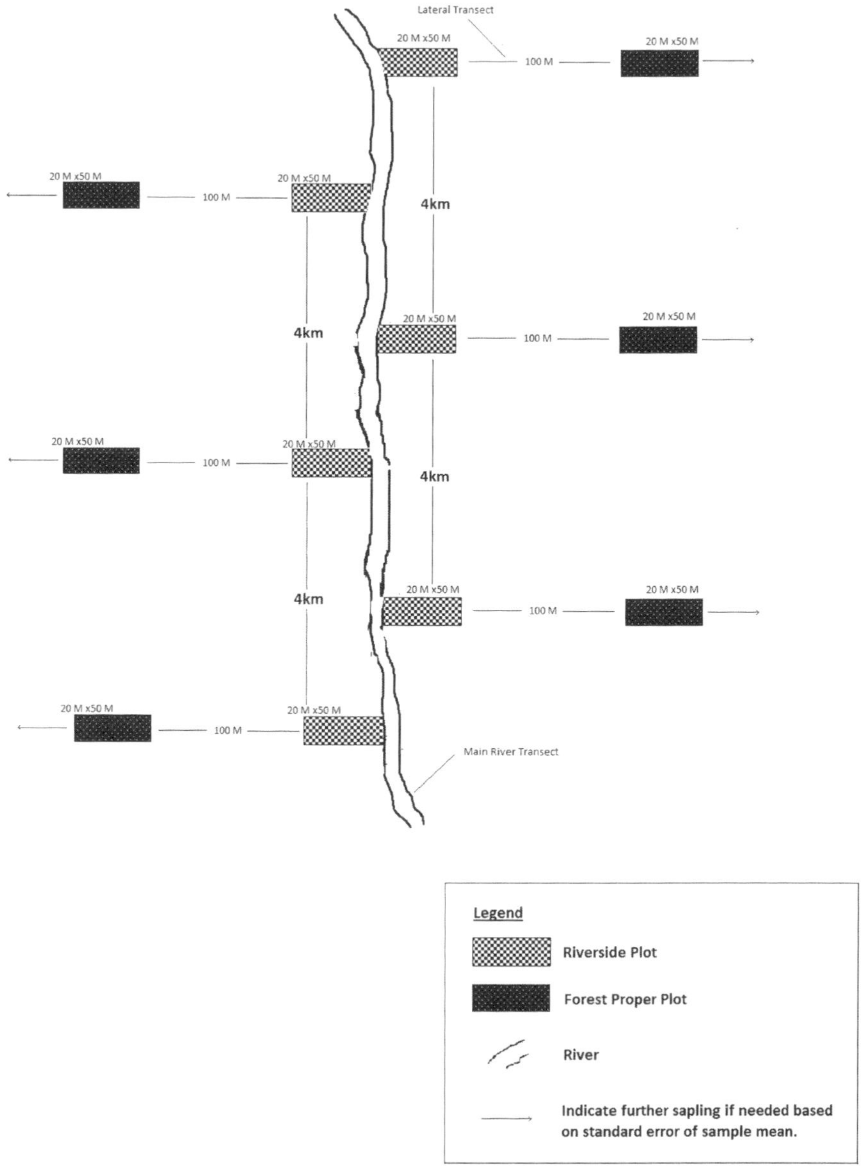

Fig. 4 Schematic diagram showing lay-out of main and lateral transects and positioning of plots in the sampling site having small streams (Design 2)

transect (river track) at a distance of 2 km (Figs. 3 and 4). To capture the maximum diversity of streamside, study plots were also established at a 100-m distance along the main transects. In the case of sampling forest proper, every first plot was positioned on the lateral transect after a 100-m distance from the edge of the stream plot to distinguish two different habitat types (i.e. streamside and forest proper).

Rectangular plots of 20 × 50 m were found manageable to record necessary biological parameters. Trees of ≥ 15 cm diameter at breast height (dbh) were measured

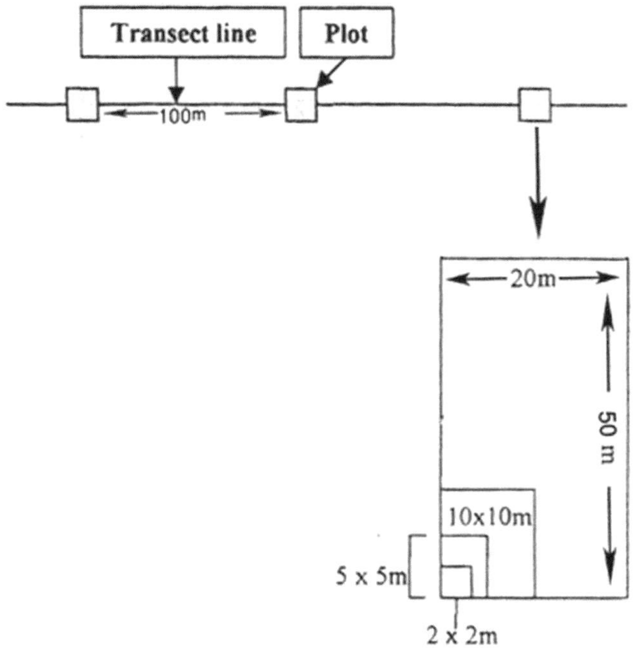

Fig. 5 Schematic diagram of nested plot design showing different shapes and sizes of plots for enumeration of various tree size classes and plants. Plants (herbs, shrubs, climbers, etc.) were enumerated within a 10 × 10-m plot

within the main plot of 20 × 50-m size and three kinds of nested subplots such as 10 × 10 m, 5 × 5 m and 2 × 2 m are distributed inside the main plot for poles (5.0 cm to < 20 cm dbh), saplings (1.5 m ht. to < 5.0 cm dbh), and seedlings (10 cm to < 1.5 m tall), respectively. The detail of the plot design is shown in Fig. 5.

2.2.2 Data Collection

Diameter at breast height (DBH) of all trees according to the diameter class selected for different plot sizes (described above) was measured with a diameter tape at 1.3 m above ground level or just above the buttress. Voucher specimens were only collected from the trees and plants that could not be identified in situ. Every such plant sample was tagged and given a unique identification code. Voucher specimens were air dried and finally identified up to species level using available literature (e.g. Tomlinson 1986; Aksornkoae et al. 1992) and matched with the collections preserved in the herbarium as well as with the images available in the website. However, few specimens could not be identified (e.g. climbers and members of the family gramineae) but preserved for identification.

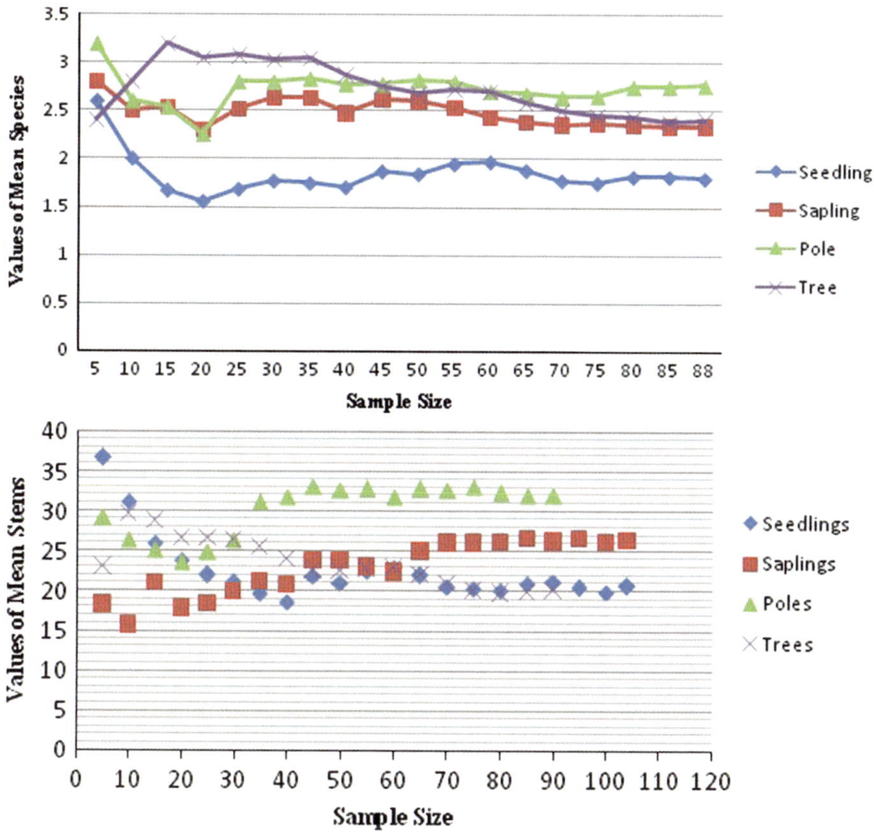

Fig. 6 Travelling mean of species (*above*) and stems (*below*) showing decrease in fluctuation of mean values with the cumulative number of study plots (sample size) indicating minimum number of samples required for all tree size classes

2.2.3 Intensity of Sampling

The intensity of sampling (i.e. number of sample required) within the study site was examined based on two measures of variability, i.e. number of stems and number of species of the plot sampled (Fig. 6). In the graphs, all four size-classes of trees showed gradual decline in fluctuation in mean values of number of stems and species with cumulative number of study plots indicating minimum required sample size for the study site. This travelling mean or performance curve is considered as analogous to a species area-curve (Brower et al. 1990), but it plots a cumulative mean rather than the cumulative number of species. In addition, to obtain desired level of precision of survey estimates as well as number of samples required, calculation of standard error (%) of the sample mean based on number of stems and species was also determined and found within the acceptable limit of precision, i.e., 6.0, 3.5, 3.0, and 3.8 % (log ten transformed data) for seedlings, saplings, poles and timber trees, respectively (Table 1).

Table 1 Precision of survey estimates in terms of standard error (SE) of sample mean based on number of individuals and number of species per plot across the three protected areas of Sundarban Reserve Forest, Bangladesh

Size-class	Sample size (N)	Based on number of stems		Based on number of species	
		Mean (SE)	SE % (log ten transformed data)	Mean (SE)	SE % (original data)
Seedlings (2 × 2 m plot)	104	20.65 (2.04)	6	1.80 (0.120)	6
Saplings (5 × 5 m plot)	104	26.15 (1.88)	3.5	2.30 (0.112)	5
Poles (10 × 10 m plot)	88	32.00 (1.81)	3.0	2.74 (0.141)	5
Timber (20 × 50 m plot)	88	20.14 (1.37)	3.8	2.41 (0.135)	5.6

2.2.4 Data Analysis

The underlying distribution of the dataset was examined by using a histogram to distinguish a symmetric from a skewed distribution. For a skewed data set, instead of using median, data were log-ten transformed to obtain symmetric distribution for calculation of mean value. The mean comparison of the variables between two groups was determined using a two-sample t-test. One way analysis of variance (ANOVA) was applied to test the differences in mean values of floristic variables between different habitat types. Statistical significance level was established at $p < 0.05$. Analysis was performed with the Minitab (Release 15 for Windows) statistical package (Minitab Inc. 1994).

The total count of species, based on the number of species per unit number of stems, was used to indicate species richness. Shannon's species diversity index was used to compare diversity of three protected areas. Multivariate cluster analysis was adopted to detect the pattern of similarity and dissimilarity in species composition among three protected areas. The percentage of canopy openness was calculated using fish-eye photographs and measured as the ratio of canopy gaps and holes including any area of the open sky that is unobscured to the whole photograph area (Brown 1990; Whitmore et al. 1993). By superimposing a dot grid to the fish-eye photograph, the proportion of the total number of dots that falls on canopy gaps and minor openings provided the estimates of canopy openness. The crown density in terms of percentage of crown closure was measured by percent of canopy gaps. Importance value index (IVI) was calculated to determine the changes in dominance structure and composition of the three protected areas. Species accumulation curves were constructed for three protected areas and different size classes of trees as well as for the study area as a whole.

3 Results and Discussion

3.1 Taxonomic Composition

3.1.1 Family Dominance

A summary of taxonomic composition of trees and plants of the study sites is provided (Table 2). A total of 31 families with 63 species and 11,619 individuals were enumerated from the study site across all size classes of trees from 1.5-m height and above including plants (herbs, shrubs, climbers, palms, etc.). However, 15 different plant specimens (herbs, grasses and climbers) could not be identified as the inventory just finished, but the vouchers specimens were preserved for identification. The family dominance was predominantly attributed to ten large or commonest families based on the number of species and individuals (Table 3). In the study site as a whole, Leguminosae was found to be a comparatively diverse family having ten species or alone 15.9 % of the total number of species recorded, followed distantly by Rhizophoraceae and Gramineae (five species each or 7.9 %), Pteridiaceae (four species or 6.3 %) and Avicenniaceae, Meliaceae and Asclepiadaceae (three species each or 4.8 %). On the other hand, Euphorbiaceae was dominated by a large number of individuals (3,778), but followed distantly by Sterculiaceae, Rhizophoraceae, Leguminosae and Palmae with 2,467, 2,152, 744 and 737 individuals, respectively. The members of the family Sterculiaceae, Euphorbiaceae, Avicenniaceae and Sonneratiaceae were successful in dominating in the overstorey canopy.

Table 3 also highlights that the dominant families by species accounted for more than half of the species richness (61.9 % or 39 species) listed from the protected area. Surprisingly, family dominance by individuals (i.e. 10 families) contributed more than 90 % of the total individual recorded (i.e., 11,271 individuals or 97.0 %).

3.1.2 Species Abundance and Dominance

Based on the list of species composition as shown in Table 2, the ten most abundant trees and plant species in the study area is given in Table 4. The dominant tree species represented 76.4 % of the total individuals recorded whereas the contribution of the dominant plant species was only 20.5 %, and together the figure reaches 97 % of the total individuals enumerated from the entire protected area. *Excoecaria agallocha* (Euphorbiaceae) and *Heritiera fomes* (Sterculiaceae) were among the two most dominant tree species in the overstorey canopy while *Ceriops decandra* (Rhizophoraceae) dominated in the understorey canopy. Among the plants, *Phoenix paludosa* (Palmae), *Derris trifoliata* (Leguminosae), *Acanthus ilicifolius* (Acanthaceae) and *Sarcolobus globosus* (Asclepiadaceae) were most abundant on the forest floor.

Table 2 Plants species list of three protected areas (wildlife sanctuaries) of the Bangladesh Sundarbans derived from stratified systematic sampling executed in the year 2012–2013

Scientific name	Family name	Local/ vernacular name	No. of individuals	Life form
Abrus precatorious	Leguminosae	Kuch lata	1	Climber, plants best known for its seeds
Acanthus ilicifolius	Acanthaceae	Hargaza	442	Scrambling, woody, thorny herb
Acrostichum aureum	Pteridiaceae	Hodo, tiger fern	14	Gregarious fern
Aegialitis rotundifolia	Plumbaginaceae	Dhalchaka	7	Small tree
Aegiceras corniculatum	Myrsinaceae	Khalisha, khalshi	26	Shrub or small tree
Amoora cucullata	Meliaceae	Amur	93	Small tree
Asplenium nidus	Pteridiaceae	–	26	Large epiphytic fern
Asplenium spp.	Pteridiaceae	–	1	Epiphytic fern
Avicennia alba	Avicenniaceae	Maricha Baen	11	Medium size tree
Avicennia marina	Avicenniaceae	Sada baen	56	Small to big tree
Avicennia officinalis	Avicenniaceae	Baen	123	Big tree
Brownlowia tersa	Tiliaceae	Sundri lota, Lota Sundri	85	Scan dent shrub
Bruguiera gymnorrhiza	Rhizophoraceae	Kankra	11	Small to large tree, red calyx 'cap'
Caesalpinia crista	Leguminosae	Kutum katta	11	Scan-dent, armed shrub
Ceriops decandra	Rhizophoraceae	Goran	1,960	Shrub or small tree, usually coppices
Clerodendrum inerme	Verbenaceae	Sitka, sitki	2	Scan-dent shrub
Cynodon dactylon	Graminae	Durba gash	Abundant	Grass
Cynometra ramiflora	Leguminosae	Shingra	77	Shrub
Cyperus javanicus	Cyperaceae	Kucha gash	Abundant	Grass-like herb (sedge)
Dalbergia candenatensis	Leguminosae	Chanda lota/Sitki	7	Scrambling climber
Dalbergia melanoxylon?	Leguminosae	Kata bohoi	2	Small tree, branch with spine
Dalbergia spinosa	Leguminosae	Chanda katta	18	Scan-dent, armed shrub
Dendrobium striolatum?	Orchidaceae	Parachula	33	Epiphytic orchid with needle like leaf
Dendrophthoe falcata	Loranthaceae	Dhoripata, Pargasa	1	Woody parasite in tree crowns
Derris indica	Leguminosae	Kali lota	1	Climber, flower pinkish white
Derris trifoliata	Leguminosae	Kali lota	594	Climber, flower whitish
Entada scandens	Leguminosae	Gila lota	4	A large woody twisted climber

Table 2 (continued)

Scientific name	Family name	Local/ vernacular name	No. of individuals	Life form
Eriochloa procera	Gramineae	Nol gash	Abundant	Grass
Eugenia fruticosa	Myrtaceae	Ban jam, jam	5	Small tree
Excoecaria agallocha	Euphorbiaceae	Gewa	3,773	Tree
Finlaysonia obovata	Asclepiadaceae	Dudhi lata	1	Climber
Flagellaria indica	Flagellariaceae	Abetaa	1	Climber
Heritiera fomes	Sterculiaceae	Sundri	2,467	Tree
Hibiscus tiliaceous	Malvaceae	Bhola	12	Shrub
Hoya species	Asclepiadaceae	Agusha, Pudipata	11	Climber
Imperata cylindrica	Gramineae	Chan gash	Abundant	Grass
Kandelia candel	Rhizophoraceae	Gura, gural, Bhatkathi	60	Small tree
Lannea coromandelica	Anacardiaceae	Jiga, Bhadi, kapila	9	Medium size tree
Lumnitzera racemosa	Combretaceae	Kirpa, kripa	4	Small tree
Myriostachya wightiana	Gramineae	Dhanshi	Abundant	Grass, common on new accretions
Nypa fruticans	Palmae	Golpata	140	Palm with underground stem
Pandanus foetidus	Pandanaceae	Kewa katta	57	Prickly succulent screw-pine
Petunga roxburghii	Rubiaceae	Narikili/Naholi	4	Small tree
Phoenix paludosa	Palmae	Hantal	597	Thorny palm
Phragmites karka	Gramineae	Nol kagra	Abundant	Grass
Pongamia pinnata	Leguminosae	Karanja	29	Small tree
Premna corymbosa	Verbenaceae	Serpoli, Setpoli, kunail	1	Shrub or small tree
Rhizophora mucronata	Rhizophoraceae	Garjan, Jhanna	91	Tree with stilt roots
Rhizophora apiculata	Rhizophoraceae	Garjan, Jhanna	30	Tree with stilt roots
Saccharum cylindricum	Gramineae	Eli ghas	Abundant	Grass
Salacia chinensis	Celastraceae	Choyt barai	4	Small tree
Sapium indicum	Euphorbiaceae	Urmui	5	Tree
Sarcolobus globosus	Asclepiadaceae	Bawali lata	365	Climber
Sonneratia caseolaris	Sonneratiaceae	Choyla, ora, soyla	1	Small tree
Sonneratia apetala	Sonneratiaceae	Keora	195	Tree
Stenochlaena palustris	Blechnaceae	Deki lota	2	Climbing fern
Taimrix indica	Tamaricaceae	Jhao, nona jhao.	10	Small tree
Typlophora spp.	Apocynaceae	Mohazani lata	24	Slender climber, leaf thin/papery, opposite with long petiole
Viscum monoicum	Loranthaceae	Shamu lota	1	Woody parasite in tree crown of *Excoecaria agallocha*

Table 2 (continued)

Scientific name	Family name	Local/ vernacular name	No. of individuals	Life form
Vittaria sp. (elongate?)	Pteridiaceae	–	18	Tape fern, common epiphytic fern
Vittaria sp.	Pteridiaceae	–	1	Epiphyte, leaf base cushion like
Xylocarpus granatum	Meliaceae	Dhundul	36	Small tree
Xylocarpus mekongensis	Meliaceae	Passur	59	Tree

Table 3 The family dominance of ten large families based on abundance of species and individuals for the overall protected area (ranked in declining order of abundance). Figures in parentheses are percentage of the total

Based on number of species		Based on number of individuals	
Family	No. of species	Family	No. of individuals
Leguminosae	10 (15.9)	Euphorbiaceae	3,778 (32.5)
Rhizophoraceae	5 (7.9)	Sterculiaceae	2,467 (21.2)
Graminae	5 (7.9)	Rhizophoraceae	2,152 (18.5)
Pteridiaceae	4 (6.3)	Leguminosae	744 (6.4)
Avicenniaceae	3 (4.8)	Palmae	737 (6.3)
Meliaceae	3 (4.8)	Acanthaceae	442 (3.8)
Asclepiadaceae	3 (4.8)	Asclepiadaceae	377 (3.2)
Euphorbiaceae	2 (3.2)	Sonneratiaceae	196 (1.7)
Loranthaceae	2 (3.2)	Avicenniaceae	190 (1.6)
Palmae	2 (3.2)	Meliaceae	188 (1.6)
Total	*39 (61.9)*	*Total*	*11,271 (97.0)*

3.2 Species Richness and Diversity

In this study, the estimated species richness, i.e. the number of species per unit area or per unit number of stems has been considered. Diversity is a combination of richness and evenness (whether the site consists of a few abundant species and many rare ones, or all species being equally frequent) (Magurran 1988). A high value of Shannon H′ indicates a large number of species with similar abundances; a low value indicates domination by a few species. Species richness has been widely used as a parameter for diversity assessment largely because it is usually the straightforward assessment based on information available. The contribution of various tree size classes and non-tree plant species towards species accumulation with respect to the collection of individuals and area sampled is provided in Table 5. Throughout the protected areas, the highest number of plant species (33) was recorded from 10 × 10-m plots with a sampled area of only 0.88 ha while contribution of tree species was only 30 with a sampled area of 8.8 ha (Table 5). The plant species richness (herbs, shrubs, climbers, epiphytes, etc.) was predominantly confined to the streamside than

Table 4 Ten dominant tree species and plants recorded from the study plots of three protected areas. Species are ranked in declining order of abundance. Species dominated in the overstorey (> 15 cm dbh) is indicated with asterisks (*). Figures in parentheses are percentage of the total

Dominant tree species by number of stems			Dominant plant species by number of individuals		
Scientific name	Family	No. stems	Scientific name	Family	No. individuals
*Excoecaria agallocha**	Euphorbiaceae	3,773	*Phoenix paludosa*	Palmae	597
*Heritiera fomes**	Sterculiaceae	2,467	*Derris trifoliata*	Leguminosae	594
Ceriops decandra	Rhizophoraceae	1,960	*Acanthus ilicifolius*	Acanthaceae	442
*Sonneratia apetala**	Sonneratiaceae	195	*Sarcolobus globosus*	Asclepiadaceae	365
*Avicennia officinalis**	Avicenniaceae	123	*Nypa fruticans*	Palmae	140
Amoora cucullata	Meliaceae	93	*Brownlowia tersa*	Tiliaceae	85
*Rhizophora mucronata**	Rhizophoraceae	91	*Cynometra ramiflora*	Leguminosae	77
Kandelia candel	Rhizophoraceae	60	*Pandanus foetidus*	Pandanaceae	57
*Xylocarpus mekongensis**	Meliaceae	59	*Dendrobium striolatum?*	Orchidaceae	33
*Avicennia marina**	Avicenniaceae	56	*Asplenium nidus*	Pteridiaceae	26
	Total	*8,877 (76.4)*		*Total*	*2,390 (20.5)*

Table 5 The contribution of various tree size classes and non-tree plant species towards species accumulation with respect to the collection of individuals and area sampled. Seedlings 10 cm to < 1.5 m height; Sapling: 1.5 m height ≤ 5 cm dbh; Poles: 5 ≤ 15 cm dbh; Timber: ≥ 15 cm dbh

Variable	Tree size classes				Plants[a]	All combined
	Seedlings	Saplings	Poles	Timber		
Sample size (n)	104	104	88	88	88	
Area surveyed in m^2 (ha)	416 (0.042 ha)	2,600 (0.26 ha)	8,800 (0.88 ha)	88,000 (8.8 ha)	8,800 (0.88 ha)	
No. species	16	20	23	25	33	63
No. individuals	2,059	2,632	2,736	1,682	2,510	11,619
Species-individual ratio	0.008	0.007	0.008	0.015	0.013	0.005

[a]Other than tree species

the 'forest proper', and the mean of these two groups was statistically significant (two-sample t-test: $t = -5.51$, $p = 0.0000$, DF = 100). This implies that biodiversity inventory of mangroves must include a sufficient number of statistically valid samples from the streamside, particularly secondary and tertiary streams. The banks of the

Table 6 Comparison of mean values of trees and plant species (species/plot) for three protected areas (wildlife sanctuaries) of Bangladesh Sundarbans

Size class	Plot size (m^2)	Protected area (wildlife sanctuaries)			ANOVA
		East PA	South PA	West PA	
Timber trees (≥ 15.0 cm dbh)	1,000 m^2	$n = 25$ Mean = 3.08 SE = 0.29	$n = 34$ Mean = 2.44 SE = 0.19	$n = 29$ Mean = 1.79 SE = 0.17	F = 8.10 $p = 0.001$**
Poles (5 cm ≤ 15 cm dbh)	100 m^2	$n = 25$ Mean = 2.80 SE = 0.34	$n = 34$ Mean = 2.67 SE = 0.18	$n = 29$ Mean = 2.75 SE = 0.24	F = 0.07 $p = 0.936$ ns
Saplings (1.5 m height ≤ 5.0 cm dbh)	25 m^2	$n = 25$ Mean = 2.52 SE = 0.32	$n = 34$ Mean = 2.38 SE = 0.20	$n = 45$ Mean = 2.11 SE = 0.11	F = 1.18 $p = 0.31$ ns
Seedlings (10 cm ≤ 1.5 m height)	4 m^2	$n = 25$ Mean = 1.68 SE = 0.29	$n = 34$ Mean = 2.21 SE = 0.17	$n = 45$ Mean = 1.56 SE = 0.16	F = 3.21 $p = 0.045$*
Plants (other than tree species)	100 m^2	$n = 25$ Mean = 3.12 SE = 0.45	$n = 34$ Mean = 2.18 SE = 0.36	$n = 45$ Mean = 2.31 SE = 0.27	F = 1.78 $p = 0.174$ ns

SE standard error, *ANOVA* analysis of variance, *ns* not significant, *n* number of plots sampled, *PA* protected area
ns $p > 0.05$; **$p < 0.05$; *Significant at $p = 0.05$

major rivers are generally washed out and eroded by wave action and deplete species richness. The species individual ratio was calculated for each tree size classes as well as for plants, and the analysis has shown that due to species dominance of few species the ratios are quite depressed (Table 5). Similarly, the species abundance distribution (i.e. domination by few species) in the three protected areas and the study site as a whole contributed to low values of Shannon H′. Shannon's index is preferred to be more useful for the purpose of comparison of different habitats. The overall Shannon's value of the study area was 2.19 while the east, south and west protected areas contributed 2.18, 1.99 and 1.95, respectively. Among the three protected areas, the mean number of species also significantly varied for seedlings and trees with dbh ≥ 15.0 cm (timber trees). Means of other two size classes (saplings and poles) as well as plants species did not vary among themselves (Table 6).

3.3 *Species Accumulation Curve*

The species accumulation curves were also constructed to observe the trend in species accumulation for different tree size classes (and also for plants) within the overall study site (Fig. 7). In the overstorey, a total of 25 tree species of ≥ 15.0 cm dbh (timber trees) was accumulated by 88 plots each of 1000-m^2 in size. However, in all tree size classes, accumulations were not regular and in different stages no substantial increase of species accumulation was observed due to domination by few tree species

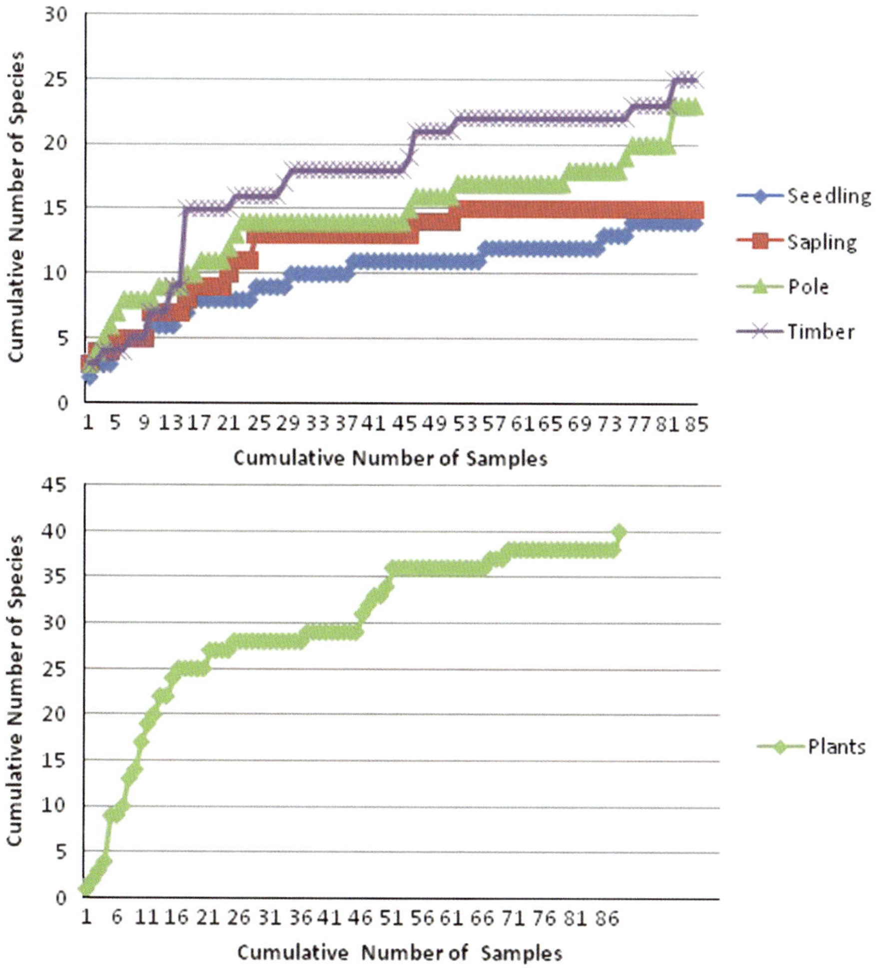

Fig. 7 Species accumulation curves of the protected area (wildlife sanctuary) of Bangladesh Sundarbans constructed for different diameter classes of trees (*above*) and plants (*below*). Seven unidentified plants were also included in the graph (*below*). Note that all curves showed no tendency to flatten out but then increased surprisingly with many plots adding not a single species

(Fig. 7). Similar observations were also made for three protected areas. On the other hand, species-area curves of plant species showed regular accumulation up to a certain point, and then increased surprisingly with many plots adding not a single species (Fig. 7). Species accumulation curves demonstrated by Hughell (1997) for three protected areas of Sundarbans showed an initial increase of species followed by complete flattening out of the curves and captured a total of 37 plant species including trees, whereas Leech and Ali (1997) demonstrated a smoother graph for all of the Sundarbans that captured 48 plant species (including trees).

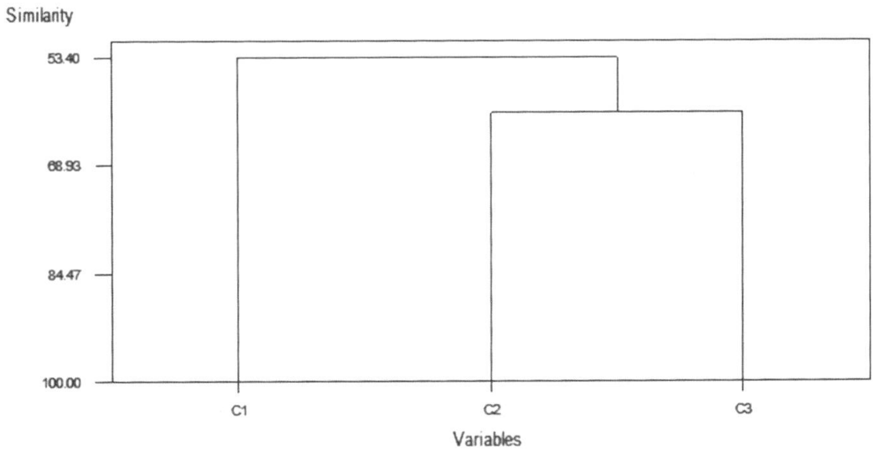

Fig. 8 A dendrogram showing the similarity and dissimilarity in species composition between the protected areas (wildlife sanctuaries) of the study site. Sunderland South Protected Area (South PA) (*C2*) and the West PA (*C3*) are the two most similar and adjacent sites clustered together while the East PA (*C1*) is quite different from the other two PAs and distantly placed. Presence/absence of species data used for cluster analysis derived from field survey (see text for detail)

3.4 Species Similarity and Composition

A hierarchical cluster analysis of the habitat-wise data was also undertaken to detect some pattern of similarity or dissimilarity in species composition among three protected areas of the study site . The single-linkage clustering was used based on presence/absence of species data for each habitat (protected area) and for trees from 1.5-m height and above. The dendrogram from the analysis is shown in Fig. 8. It shows distinct dissimilarity between the East Protected Area (East PA) and the other two habitats, and close similarity between South PA and West PA that clustered together. The dissimilarity of the East Protected Area (East PA) from the other two habitats validated environmental factors such as fresh water flow and low salinity of the East PA. The overall preferences of most of the species to a particular site determined the representative floristic variation.

3.5 Stand Dynamics and Changes in Dominance Structure

3.5.1 Stand Dynamics

Among the three protected areas, mean density (stems/plot) significantly varied for trees with dbh $\geq$ 15.0 cm, and higher number of trees was recorded in the East Protected Area and lowest in the West Protected Area (Table 7). The mean density of both poles and saplings of the East Protected Area also significantly varied but with lower number of individuals than the other two protected areas. This low number of

Table 7 Comparison of mean density of trees (stems/plot) for three protected areas (wildlife sanctuaries) of Bangladesh Sundarbans

Size class	Plot size (m^2)	Protected Area (wildlife sanctuaries)			ANOVA
		East PA	South PA	West PA	
Timber trees ($\geq$ 15.0 cm dbh)	1,000 m^2	$n = 25$ Mean = 26.52 SE = 2.72	$n = 34$ Mean = 19.79 SE = 1.82	$n = 29$ Mean = 15.03 SE = 2.35	F = 5.97 $p = 0.004$**
Poles (5 cm $\leq$ 15 cm dbh)	100 m^2	$n = 25$ Mean = 24.88 SE = 2.96	$n = 34$ Mean = 37.32 SE = 3.17	$n = 29$ Mean = 31.90 SE = 2.79	F = 4.14 $p = 0.02$*
Saplings (1.5 m height $\leq$ 5.0 cm dbh)	25 m^2	$n = 25$ Mean = 18.32 SE = 3.73	$n = 34$ Mean = 24.24 SE = 3.16	$n = 45$ Mean = 33.39 SE = 2.77	F = 5.55 $p = 0.005$**
Seedlings (10 cm $\leq$ 1.5 m height)	4 m^2	$n = 25$ Mean = 21.88 SE = 4.33	$n = 34$ Mean = 23.38 SE = 3.78	$n = 45$ Mean = 17.91 SE = 2.91	F = 0.72 $p = 0.488$ ns

SE standard error, *ANOVA* analysis of variance, *ns* not significant, *n* number of plots sampled, *PA* protected area
ns $p > 0.05$; **$p < 0.05$; * Significant at $p = 0.05$

individuals (poles and saplings) in the East Protected Area could be due to effects of cyclone Sidr in the year 2007 with delayed recovery owing to severe injury. However, most of the damaged canopy trees (crown damage) by cyclone Sidr within the East Protected Area have been successfully sprouted. Unlike overstorey canopy trees, both poles and saplings were possibly confined to stem or bole injury and also crushed down by fallen tree crowns.

The majority of the forest is two storied with scattered emergent trees attaining a height of up to 20 m. The mean stem diameter of canopy trees at breast height across the three protected areas was 19.75 ($n = 1{,}662$, SE = 0.166). As per ocular estimation, the dominant vegetation type in the East Protected Area was generally $\geq$ 15 m in height with a gradual decrease observed towards the west (Table 8).

Canopy photography using fish-eye lens was the only way to measure the canopy structure, and it supports more general impressions of the forest canopy of the three protected areas (Fig. 9). However, percent canopy closure of the three protected areas was measured and the mean was statistically significant (Table 8). As shown in Table 8, the most dominant vegetation type (*Heritiera-Excoecaria or Excoecaria-Heritiera*) in the East Protected Area was mostly with about 90 % in mean canopy closure, while in the South and West Protected Area, the mean canopy closure was 83 % and 75 %, respectively.

3.5.2 Dominance Structure: Importance Value Index

The importance value index (IVI) was calculated as the sum of the relative frequency, relative density and relative basal area for each species in three protected areas. In terms of contribution of species to IVI, all protected areas were dominated by

Table 8 Percent canopy closure of the three protected areas of Bangladesh Sundarbans

Protected area	Vegetation types	Canopy height	Sample size (n)	Mean canopy closure (%)	ANOVA
East protected area	*Heritiera-Excoecaria or Excoecaria-Heritiera*	≥ 15 m	30	Mean = 89.0 (SE = 1.37)	F = 12.03 $p = 0.000$***
South protected area	*Excoecaria- Ceriops*	10 m $\leq$ 15 m	30	Mean = 83.0 (SE = 1.76)	
West protected area	*Ceriops- Excoecaria*	5 m $\leq$ 10 m	30	Mean = 75.0 (SE = 2.60)	

SE standard error, *ANOVA* analysis of variance, *n* number of plots sampled
***$p < 0.001$

Fig. 9 Fish-eye photography of the three protected areas of Bangladesh Sundarbans showing general impressions of the forest canopy. Note vegetation density and canopy openness

Excoecaria agallocha (Euphorbiaceae) with highest IVI followed by *Heritiera fomes* (Sterculiaceae) and *Sonneratia apetala* (Sonneratiaceae) (Table 9). But in the West Protected Area, *Excoecaria agallocha* was followed by *Xylocarpus mekongensis* (Meliaceae) *and Xylocarpus granatum* (Meliaceae), and *Heritiera fomes* (Sterculiaceae) was in the fourth position in IVI ranking (Table 9). Apart from

Table 9 Dominant species in terms of contribution to important value index (IVI) in three protected areas of Bangladesh Sundarbans (ranked in declining order). The index is calculated as the sum of the relative frequency (rF), relative density (rD) and relative basal area (rBA) for each species. Only tree species with dbh ≥ 15.0 cm are taken into consideration

Family	Scientific name	Local name	IVI	Ranking
A. East protected area				
Euphorbiaceae	*Excoecaria agallocha*	Gewa	35.932	1
Sterculiaceae	*Heritiera fomes*	Sundri	35.656	2
Sonneratiaceae	*Sonneratia apetala*	Keora	9.521	3
Avicenniaceae	*Avicennia officinalis*	Baen	5.258	4
Meliaceae	*Xylocarpus mekongensis*	Passur	4.277	5
Leguminosae	*Pongamia pinnata*	Karamja	1.289	6
Leguminosae	*Cynometra ramiflora*	Singra	1.283	7
Euphorbiaceae	*Sapium indicum*	Urmui	1.078	8
Meliaceae	*Xylocarpus granatum*	Dhundul	1.074	9
Anacardiaceae	*Lannea coromandelica*	Kapla/Jiga	1.073	10
Myrtaceae	*Eugenia fruticosa*	Jam	0.856	11
Meliaceae	*Amoora cucullata*	Amoor	0.539	12
Rhizophoraceae	*Kandelia candel*	Bhatkhati	0.539	13
Verbenaceae	*Premna corymbosa*	Kunail	0.539	14
Tamaricaceae	*Taimrix indica*	Nuna jhau	0.539	15
Leguminosae	*Dalbergia melanoxylon?*	Kata bahoi	0.539	16
B. South protected area				
Euphorbiaceae	*Excoecaria agallocha*	Gewa	35.649	1
Sterculiaceae	*Heritiera fomes*	Sundri	24.173	2
Sonneratiaceae	*Sonneratia apetala*	Keora	15.978	3
Avicenniaceae	*Avicennia marina*	Sada baen	8.299	4
Rhizophoraceae	*Rhizophora mucronata*	Garjan, jhana	4.530	5
Meliaceae	*Xylocarpus granatum*	Dhundul	2.405	6
Meliaceae	*Xylocarpus mekongensis*	Passur	2.098	7
Avicenniaceae	*Avicennia alba*	Maricha baen	1.941	8
Avicenniaceae	*Avicennia officinalis*	Baen	1.605	9
Leguminosae	*Pongamia pinnata*	Karamja	1.108	10
Rhizophoraceae	*Bruguiera gymnorrhiza*	Kakra	0.995	11
Rhizophoraceae	*Rhizophora apiculata*	Garjan, jhana	0.711	12
C. West protected area				
Euphorbiaceae	*Excoecaria agallocha*	Gewa	73.217	1
Meliaceae	*Xylocarpus mekongensis*	Passur	10.866	2
Meliaceae	*Xylocarpus granatum*	Dhundul	4.978	3
Sterculiaceae	*Heritiera fomes*	Sundri	4.541	4
Anacardiaceae	*Lannea coromandelica*	Kapla, jiga	1.839	5
Avicenniaceae	*Avicennia officinalis*	Baen	1.808	6
Rhizophoraceae	*Bruguiera gymnorrhiza*	Kakra	1.151	7
Myrtaceae	*Eugenia fruticosa*	Jam	0.804	8
Sonneratiaceae	*Sonneratia apetala*	Keora	0.793	9

the individual species, both the East and South Protected Areas were more similar in dominance structure and composition than the West Protected Area.

Thus in terms of species composition, diversity and forest structure, this study shows that the forest of the protected areas of Bangladesh Sundarbans seem to be largely controlled by fresh water flow from the upstream in the eastern side and high salinity towards the west, along with other associated microenvironments.

3.6 Comparison of Species Diversity with Other Studies

There are limitations in comparative studies due to differences in method of study, intensity of sampling, and plot positioning and diameter class of trees. The floristic composition of the Sundarbans is rich compared to many other mangroves of the world. Prain (1903) recorded 334 species of plants belonging to 245 genera and 75 families for the entire Sundarbans and adjoining areas (Bangladesh and India). This was possibly a taxonomic survey to capture a great number of species. Heining (1892) reported 70 species from 34 families for the entire Sundarbans (India and Bangladesh). Chaffey and Sandom (1985) listed 66 plant species from Bangladesh Sundarbans with 37 families. Leech and Ali (1997) recorded 48 plant species from the Sundarban Reserve Forest. A study made by Bangladesh Center for Advance Studies (BCAS) registered 37 plant species from three protected areas (wildlife sanctuaries) of Sundarbans (Rosario 1997). However, this study recorded 63 identified plant species with 15 unidentified from the three protected areas. As this study has just finished, these 15 unidentified species have been preserved for identification. It is already stated that streamside, particularly secondary and tertiary streams, were much higher in plants species richness than the 'forest proper', and the species area curve was progressively going upward without flattening off. This indicates more samples are required to capture the maximum number of plant species from the streamside.

4 Conclusions and Management Perspective

Three existing wildlife sanctuaries are all at the sea face (extreme south) and it would therefore seem appropriate to protect an area of the Sundarbans at the northern edge; an area of which others suggest would have generally lower salinity levels. It is recommended that compartments 1 and 2, adjacent to the northern boundary of the East Protected Area, be protected. This would broaden the variants of the general mangrove ecosystem that are included in sanctuary area. These two compartments have a higher stocking, basal area and species composition than the Sundarbans as a whole. Similarly, the South Protected Area should be extended northwards for better protection of the forest. Management of these areas should incorporate research activities to determine successional changes due to dynamic ecological processes occurring quite rapidly.

The structure and composition of the East Protected Area have shown quite different from other the two protected areas as revealed from different parameters of the study. However, all these protected areas are representative of the particular environmental conditions such as salinity and silt deposition. Vegetation height and density also differentiated the three protected areas. As the Sundarbans is naturally regenerated, there is no need for an artificial plantation inside even in the vacant areas. Previous attempts to increase population size by planting additional individuals of mangrove species showed unsuccessful and was not environmental friendly. A homogenous flora (such as plantation) does not support a diversified terrestrial fauna due to limited number of ecological niches. Most of the injured canopy trees (crown damage) by cyclone Sidr (in 2007) within the protected area have been successfully sprouted and vacant areas are also being gradually replenished by the original species. Fortunately, all these protected areas of Sundarbans are located out-of-the-way and thus free from any encroachment and other major threats. But management of these protected areas in terms of capacity building (housing, marine transports, wireless communication and skill development) is imperative.

Acknowledgements We appreciate the financial assistance provided by the European Union under Sundarban Environmental And Livelihoods Security (SEALS) Project and cooperation received from the Forestry Department, Bangladesh for field study. The field staff of Forest Ranges inside Sundarbans is also greatly acknowledged. This study was performed under the Technical Assistance Programme of the SEALS project.

References

Aksornkoae S, Maxwell GS, Havanond S, Panichsuko S (1992) Plants in mangroves. Chalongrat, Bangkok

Anderson JE (1991) A conceptual framework for evaluating and quantifying naturalness. Conser Biol 5(3):347–352

Anon. (1998) The Sundarbans world heritage site. Forest Department and IUCN, Dhaka, p 8

Anon. (2012) Life, lives, livelihoods: the European Commission's work on biodiversity and development. DG EuropeAid, European Commission

Brower JE, Zar JH, Von Ende CN (1990) Field and laboratory methods for general ecology, 3rd edn. Wm. C. Brown Publishers, Dubuque

Brown ND (1990) Dipterocarp regeneration in tropical rain forest gaps of different sizes. DPhil Thesis, University of Oxford

Callow P (1992) Can ecosystems be healthy? Critical consideration of concepts. J Aq Ecosys Health 1:1–5

Canonizado JA, Hossain MA (1998) Integrated forest management plan for the Sundarbans reserved forest. GOB/WB Forest Resources Management Project, TA Component. Mandala Agricultural Development Corporation, Dhaka, Bangladesh

Chaffey DR, Sandom JH (1985) Sundarbans forest inventory project, Bangladesh. A glossary of vernacular plant names and a field key to the trees. Overseas Development Administration, England, p 23

Chaudhuri AB, Naithani HB (1985) A comparative survey of tropical mangrove forests of Sundarbans and Andamans, part 1. International Book Distributors, Dehra Dun, p 41

Choudhury AM (1968) Working plan of the Sundarbans forest division for the period from 1960–61 to 1979–80. East Pakistan Government Press, Dhaka, p 82

Curtis SJ (1933) Working Plan for the Forest of the Sundarbans Divisions (April 1931 to March 1951), vol 1. Bengal Government Press, Calcutta, p 175

De Leo GA, Levin S (1997) The multifaceted aspects of ecosystem integrity. Conserv Ecol 1(1):3. http://www.consecol.org/vol1/iss1/art3

FAO (1994) Mangrove forest management guidelines. FAO Forestry Paper 117

Forestal (1960) Forest inventory 1958–59 Sundarban forest. Forestal International Incorporated, Canada, Oregon

Heining RL (1892) Working plan of Sundarbans government forest, Khulna and 24-Pargana District. Bengal Secretariat Press, Bengal

Hughell DA (1997) Nature resources survey. Forest Resources Management Project, Bangladesh, GOB/WB

Iftekhar MS, Saenger P (2007) Vegetation dynamics in the Bangladesh Sundarbans mangroves: a review of forest inventories. Wetlands Ecol Manage. Springer Science+Business Media B.V.

Leech J, Ali SS (1997) Extended natural resources survey: part III- report. Forest Resources Management Project, GOB/WB

Magurran AE (1988) Ecological diversity and its measurement. Chapman and Hall, New York

Meyer WB, Turner IIBL (1992) Human population growth and global land-use/cover change. Ann Rev Ecol Syst 23:39–61

Minitab Inc. (1994) Minitab user's guide. Release 15 for Windows

Prain D (1903) Flora of Sundarbans. Records of the Botanical Survey of India. Periodical Expert Book Agency (reprinted in 1979), Delhi, pp 231–370

Rapport DJ (1995) Ecosystem health: exploring the territory. Ecosys Health 1(1):5–13

Rapport DJ (1998) Defining ecosystem health. In: Rapport DJ, Costanza R, Epstein PR, Gaudet CL, Levins R (eds) Ecosystem health. Blackwell Science, Malden, pp 18–33

Revilla JAV, Ahmad IU, Hossain MA (1998) Final report: forest inventory of the Sundarbans reserved forest, vol 2. Forest Resources Management Project. Ministry of Environment and Forest, Dhaka

Rosario EA (1997) Draft conservation management plan of the wildlife sanctuaries in the Sundarbans forests. GOB/WB Forest Resources Management Project, Mandala Agricultural Development Corporation, Dhaka, Bangladesh

Siddiqi NA (2001) Mangrove forestry in Bangladesh. Institute of Forestry & Environmental Sciences, University of Chittagong, Chittagong, p 201

Sutherland WJ (2000) The conservation handbook: research, management and policy. Blackwell Science, UK

Tomlinson PB (1986) The botany of mangroves. Cambridge University Press, Cambridge

Whitmore TC, Brown ND, Swaine MD, Kennedy D, Goodwin-Bailey CI, Gong WK (1993) Use of hemispherical photographs in forest ecology: measurement of gap size and radiation totals in a Bornean tropical rain forest. J Trop Ecol 9:131–151

Wood GB (1990) Ground-sampling methods used to inventory tropical mixed moist forests. Forest Ecol Manag 35:199–206

Mangrove Fauna of Asia

Muhammad Nawaz Rajpar and Mohamed Zakaria

Contents

Abstract Mangrove is a plant community of salt tolerant plant species which grow within transitional or inter-tidal zones of coastal, estuary and riverine areas of tropical and subtropical regions where rivers drain into the sea. They are highly productive habitat for a variety of fauna such as birds, fishes, reptiles, amphibians, mammals and aquatic as well as terrestrial invertebrates. The occurrence of higher diversity of fauna could be due to richness of food resources and diversity of vegetation, i.e. they provide ideal foraging and breeding sites and also shelter for these wide array of animals. Mangrove fauna are an important component of the food web and play a significant role in the mangrove ecosystem. Unfortunately, despite such a richness in animal communities, mangrove areas are still declining at an alarming rate day by day due to human activities. The habitat loss has seriously caused threats to different mangrove dependent animals such as birds, mammals, reptiles and amphibians, i.e., extinct

M. N. Rajpar · M. Zakaria (✉)
Faculty of Forestry, Universiti Putra Malaysia,
43400 UPM Serdang, Selangor Darul Ehsan, Malaysia
e-mail: mzakaria@putra.upm.edu.my

I. Faridah-Hanum et al. (eds.), *Mangrove Ecosystems of Asia*,
DOI 10.1007/978-1-4614-8582-7_8,

and critically endangered species. The current information on the various fauna such as reptiles, mammals, invertebrates and fishes in Asia's mangrove ecosystem is not sufficient. In the future, more research is required to determine the various aspects of fauna such as species richness, diversity, distribution and the association of fauna with water quality, food resources and habitats to explore the ways and means to conserve the fauna in and around mangrove areas.

1 Introduction

Mangrove is a plant community of salt tolerant species such as trees, shrubs, palms, and ferns which grow within transitional or inter-tidal zones of coastal, estuary and riverine areas of tropical and subtropical regions where rivers drain into the sea (Macintosh and Ashton 2002; FAO 2007; Rajkumar et al. 2009; Naidoo 2009; Wan Juliana et al. 2010; Zhou et al. 2010). Worldwide, mangrove vegetation covers an area of 137,760 km^2 (Giri et al. 2011) and the higher percent of mangrove vegetation occurs between 5 N and 5 S, 32 N and 38 S (Morrisey et al. 2010; Friess et al. 2012). Out of the total mangrove areas, around 42.0 % occur in Asia, 21.0 % in Africa, 15.0 % in North/Central America, 12.0 % in Oceania, and 11.0 % in South America (Giri et al. 2010). Mangrove areas are considered as a wetland that include waterways such as estuaries, creeks, canals, lagoons, backwaters, mudflats, salt pans and islands (Kjerfve 1990; Wan Juliana et al. 2010). However, vegetation composition and structure of mangrove areas may vary from area to area or region to region depending on soil condition, rainfall pattern, and inflow of river water into the sea. Mangrove areas are rich in tree diversity that comprise about 69 true tree species that represent 27 genera and 20 families (Selvam et al. 2004). Mangrove trees are divided into three categories such as (i) true mangroves or mangrove exclusive, (ii) mangrove non-exclusive and (iii) mangrove associated (Wan Juliana et al. 2010).

1.1 Mangrove Exclusive or Major Mangroves

Tree species that are mainly restricted to the intertidal zone within deep water and high salinity include *Avicennia lanata*, *A. marina*, *A. officinalis*, *Bruguiera cylindrical*, *B. gymnorrhiza*, *B. parviflora*, *B. sexangula*, *Ceriops decandra*, *C. tagal*, *Kandelia candel*, *Lumnitzera littorea*, *Nypa fruticans*, *Rhizophara apiculata*, *R. stylosa*, *Sonneratia alba*, and *S. caseolaris*, etc. (Saenger et al. 1983; Tomlinson 1986; Rotaquio et al. 2007).

1.2 Mangrove Non-exclusive or Minor Mangroves

Tree species that tolerate low salinity and are restricted to shallow water where salinity fluctuates from time to time include *Acrostichum aureum*, *A. speciosum*, *Aegiceras corniculatum*, *A. floridum*, *Excoecaria agallocha*, *Heritiera littoralis*, *Osbornia octodonta*, *Pemphis acidula*, *Planchonella obovata*, *Scyphiphora hydrophyllacea*, and *Xylocarpus granatum*, etc. (Saenger et al. 1983; Tomlinson 1986; NTG 2002; Rotaquio et al. 2007).

1.3 Mangrove Associated

Plant species that grow with mangrove tree species include grasses, epiphytes, pteridophytes, bryophytes, and parasitic plants, e.g. *Acanthus ilicifolius*, *A. volubilis*, *Barringtonia asiatica*, *B. racemosa*, *Brownlowia tersa*, *Cerbera odallam*, *C. manghas*, *Clerodendrum inerme*, *Crinum asiaticum*, *Dolichandrone spathacea*, *Inocarpus edulis*, *Hibiscus titiaceus*, *Morinda citrifolia*, etc. (Tomlinson 1986; Rotaquio et al. 2007).

2 Threats to Mangrove Fauna

Being an important habitat for wildlife species, about one-third of the mangrove area has been lost over the past two decades due to land reclamation, conversion into agricultural fields, deforestation, aquaculture, and urbanization, i.e. coastal development (Macintosh and Ashton 2002; Penha-Lopes et al. 2011). The habitat loss and degradation have caused serious threats to wildlife species, particularly bird species, i.e. 40.0 % of the bird population has been decreased in mangrove areas (Sandilyan et al. 2010). In addition, 100 % turtles species, 43 % crocodiles species, 20 % fish species, 37 % mammal species, 21 % bird species and 43 % amphibians that directly or indirectly depend on mangroves, mudflats and estuarine habitats are globally critically endangered (Millennium Ecosystem Assessment 2005).

The major driven factors that cause population decrease of mangrove fauna are habitat loss (Tidwell and Allan 2001), over exploitation (FAO 2009), coastal degradation and climate change (Gracia and Rosenberg 2010), organic pollution and toxic contamination (Naylor et al. 2000; Gracia and Rosenberg 2010). These factors cause habitat degradation, reduced food resources, and destroy nursery grounds that ultimately affect the fauna population of mangrove habitats.

3 Economic Importance of Mangrove

The diverse vegetation structure and composition of mangroves with denser foliage (Stafford-Deitsch 1996) has created different layers of vegetation that offer heterogeneous habitats which support a variety of marine, freshwater and terrestrial wildlife species. The mangrove vegetation interacts with aquatic, inshore, upstream and terrestrial ecosystems that also form intertidal habitats for birds, fishes, reptiles, amphibians, mammals and a variety of aquatic invertebrates such as insects, mollusk, i.e. gastropods (snails) and bivalves (mussels), crabs, shrimps, oysters, sponges, barnacles, and polychaetes (worms).

Some of the animals depend on mangrove areas their whole lives while others utilize them only during specific periods such foraging, shelter and breeding (Hutchings and Recher 1982; Hutchings and Saenger 1987; Yaez-Arancibia et al. 1988; Macintosh and Ashton 2002; Northern Territory Government 2002; Thu and Populus 2007; Han 2011; Talaat et al. 2011; Nyanti et al. 2012).

Mangrove fauna can be divided into three inhabitants such as (i) aquatic animals, i.e., fishes, amphibians, (ii) semi-aquatic animals (i.e., reptiles, amphibians and birds) and (iii) terrestrial animals based on their living behaviour (i.e., mammals and birds). These animal communities utilize mangrove areas for their daily activities such as foraging, breeding, and loafing. These animals play a significant role in the management of mangrove forests and in balancing nature in and around the mangrove areas (Spalding et al. 2010; Nyanti et al. 2012).

3.1 Mangroves as a Habitat for Avifauna

Mangrove areas are a favorable habitat for a variety of waterbirds (i.e., the bird species that entirely depend on water for a variety of activities such as foraging, nesting, loafing and moulting) as well as terrestrial birds (i.e., bird species that do not entirely depend on water but may visit some time in search of food, shelter and perch) (Table 1). This is due to the diversity of habitats such as mangroves, mudflats, estuaries and richness of food resources which includes fishes (Blaber 2000; White and Potter 2004; Martin 2005), turtles (Blanco et al. 1991), snake (Guinea et al. 2004), amphibians (Kathiresan and Bingham 2001; Nagelkerken et al. 2008), mammals (Nijman 2000; Angelici et al. 2005; Bordignon 2006), and invertebrates such as gastropods (Plaziat 1984; Jiang and Li 1995), bivalves (Lebata and Primavera 2001), prawn (Kenyon et al. 2004), nekton (Minello et al. 2003), crabs (Ashton 2002; Skov and Hartnoll 2002) and insects (Nagelkerken et al. 2008).

Noske (1996) reported that mangroves support more than 200 bird species that utilize mangrove forest, mudflats, estuaries and adjacent areas. Avifauna of mangrove can be divided into four categories including (i) aerial feeders, (ii) waders, (iii) surface/diving foragers and (iv) foliage gleaners.

3.1.1 Aerial Feeders or Sallying Birds

The bird species that catch their prey on wing i.e. Fish Eagles (Fig. 1) and Kites (Fig. 2) (Accipitridae), Wood Swallow (Artamidae) (Fig. 3), Swallows (Hirundinidae), Bee-eaters (Meropidae) (Fig. 4), Kingfisher (Alcidinidae) (Fig. 5), and Swiftlet (Apodidae) always hovers on mudflats and mangrove areas in search of

Table 1 List of bird species recorded in different areas of mangrove and adjacent habitat

Family	Scientific name	Common name	Habitat	Authors
Accipitridae	*Accipter cooperii*	Cooper's Hawk	Mangrove	Gough et al. 1998; South Florida Aquatic Environments 2013
Accipitridae	*Buteo jamaicensis*	Red-tailed Hawk	Mangrove	Gough et al. 1998; SFAE 2013
Accipitridae	*Buteo lineatus*	Red-shouldered Hawk	Mangrove	Gough et al. 1998; SFAE 2013
Accipitridae	*Cathartes aura*	Turkey Vulture	Mangrove	Gough et al. 1998; SFAE 2013
Accipitridae	*Circus cyaneus*	Marsh Hawk	Mangrove and adjacent areas	Gough et al. 1998; SFAE 2013; Rajpar and Zakaria 2010
Accipitridae	*Coragyps atratus*	Black Vulture	Mangrove	Gough et al. 1998; SFAE 2013
Accipitridae	*Haliaeetus leucocephalus*	Bald Eagle	Mangrove	Gough et al. 1998; SFAE 2013
Accipitridae	*Haliaeetus leucogaster*	White-bellied Fish-Eagle	Mangrove and adjacent areas	Rajpar and Zakaria 2010
Accipitridae	*Haliastur Indus*	Brahminy Kite	Mangrove and adjacent areas	Lim et al. 2001; Norhayati et al. 2009; Rajpar and Zakaria 2010
Accipitridae	*Pandion haliaetus*	Osprey	Mangrove	Gough et al. 1998; SFAE 2013
Alcedinidae	*Ceyx erithacus*	Oriental Dwarf Kingfisher	Mangrove	Norhayati et al. 2009
Alcedinidae	*Halcyon chloris*	Collared Kingfisher	Mangrove and adjacent areas	Rajpar and Zakaria 2010
Alcedinidae	*Pelargopsis capensis*	Stork-billed Kingfisher	Mangrove and adjacent areas	Rajpar and Zakaria 2010
Alcedinindae	*Halcyon pileata*	Black-capped Kingfisher	Mangrove	Norhayati et al. 2009
Alcidinidae	*Todirhamphus chloris*	Collard Kingfisher	Mangrove	Lim et al. 2001
Anatidae	*Anas Penelope*	Eurasian Wigeon	Mangrove	Sabir 2011
Anatidae	*Anas platyrhynchos*	Mallard	Mangrove	Gough et al. 1998; SFAE 2013
Anatidae	*Anas acuta*	Pintail	Mangrove	Gough et al. 1998; SFAE 2013
Anatidae	*Aythya affinis*	Lesser Scaup	Mangrove	Gough et al. 1998; SFAE 2013
Anatidae	*Aythya valisineria*	Canvasback	Mangrove	Gough et al. 1998; SFAE 2013
Anatidae	*Tadorna tadorna*	Common Shelduck	Mangrove	Sabir 2011
Anhingidae	*Anhinga anhinga*	Snakebird/Darter/Water Turkey	Mangroves	Gough et al. 1998; SFAE 2013
Anhingidae	*Anhinga melanogaster*	Oriental Darter	Mangrove and adjacent areas	Rajpar and Zakaria 2010
Apodidae	*Collocalia esculenta*	White-bellied Swiftlet	Mangrove and adjacent areas	Rajpar and Zakaria 2010
Aramidae	*Aramus guarauna*	Limpkin	Mangrove	Gough et al. 1998; SFAE 2013
Ardeidae	*Ardea cinerea*	Grey Heron	Mangrove	Norhayati et al. 2009; Sabir 2011

Table 1 (continued)

Family	Scientific name	Common name	Habitat	Authors
Ardeidae	*Ardea herodias*	Great Blue Heron	Mangrove	Gough et al. 1998; SFAE 2013
Ardeidae	*Ardea purpurea*	Purple Heron	Mangrove	Lim et al. 2001; Han 2011; Sabir 2011
Ardeidae	*Ardeola grayi*	Indian Pond Heron	Mangrove	Sabir 2011
Ardeidae	*Bubulcus ibis*	Cattle Egret	Mangrove and adjacent areas	Macintosh and Ashton 2002
Ardeidae	*Butorides striatus*	Little Heron	Mangrove and adjacent areas	Lim et al. 2001; Norhayati et al. 2009; Rajpar and Zakaria 2010
Ardeidae	*Butorides virescens*	Green Heron	Mangrove	Acevedo and Aide 2008
Ardeidae	*Casmerodius albus*	Great Egret	Mangroves	Gough et al. 1998; Han 2011; Sabir 2011; SFAE 2013
Ardeidae	*Egretta alba*	Great Egret	Mangrove and adjacent areas	Rajpar and Zakaria 2010
Ardeidae	*Egretta eulophotes*	Chinese Egret	Mangrove and adjacent areas	Rajpar and Zakaria 2010
Ardeidae	*Egretta garzetta*	Little Egret	Mangrove and adjacent areas	Lim et al. 2001; Rajpar and Zakaria 2010; Han 2011; Sabir 2011
Ardeidae	*Egretta gularis*	Western Reef Egret	Mangrove	Sabir 2011
Ardeidae	*Egretta intermedia*	Intermediate Egret	Mudflats, mangrove and adjacent areas	Han 2011; Rajpar and Zakaria 2010
Ardeidae	*Egretta rufescens*	Reddish Egret	Mangroves	Gough et al. 1998; SFAE 2013
Ardeidae	*Egretta thula*	Snowy Egret	Mangrove mudflats	Macintosh and Ashton 2002
Ardeidae	*Ixobrychus eurhythmus*	Schrenck's Bittern	Mangrove and adjacent areas	Rajpar and Zakaria 2010
Ardeidae	*Nyctanassa violacea*	Yellow Crowned Night Heron	Mangroves	Gough et al. 1998; SFAE 2013
Ardeidae	*Nycticorax caledonicus*	Rufous Night-Heron	Mangrove and adjacent areas	Rajpar and Zakaria 2010
Ardeidae	*Nycticorax nycticorax*	Black-crowned Nightheron	Mangrove	Lim et al. 2001; Sabir 2011
Artamidae	*Artamus leucorhynchus*	White-breasted Wood-Swallow	Mangrove and adjacent areas	Rajpar and Zakaria 2010
Burhinidae	*Burhinus recurvirostris*	Great Thick-knee	Mangrove	Sabir 2011
Campephagidae	*Lalage nigra*	Pied Thriller	Mangrove and adjacent areas	Rajpar and Zakaria 2010
Charadriidae	*Charadrius alexandrinus*	Kentish Plover	Mangrove	Sabir 2011
Charadriidae	*Charadrius dubius*	Little Ringed Plover	Mangrove and Adjacent areas	Rajpar and Zakaria 2010; Sabir 2011
Charadriidae	*Charadrius hiaticula*	Common Ringed Plover	Mangrove and adjacent areas	Rajpar and Zakaria 2010; Sabir 2011
Charadriidae	*Charadrius leschenaultia*	Greater Sand Plover	Mangrove	Sabir 2011

Table 1 (continued)

Family	Scientific name	Common name	Habitat	Authors
Charadriidae	*Charadrius mongolus*	Mongolian Plover	Mangrove	Sabir 2011
Charadriidae	*Pluvialis fulva*	Pacific Golden-Plover	Mudflats, mangrove and adjacent areas	Lim et al. 2001; Rajpar and Zakaria 2010
Charadriidae	*Pluvialis squatarola*	Grey Plover	Mangrove	Sabir 2011
Charadriidae	*Vanellus indicus*	Red-wattled Lapwing	Mangrove	Sabir 2011
Chloropseidae	*Aegithina tiphia*	Common Iora	Mangrove and adjacent areas	Rajpar and Zakaria 2010
Ciconiidae	*Leptoptilos javanicus*	Lesser Adjutant	Mangrove and adjacent areas	Norhayati et al. 2009; Rajpar and Zakaria 2010
Ciconiidae	*Mycteria cinerea*	Milky Stork	Mangrove	Macintosh and Ashton 2002
Columbidae	Ptilinopus jambu	Jambu Fruit-Dove	Mangrove and adjacent areas	Rajpar and Zakaria 2010
Columbidae	*Treron curvirostra*	Thick-billed Green Pigeon	Mangrove	Norhayati et al. 2009
Columbidae	*Treron olax*	Little Green-Pigeon	Mangrove and adjacent areas	Rajpar and Zakaria 2010
Columbidae	*Treron vernans*	Pink-necked Green Pigeon	Mangrove and adjacent areas	Rajpar and Zakaria 2010
Coraciidae	*Eurystomus orientalis*	Dollar Bird	Mangrove and adjacent areas	Rajpar and Zakaria 2010
Cuculidae	*Centropus bengalensis*	Lesser Coucal	Mangrove and adjacent areas	Rajpar and Zakaria 2010
Cuculidae	*Centropus sinensis*	Greater Coucal	Mangrove and adjacent areas	Rajpar and Zakaria 2010
Cuculidae	*Chrysococcyx minutillus*	Malayan Bronze Cuckoo	Mangrove and adjacent areas	Rajpar and Zakaria 2010
Cuculidae	*Coccyzus minor*	Mangrove Cuckoo	Mangrove	Acevedo and Aide 2008
Cuculidae	*Phaenicophaeus diardi*	Black-bellied Malkoha	Mangrove and adjacent areas	Rajpar and Zakaria 2010
Cuculidae	*Phaenicophaeus sumatranus*	Chestnut-bellied Malkoha	Mangrove and adjacent areas	Rajpar and Zakaria 2010
Dicaeidae	*Dicaeum curentatum*	Scarlet-backed Flowerpecker	Mangrove and adjacent areas	Rajpar and Zakaria 2010
Dormatidae	*Dromas ardeola*	Crab Plover	Mangrove	Sabir 2011
Estrilidae	*Lonchura fuscans*	Dusky Munia	Mangrove and adjacent areas	Rajpar and Zakaria 2010
Falconidae	*Falco columbarius*	Peregrine Falcon	Mangrove	Gough et al. 1998; SFAE 2013
Haematopodidae	*Haematopus ostralegus*	Eurasian Oystercatcher	Mangrove	Sabir 2011
Hemiprocnidae	*Hemiprocne comata*	Whiskered Treeswift	Mangrove and adjacent areas	Rajpar and Zakaria 2010
Hirundinidae	*Hirundo tahitica*	Pacific Swallow	Mangrove and adjacent areas	Lim et al. 2001; Rajpar and Zakaria 2010
Icteridae	*Lonchura punctulata*	Nutmeg Mannikin	Mangrove	Acevedo and Aide 2008
Icteridae	*Quiscalus niger*	Great Antillean Grackle	Mangrove	Acevedo and Aide 2008

Table 1 (continued)

Family	Scientific name	Common name	Habitat	Authors
Laniidae	*Hemipus hirundinaceus*	Black-winged Flycatcher-Shrike	Mangrove and adjacent areas	Rajpar and Zakaria 2010
Laniidae	*Hemipus picatus*	Bar-winged Flycatcher-Shrike	Mangrove and adjacent areas	Rajpar and Zakaria 2010
Laridae	*Chroicocephalus genei*	Slender-billed Gull	Mangrove	Sabir 2011
Laridae	*Chroicocephalus ridibundus*	Black-headed Gull	Mangrove	Sabir 2011
Laridae	*Ichthyaetus ichthyaetus*	Greater Black-headed Gull	Mangrove	Sabir 2011
Laridae	*Larus cachinnans*	Caspian Gull	Mangrove	Sabir 2011
Laridae	*Larus canus*	Common Gull	Mangrove	Sabir 2011
Laridae	*Larus heuglini*	Heuglini Gull	Mangrove	Sabir 2011
Laridae	*Sterna albifrons*	Little Tern	Mangrove	Lim et al. 2001
Meropidae	*Merops philippinus*	Blue-tailed Bee-eater	Mangrove and adjacent areas	Rajpar and Zakaria 2010
Meropidae	*Merops viridus*	Blue-throated Bee-eater	Mangrove and adjacent areas	Rajpar and Zakaria 2010
Muscicapidae	*Cyornis rufigastra*	Mangrove Blue-Flycatcher	Mangrove and adjacent areas	Norhayati et al. 2009; Rajpar and Zakaria 2010
Muscicapidae	*Muscicapa danurica*	Asian Brown Flycatcher	Mangrove	Norhayati et al. 2009
Nectariniidae	*Anthreptes malacensis*	Brown-throated Sunbird	Mangrove and adjacent areas	Norhayati et al. 2009; Rajpar and Zakaria 2010
Nectariniidae	*Anthreptes simplex*	Plain Sunbird	Mangrove and adjacent areas	Rajpar and Zakaria 2010
Nectariniidae	*Nectarinia calcostetha*	Copper-throated Sunbird	Mangrove	Norhayati et al. 2009
Nectariniidae	*Nectarinia jugularis*	Olive-backed Sunbird	Mangrove and adjacent areas	Rajpar and Zakaria 2010
Nectariniidae	*Nectarinia sperata*	Purple-throated Sunbird	Mangrove and adjacent areas	Rajpar and Zakaria 2010
Pachycephalidae	*Pachycephala grisola*	Mangrove Whistler	Mangrove and adjacent areas	Norhayati et al. 2009; Rajpar and Zakaria 2010
Parulidae	*Dendroica discolor*	Prarie Warbler	Mangrove	Acevedo and Aide 2008
Parulidae	*Mniotilta varia*	Black-and-white Warbler	Mangrove	Acevedo and Aide 2008
Parulidae	*Protonotaria citrea*	Prothonotary Warbler	Mangrove	Acevedo and Aide 2008
Parulidae	*Seiurus aurocapilla*	Ovenbird	Mangrove	Acevedo and Aide 2008
Parulidae	*Seiurus noveboracensis*	Northern Waterthrush	Mangrove	Acevedo and Aide 2008
Parulidae	*Setophaga ruticilla*	American Redstart	Mangrove	Acevedo and Aide 2008

Table 1 (continued)

Family	Scientific name	Common name	Habitat	Authors
Pelicanidae	*Pelecanus crispus*	Dalmation Pelican	Mangrove	Sabir 2011
Pelicanidae	*Pelecanus onocrotalus*	White Pelican	Mangrove	Sabir 2011
Pelicanidae	*Pelecanus occidentalis*	Brown Pelican	Mangrove	Gough et al. 1998; SFAE 2013
Phalacrocoracidae	*Phalacrocorax carbo*	Great Cormorant	Mangrove and adjacent areas	Rajpar and Zakaria 2010; Sabir 2011
Phalacrocoracidae	*Phalacrocorax niger*	Little Cormorant	Mangrove and adjacent areas	Rajpar and Zakaria 2010
Phalacrocoracidae	*Phalacrocorax nigrogularis*	Socotra Cormorant	Mangrove	Sabir 2011
Phoenicorpteridae	*Phoenicopterus ruber roseus*	Greater Flamingo	Mangrove	Sabir 2011
Picidae	*Blythipicus rubiginosus*	Maroon Woodpecker	Mangrove and adjacent areas	Rajpar and Zakaria 2010
Picidae	*Dendrocopus moluccensis*	Sunda Woodpecker	Mangrove	Lim et al. 2001
Picidae	*Dinopium javanense*	Common Flameback	Mangrove and adjacent areas	Norhayati et al. 2009; Rajpar and Zakaria 2010
Picidae	*Dryocopus javensis*	White-bellied Woodpecker	Mangrove and adjacent areas	Rajpar and Zakaria 2010
Picidae	*Picoides moluccensis*	Brown-capped Woodpecker	Mangrove and adjacent areas	Rajpar and Zakaria 2010
Picidae	*Picumnus innominatus*	Speckled Piculet	Mangrove and adjacent areas	Rajpar and Zakaria 2013
Podiepedidae	*Pdiceps nigricollis*	Black-necked Grebe	Mangrove	Sabir 2011
Podiepedidae	*Pediceps cristatus*	Great Crested Grebe	Mangrove	Sabir 2011
Psittacidae	*Loriculus galgulus*	Blue-crowned Hanging Parrot	Mangrove and adjacent areas	Rajpar and Zakaria 2010
Rallidae	*Eulabeornis castaneoventris*	Chestnut Rail	Mangrove	MNS-BCC 2005
Rallidae	*Gallinule chloropus*	Eurasian Moorhen	Mangrove	Gough et al. 1998; SFAE 2013
Recurvirostridae	*Himantopus himantopus*	Black-winged Stilt	Mangrove	Sabir 2011
Recurvirostridae	*Recurvirostra himantopus*	Avocet	Mangrove	Sabir 2011
Rhipiduridae	*Rhipidura javanica*	Pied Fantail	Mangrove and adjacent areas	Lim et al. 2001; Norhayati et al. 2009; Rajpar and Zakaria 2010
Scolopacidae	*Actitis hypoleucos*	Common Sandpiper	Mangrove	Sabir 2011
Scolopacidae	*Arenaria interpres*	Ruddy turnstone	Mangrove	Sabir 2011
Scolopacidae	*Calidris alba*	Sanderling	Mangrove	Sabir 2011
Scolopacidae	*Calidris alpina*	Dunlin	Mangrove	Sabir 2011
Scolopacidae	*Calidris ferruginea*	Curlew Sandpiper	Mangrove	Sabir 2011
Scolopacidae	*Calidris minuta*	Little Stint	Mangrove	Sabir 2011

Table 1 (continued)

Family	Scientific name	Common name	Habitat	Authors
Scolopacidae	*Limicola falcinellus*	Broad-billed Sandpiper	Mangrove	Sabir 2011
Scolopacidae	*Limosa lapponica*	Bar-tailed Godwit	Mangrove	Sabir 2011
Scolopacidae	*Limosa limosa*	Black-tailed Godwit	Mangrove	Sabir 2011
Scolopacidae	*Numenius arquata*	Eurasian Curlew	Mangrove	Sabir 2011
Scolopacidae	*Numenius phaeopus*	Whimbrel	Mudflats, mangrove and adjacent areas	Lim et al. 2001; Rajpar and Zakaria 2010; Sabir 2011
Scolopacidae	*Philomachus pugnax*	Ruff	Mangrove	Sabir 2011
Scolopacidae	*Tringa brevipes*	Grey-tailed Tattler	Mangrove and Adjacent areas	Rajpar and Zakaria 2010
Scolopacidae	*Tringa guttifer*	Nordmann's Greenshank	Mangrove and adjacent areas	Rajpar and Zakaria 2010
Scolopacidae	*Tringa hypoleucos*	Common Sandpiper	Mudflats, Mangrove and adjacent areas	Lim et al. 2001; Rajpar and Zakaria 2010; Han 2011
Scolopacidae	*Tringa nebularia*	Greenshank	Mangrove	Sabir 2011
Scolopacidae	*Tringa stagnatilis*	Marsh Sandpiper	Mangrove	Sabir 2011
Scolopacidae	*Tringa totanus*	Common Redshank	Mudflats, Mangrove and adjacent areas	Lim et al. 2001; Rajpar and Zakaria 2010; Sabir 2011
Scolopacidae	*Xenus cinereus*	Terek Sandpiper	Mangrove and adjacent areas	Rajpar and Zakaria 2010; Sabir 2011
Sittidae	*Sitta frontalis*	Velvet-fronted Nuthatch	Mangrove and adjacent areas	Rajpar and Zakaria 2010
Sternidae	*Chlidonias hybridus*	Whiskered Tern	Mangrove	Sabir 2011
Sternidae	*Chlidonias leucopterus*	White-winged Black Tern	Mangrove	Sabir 2011
Sternidae	*Gelochelidon nilotica*	Gull-billed Tern	Mangrove	Sabir 2011
Sternidae	*Hydroprogne caspia*	Caspian Tern	Mangrove	Sabir 2011
Sternidae	*Onychoprion anaethetus*	Bridled Tern	Mangrove	Sabir 2011
Sternidae	*Sterna hirundo*	Common Tern	Mangrove	Sabir 2011
Sternidae	*Sterna repressa*	White-cheeked Tern	Mangrove	Sabir 2011
Sternidae	*Sternula saundersi*	Saunders Little Tern	Mangrove	Sabir 2011
Sternidae	*Thalasseus bengalensis*	Lesser Crested Tern	Mangrove	Sabir 2011
Sternidae	*Thalasseus bergii*	Great Crested Tern	Mangrove	Sabir 2011
Sternidae	*Thalasseus sandvicensis*	Sandwich Tern	Mangrove	Sabir 2011
Strigidae	*Bubo virginianus*	Great Horned Owl	Mangrove	Gough et al. 1998; SFAE 2013

Table 1 (continued)

Family	Scientific name	Common name	Habitat	Authors
Strigidae	*Strix varia*	Barred Owl	Mangrove	Gough et al. 1998; SFAE 2013
Strigidae	*Tyto alba*	Barn Owl	Mangrove	Gough et al. 1998; SFAE 2013
Sturnidae	*Aplonis panayensis*	Philippine Glossy Starling	Mangrove and adjacent areas	Rajpar and Zakaria 2010
Sturnidae	*Gracula religiosa*	Hill Myna	Mangrove and adjacent areas	Rajpar and Zakaria 2010
Sylviidae	*Orthotomus ruficeps*	Ashy Tailorbird	Mangrove and adjacent areas	Lim et al. 2001; Norhayati et al. 2009; Rajpar and Zakaria 2010
Sylviidae	*Orthotomus sericeus*	Rufous-tailed Tailorbird	Mangrove and adjacent areas	Rajpar and Zakaria 2010
Thraupidae	*Spindalis portoricensis*	Puerto Rican Spindalis	Mangrove	Acevedo and Aide 2008
Threskiornithidae	*Ajaia ajaja*	Roseate Spoonbill	Mangrove	Gough et al. 1998; Macintosh and Ashton 2002; SFAE 2013
Threskiornithidae	*Eudocimus albus*	White Ibis	Mangrove	Gough et al. 1998; SFAE 2013
Threskiornithidae	*Eudocimus ruber*	Scarlet Ibis	Mangrove mudflats	Macintosh and Ashton 2002
Threskiornitidae	*Platalea leucordia*	Spoonbill	Mangrove	Sabir 2011
Timaliidae	*Macronus gularis*	Striped Tit-Babbler	Mangrove and adjacent areas	Rajpar and Zakaria 2010
Trochilidae	*Anthracothorax viridis*	Green Mango	Mangrove	Acevedo and Aide 2008
Turdidae	*Copsychus malabaricus*	White-rumped Shama	Mangrove and adjacent areas	Rajpar and Zakaria 2010
Turdidae	*Copsychus saularis*	Oriental Magpie Robin	Mangrove and adjacent areas	Rajpar and Zakaria 2010
Turdidae	*Turdus plumbeus*	Red-legged Trush	Mangrove	Acevedo and Aide 2008
Tyrannidae	*Tyrannus caudifasciatus*	Loggerhead Kingbird	Mangrove	Acevedo and Aide 2008
Tyrannidae	*Tyrannus dominicensis*	Gray Kingbird	Mangrove	Acevedo and Aide 2008
Virenidae	*Vireo altiloquus*	Black-whiskered Vireo	Mangrove	Acevedo and Aide 2008
Zosteropidae	*Zosterops palpebrosa*	Oriental White-eye	Mangrove	Norhayati et al. 2009

Fig. 1 White-bellied Fish Eagle—*Haliaeetus leucogaster* hovers over mangrove area in search of food

Fig. 2 Brhaminy Kite—*Haliaster indus* hovers over mangrove area in search of food

Fig. 3 White-breasted Woodswallow—*Artamus leucorynchus* perching on the dead branches of *Rhizophora* sp

Fig. 4 Blue-throated Bee-eater—*Merops viridis* sitting in wire near mangrove area

food such as fishes, birds, monkeys, snakes, and insects. Raptor birds such eagles, hawks, kites and osprey extensively prey on fishes, small birds, small mammals, reptiles, amphibians and large invertebrates (Ridgely and Greenfield 2001; Solano-Ugalde et al. 2009; Alava et al. 2011) while swallows, bee-eaters and swiftlets catch flying insects on the wing over mangrove areas and roost within mangrove areas.

3.1.2 Wader Birds

These are a group of waterbird species that wade in shallow water (Fig. 6) to catch different food resources such as fishes, prawns, mollusks, crustaceans, polychaetes and other invertebrates during low tides or in soft mud such as Egrets, Herons, Bitterns (Ardeidae), Finfoots (Heliornithidae), Plovers (Charadriidae), Oystercatchers (Haematopodidae), Sandpipers, Curlews, Shanks (Fig. 7), Tattlers (Fig. 8), Stints, Ruffs, Godwits, Knots, Dowitchers, Turnstones, Whimbrel, Snipes, Oystercatchers (Scolopacidae), Stilts and Avicets (Recurvirostridae), Phalaropes (Phalaropidae), Gulls, Terns and Noddys (Laridae), Spoonbills, Ibis, and Storks (Ciconiidae), Frigate birds (Fregatidae) and Famingos (Phoenicopteridae) (Fig. 9). These bird species utilize mangrove areas for foraging, roosting, nesting and shelter from harsh weather and hide cover from predators. Habitat selection among bird species may vary depending upon the nature of food selection, shape of the bill and location of food resources. It has been reported that mangrove forests may harbour a variety of water as well as terrestrial bird through offering safe habitats, foraging and loafing sites (Jayson 2001; Laakkonnen 2003; Berg and Angel 2006; Carvajal and Alava 2007; Saari and Ibrahim 2001).

3.1.3 Surface/Diving Foragers

Some bird species forage on the surface of water and sometimes dive into deep water to catch their prey especially fishes, amphibians, aquatic invertebrates, and

Fig. 5 Different Kingfisher species loafing in mangrove area

vegetable matter. For example, Pelicans (Pelecanidae) (Fig. 10), Ducks and Goose (Anatidae) mostly swim on the surface of water to forage small fishes, amphibians, aquatic invertebrates, and vegetable matter while Cormorants (Phalacrocoracidae), Darters (Anhingidae) (Fig. 11), Loons (Gaviidae), and Grebes (Podicipedidae) dive into deep water, particularly river beds, in search of food, mainly fishes and aquatic invertebrates such as mollusks.

Fig. 6 **a** Different wading birds searching for food in mudflat area. **b** Whimbrel—*Numenius phaeopus* foraging in shallow water during low tides. **c** Whimbrel—*Numenius phaeopus* loafing on mangrove tree

3.1.4 Foliage and Bark Gleaners

These are mostly terrestrial bird species which prefer to use mangrove vegetation, i.e. trees, shrubs, palms, and ferns for foraging, perching, nesting and roosting

Fig. 7 Shank species—*Tringa spp*. perch on soft mud under Rhizophora sp. during low tides

Fig. 8 Grey–tailed Tattler—*Tringa brevipes* loaf on the roots of Rhizophora sp. during low tides

Fig. 9 Greater Flamingo—*Phoenicopterus roseus* foraging in shallow water of estuary

Fig. 10 Juvenile of Great White Pelican—Pelecanus onocrotalus loafing on dead wood fallen into water

Fig. 11 Oriental Darter—*Anhinga melanogaster* perching on the dead branches of *Rhizophora* sp

(Fig. 12) such as woodpeckers (Picidae), Tailorbirds, Warblers, Flyeaters (Sylviidae), Flycatchers (Muscipcapidae), Trush, Shama and Robins, (Turdidae), Nuthatch (Sittidae), Sunbirds Spiderhunters (Nectariniidae), Pigeons (Columbidae), Owls (Strigidae), Cuckoos and Malkohas (Cuculidae), Parrots (Psittacidae), Tits (Paridae), Orioles (Oriolidae), Drongos (Dicruridae), Ioras (Chloropseidae), Flycatcher Shrikes (Campephagidae) and Pittas (Pittidae). Some of them are frugivorous birds that feed on fruits such as pigeons and parrots, insectivorous birds that feed on insects such as woodpeckers, robins, warblers, tits, and nectarivorous birds that nip on the nectar such as sunbirds and spider-hunters, and carnivorous birds that forage on other animals such as owls.

Bird species are a bioindicator of a mangrove ecosystem and play a significant role in the management of vegetation. They control the population of insect pest that cause the defoliation among trees and reduce their growth and also cause damage to the seeds. For example, insect eating birds such as tailorbirds, shrikes, flycatchers, iroas, and robins prey on different insect species such as caterpillars, beetles, bugs, and aphids that may cause the defoliation, bark damage that vigorously decreased

Fig. 12 Different foliage and bark gleaning birds resting in mangrove area

the productivity and health of trees. Sunbirds and spider-hunters play a vector role in pollination, i.e., they transfer pollen from one flowering tree to another thus increasing the process of pollination that ultimately increases the seed production. Raptors such as eagles, falcons, and hawks prey on mammals such as monkeys and squirrels that foraged on fruits and tender leaves in the mangrove. Waterbirds are predators of fishes, amphibians, reptiles and a variety of aquatic invertebrates. They control their population and balance the mangrove and mudflat ecosystem. In addition, they are also an important source of food for other animals such as snakes, lizards, fishes, and crocodiles.

3.2 Mangrove as a Habitat for Reptiles

Mangroves are an ideal habitat and are rich in reptile fauna, which include snakes, turtles, crocodiles and alligators. The turtle species found in the mangrove area include Loggerhead Turtle—*Caretta caretta,* Green Sea Turtle—*Chelonia mydas,* Ornate Diamondback Terrapin—*Malaclemys terrapin macrospilota* (Laakkonnen 2003; Boykin 2004; SPGMEC 2013), Mangrove Diamondback Terrapin—*Malaclemys terrapin rhizophorarum* (Burke 2000; Laakkonen 2003; Boykin 2004), Hawksbill Sea Turtle—*Eretmochelys imbricate,* Atlantic Ridley Sea Turtle—*Lepidochelys kempii* (Laakkonnen 2003), Olive Ridley—*Lepidochelys olivacea*, and Leatherback turtle—*Dermochelys coriacea* (SPGMEC 2013). These turtle species utilize mangrove areas, estuaries and creeks for foraging and breeding purposes due to the richness and diversity of plankton and benthic food resources. They use sandy beaches for breeding purpose.

Mangrove areas are also rich and diverse in snake fauna which include, e.g., Dog-faced Water Snake—*Cerebus rynchops* (Lim et al. 2001; Han 2011), File Snake—*Acrochordus granulates* (Lim et al. 2001), Mangrove Snake—*Boiga dendrophila* (Norhayati et al. 2009), Mangrove Pit-Viper—*Trimeresurus purpureomaculatus* (Lim et al. 2001), Mangrove Skin—*Emoia atrocostata* (Lim et al. 2001; Norhayati et al. 2009; Han 2011) and Green Pit Viper—*Vipera trimeresurus* (Macintosh and Ashton 2002). These snake species prey on a variety of animals such as birds, amphibians, small mammals and are also eaten by fishes, crocodiles and eagles.

Only a few crocodile species exist in mangroves, estuarine, and adjacent rivers, e.g., Saltwater/Estuary Crocodile—*Crocodylus porosus* (Fig. 13) (Macintosh and Ashton 2002; Foote 2013), Common Caiman—*Caiman crocodylus* (Macintosh and Ashton 2002) and Marsh Crocodile—*Crocodylus palustris* (Fig. 14) (SPGMEC 2013). These crocodile species prey on a wide array of animals such as birds, fishes, snakes and mammals.

Mangrove areas are also home to a few lizard species such as Mangrove Monitor Lizard—*Varanus indicus* (Fig. 15) and Malaysian Water Monitor Lizard—*Varanus salvator* (Fig. 16) (Lim et al. 2001; Macintosh and Ashton 2002; NTG 2002; Norhayati et al. 2009). These are predators of different animals such as birds, amphibians and small reptiles.

Fig. 13 Saltwater/Estuary Crocodile—*Crocodylus porosus* resting in mangrove roosts in shallow water

Fig. 14 Marsh Crocodile—*Crocodylus palustris* taking a sunbath on soft mud

Fig. 15 Mangrove Monitor Lizard—*Varanus indicus* search for food near mangrove roots in shallow water

Fig. 16 Malaysian Water Monitor Lizard—*Varanus salvator* resting in grasses near a mangrove area

Fig. 17 Giant Toad—*Bufo marinus* resting on soft soil

Fig. 18 Mangrove Frog—*Fejervarya cancrivora* resting on mangrove root

3.3 Mangrove as a Habitat for Amphibians

Only a few species of frogs occur in mangrove forests including Giant Toad—*Bufo marinus* (Fig. 17), Tree Frog—*Osteopilus septentrionalis* and Mangrove Frog—*Fejervarya cancrivora* (Fig. 18) (Dicroglossidae) Macintosh and Ashton 2002;

Fig. 19 Long-tailed Macaque—*Macaca fascicularis* sitting on mangrove tree

Fig. 20 Crab-eating Macaque—*Macaca fascicularis* swimming in mangrove area

Wright et al. 2004; Satheeshkumar 2011). The occurrence of such a few number of amphibians could be due to high salt contents of the water. Mangrove frogs are predators that may eat almost every small living thing such as insects (e.g., beetles, bees, ants, termites, crickets and bugs), snails, smaller toads, prawns, and fishes.

3.4 Mangrove as a Habitat for Mammals

Mangrove forests are rich in mammal species such as White-tailed Deer—*Odocoileus virginianus*, Key Deer—*Odocoileus virginianus clavium*, Bengal Tiger—*Panthera tigris*, Leopard—*Panthera pardus*, Spotted Deer—*Axis axis*, Long-tailed Macaque—*Macaca fascicularis* (Fig. 19), Crab-eating Macaque—*Macaca fascicularis* (Fig. 20), White-faced monkey—*Cebus capucinus*, Malaysian Proboscis Monkey—*Nasalis larvatus* (Fig. 21), Wild Pigs—*Sus scrofa* and Mousedeer—*Tragulus* sp., Long-tongued Nectar Bat—*Macroglossus minimus*, Lesser Dog-faced

Fig. 21 Malaysian Proboscis Monkey—*Nasalis larvatus* sitting on trees

Source: http://borneoclimbanddive.com/wp-content/uploads/2013/03/Proboscis-Monkey-18.jpg

Fruit Bat—*Cynopterus brachyotis*, Marsh Rabbit—*Sylvilagus palustrias*, Cotton Rat—*Sigmodon hispidus*, and Marsh Rat—*Oryzomys palustris* (Table 2). In addition, Bottle-nosed Dolphin—*Tursiops truncates*, Gangetic Dolphin—*Platanista gangetica*, Common Dolphin—*Delphinus delphis,* Manatees—*Trichechus* spp., Smooth Otter—*Lutrogale perspicillata,* Small-clawed Oriental Otter—*Amblonyx cinereus* and Manatees—*Trichechus manatus* are often observed swimming in canals, coastal rivers, and other waters close in proximity to mangroves (FAO 1984, 1994; Hogarth 1999; Ng and Sivasothi 2001; Laakkonnen 2003; Warne 2013).

Mammals are a major source of food for a variety of animals such as raptor birds, snakes, crocodiles, and a significant component of mangrove ecosystems. Frugivore mammals such as monkeys, squirrels, and bats are also important as seed dispersal agents. Herbivorous mammals browse on young shoots of trees, shrubs and other vegetation; hence, they control the growth of shrubs and bushes that may compete for nutrition with trees.

3.5 Mangrove as a Habitat for Fish

Mangrove areas are rich in fish fauna (Table 3). For example, a total of 128 fish species were sampled in mangroves of Paglibao, Philippines (Pinto 1988), 119 fish species have been recorded in the mangrove of Selangor, Malaysia (Chong et al. 1990), 135 fish species in the mangrove estuary of Sikao Creek, Trang Province, Thailand (Prasert et al. 2002), 33 fish species in the mangrove river of Sarawak, Malaysia (Nyanti et al. 2012), and 105 fish species in the mangrove of India (Naik et al. 2013). In addition, mud skippers are one of the fish which live on the mud

Table 2 List of mammal species recorded in different areas of mangrove and adjacent habitat

Family	Scientific Name	Common Name	Habitat	Authors
Cercopithecidae	*Macaca fascicularis*	Lon-tailed Macaque	Mangrove forest	Lim et al. 2001; Norhayati et al. 2009; Han 2011
Cercopithecidae	*Nasalis larvatus*	Proboscis/Long-nose Monkey	Mangrove	Ecology Asia 2001; Macintosh and Ashton 2002; Meijaard et al. 2008
Cercopithecidae	*Macaca fascicularis*	Crab-eating Macaque	Mangrove	Ong and Richardson 2008
Cercopithecidae	*Trachypithecus obscurus*	Dusky Leaf Monkey	Mangrove	Norhayati et al. 2009
Cercopithecidae	*Cercopithecus sp.*	Vervet Monkey	Mangrove	Macintosh and Ashton 2002
Felidae	*Panthera tigris*	Bengal Tiger	Mangrove	Macintosh and Ashton 2002
Mustelidae	*Amblonyx cinereus*	Small-clawed Oriental Otter	Coastal areas	Lim et al. 2001
Mustelidae	*Lutrogale perspicillata*	Smooth Otter	Coastal areas	Lim et al. 2001
Pteropodidae	*Pteropus vampyrus*	Flying Fox	Mangrove Island	Bates et al. 2008; Norhayati et al. 2009
Pterpodidae	*Cynopterus brachyotis*	Lesser Dog-faced Fruit Bat	Mangrove	Lim et al. 2001; Macintosh and Ashton 2002
Pterpodidae	*Eonycteris spelaea*	Cave Fruit Bat	Mangrove	Macintosh and Ashton 2002
Pterpodidae	*Macroglossus minimus*	Long-tongued Nectar Bat	Mangrove	Lim et al. 2001; Norhayati et al. 2009
Sciuridae	*Callosciurus notatus*	Plantain Squirrel	Mangrove	Norhayati et al. 2009
Bovidae	*Tragelaphus spekii*	Swamp Antelope	Mangrove adjacent	Macintosh and Ashton 2002; IUCN 2008
Suidae	*Sus scrofa*	Wild Pig	Mangrove adjacent	Bhattacharya 2011; Macintosh and Ashton 2002

Table 3 List of fish species recorded in different areas of mangrove and adjacent habitat

Family	Scientific Name	Common Name	Habitat	Authors
Acanthuridae	*Acanthurus chirugus*	Doctor Fish	Mangrove root	Osorio et al. 2011
Achiridae	*Trinectes paulistanus*	Slipper Sole Fish	Mangrove root	Osorio et al. 2011
Adrianichthyidae	*Aplocheilus panchax*	Whitespot	Mangrove	Lim et al. 2001
Adrianichthyidae	*Etroplus suratensis*	Green Chromide	The mangrove-lined estuaries	Lim et al. 2001
Ambassidae	*Ambassis gymnocephalus*	Glass Perchlet	Mangrove	Chong et al. 1990
Ambassidae	*Ambassis* interrupta	Long-spined Glass Perchlet	Mangrove river	Nyanti et al. 2012
Ambassidae	*Ambassis kopsii*	Freckled Hawkfish	Mangrove river	Nyanti et al. 2012
Ambassidae	*Ambassis urotaenia*	Banded-tail Glassy perchlet	Mangrove river	Nyanti et al. 2012
Anabantidae	*Anabas* testudineus	Gouramies Fish	Mangrove river	Nyanti et al. 2012
Apogonidae	*Apogon hyalsoma*	Mangrove Cardinalfish	Mangrove roots	Lim et al. 2001
Apogonidae	*Apogon quadrifasciatus*	Cardinal Fish	Mangrove	Chong et al. 1990
Apogonidae	*Sphaeraemia orbicularis*	Orbi Cardinalfish	Mangrove river	Huxham et al. 2004
Ariidae	*Arius sagor*	Engraved Catfish	Mangrove	Chong et al. 1990; Nyanti et al. 2012
Ariidae	*Arius tonggol*	Catfish	Mangrove	Chong et al. 1990
Ariidae	*Arius venosus*	Veined Catfish	Mangrove	Chong et al. 1990
Ariidae	*Osteogeneiosus militaris*	Solider/Sea Catfish	Mangrove	Chong et al. 1990
Atherinidae	*Atherinomorus duodecimalis*	Tropical Silverside	Mangrove	Lim et al. 2001
Atherinidae	Hypoatherina temminckii	Samoan Silverside	Mangrove river	Huxham et al. 2004
Bagridae	*Mystus gulio*	Estuarine Catfish	Mangrove	Lim et al. 2001; Nyanti et al. 2012
Bagridae	*Mystus sp.*	Long Whiskers Catfish	Mangrove	Chong et al. 1990
Batrachoididae	*Batrachoides surinamensis*	Pacuma Toadfish	Mangrove root	Osorio et al. 2011
Belonidae	*Strongylura strongylura*	Spot-tail Needlefish	Mangrove	Lim et al. 2001
Bleniidae	*Antennablennius australis*	Moustached Rockskipper	Mangrove river	Huxham et al. 2004
Carangidae	*Caranx ignobilis*	Giant Trevally	Mangrove river	Chong et al. 1990
Centropomidae	*Lates calcarifer*	Sea Bass	Mangrove	Chong et al.1990; Lim et al. 2001; Nyanti et al. 2012
Centropomidae	*Centropomus spp.*	Snook Fish	Mangrove root	Osorio et al. 2011
Chanidae	*Chanos chanos*	Milkfish	Mangrove river	Huxham et al. 2004
Clariidae	*Claris* sp.	Walking Catfish	Mangrove river	Nyanti et al. 2012
Clupeidae	*Sardinelia melanura*	Tamban	Mangrove	Chong et al. 1990; Lim et al. 2001
Clupeidae	*Tenualosa sinensis*	Hilsa	Mangrove	Chong et al. 1990

Table 3 (continued)

Family	Scientific Name	Common Name	Habitat	Authors
Cynoglossidae	*Cynooglosus macrolepidotus*	Flat Fish	Mangrove	Chong et al. 1990
Elopidae	*Elops machnata*	Elops Machnata	Mangrove river	Nyanti et al. 2012
Engraulidae	*Engraulis japonicus*	Japanese anchovy	Mangrove river	Huxham et al. 2004
Engraulidae	*Stolephorus dubiosus*	Indian Anchovy	Mangrove	Nyanti et al. 2012
Engraulidae	*Stolephorus indicus*	Indian Anchovy	Mangrove	Chong et al. 1990
Engraulidae	*Stolephorus tri*	Spined Anchovy	Mangrove	Chong et al. 1990
Engraulidae	*Thryssa hamiltonii*	Thryssa Fish	Mangrove	Chong et al. 1990
Engraulidae	*Thryssa kammalensis*	Kammal Thryssa	Mangrove	Chong et al. 1990
Ephippidae	*Chaetodipterus faber*	Spade Fish	Mangrove root	Osorio et al. 2011
Gerreidae	*Gerres filamentosus*	Whipfin Silver-biddy	Mangrove river	Nyanti et al. 2012
Gerreidae	*Gerres oyena*	Common Silver-biddy	Mangrove river	Huxham et al. 2004
Gerreidae	*Diapterus sp.*	Mojarra Fish	Mangrove root	Osorio et al. 2011
Gerreidae	*Eucinostomus melanopterus*	Flagfin Mojarra	Mangrove root	Osorio et al. 2011
Gobiidae	*Bathygobius cyclopterus*	Spotted Frillgoby Fish	Mangrove river	Nyanti et al. 2012
Gobiidae	*Boleophthalmus boddarti*	Blue-spotted Mudskipper	Mangrove and mudflats	Lim et al. 2001; Han 2011
Gobiidae	*Callogobius maculipinnis*	Ostrich Goby	Mangrove river	Huxham et al. 2004
Gobiidae	*Glossobitus aureus*	Golden Tank Goby	Mangrove river	Nyanti et al.2012
Gobiidae	*Glossogobius givris*	Goby Fish	Mangrove	Chong et al. 1990
Gobiidae	*Gobius nebulosus*	Shadow Goby	Mangrove river	Huxham et al. 2004
Gobiidae	*Periophthalmodon schlosseri*	Giant Mudskipper	Mangrove and mudflats	Lim et al. 2001; Ecology Asia 2001
Gobiidae	*Periophthalmus novemradiatus*	Dwarf Indian Mudskipper	Estuarine mangrove swamp	Lim et al. 2001; NTG 2002
Gobiidae	*Periophthalmus variabilis*	Dusky-gilled Mudskipper	Landward mangrove	Lim et al. 2001
Gobiidae	*Bathygibius soporator*	Frillfin Goby	Mangrove root	Osorio et al. 2011
Haemulidae	*Pomadasys hasta*	Grunt Fish	Mangrove	Chong et al. 1990
Haemulidae	*Haemulon parra*	Sailor's Grunt	Mangrove root	Osorio et al. 2011
Hemiramphidae	*Hemiramphus far*	Barred Garfish	Mangrove river	Huxham et al. 2004
Hemiramphidae	*Zenarchopterus buffonis*	Striped-nose Halfbeaks	Mangrove	Lim et al. 2001
Hemiramphidae	*Zenarchopterus candovittatus*	Striped-nosed Halfbeak	Mangrove	Chong et al. 1990
Lacepede	*Scomberoides commersonianus*	Queen Fish	Mangrove	Chong et al. 1990

Table 3 (continued)

Family	Scientific Name	Common Name	Habitat	Authors
Leiognathidae	*Leiognathus brevirostris*	Short-nosed Pony Fish	Mangrove	Chong et al. 1990
Leiognathidae	*Leiognathus daura*	Goldstripe Ponyfish	Mangrove	Chong et al. 1990
Leiognathidae	*Secutor insidiator*	Ponyfish	Mangrove	Chong et al. 1990
Lutjanidae	*Lutjanus argentimaculatus*	Mangrove Red Snapper	Mangrove river	Huxham et al. 2004; Nyanti et al. 2012
Lutjanidae	*Lutjanus bohar*	Two-spot Red Snapper	Mangrove river	Huxham et al. 2004
Lutjanidae	*Lutjanus ehrenbergi*	Blackspot Snapper	Mangrove river	Huxham et al. 2004
Lutjanidae	*Lutjanus alexandrei*	Brazilian Snapper	Mangrove root	Osorio et al. 2011
Lutjanidae	*Lutjanus cyanopterus*	Cubera Snapper	Mangrove root	Osorio et al. 2011
Lutjanidae	*Lutjanus jocu*	Dog Snapper	Mangrove root	Osorio et al. 2011
Megalopidae	*Megalops scyprinoides*	Indo-Pacific Tarpon	Mangrove river	Nyanti et al. 2012
Monodactylidae	*Monodactylus argenteus*	Silver Moony	Mangrove	Lim et al. 2001
Monodactylidae	*Monodactylus argenteus*	Silver Moony	Mangrove river	Huxham et al. 2004
Mugilidae	*Liza melinopterus*	Mullet Fish	Mangrove	Chong et al. 1990; Nyanti et al. 2012
Mugilidae	*Liza subviridis*	Greenback Mullet	Mangrove	Chong et al. 1990
Mugilidae	*Mugil cephalus*	Flathead Mullet	Mangrove river	Nyanti et al. 2012
Mugilidae	*Valamugil buchanani*	Blue-tail Mullet	Mangrove river	Nyanti et al. 2012
Mugilidae	*Valamugil seheli*	Bluespot mullet	Mangrove river	Huxham et al. 2004
Mullidae	Upeneus sulphurous	Sulphur Goatfish	Mangrove river	Chong et al. 1990; Huxham et al. 2004
Narrkidae	*Narke capensis*	Onefin Electric ray	Mangrove river	Huxham et al. 2004
Phallostethidae	*Neostethus sp.*	Pariapus Fish	Mangrove	Lim et al. 2001
Platycephalidae	*Platycephalus scaber*	Flathead Fish	Mangrove	Chong et al. 1990
Platychepalidae	*Sorsogona prionata*	Flathead Fish	Mangrove river	Huxham et al. 2004
Plotosidae	*Paraplotosus albilabris*	White-lipped Eeltail Catfish	Mangrove river	Nyanti et al. 2012
Pristigasteridae	*Ilisha megaloptera*	Shad Fish	Mangrove	Chong et al. 1990
Scaridae	*Sparisoma sp.*	Parrot Fish	Mangrove root	Osorio et al. 2011
Scatophagidae	*Scatophagus argus*	Green/Ruby/Red Scat	Mangrove river	Nyanti et al. 2012
Scatophagidae	*Scatophagus sp.*	Scat Fish	Mangrove	Chong et al. 1990

Table 3 (continued)

Family	Scientific Name	Common Name	Habitat	Authors
Sciaenidae	*Johnius belangerii*	Belanger's Croaker	Mangrove	Chong et al. 1990
Sciaenidae	*Johnius trachycephalus*	Leaf-tailed Croaker	Mangrove river	Nyanti et al. 2012
Sciaenidae	*Johnius soldado*	Solider Croaker	Mangrove	Chong et al. 1990
Sciaenidae	*Johnius coiter*	Coiter Croaker	Mangrove	Chong et al. 1990
Sciaenidae	*Otolithes ruber*	Drums or Croaker Fish	Mangrove	Chong et al. 1990
Serranidae	*Epinephelus coioides*	Orange-spotted Grouper	Mangrove	Lim et al. 2001
Serranidae	*Mycteroperca bonaci*	Black Grouper	Mangrove root	Osorio et al. 2011
Sparidae	*Archosargus probatocephalus*	Sheep-head Fish	Mangrove root	Osorio et al. 2011
Sphyraenidae	*Sphyraena putnamiae*	Chevron Barracuda	Mangrove river	Huxham et al. 2004
Sphyraenidae	*Sphyraena sp.*	Barracuda Fish	Mangrove	Chong et al. 1990
Sphyraenidae	*Sphyraena barracuda*	Great Barracuda	Mangrove root	Osorio et al. 2011
Syngnathidae	*Hippocampus reidi*	Slender Seahorse	Mangrove root	Osorio et al. 2011
Synodontidae	*Harpodon nehereus*	Bombay Duck Fish	Mangrove	Chong et al. 1990
Synodontidae	*Saurida undosquamis*	Brushtooth Lizardfish	Mangrove river	Huxham et al. 2004
Tetradontidae	*Arothron reticularis*	Reticulated Putterfish	Mangrove river	Nyanti et al. 2012
Tetraodontidae	*Chelondon fluciatilis*	Green Puffer Fish	Mangrove	Chong et al. 1990
Tetraodontidae	*Tetraodon nigroviridis*	Spotted Green Puffer	Mangrove	Lim et al. 2001
Tetraodontidae	*Sphoeroides testudineus*	Checkered Puffer	Mangrove root	Osorio et al. 2011
Theraponidae	*Terapon jarbua*	Crescent Perch, Tiger Bass	Mangrove river	Nyanti et al. 2012
Toxotidae	*Toxotes sp.*	Archerfish	Mangrove	Chong et al. 1990

Fig. 22 Mullet—*Liza* sp. caught in mangrove area

Fig. 23 Black Snapper—*Apsilus dentatus* caught in mangrove area

flats associated with mangrove shores. This indicates that a variety of fish species use mangrove areas for foraging, i.e. feed on amphipods, isopods, crabs, snails, insects, spiders, copepods, shrimp, and organic matter (Sasekumar et al. 1992; Ewel et al. 1998; Clayton 1993; Macintosh and Ashton 2002; Nagelkerken et al. 2008). Many scientists have reported that an array of fish species extensively use mangrove areas as breeding and nursery sites especially during early juvenile stages (Robertson and Duke 1987; Morton 1990; Chong et al. 1990; Laegdsgaard and Johnson 1995; Dorenbosch et al. 2007; Jaxion-Harm et al. 2012). This could be due to the abundance and richness of food resources (Nyanti et al. 2012) such as invertebrates that inhabit the vegetated area (Lubbers et al. 1990; Schneider and Mann 1991), and richness of benthic fauna (Laegdsgaard and Johnson 2001; Marlena 2005).

Fishes utilize a variety of aquatic habitats (Gratwicke et al. 2006) such as fresh water, brackish water and salt water. The fish fauna of mangrove areas include mud skippers, carangids, clupeids, serranids, mullets, hilsa, seabass, and milkfish (Naik et al. 2013). Some of the common fishes that may occur in mangrove area of southeast asia such as Malaysia include Mullet—*Liza* sp. (Fig. 22), Black Snapper—Apsilus dentatus (Fig. 23), Spottail Needle Fish—*Strongylura*

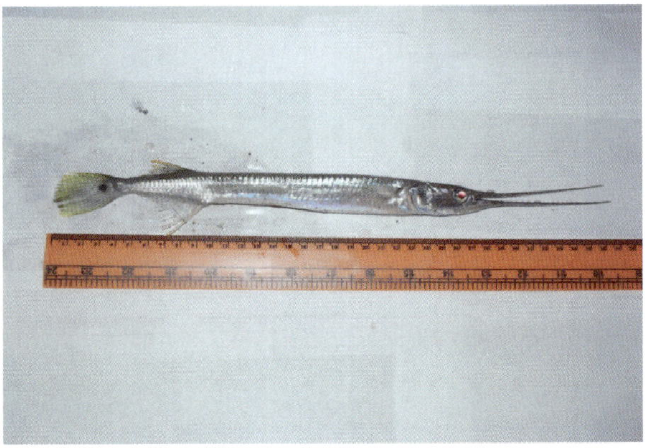

Fig. 24 Spottail Needle Fish—*Strongylura strongylura* caught in mangrove area

Fig. 25 One Spot Snapper—*Lutjanus monostigma* caught in mangrove area

Fig. 26 Orange Spotted Grouper—*Epinephelus coioides* caught in mangrove area

strongylura (Fig. 24), One Spot Snapper—*Lutjanus monostigma* (Fig. 25), Orange Spotted Grouper—Epinephelus coioides (Fig. 26), Snapper Fish—*Lutjanus* sp. (Fig. 27), Cloudy Grouper Fish—Epinephelus erythrurus (Fig. 28), Mangrove Red Snapper—*Lutjanus* sp. (Fig. 29), Garfish—*Hemiramphus* sp. (Fig. 30) and Whipfin Silver-biddy—*Whipfin mojarra* (Fig. 31).

Fig. 27 Snapper Fish—*Lutjanus* sp. caught in mangrove area

Fig. 28 Cloudy Grouperfish—*Epinephelus erythrurus* caught in mangrove area

Fig. 29 Mangrove Red Snapper—*Lutjanus* sp. caught in mangrove area

Fig. 30 Garfish—*Hemiramphus* sp. caught in mangrove area

Fig. 31 Whipfin Silver-biddy—*Whipfin mojarra* caught in mangrove area

However, the fish species composition and distribution may vary from area to area depending on the water quality, aquatic vegetation structure and composition (Pittman et al. 2004), richness of food resources, i.e. invertebrates, vertebrates and vegetable matter (Verweij et al. 2006), detritus (Naik et al. 2013), suitability of breeding sites (Almany 2004), habitat connectivity (Paris et al. 2007; Nakamura et al. 2008) and rate of predation (Chittaro et al. 2005).

The occurrence of a higher diversity of fish species in the mangrove area might be due to the richness and diversity of food resources (Nagelkerken et al. 2002; Aguilar-Perera and Appledoom 2007). The other reason could be due to the complex and extensive root system that reduces the risk of predation and provides safe breeding and nursery grounds, i.e. fishes lay their eggs in extensive roots of mangrove trees (Robertson and Duke 1987; Thayer et al. 1987; Laegdsgaard and Johnson 2001; Huxham et al. 2004; Naik et al. 2012), and after hatching they feed on detritus and other food resources which are easily available in mangrove areas. The water in mangrove areas is turbid and rich in detritus which provide instant food material for juvenile fishes and also reduced predator's vision (Abrahams and Kattenfeld 1997).

Fish is highly nutritious and a major source of human diet, i.e. proteins, vitamins and micronutrients, particularly for low income rural communities (Gracia and Rosenberg 2010). Mangrove supports 75–90 % of the commercial and subsistence fish industries (Lee 1999; Hoyle and Gibbons 2000; NTG 2002). It has been reported that almost 400 million low income people depend on fish as their food (Hortle 2007; Laurenti 2007). Globally, fishes provide around 16 % of animal protein for human beings (Tidwell and Allan 2001). They are economically important for humans, i.e. more than 200 million people directly or indirectly obtain income from the fish industry (Gracia and Newton 1997; FAO 2009).

In addition, fishes are sources of food for a variety of wildlife species such as birds, reptiles and amphibians, mammals, carnivore fishes and invertebrates (Battley et al. 2003; Yu-Seong et al. 2008; Liordos 2010).

3.6 Mangrove as a Habitat for Invertebrate Fauna

Mangrove vegetation has attracted diverse insect species (Macintosh and Ashton 2002) (Table 4) such as Tide-watching Moth—*Aucha velans,* Avicennia Seed Moth—*Autoba alabstrata*, Pneumatophore Moth—*Hymenoptychis sordid,* Common Aquatic Moth—*Erisena mangalis,* Mangrove Moth—*Odites* spp., Avicennia leaf Beetle—*Monolepta* spp., Rhizophora Root Borer—*Coccotrypes rhizophorae*, Sonneratia Weevil—*Rhynchites* sp., Ants—*Crematogaster* sp., and Mangrove Cricket—*Apterombius asahinai* (Lim et al. 2001). These insects play a significant role in the mangrove ecosystem such as pollinator and detritivore, and are a major source of food for birds, fishes and amphibians. It has been reported that *Rhizophora mucronata* and *Avicennia marina* support a higher abundance of crabs (Macintosh et al. 2002; Bosire et al. 2004; Walton et al. 2007).

Mangroves are also ideal habitats for a variety of crustaceans, e.g., prawn species such as Mangrove Snapping Prawn—*Alpheus* spp., Fiddler Shrimp—*Macrobrachium* sp., Glass Shrimp—*Palaemon stylifera,* Red-tailed Prawn—*Penaeus penicillatus*, Edible Prawn—*Metapenaeopsis affinis*, Small White Prawn—*Metapenaeus lysianassa*, Giant Tiger Prawn—*Penaeus monodon* and Mangrove Mud Shrimp—*Wolffogebia* sp. (Chong et al. 1990; Lim et al. 2001; Penha-Lopes et al. 2011; Nyanti et al. 2012) and crab species such as Mud Crab—*Scylla tranquebarica*, Mangrove Mud-hopper—*Microrchestia* sp., Sentinel Crabs—*Macrophthalmus* spp., Shen Crab—*Shenius anomalum*, Fiddler crab—*Uca* sp. (Fig. 32), Orange Mud Crab—*Scylla* spp., Tree Climbing Crab—*Episesarma* spp. and Mangrove Tree-dwelling (Lim et al. 2001; Skov et al. 2002; Penha-Lopes et al. 2009; Han 2011; Nyanti et al. 2012).

Aquatic invertebrates play an important role in the ecology of mangrove because they break down leaf litter that act as fertilizer (Robertson 1986; Smith 1987; Slim et al. 1997), increase surface area of mud through burrowing (Botto and Iribarne 2000; Macintosh and Ashton 2002; Kristensen 2008; Penha-Lopes et al. 2009) and increasing the diffusion rate of gases (Lee 1998; Gribsholt et al. 2003) that ultimately affect the growth and productivity of the mangrove vegetation (Smith et al. 1991; Nielsen et al. 2003; Kristensen and Alongi 2006). In addition, aquatic invertebrates are a major source of food for different animals such as monkeys, birds, snakes, fishes, and even for humans such as oysters and mussels (Macintosh and Ashton 2002) (Fig. 33).

4 Management of Mangrove Fauna

Mangrove is considered as the most productive natural wetland ecosystem on the earth (Ahmed 2008; Jusoff 2008) due to the richness of nutrients, as well as diversity of flora and fauna. They are rich and diverse in fauna species such as birds, mammals, reptiles, amphibians, fishes and aquatic invertebrates (gastropods, bivalves, echinoderms, arthropods, crustaceans, flatworms, etc.). The majority of fauna species

Table 4 List of invertebrate species recorded in different areas of mangrove and adjacent habitat

Family	Scientific name	Common name	Habitat	Authors
Moths				
Gracilariidae	*Caloptia scaedesma*	Kacang Putih Moth	Mangrove	Lim et al. 2001
Lyonetidae	*Lyonetia sp.*	Hammock Moth	Mangrove	Lim et al. 2001
Noctuidae	*Aucha velans*	Tide-watching Moth	Mangrove	Lim et al. 2001
Noctuidae	*Autoba alabstrata*	Avicennia Seed Moth	Mangrove	Lim et al. 2001
Phyllocnistidae	*Phyllocnistis spp.*	Leaf Minor Moth	Mangrove	Lim et al. 2001; Han 2011
Pyralidae	*Hymenoptychis sordida*	Pneumatophore Moth	Mangrove	Lim et al. 2001
Pyralidae	*Erisena mangalis*	Common Aquatic Moth	Mangrove	Lim et al. 2001
Torticidae	*Capua endocypha*	Leaf Binder Moth	Mangrove	Lim et al. 2001
Tortricidae	*Euopicilia sp.*	Bud Worms	Mangrove	Lim et al. 2001
Xyloryctidae	*Odites spp.*	Mangrove Moth	Mangrove	Lim et al. 2001
Beetles				
Attelabidae	*Rhynchites sp.*	Sonneratia Weevil	Mangrove	Lim et al. 2001
Cerambycidae	*Aeolesthes holoserices*	Mangrove Longicorn Beetle		Macintosh and Ashton 2002
Chrysomelidae	*Rhyparida wallacei*	Wallace's Leaf beetle	Mangrove	Lim et al. 2001
Chrysomelidae	*Monolepta spp.*	Avicennia leaf Beetle	Mangrove	Lim et al. 2001; Han 2011
Scolyridae	*Coccotrypes rhizophorae*	Rhizophora Root Borer	Mangrove	Lim et al. 2001
Bugs				
Margarodidae	*Icerya sp.*	Mealy Bug	Mangrove	Lim et al. 2001
Pentatomidae	*Calliphara nobilis*	Shield/Stink Bug	Mangrove	Lim et al. 2001
Pyrrhocoridae	*Dysdercus decussates*	Cotton Stainer Bug	Mangrove	Lim et al. 2001
Other insects				
Formicidae	*Crematogaster sp.*	Ant	Mangrove	Lim et al. 2001
Formicidae	*Oecophylla smaragdina*	Weaver Ants	Mangrove	Macintosh and Ashton 2002
Formicidae	*Polrachis sokolova*	Mangrove Ant	Mangrove	Nielsen, 1997
Gryllidae	*Apterombius asahinai*	Mangrove Cricket	Mangrove	Lim et al. 2001
Hebridae	*Hebrus mangrovensis*	Ground Feeding Insect	Mangrove	Lim et al. 2001
Lampyridae	*Pteroptyx spp*	Firefly	Mangrove	Macintosh and Ashton 2002
Mesoveliidae	*Nereivelia sp.*	Ground Feeding Insect	Mangrove	Lim et al. 2001

Table 4 (continued)

Family	Scientific name	Common name	Habitat	Authors
Pselaphidae	*Mangalobythus furcifer*	Ground Feeding Insect	Mangrove	Lim et al. 2001
Pselaphidae	*Berlara bella*	Ground Feeding Insect	Mangrove	Lim et al. 2001
Saldidae	*Saldoida arrnata*	Ground Feeding Insect	Mangrove	Lim et al. 2001
Terphritidae	*Elleipsa quadrifasciata*	Excocaria Fruit Fly	Mangrove	Lim et al. 2001
Veliidae	*Xenobates sp.*	Mangrove Water Skater	Mangrove	Lim et al. 2001
Ecinoderms				
Amphiuridae	Ophiactis sp.	Mangrove Brittle Star	Mangrove/Mudflats	Lim et al. 2001
Echinasteridae	Echinaster spinulosus	Orange/Sea Starfish	Mangrove roots	Gough et al. 1998
Prawns				
Alpheidae	*Alpheus spp.*	Mangrove Snapping Prawn	Mangrove	Lim et al. 2001
Atyidae	*Caridina sp.*	Mangrove hairy-handed Prawn	Mangrove	Lim et al. 2001
Palaemonidae	*Macrobrachium sp*	Estuarine Prawn	Mangrove	Lim et al. 2001
Palaemonidae	*Macrobrachium sp.*	Fiddler/Tank Shrimp	Mangrove	Chong et al. 1990
Palaemonidae	*Palaemon stylifera*	Glass Shrimp	Mangrove	Chong et al. 1990; Penha-lopes et al. 2011
Penaeidae	*Penaeus penicillatus*	Red-tailed Prawn	Mangrove river	Nyanti et al. 2012
Penaeidae	*Metapenaeopsis stridulans*	Marbled Prawn	Mangrove	Chong et al. 1990
Penaeidae	*Metapenaeopsis affinis*	Edible Prawn	Mangrove	Chong et al. 1990
Penaeidae	*Metapenaeus brevicornis*	Yellow Prawn	Mangrove	Chong et al. 1990
Penaeidae	*Metapenaeus lysianassa*	Small White Prawn	Mangrove	Chong et al. 1990
Penaeidae	*Parapenaeopsis coramandelica*	Red Prawn	Mangrove	Chong et al. 1990
Penaeidae	*Parapenaeopsis gracillima*	Red Prawn	Mangrove	Chong et al. 1990
Penaeidae	*Parapenaeopsis hardwickii*	Spear Shrimp	Mangrove	Chong et al. 1990
Penaeidae	*Parapenaeopsis hungerfordii*	Torpedo Shrimp	Mangrove	Chong et al. 1990
Penaeidae	*Parapenaeopsis maxillipedo*	Sharp Rostrum Prawn	Mangrove	Chong et al. 1990
Penaeidae	*Parapenaeopsis sculptilis*	Rainbow Shrimp	Mangrove	Chong et al. 1990
Penaeidae	*Penaeus indicus*	Indian Prawn	Mangrove	Chong et al. 1990
Penaeidae	*Penaeus merguiensis*	Banana Prawn	Mangrove	Chong et al. 1990
Penaeidae	*Penaeus monodon*	Giant Tiger Prawn	Mangrove	Chong et al. 1990; Nyanti et al. 2012
Penaeidae	*Solenocera subnuda*	Deepwater peneid Prawn	Mangrove	Chong et al. 1990
Penaeidae	*Trachypenaeus fulvus*	Penaeid Shrimp	Mangrove	Chong et al. 1990
Upogebiidae	*Wolffogebia sp.*	Mangrove Mud Shrimp	Mangrove	Lim et al. 2001

Table 4 (continued)

Family	Scientific name	Common name	Habitat	Authors
Crabs				
Diogenidae	*Diogenes sp.*	Mangrove Hermit Crab	Mangrove	Lim et al. 2001
Gecarcinidae	Gecarcinus quadrates	Halloween/Red Land Crab	Mangrove roots	Gough et al. 1998
Grapsidae	*Metopograpsus spp.*	Purple Climber Crab	Mangrove	Lim et al. 2001
Grapsidae	*Sarmatium germaini*	Searmine Crab	Mangrove	Lim et al. 2001
Grapsidae	*Episesarma spp.*	Tree Climbing Crab	Mangrove	Lim et al. 2001
Grapsidae	*Selatium brockii*	Mangrove Tree-dwelling Crab	Mangrove	Lim et al. 2001
Grapsidae	*Metaplax elegans*	Orange Signaller Crab	Mudflats	Lim et al. 2001
Grapsidae	*Scylla* spp.	Orange Mud Crab	Mudflats	Lim et al. 2001
Ligiidae	*Ligia hawaiiensis*	Mangrove Sea Slater	Mangrove	Lim et al. 2001
Limulidae	*Tachypleus tridentatus*	Horseshoe Crab	Mangrove river	Nyanti et al. 2012
Ocypodidae	*Uca burgersi*	Fiddler crab	Mangrove roots	Lim et al. 2001; Skov et al. 2002; Penha-lopes et al. 2009; Han 2011
Ocypodidae	*Dotilla myctiroides*	Solider Crab	Mangrove fringe	Lim et al. 2001
Ocypodidae	*Macrophthalmus spp.*	Sentinel Crabs	Mudflats	Lim et al. 2001
Ocypodidae	*Shenius anomalum*	Shen Crab	Mudflats and mangrove roots	Lim et al. 2001
Portunidae	*Scylla olvacea*	Mud Crab	Mangrove river	Nyanti et al. 2012
Portunidae	*Scylla tranquebarica*	Mud Crab	Mangrove river	Nyanti et al. 2012
Talitridae	*Microrchestia* sp.	Mangrove Mud-hopper	Mangrove	Lim et al. 2001
Flatworms				
Stylochidae	*Limnostylochus sp,*	Reddish-brown Mangrove Flatworm	Wet areas	Lim et al. 2001
Stylochidae	*Meixneria furva*	Large Grey Mangrove Flatworm	Mud lobster mounds	Lim et al. 2001

Fig. 32 Fiddler crab—*Uca burgersi* feeding in mudflat area

Fig. 33 Bivalve mollusc in muddy soil of mangrove area

depend exclusively on mangrove for their whole life while others utilize this area in search of food, shelter and breeding purposes.

Nowadays, mangrove areas are decreasing at an alarming rate due to anthropogenic activities such as over-exploitation for fuel wood and fodder, conversion into urbanization, agricultural fields, aquaculture and fish farming, diversion of rivers due to construction of the water reservoirs which decrease inflow of fresh water into mangrove areas, pollution (oil spills, domestic and industrial sewage) and reclamation of inter-tidal areas (Barter 2002; Barbier and Cox 2004; Mineau et al. 2005). These activities have negatively affected the fauna population of the mangrove. In addition, natural causes can also affect the population of wildlife species such as global warming (Robinson et al. 2009) and diseases outbreak (Rocke et al. 2005; Boyce et al. 2009). Pullin et al. (2013) argued that species extinction and vulnerability is associated with habitat loss and over-exploitation that may cause the loss of ecosystem functions. Mangrove fauna are under severe pressure, and therefore they need protection and proper management to sustain their population in the future.

4.1 Management Through Habitat Restoration

The regeneration of mangrove vegetation in previously exited areas which have been degraded or destroyed can be done through artificial plantations by the relevant agencies (e.g. Forestry Department and NGOs). This will successfully restore the disturbed mangrove ecosystem into its preexisting condition and also strengthen its capacity to adapt change over time. In addition, the areas devoid of vegetation should be planted with economic and ecologically important mangrove tree species on a large scale to compensate the loss of vegetated areas. This will trap sediments, improve water quality and provide a crucial habitat for a variety of fauna.

4.2 Management Through the Involvement of Local Communities

Involvement of local communities residing near the vicinity of mangrove areas and directly dependent on mangrove goods and service for their livelihood is an essential element in sustainable management and conservation of mangrove fauna. A mass awareness programme should be launched in local communities to create awareness among the people about the benefits, economic and social importance of mangrove fauna. Local communities should be involved in the decision of restoration and management activities. Their involvement and collaboration with stakeholder and government agencies will be fruitful and effective for conservation and management of mangrove fauna in the future.

5 Conclusion and Future Perspectives

The current review indicated that mangrove areas are ideal habitat for a variety of fauna such as birds, fishes, reptiles, amphibians, mammals and aquatic as well as terrestrial invertebrates. These fauna are an important component of the food web and play a significant role in the mangrove ecosystem. In this review we focused on the various fauna of mangrove and adjacent area, threats and their important roles in the ecosystem. We have found that these fauna species are facing overwhelming pressure due to habitat loss and degradation. Furthermore, the current information on the various fauna such as reptiles, mammals, invertebrates, and fishes is not sufficient; thus, there is a need to conduct a more detailed research on various aspects of fauna such as species richness, diversity, distribution and the association of fauna with water quality, food resources and habitats. We hope the findings will provide the ways and means to conserve the fauna in and around mangrove areas.

References

Abrahams M, Kattenfeld M (1997) The role of turbidity as constraints. Beh Ecol Soci 40:169–174

Acevedo MA, Aide TM (2008) Bird community dynamics and habitat association in Karst, mangrove and Pterocarpus forest fragments in an urban zone in Puerto Rico. Cari J Sci 44:402–416

Aguilar-Perera and Appledoom (2007) Variation in juvenile fish denssity along the mangrove-seagrass coral reef continuum in SW Puerto Rico. Inter-Res Mari Ecol Pro Seri 348:139–148

Ahmed AIMU (2008) Underlying causes of deforestation and forest degradation in Bangladesh. A report submitted to the Global Forest Coalition (GFC), The Netherlands. University of Dhaka, Bangladesh, p 76

Alava JJ, Saavedra M, Arosemena X, Calle M, Vinueza C, Jimenez PJ, Carvajal R, Vargas FH (2011) Distributional records and potential threats to the common (mangrove) black hawk (*Buteogallus anthracinus subtilis*) in southwestern Ecuador. Bol SAO 20:18–28

Almany GR (2004) Does increased habitat complexity reduce predation and competition in coral reef fish assemblages? Oik 106:275–284

Angelici FM, Politano E, Bogudue AJ, Luiselli L (2005) Distribution and habitat of otters (*Aonyx capensis* and *Lutra maculicollis*) in southern Nigeria. Ital J Zool 72:223–227

Ashton EC (2002) Mangrove sesarmid crab feeding experiments in Peninsular Malaysia. J Exp Mari Biol Ecol 273:97–119

Barbier FB, Cox M (2004) An economic analysis of shrimp farm expansion and mangrove conversion in Thailand. Land Econ 80:389–407

Barter MA (2002) Shorebirds in yellow sea: importance, threats and conservation status. Wetland international global series 9, international wader studies 12, Canberra, Australia

Bates P, Francis C, Gumal M, Bumrungsri S, Walston J, Heaney L, Mildenstein T (2008) *Pteropus vampyrus*. In: IUCN 2012. IUCN red list of threatened species. Version 2012.2. www.iucnredlist.org. Accessed 02 April 2013

Battley PF, Poot M, Wiersma P, Gordon C, Ntiamoa-Baidu Y, Piersma T (2003) Social foraging by waterbirds in shallow coastal lagoons in Ghana. Waterbirds 26:26–34

Berg KS, Angel R (2006) Seasonal roosts of red-lored amazons in Ecuador provide information about population size and structure. J F Ornit 77:95–103

Bhattacharya S (2011) Statue of faunal status with special reference to Bengal Tiger (*Panthera tigris tigris*) in Sundarban Delta. Exploratory animal and medical research, West Bengal Veterinary Alumni Association, West Bengal. India 1:41–52

Blaber SJM (2000) Tropical estuarine fishes: ecology, exploitation and conservation. Blackwell, Oxford

Blanco S, Behler JL, Kostel F (1991) Propagation of the Batagurine turtles batagur baska and Callagur borneoensis at the Bronx zoo. In: Beaman KR, Caporaso F, McKeown S, Graff MD (ed) Proceedings of the 1st international symposium on turtles and tortoises: conservation and captive husbandry, pp 63–65

Bordignon MO (2006) Diet of the fishing bat *Noctilio leporinus* (Linnaeus) (Mammalia, Chiroptera) in a mangrove area of southern Brazil. Revi Brasi de Zool 23:256–260

Bosire JO, Dahdouh-Guebas F, Kairo JG, Cannicci S, Koedam N (2004) Spatial macrobenthic variations in a tropical mangrove bay. Biod Conse 13:1059–1074

Botto F, Iribarne O (2000) Contrasting effects of two burrowing crabs (*Chasmagnathus granulata* and *Uca uruguayensis*) on sediment composition and transport in estuarine environments. Est Coast Shel Sci 51:141–151

Boyce WM, Sandrock C, Kreuder-Johnson C, Kelly T, Cardona C (2009) Avian influenza virus in wild birds: a moving target. Comp Immu Micr Infe Dise 32:275–286

Boykin C (2004) The status and demography of the ornate diamondback terrapin (*Malaclemys terrapin macrospilota*) within the Saint Martins marsh aquatic preserve. http://www.dep.state.fl.us/coastal/sites/stmartins/pub/SMM_Terrapin_Report.pdf. Accessed 28 July 2013

Burke VJ (2000) Diamondback terrapin: a coastal species. Turtle conservation. Mithsonian Institution Press, Washington, DC, pp 168–170

Carvajal R, Alava JJ (2007) Mangrove wetlands conservation project and the shrimp farming industry in Ecuador: lessons learned. World Aqua 38:14–17
Chittaro PM, Usseglio P, Sale P (2005) Variation in fish density, assemblage composition and relative rates of predation among mangrove, seagrass and coral reef habitats. Envir Biol Fish 72:175–187
Chong VC, Sasekumar A, Leh MUC, D'Cruz R (1990) The fish and prawn communities of a Malaysian coastal mangrove system, with comparison to adjacent mud flats and inshore waters. Estu Coast Shel Sci 31:703–722
Clayton DA (1993) Mudskippers. Oceano Mari Biol Ann Rev 31:507–577
Das S, Siddiqi NA (1985) The mangroves and mangrove forests of Bangladesh. Magrove silviculture division bulletin no. 2, Bangladesh forest research institute and UNDP/FAO Project BGD/79/017. Chittagong, Bangladesh
Dorenbosch M, Verberk W, Nagelkerken I, van der Velde G (2007) Influence of habitat configuration on connectivity between fish assemblages of caribbean seagrass beds, mangroves and coral reefs. Inter-Res Mari Ecol Prog Seri 334:103–116
Ecology Asia (2001) Brunei bay, home of monkey with long nose. http://www.ecologyasia.com/html-loc/brunei-bay.htm. Accessed 1 April 2013
Ellis WL, Bell SS (2013) Intertidal fish communities may make poor indicators of environmental quality: lesson from a study of mangrove habitat modification. Ecol Indi 24:421–430
Ewel KC, Twilley RR, Ong JE (1998) Different kinds of mangrove forests provides different goods and services. Glob Ecol Biogeo Lett 7:83–94
FAO (1982) Management and utilization of mangroves in Asia and the Pacific. FAO environmental paper 3. Rome
FAO (1994) Mangrove fauna. http://www.fao.org/forestry/mangrove/3645/en/. Accessed 28 July 2013
Food and Agriculture Organization (2007) The world's mangroves, 1980–2005: a thematic study in the framework of the Global Forest Resources Assessment 2005. FAO of United Nation Rome, Italy. Forestry Paper 153. ISBN: 978-92-5-10586-5. Edition: Illustrated, pp 1–5
Food and Agriculture Organization (2009) The state of world fisheries and aquaculture 2008. pp 176. Rome, Italy. http://www.fao.org/fishery/sofia/en. Accessed 28 July 2013
Foote J (2013) Animals in the mangrove ecosystem. http://www.ehow.com/list_7596229_animals-mangrove-ecosystem.html#page=0. Accessed 10 April 2013
Friess DA, Krauss KW, Horstman EM, Balke T, Bouma TJ, Galli D, Webb EL (2012) Are all intertidal wetlands naturally created equal? Bottlenecks, thresholds and knowledge gaps to mangrove and saltmarsh ecosystems. Biol Revi 87:346–366
Giri C, Ochieng E, Tieszen L, Zhu Z, Singh A, Loveland T, Masek J, Duke N (2010) Status and distribution of mangrove forests of the world using earth observation satellite data. Glob Ecol Biogeog 20:154–159
Giri C, Ochieng E, Tieszen LL, Zhu Z, Singh A, Loveland T, Masek J, Duke N (2011) Status and distribution of mangrove forests of the world using earth observation satellite data. Glob Ecol Biogeo 20:154–159
Gough GA, Sauer JR, Iliff M (1998) *Patuxent bird identification infocenter*. Version 97.1. Patuxent Wildlife Research Center, Laurel, MD. http://www.mbr-pwrc.usgs.gov/id/framlst/infocenter.html. Accessed 28 July 2013
Gracia S, Newton C (1997) Current situation, trends and prospects in world capture fisheries. In: Pikitch E, Hubert D, Sissenwine (ed) Global trends in fisheries management. American fisheries society monograph series, Bethesda, MD, pp 352
Gracia SM, Rosenberg AA (2010) Food security and marine capture fisheries: characteristics, trends, drivers and future perspective. Philos Trans Roy Soci Lond Biol Sci 365:2869–2880
Gratwicke B, Petrovic C, Speight MR (2006) Fish distribution and ontogenetic habitat preferences in non-estuarine lagoons and adjacent reefs. Envi Biol Fish 76:191–210
Gribsholt B, Kostka JE, Kristensen E (2003) Impact of fiddler crabs and plant roots on sediment biogeochemistry in a Georgia saltmarsh. Inter-Res Mari Ecol Prog Seri 259:237–251

Guinea ML, Limpus CJ, Whiting SD (2004) Marine snakes. In: National oceans office. Description of key species groups in the northern planning area. National Oceans Office, Hobart, Australia
Han KL (2011) Mangrove fauna and their adaptations in the Kota Kinabalu wetland. Sab Soci J 28:15–22
Hogarth PJ (1999) The biology of mangroves. Oxford University Press, pp 228
Hortle KG (2007) Consumption and the yield of fish and other aquatic animals from the lower Mekong Basin. Mekong River Commission (MRC) technical paper 16, pp 87
Hoyle ME, Gibbons JW (2000) Use of a marked population of diamondback terrapins (*Malaclemys terrapin*) to determine impacts of recreational crab pots. Chel Conser Biol 3:735–737
Hutchings PA, Recher HF (1982) The fauna of Australian mangroves. Proceedings of Linnean Society of New South Wales 106:83–121
Hutchings P, Saenger P (1987) Ecology of mangroves. University of Queensland Press, Queensland, Australia, p 388
Huxham M, Kimani E, Augley J (2004) Mangrove fish: a comparison of community structure between forested and cleared habitats. Estu Coas Shel Sci 60:637–647
Jaxion-Harm J, Saunders J, Sepight R (2012) Distribution of fish in seagrass, mangroves and coral reefs: life-stage dependent habitat use in Honduras. International J Tropi Biol 60:683–698
Jayson EA (2001) Structure, composition and conservation of birds in Mangalavanam mangroves, Cochin, Kerala. Zoos J 16:471–478
Jiang JX, Li RG (1995) An ecological study of the Mollusca in mangrove areas in the estuary of the Jiulong river. Hydrobiologia 295:213–220
Jussof K, Taha D (2008) *Managing sustainable mangrove forests in Peninsular Malaysia*. J Sust Devel 1:88–96
Kathiresan K, Bingham BL (2001) Biology of mangroves and mangrove ecosystems. Adva Mari Biol 40:81–251
Kenyon RA, Loneragan NR, Manson FJ, Vance DJ, Venables WN (2004) Allopatric distribution of juvenile red-legged banana prawns (*Penaeus indicus* H. Milne Edwards 1837) and juvenile white banana prawns (*Penaeus merguiensis* De Man 1888), and inferred extensive migration, in the Joseph Bonaparte gulf, northwest Australia. J Experi Mari Biol Ecol 309:79–108
Kjerfve B (1990) Manual for investigation of hydrological processes in mangrove ecosystems. UNESCO/UNDP, New Delhi, pp 79
Kristensen E (2008) Mangrove crabs as ecosystem engineers; with emphasis on sediment processes. J Sea Res 59:30–43
Kristensen E, Alongi DM (2006) Control by fiddler crabs (Uca vocans) and plant roots (Avicennia marina) on carbon, iron, and sulfur biogeochemistry in mangrove sediment. Limnol and Oceanog 51:1557–1571
Laakkonnen K (2003) Crystal river state buffer preserve draft ten-year land management plan (2004–2014). Office of coastal and aquatic managed areas. Florida Department of Environmental Protection p 95
Laegdsgaard P, Johnson CR (1995) Mangrove habitat as nurseries: unique assemblages of juvenile fish in subtropical mangroves in eastern Australia. Mar Ecol Prog Ser 126:67–81
Laegdsgaard P, Johnson C (2001) Why do juvenile fish utilize mangrove habitats? J Experi Mari Biol and Ecol 257:229–253
Laurenti G (2007) 1961–2003 Fish and fishery products: world apparent consumption statics based on food balance sheets. FAO Fisheries Circular 821 (9), 429 Italy: FAO
Lebata MJHL, Primavera JH (2001) Gill structure, anatomy and habitat of *Anodontia edulenta*: evidence of endosymbiosis. J Shell Res 20:1273–1278
Lee SY (1998) Ecological role of grapsid crabs in mangrove ecosystems: a review. Mari Fresh Res 49:335–343
Lee SY (1999) Tropical mangrove ecology: physical and biotic factors influencing ecosystem structure and function. Aust J Ecol 24:355–366
Lim KKP, Murphy DH, Morgany T, Sviasothi N, Ng PKL, Soong BC, Tan HTW, Tan HS, Tan TK (2001) A guide to the mangroves of Singapore. Raffles Museum of Biodiversity Research,

National University of Singapore, and Singapore Science Centre, Singapore. Volume 1, animal diversity

Liordos V (2010) Foraging guilds of waterbirds wintering in a Mediterranean coastal wetland. Zool Stud 49:311–323

Lubbers L, Boynton WR, Kemp WM (1990) Variation in structure of estuarine fish communities in relation to abundance of submersed vascular plants. Mar Eco Prog Ser 65:1–14

Macintosh DJ, Ashton EC (2002) A Review of mangrove biodiversity conservation and management. Centre for Tropical Ecosystems Research, University of Aarhus, Denmark. Final Report 10/06/2002. http://mit.biology.au.dk/cenTER/MCB_Files-/2002_Review_WB_MCB_Final.pdf. Accessed on 15 June 2013

Macintosh DJ, Ashton EC, Havanon S (2002) Mangrove rehabilitation and intertidal biodiversity: a study in the Ranong mangrove ecosystem, Thailand. Estu Coas Shel Sci 55:331–345

Marlena A (2005) Benthic diatom flora of Paloh mangrove area, Kuala Rajang, Sarawak. Final year project report, Universiti Malaysia Sarawak

Martin RA (2005) Conservation of freshwater and euryhaline elasmobranchs: a review. J Mari Biol Associ U K 85:1049–1073

Meijaard E, Nijman V, Supriatna J (2008) *Nasalis larvatus*. In: IUCN 2012. IUCN red list of threatened species. Version 2012.2. www.iucnredlist.org. Accessed 02 April 2013

Millennium Ecosystem Assessment (2005) Ecosystem and human well-being: wetlands and water synthesis. World Resources Institute, Washington, DC

Mineau P, Downes CM, Kirk DA, Bayne E, Csizy M (2005) Patterns of bird species abundance in relation to granular insecticides use in the Canadian prairies. EcoSci 12:267–278

Minello TJ, Able KW, Weinstein MP, Hays CG (2003) Salt marshes as nurseries for nekton: testing hypotheses on density, growth and survival through meta-analysis. Inter-Res Mari Ecol Prog Seri 246:39–59

MNS–BCC (2005) Checklist of the Malaysian birds with information about occurrence and status (Malaysian Nature Society). http://www.birdinginmalaysia.com-/lists.php. Accessed 06 August 2012

Morrisey DJ, Swales A, Dittmann S, Morrison MA, Lovelock CE, Beard CM (2010) The ecology and management of temperate mangroves. Ocean & Mari Biol: Annu Revi 48:43–160

Morton RM (1990) Community structure, density and standing crop of fishes in subtropical Australian mangrove area. Mari Biol 105:385–394

Nagelkerken I, Roberts CM, van der Velde G, Dorenbosch M, van Riel MC, Cocheret de la ME, Nienhuis PH (2002) How important are mangroves and seagrass beds for coral reef fish? The nursery hypothesis tested on an island scale. Inter-Res Mari Ecol Prog Seri 244:299–305

Nagelkerken I, Blaber SJM, Bouillon S, Green P, Haywood M, Kirton LG, Meynecke JO, Pawlik J, Penrose HM, Sasekumar A, Somerfield PJ (2008) The habitat function of mangroves for terrestrial and marine fauna: a review. Aqua Bot 89:155–185

Naidoo G (2009) Different effects of nitrogen and phosphorus enrichment on growth of dwarf *Avicennia marina mangroves*. Aquat Bot 90:184–190

Naik NR, Kamalendra V, Naik J (2013) Mangrove ecosystem and its importance in fisheries. http://aquafind.com/articles/Mangrove_Ecosystem.php. Accessed 12 June 2013

Nakamura Y, Horinouchi M, Shibuno T, Tanaka Y, Miyajima T, Koike I, Kurokura H, Sano M (2008) Evidence of ontogenetic migration from mangroves to coral reefs by black-tail snapper *Lutjanus fulvus*: stable isotope approach. Inter-Res Mari Ecol Prog Seri 355:257–266

Naylor RL, Goldberg RJ, Primavera JH, Kautsky N, Beveridge MC, Clay J, Folk C, Lubchenco J, Mooney H, Troell M (2000) Effect of aquaculture on world fish supplies. Nature 405:1017–1024

Ng PKL, Sivasothi N (eds) (2001) A guide to mangroves of Singapore. Raffles museum of Biodiversity research, National University of Singapore and Singapore Science Centre. http://mangrove.nus.edu.sg/guidebooks/contents.htm

Nielsen MG (1997) Nesting biology of the mangrove mud-nesting ant *Polyrhachis sokolova* Forel (Hymenoptera: Formicidae) in northern Australia. Insec Soci 44:15–21

Nielsen OI, Kristensen E, Macintosh DJ (2003) Impact of fiddler crabs (*Uca* spp.) on rates and pathways of benthic mineralization in deposited mangrove shrimp pond waste. J Experi Mari Biol Ecol 289:59–81

Nijman V (2000) Geographic distribution of ebony leaf monkey Trachypithecus auratus (E. Geoffroy Saint-Hilaire, 1812) (Mammalia: Primates: Cercopithecidae). Contri Zool 69:157–177

Norhayati A, Shukor MN, Juliana S, Wan Juliana WA (2009) Mangrove flora and fauna of Klang islands mangrove forest reserves, Selangor, Malaysia. Malaysian J Sci 28:275–288

Northern Territory Government (2002) Mangrove management in the northern territory. http://lrm.nt.gov.au/_data/assets/pdf_file/0018/10395/mangrove_mngt_inthe-NT_2002.pdf

Noske RA (1996) Abundance, zonation and feeding ecology of birds in mangroves of Darwin Harbour, Northern Territory. Wild Res 23:443–474

Nyanti L, Ling TY, Jongkar G (2012) Fish and crustacean communities and fish length-weight relationship of Lutong river, Miri, Sarawak, Malaysia. World J Fish Mari Sci 4:102–110

Ong P, Richardson M (2008) *Macaca fascicularis*. In: IUCN 2012. IUCN Red list of threatened species. Version 2012.2. www.iucnredlist.org. Accessed 02 April 2013

Osorio FM, Godinho WO, Lotufo TMC (2011) Fish fauna associated to mangrove roots at the Pacoti river estuary. Biota Neot 11:415–420

Paris CB, Chérubin LM, Cowen RK (2007) Surfing, spinning, or diving from reef to reef: effects on population connectivity. Inter-Res Mari Ecol Prog Seri 347:285–300

Penha-Lopes G, Bartolini F, Limbu S, Cannicci S, Kristensen E, Paula J (2009) Are fiddler crabs potentially useful ecosystem engineers in mangrove wastewater wetlands? Mari Poll Bull 58:1694–1703

Penha-Lopes G, Torres P, Cannicci S, Narciso L, Paula J (2011) Monitoring anthropogenic sewage pollution on mangrove creeks in southern mozambique: a test of palaemon concinnus dana, 1852 (Palaemonidae) as a biological indicator. Envir Pollu 159:636–645

Pinto L (1988) Population dynamics and community structure of fish in the mangroves of pagbilao, Philippines. J Fish Biol 33:35–43

Pittman SJ, McAlpine CA, Pittman KM (2004) Linking fish and prawns to their environment: a hierarchical landscape approach. Inter-Res Mari Ecol Prog Seri 283:233–254

Plaziat JC (1984) Mollusc distribution in the Mangal. In: Por FD, Dor I (ed) Hydrobiology of the Mangal—the ecosystem of the mangrove forests. Devel Hydrob 20:111–143

Prasert T, Kou I, Takeshi Y, Masahiro H, Tomon M, Mitsuhiko S, Hisashi K, Toru T (2002) Fish fauna of the Sikao creek mangrove estuary, Trang, Thailand. Fish Sci 68:10–17

Pullin AS, Sutherland W, Gradner T, Kapos V, Fa JE (2013) Conservation priorities: identifying need, taking action and evaluating success. In: Macdonald DW, Willis KJ (ed) Key topics in conservation biology 2. John Wiley & Sons, Ltd. First Edition

Rajkumar M., Perumal P, Prabu V, Perumal N, Rajasekar K (2009) Phytoplankton diversity in Pichavaram mangrove waters from south-east coast of India. J Environ Biol 30:489–498

Rajpar MN, Zakaria M (2010) Avifauna of marudu bay mangrove forest. (unpublished data)

Ridgely RS, Greenfield PJ (2001) The birds of Ecuador, status, distribution and taxonomy, vol 1. Cornell Univ. Press, Ithaca

Robertson AI (1986) Leaf-burying crabs: their influence on energy flow and export from mixed mangrove forests (*Rhizophora* spp.) in northeastern Australia. J Exp Mari Biol Ecol 102:237–248

Robertson AI, Duke NC (1987) Mangrove as nursery sites: comparison of the abundance and species composition of fish and crustaceans in mangrove and other near shore habitats in tropical Australia. Mari Biol 96:193–205

Robinson RA, Julliard R, Saracco JF (2009) Constant efforts: studying avian population processes using standardized ringing. Ring Migra 24:199–204

Rocke T, Converse K, Meteyer C, Mclean B (2005) The impact of disease in the American white pelican in North America. Waterbirds 28:87–94

Rotaquio Jr EL, Nakagoshi N, Rotaquio RL (2007) Species composition of mangrove forests in Aurora, Philippines-A special reference to the presence of Kandelia Candel (L.) Druce. J Inter Devel Coop 13:61–78

Saari MMS, Ibrahim S (2001) Review on Mangroves restoration effort focused on encasement replanting methods. Proceeding of Universiti Malaysia Terengganu (UMTAS 2011). International annual symposium on life science, 11–13th July, 2011 Kuala Terengganu Malaysia; empowering science, technology and innovation towards a better tomorrow. LSP 48:450–456
Sabir G (2011) Population of waterbirds in two mangrove types, Avicennia and Rhizophora in Iran. PhD Thesis, Faculty of Forestry, University Putra Malaysia
Saenger P, Hegerl EJ, Davie JDS (1983) Global status of mangrove ecosystems. The Envir 3:1–88
Sandilyan S, Thiyagesan K, Nagarajan R (2010) Major decline in species-richness of waterbirds in the Pichavaram mangrove wetlands, southern India. Wad Stu Gro Bull 117:91–98
Sasekumar A, Chong VC, Leh MU, D'Cruz R (1992) Mangroves as a habitat for fish and prawns. Hydrobiologia 247:195–207
Satheeshkumar P (2011) First record of a mangrove frog Fejervarya cancrivora (Amphibia: Ranidae) in the Pondicherry mangroves, Bay of Bengal-India. Worl J Zool 6:328–330
Schneider FI, Mann KH (1991) Species specific relationships of invertebrates to vegetation in seagrass bed. I. Correlational studies. J Expe Mari Biol Ecol 145:101–117
Selvam V, Ravichandran KK, Karunagaran VM, Mani KG, Evanjalin JBeulaG (2004). Joint mangrove management in Tamil Nadu: processes, experiences and prospects: Part 1 to 4. M. S. Swaminathan Research Foundation, Chennai, India
Skov MW, Hartnoll RG (2002) Paradoxical selective feeding on a low nutrient diet: why do mangrove crabs eat leaves? Oecologia 131:1–7
Skov MW, Vannini M, Shunula JP, Hartnoll RG, Cannicci S (2002) Quantifying the density of mangrove crabs. Ocypodidae and Grapsidae. Mari Biol 141:725–732
Slim FJ, Hemminga MA, Ochieng C, Jannink NT, Cocheret delaME, van der Velde G (1997) Leaf litter removal by the snail *Terebralia palustris* (Linneaus) and sesarmid crabs in an East African mangrove forest. J Experi Mari Biol Ecol 215:35–48
Smith TJ (1987) Seed predation in relation to tree dominance and distribution in mangrove forests. Ecology 68:266–273
Smith TJ, Boto KG, Frusher SD, Giddins RL (1991) Keystone species and mangrove forest dynamics: the influence of burrowing by crabs on soil nutrient status and forest productivity. Estu Coast Shel Sci 33:19–32
Solano-Ugalde A, Freile AJ, Moscoso P, Prieto-Albuja F (2009) New and confirmative bird records from northern Esmeraldas province, Ecuador. Cotinga 31:115–118
Soonabai Pirojsha Godrej Marine Ecology Centre (2013) Mangrove diversity. http://www.mangroves.godrej.com/Mangrove-Vegetation.htm. Accessed 5 April 2013
South Florida Aquatic Environments (2013) Mangrove life: birds. http://www.flmnh.ufl.edu/fish/southflorida/mangrove/lifebirds.-html. Accessed 1 April 2013
Spalding M, Kainuma M, Collins L (2010) World atlas of mangrove. EarthScan, UK.
Stafford–Deitsch J (1996) Mangrove: the forgotten habitat. IMMEL Publishing Limited, London, p 277
Talaat WAW, Tahir NM, Husain ML (2011) Mangrove biodiversity management and conservation in Malaysia: The present legislative, administrative and policy framework. Proceeding of Universiti Malaysia Terengganu (UMTAS 2011). International annual symposium on life science, 11–13th July, 2011 Kuala Terengganu Malaysia; empowering science, technology and innovation towards a better tomorrow. LSP 36:399–404
Thayer GW, Colby DR, Hettler WF (1987) Utilization of the red mangrove pro root habitat by fishes in south Florida. Inter-Res Mari Ecol Prog Seri 35:25–38
Thu PM, Populus J (2007) Status and changes of mangrove forest in mekong delta: Case study in Tra Vinh, Vietnam. Estu, Coast and Shel Sci 71:98–109
Tidwell JH, Allan GL (2001) Fish as food: aquaculture's contribution: ecological and economic impacts and contributions of fish farming and capture fisheries. EMBO Rep 15:958–963
Tomlinson PB (1986) The botany of mangroves. University Press, Cambridge, p 419

Verweij MC, Nagelkerken I, Wartenbergh SLJ, Pen IR, van der Velde G (2006) Caribbean mangroves and seagrass beds as daytime feeding habitats for juvenile French grunts, *Haemulon flavolineatum*. Mari Biol 149:1291–1299

Walton MEM, Le Vay L, Lebata JH, Binas J, Primavera JH (2007) Assessment of the effectiveness of mangrove rehabilitation using exploited and non-exploited indicator species. Biol Conserv 138:180–188

Wan Juliana WA, Damanhuri A, Razali MS, Norhayati A, Latiff A (2010) Mangrove flora of Langkawi. Langkawi research centre, Institute for environment and development (LESTRI), Universiti Kebagsaan Malaysia, and Langkawi development (LDA)

Wan Juliana WA, Maimon A, Norhayati A, Razali SM, Mohd Fauzi E, Izwahanie I, Nurhanim MN (2011) Mangrove flora and fauna of Manjung riparian reserves, Perak, Malaysia. Proceeding of Universiti Malaysia Terengganu (UMTAS 2011). International annual symposium on life science, 11–13th July, 2011 Kuala Terengganu Malaysia; empowering science, technology and innovation towards a better tomorrow. LSP 39:412–417

Warne K (2013) Forest of the tide. http://ngm.nationalgeographic.com/2007/02/mangroves/warne-text. Accessed 10 April 2013

White WT, Potter IC (2004) Habitat partitioning among four elasmobranch species in near shore, shallow waters of a subtropical embayment in Western Australia. Mari Biol 145:1023–1032

Wright P, Anderson P, Weng L, Frick N, Wong WP, Ip YK (2004) The crab-eating frog, *Rana cancrivora*, up-regulates hepatic carbamoyl phosphate synthetase I activity and tissue osmolyte levels in response to increased salinity. J Exp Zool 301A:559–568

Yaez-Arancibia A, Lara-Domínguez AL, Rojas-Galaviz JL, Sánchez-Gil P, Day JW, Madden CJ (1988) Seasonal biomass and diversity of estuarine fishes coupled with tropical habitat heterogeneity (southern Gulf of Mexico). J Fish Biol 33:191–200

Yu-Seong C, Kwon I, Yoo J (2008) A Study of feeding methods in five species of herons and egrets in Korea. J Ecol Fiel Biol 31:147–151

Zhigang Y, Ermi Z, Haitao S, Diesmos A, Alcala A, Brown R, Afuang L, Gee G, Sukumaran J, Yaakob N, Ming LT, Chuaynkern Y, Thirakhupt K, Das I, Mumpuni ID, Robert Inger R (2004) *Fejervarya cancrivora*. In: IUCN 2012. IUCN Red list of threatened species. Version 2012.2. www.iucnredlist.org. Accessed 11 April 2013

Zhou Y, Zhao B, Peng Y, Chen G (2010) Influence of mangrove reforestation on heavy metal accumulation and speciation in intertidal sediments. Mari Pollu Bull 6:1319–1324

Mangrove Forests of Timor-Leste: Ecology, Degradation and Vulnerability to Climate Change

Daniel M. Alongi

Contents

Abstract Mangrove forests occupy a relatively small area (1,300 ha) of the coastal zone of Timor-Leste, being composed of fringing stands of relatively few species (a total of 19 true mangrove species) inhabiting sand-dominated deposits in small lagoons along the south coast and sheltered embayments along the north coast. Despite their small size and disjunct distribution these mangroves are heavily used as a source of food, and wood for housing and fuel, and have been used as burial sites during past episodes of violence during occupation. Links between mangroves and fisheries offshore are uncertain, but it is clear that net canopy production is low, equivalent to the mangrove forests in other dry tropical zones. Timorese mangroves face a very uncertain future in light of their small, fragmented distribution, heavy human encroachment, and forecasted rise in sea level.

1 Introduction

Timor-Leste, an independent nation since 2002, lies on the eastern half of the island of Timor and also consists of the Oecussi-Ambeno enclave within West Timor (part of the Indonesian province of Nusa Tenggara Timur), Atuaro Island off the capital of

D. M. Alongi (✉)
Australian Institute of Marine Science, PMB 3 Townsville MC,
4810 Queensland, Australia
e-mail: D.Alongi@aims.gov.au

I. Faridah-Hanum et al. (eds.), *Mangrove Ecosystems of Asia*,
DOI 10.1007/978-1-4614-8582-7_9,

Dili, and the small island of Jaco off the eastern tip of the island. Timor-Leste covers an area of 14,610 km^2 and has a population of just over one million (Macauley 2003).

Unlike the Lesser Sunda Islands and even Atuaro Island, Timor is not of volcanic origin and its position between the Asian and Australian plates has resulted in several million years of relative isolation (Audley-Charles 1968; Hall 2012). Partly for this reason, Timor is closer to the Melanesian than the Asiatic biogeographical zone, lying east of the Wallace Line. Thus, the flora and fauna of the island reflect both Melanesian and Austronesian origins (Richardson et al. 2012). Geologically, Timor is one of the most complex islands in Southeast Asia, consisting of a mountainous spine, the Ramelau range, which divides the island into two distinct climatic zones (Durand 2007; Penny 2012). The south coast and mountainous center receive more rain (1,500–2,000 mm yr^{-1}) than both the north coast ($<$ 1000 mm yr^{-1}) and the northern highlands and slopes of the central mountain chain (1,000–1,500 mm yr^{-1}). There is a distinct dry season from July to October along the north coast and two wet seasons (November–April, May–July) along the south coast (Durand 2007). Vegetation types reflect the geographical differences in climate with tropical humid forests south of the central mountain chain and savanna woodlands and fragments of scrub forest along the north coast.

The Timorese coastline stretches for 706 km and is influenced partly by the Indonesian Throughflow (Hantoro et al. 1997; Alongi et al. 2013). The direction of the main currents offshore switch seasonally from flow directed towards Timor from New Guinea at mid-year, to water flow from the direction of mainland Asia at the beginning of the year. The north coast is characterized by karst geology and uplifted ancient coral reefs (Chappell and Veeh 1978; Audley-Charles 2004; Boggs et al. 2009); the narrow continental margin consists of fringing reefs, seagrass beds, rocky intertidal outcrops, sandflats, sandy beaches, and fringing mangroves bathed in very clear water. The south coast, in contrast, is very turbid and open with long stretches of sandy beach, uplifted ancient reefs, and numerous small lagoons fronted by large sand bars (Sandlund et al. 2001; Wyatt 2004).

In this chapter, I review various aspects of the structure and function of the mangroves inhabiting pockets of both the north and south coast of Timor-Leste. While few published studies exist, there are a number of extensive unpublished reports detailing the size, structure and human encroachment of these small, but heavily used, fringing mangroves.

2 Forest Area, Distribution, and Species Composition

Estimates of the total mangrove area in Timor-Leste are few and, until recently, unreliable (Table 1). These data show an apparently severe loss of mangroves from about 9,000 ha in 1940 to only about 1,000 ha in 2010. And while there is no doubt that much mangrove forest has been lost since 1940, a deforestation rate cannot be accurately determined, as the early references do not describe how their estimates were calculated or what methods were used. Even my estimate of 1,300 ha

Table 1 Estimates of mangrove coverage (hectares) in Timor-Leste

Area (ha)	Year	Reference
9,000	1940	MacKinnon et al. 1982
4,000	1982	MacKinnon et al. 1982
3,035	2000	Wilkie et al. 2003
1,802	2000	FAO 2007
899 (north coast)	2009	Boggs et al. 2009
1,300	2013	This study

(Table 1) is imprecise even though it is derived from the Boggs et al. (2009) study and from several ground-truth surveys along the south coast. Most of the previous estimates overestimated mangrove area as it is difficult to distinguish mangroves from freshwater swamps in the photo reconnaissance conducted in the last century. *Lontar* palm and other freshwater palm and swamp species commonly occur landward of fringing mangroves, especially along the south coast.

The current distribution of mangroves along the East Timorese coast (Fig. 1) indicates small patches of fringing mangrove forest along the north coast, usually near river mouths or small inlets that provide sufficient shelter for forest development. Along the south coast, mangroves are few and all are located in small channels or lagoons behind sand bars (Fig. 2) that are usually located east of river mouths; the south coast is much more open to wave action than the north coast.

Nineteen true mangrove and 13 mangrove-associated species are found in Timor-Leste (Table 2), with *Avicennia marina*, *Sonneratia alba*, *Rhizophora stylosa*, *Ceriops tagal,* and *Lumnitzera racemosa* being the most common. Most trees are stunted or < 6–8 m in height, although a few tall (> 40 m) *Sonneratia alba* stands (Fig. 3)

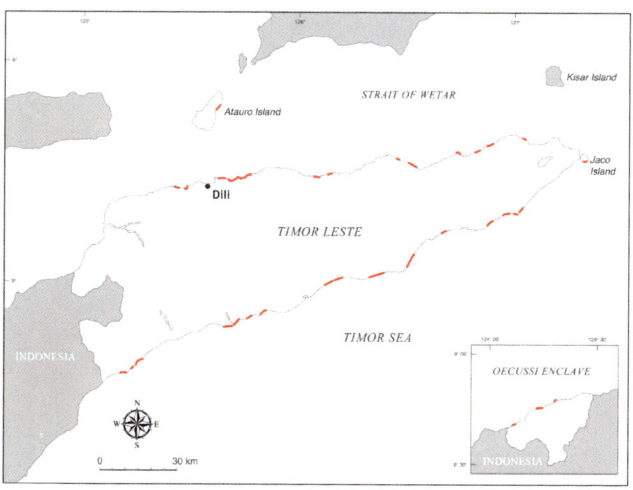

Fig. 1 Location of mangrove forests (in *red*) fringing the entire coastline of Timor-Leste, 2007–2010. The length of the red line is not indicative of the size of the forest

Fig. 2 Typical mangrove forest in a small coastal lagoon behind large sand dunes, south Timor coast

Table 2 True mangrove and mangrove associate species recorded in Timor-Leste

True mangroves	Mangrove associates
Acanthus ilicifolius	*Barrintonia asiatica*
Acrostichum aureaum	*Callophyllum inophyllum*
Aegiceras corniculatum	*Cerbera manghas*
Avicennia marina	*Derris scandens*
Bruguiera cylindrica	*Derris trifoliata*
Bruguiera gymnorrhiza	*Hibiscus tiliaceus*
Bruguiera parviflora	*Ipomoea pes-caprae*
Ceriops australis	*Pemphis acidula*
Ceriops tagal	*Premna serratifolia*
Excoecaria agallocha	*Scaevola taccada*
Heritiera littoralis	*Suriana maritima*
Lumnitzera racemosa	*Terminalia catappa*
Rhizophora apiculata	*Ryssopterys timoriensis*
Rhizophora mucronata	
Rhizosphora stylosa	
Sonneratia alba	
Sonneratia caseolaris	
Sonneratia urama	
Xylocarpus granatum	

exist near Metinaro, the largest contiguous forest in Timor-Leste. There is distinct zonation of the Metinaro mangroves with stands of *Avicennia marina, Sonneratia alba* and a mixture of *R. apiculata, R. mucronata* and *R. stylosa* dominating the more seaward zones, and stunted *Avicennia marina, Ceriops tagal* and *Lumnitzera racemosa* dominating further inland (Boggs et al. 2009). Along the entire north coast, Boggs et al. (2009) recognized at least nine species associations as well as pure stands of *S. alba*, *A. marina* and *C. tagal* (Table 3). Along the south coast, there are

Fig. 3 Not all mangrove trees in Timor-Leste are small. This *Sonneratia alba* tree, located within the Metinaro mangroves, is 1.6 m in diameter and 45 m tall

Table 3 Remote sensing and ground-truth estimates of mangrove forest types and their coverage along the entire north coast of Timor-Leste. (Adapted from Boggs et al. 2009)

Dominant/subdominant forest type	Area (ha)
Sonneratia alba/Rhizophora mucronata/R. apiculata	285
S. alba	103
Ceriops tagal/Avicennia marina/Lumnitzera racemosa	53
Mixed C. tagal/R. apiculata/A. marina	26
A. marina	26
L. racemosa/A. marina/Excoecaria agallocha (low open canopy)	17
L. racemosa/A.marina/S. alba (low open canopy)	14
C. tagal	8
R. stylosa/A. marina (low open canopy)	7
Mixed A. marina/C. tagal	7
A. marina/R. stylosa (low open canopy)	5
R. apiculata/S. caseolaris/Aegiceras corniculatum (low open canopy)	3
Ceriops australis sparse low canopy	2
L. racemosa/Azima sparse shrub land-saltpan	1
Other (saltpan/hinterland/shrub land/beach/mixed mangrove and palm forest)	343

five species groups: (1) *E. agallocha/L. racemosa,* (2) *Sonneratia urama*, (3) mixed *Rhizophora*, (4) *R. stylosa/A. marina/E. agallocha*, and (5) *R. mucronata* (Alongi et al. 2012). In Liquisa district, near the northern border with Indonesian West Timor, de Jesus (2012) found isolated stands of *S. alba* (967 trees ha^{-1}), mixed *R. apiculata/R. mucronata* (967 trees ha^{-1}) and *B. cylindrica* (1,333 trees ha^{-1}). Near Com at the northeastern end of the island, mangrove communities are dominated by *A. marina, S. alba* and *L. racemosa*; on the southeast coast between Lore and the Mamalutu River, small lagoons behind coastal dunes are composed of stands of *E. agallocha, A. marina* and *L. racemosa*, and at river mouths, *H. littoralis* and an occasional *S. alba* are present on the ocean side.

Mangrove associates are commonly found throughout Timor-Leste as sand dune and rocky shore communities of low herbs, trailing vines and grasses. These assemblages include *Ipomoea pes-caprae*, *Barringtonia asiatica*, *Callophyllum inophyllum*, *Terminalia catappa*, *Pemphis acidula* and *Suriana maritima* (Cowie 2006). And there are closed forest community associates ('*Barringtonia* formation', sensu Cowie 2006) that form a narrow ridge between coastal lowland and the sea along the southeast coast between Lore and the Namalutu River, composed of the tree species *C. inophyllum, Cerbera manghas, Hibiscus tiliaceus,* and *Terminalia catappa*, the shrubs *Premna serratifolia* and *Scaevola taccada*, and the vines *Derris scandens, D. trifoliata* and *Ryssopterys timoriensis.*

Structural characteristics of Timorese mangrove forests vary greatly (Table 4) but most stands on average have standing biomass (mean = 237 t DW ha^{-1}) and leaf area indices (mean = 4.9 m^2 leaf area m^{-2} ground area) that are well within the range of values expected from mangroves at 8–9° latitude (Saenger and Snedaker 1993; Alongi and Dixon 2000; Clough et al. 2000). The biomass of a mangrove forest depends upon a number of interrelated factors, such as soil fertility, precipitation, species composition, and frequency of tidal inundation and presumably these same drivers operate in Timor-Leste. Indeed, the smallest forests are located on very dry soils in the high intertidal zone along the arid north coast, while some of the largest (by weight) stands are located at the sea edge along the north (Tibar and Metinaro) and south (near Suai) coast.

3 Deforestation and Overexploitation

The rate of mangrove deforestation is unknown but few, if any, of East Timor's remaining mangroves are pristine. An idea of the probable extent of mangrove deforestation in Timor-Leste can be surmised from the data on the destruction of the island's terrestrial forests. Shifting agriculture, settled agriculture, logging, and fire have substantially altered the composition and extent of vegetation cover (Benevides 2003; Bouma and Kobryn 2004). Closed canopy forests occur now only on hilltops and in deep ravines. Changes in climate have undoubtedly altered vegetation, in addition to human impacts that have increased within the last few millennia (Benevides 2003).

Table 4 Mean stand characteristics of some mangrove forests in Timor-Leste. (adapted from Alongi and de Carvalho 2008; Alongi et al. 2012)

Site Designation	Dominant species	Location	AGB (t DW ha^{-1})	Mean dbh (cm)	Canopy cover (η)	Leaf area index (m^2 leaf area m^{-2} ground area)	Stem density (stem ha^{-1})	Basal area (m^2 ha^{-1})
TL1	*B. gymnorrhiza/R. apiculata*	Metinaro, north coast	221.5	12.7	0.73	4.9	3,633	32
TL2	*C. tagal*	Metinaro, north coast	194.8	13.1	0.72	5.2	7,606	34
TL3	*A. marina/C. tagal/R. apiculata*	Metinaro, north coast	51.1	9.9	0.61	5.4	9,610	13
TL4	*E. agallocha/L. racemosa*	Betano, south coast	136.1	6.1	0.72	2.2	11,447	11
TL5	*R. apiculata/R. stylosa*	Betano, south coast	375.2	7.1	0.91	4.4	11,217	17
TL6	*R. mucronata/R. stylosa/S. alba*	Metinaro, north coast	300.0	24.1	0.89	5.4	15,269	27
SL1	*Sonneratia urama*	Suai, south coast	108.8	12.2	0.52	1.4	3,316	16
SL2	*R. stylosa/A. marina/ E.agallocha*	Suai, south coast	171.6	11.2	0.63	2.5	7,985	18
BC1	*R. mucronata*	40 km east of Suai, south coast	315.4	7.5	0.85	4.1	20,410	46
TB1	*S. alba*	Tibar, North coast	492.9	44.7	0.89	4.1	7,448	32

The only documented case of mangrove deforestation (Alongi et al. 2008) occurred in Metinaro in 2007 as the direct result of the settlement of over 6,000 internally displaced persons (IDPs) fleeing riots and disturbances in the nearby capital of Dili. During this crisis, large numbers of people harvested mangrove timber for firewood and fuel, and mangrove invertebrates, especially various shellfish, for food. Approximately one year after the start of harvesting, the Metinaro forests experienced a 30–50 % decline in live stems and a 46–86 % loss of above-ground biomass with more canopy gaps between less dense, smaller trees (Fig. 4). Harvesting was size-selective, probably as a trade-off between cutting trees small enough for women and children to carry, and being large enough to warrant the cost of carrying for selling as

Fig. 4 The result of unsustainable tree harvesting by displaced persons in the Metinaro mangroves, June 2007

firewood. Most cut trees were within 5–15 cm diameter-at-breast height in size. These harvesting operations altered the size and structure of the remaining canopy; canopy cover declined from 60–73 % to 33–39 % and leaf area index declined from 4.9–5.4 to 3.5–3.9 in three stands. A decline in forest canopy cover resulted in changes to soil chemistry with significant increases in interstitial salinity, and nutrient, metal, and trace element concentrations. Rates of anaerobic soil metabolism decreased after logging, with the decline greatest near the soil surface.

On a daily basis, people, cattle, and other domestic animals routinely enter the mangroves to harvest leaves, fruit, fallen wood, and algae growing on root and stem surfaces; goats commonly feed on mangrove tree parts, especially *Sonneratia* pneumatophores (Fig. 5). It is common to see people and animals in any given Timorese mangrove forest and very common to see small fishing boats and canoes in close proximity to mangroves. Women especially are commonly encountered harvesting large basketfuls of small benthic invertebrates such as crabs and molluscs (Fig. 6). Whether or not such harvesting is unsustainable is unclear, but there is evidence that humans have been harvesting marine resources in Timor-Leste for at least 42,000 years (O'Connor et al. 2011).

Cultural rules and traditions in some villages and coastal regions probably foster some degree of sustainable management. For example, in the city of Manatuto in northern Timor-Leste, traditional tidal (stone) fish traps (*hatu meti ian*) is the common property of particular clans and their use is managed through customary gathering arrangements. There are also customary seasonal prohibitions on harvesting or utilization of designated resources. These customs, known locally as *tara bandu* (Tetum for 'to raise a prohibition'), represent indigenous practices to limit and manage common resources. Injunctions can include prohibitions on timber cutting and wild food

Fig. 5 Goats commonly eat pnematophores and macroalgae in mangroves along the north coast. Photo taken in a *Sonneratia alba* stand in Tibar Bay, west of Dili, October 2008

Fig. 6 Women do all of the animal harvesting in Timorese mangroves. This photo was taken in the Metinaro mangroves during the June 2007 crisis

harvesting for certain periods of time. Violating these proscriptions usually results in fines and other sanctions for the perpetrators. These practices are widely reported, but there has been very little research into the scope and persistence of these local management systems either within the Manututo area or more widely across Timor Leste (McWilliam 2001, 2003).

4 Food Webs and Links to the Coastal Zone

Located within the Coral Triangle, Timor-Leste belongs to a region with the world's richest and most diverse marine life (Wever 2008); the country's coastal and open ocean fauna consist of a comparatively large diversity of whales, sharks, dugongs, dolphins, turtles, sea snakes, tuna, mackerel, and snapper. The invertebrate fauna of the shallow-water coral platforms along the south coast also appear to be abundant and species-rich (Wyatt 2004).

The mangrove benthic and pelagic fauna, in contrast, have not been examined so there is no information on species diversity or trophic relationships, despite the presence of eatable epibenthos whose apparent abundance can then be surmised from the large number of women harvesters. Other than one record of the common gastropod *Littoraria scabra* occurring in Dili Bay (Reid et al. 2010), there is no ecological or systematic knowledge of the mangrove fauna. One record does document a diverse ant fauna of predominantly Indo-Malayan affinities, with ants of the mangrove-inhabiting genus *Polyrhachis* apparently highly endemic to the island (Andersen et al. 2013).

Waterbirds and coastal seabirds are abundant (Trainor 2005; Trainor et al. 2007) with at least 59 species found in mangroves and adjacent mudflats. This is unsurprising considering that the island of Timor is known to be on the East Asian flyway for migratory shorebirds (Mayr 1944). These birds are likely to feed in the mangroves but the impact their feeding activities have on Timor's mangroves is unknown.

Fish undoubtedly form a trophic link between Timor's mangroves and the adjacent coastal ocean. There are little fisheries data, but coastal villagers routinely fish near and in mangroves at high tide along the entire coast; many species caught are commonly associated with adjacent reefs and seagrasses (Alongi et al. 2009). In a series of interviews with coastal communities on both the north and south coasts, villagers repeatedly stated that they were dependent on mangroves for food—fish and penaeid shrimp—especially during holidays and special celebrations (Alongi et al. 2009) supporting the results of earlier anthropological studies (McWilliam 2001, 2003) on fishing traditions in Timor Leste.

Another link to the coastal zone is the possible exchange of particulate and dissolved nutrients with mangroves. In a detailed biogeochemical study, it was hypothesized that most organic carbon decomposed by microbes in mangrove soils appears to seep out of the forest via groundwater (Alongi et al. 2012). There was little net algal production on the forest floor, but there were large disparities between rates of surface and subsurface respiration. These data suggest that the main chemical link between the mangroves and the coastal ocean off Timor Leste is the export of dissolved inorganic carbon, which may play a role in supporting phytoplankton production or enhancing coastal respiration, or both. Our analysis of water and sediment samples from a number of rivers suggest that terrestrial material, including from mangroves, is rapidly exported offshore and, in the case of the narrow shelf margin along the north coast, deposited into the deep ocean (Alongi et al. 2009). Along the south coast where the continental shelf is wider, sediment δ^{13} °C analyses revealed that terrestrial organic carbon (including mangrove material) accumulates offshore in zones of benthic enrichment and high biological activity (Alongi et al. 2013). The rapid regeneration of nutrients returned to the water-column supports high plankton production and possibly pelagic fish production, forming another link between mangroves and coastal Timorese waters.

5 Climate Change Forecasts

Climate change is predicted to alter the coastal environment of Timor-Leste (Barnett 2007; Kirono 2010) in a number of ways:

- Annual air temperature will increase by 0.8 °C by 2020 and 2.2 °C by 2080
- Annual rainfall will increase by 2 % by 2020 and 6 % by 2080
- Annual potential evaporation will decrease up to 5 mm d^{-1} by 2090
- The heat wave duration index (number of consecutive days when temperatures are > 5 °C above the normal maximum) will increase 2 d yr^{-1} by 2050
- Sea surface temperatures will increase by 0.6–0.8 °C by 2030 and 1.0–1.5 °C by 2050
- Sea level is forecast to rise by 3.2–10.0 cm by 2020, 8.9–27.8 cm by 2050, and 18–79 cm by 2095
- Ocean acidification is projected to increase
- The inter-annual variability of the Asian monsoon will increase but it is impossible to predict whether ENSO activity will increase or decrease.

If these predictions come to fruition, mangroves in Timor-Leste will experience warmer temperatures, increasing atmospheric CO_2concentrations, more rain and land runoff, increases in terrestrial sedimentation, exposure to more acidic tidal water, and an increase in sea-level of > 0.5 m by the end of century. Mangroves worldwide are predicted to respond positively in relation to future increases in rainfall, temperature, and increasing amounts of CO_2, and negatively to sea-level rise; they are unlikely to exhibit any discernible effects from ocean acidification (Alongi 2008).

Along the arid north coast of Timor-Leste, mangroves living at maximum tidal height may well be negatively affected by warmer temperatures as they currently appear to be under high stress due to dry, high salinity conditions; however, the negative temperature impact will probably be highly interdependent on the extent of the predicted increase in rainfall. If rainfall becomes much more frequent, soil salinities may decrease despite warmer annual temperatures. Mangroves thus may not be severely impacted by increased temperatures and rainfall; in fact, these mangroves may benefit from increasing CO_2 concentrations.

The most likely negative impact of climate change on Timor-Leste's mangroves is the predicted increase in sea-level (Kirono 2010) with the degree of impact directly dependent on the rate and magnitude of the eventual rise. If the rise is within the low range of the forecast (3.2–18 cm), most of Timor's mangroves will likely survive by migrating further up onto land. If the rise is closer to 80 cm, most mangroves will not have enough time and space to migrate landwards as roads, agricultural fields, and rocky headlands lie next to mangroves along most stretches of the coast. The slope of the land-intertidal interface is steeper on average along the south coast than on the north coast (as well as less people), so there will be geographical differences in mangrove survival in Timor-Leste.

Under climate change, the mangroves of this region are forecast to suffer a decline in species diversity (Record 2013). Such a decline may be exacerbated by the fragmented nature of the country's small mangrove stands; recruitment of new propagules may even now be limited. Most stands visited over the past eight years on the north and south coast appear to consist of mature adults with little, if few, immature trees. Saplings have rarely been observed, even at the seaward edge, where one would most expect to see new recruits colonizing accreting mudflats. In fact, during coring operations in several coastal lagoons on the south coast we found mangrove peat deposits beneath the sand dunes suggesting that mangroves have receded over time, perhaps buried by the increasing amounts of sand exiting rivers from catchments that have long experienced high erosion rates. Regardless of the true cause, it is clear that no new forests have developed along either coast in Timor-Leste in the recent past.

6 Conclusion and Future Perspectives

The mangrove forests of Timor-Leste are not pristine, being heavily used by the Timorese; there is evidence that they have suffered a severe decline since the 1940s. Compared to the mangroves of other Southeast Asian nations, the Timorese mangroves remain little studied. With few exceptions mangroves occur along the coast in small, discontinuous stands that fringe the arid coast on the north side of the island, and are restricted to small, isolated lagoons on the more exposed south coast.

Timorese mangroves face a very uncertain future in light of the small, fragmented nature of their fringing habitat, heavy human encroachment, and the forecasted rise in sea-level. Mangroves along the north and south coast constitute isolated fragments to the extent that they are susceptible to the problem of being at or below critical patch size for recruitment of new seedlings. Even now, such recruitment may be limited; exceedingly few recruits were observed in any of the stands visited from 2005 to 2012. For these reasons, the mangroves require urgent management and conservation, especially the largest contiguous mangroves at Metinaro. Roughly one-half of the country's mangroves occur in this north coastal area, and may be just large enough to remain self-sustaining if conservation plans are put in place very soon.

References

Alongi DM (2008) Mangrove forests: resilience, protection from tsunamis, and responses to global climate change. Estuar Coast Shelf Sci 76:1–13

Alongi DM, Carvalho NA de (2008) The effect of small scale logging on stand characteristics and soil biogeochemistry in mangrove forests of Timor-Leste. Forest Ecol Manage 255:1359–1366

Alongi DM, Dixon P (2000) Mangrove primary production and above and below ground biomass in Sawi Bay, southern Thailand. Phuket Mar Biol Cent Sp Publ 22:31–38

Alongi DM, Amaral A, Carvalho N, McWilliam A, Rouwenhorst J, Tirendi F, Trott L, Wasson RJ (2009) River catchments and marine productivity in Timor-Leste: caraulun and laclo catchments; South and North coasts. Project report to ATSEF on the Timor Leste coastal/marine habitat mapping for tourism and fisheries development project, Darwin, Australia

Alongi DM, Carvalho NA de, Amaral AL, Costa A da, Trott LA, Tirendi F (2012) Uncoupled surface and below-ground soil respiration in mangroves: implications for estimates of dissolved inorganic carbon export. Biogeochemistry 109:151–159

Alongi DM, Brinkman R, Trott LA, Silva F da, Pereira F, Wagey T (2013) Enhanced benthic response to upwelling of the Indonesian throughflow onto the southern shelf of Timor-Leste, Timor Sea. J Geophys Res 118:1–13. doi: 10.1029/2012JG002150

Andersen AN, Kohout RJ, Trainor CR (2013) Biogeography of Timor and surrounding Wallacean Islands: endemism in ants of the genus *Polyrhachis* Fr. Smith. Diversity 5:139–148

Audley-Charles MG (1968) The geology of Portuguese Timor. Mem Geol Soc London 4:1–76

Barnett J (2007) Vulnerability and adaptation to climate change in Timor-Leste. Ambio 36:372–378

Benevides FT (2003) Cropping systems in East Timor. In: Costa H, Piggin C, Cruz CJ, Fox JJ (eds) Agriculture: new directions for a new nation-East Timor. ACAIR Proceedings, Canberra, pp. 17–19

Boggs G, Edyvane K, Carvalho N de, Penny S, Rouwenhorst J, Brocklehurst P, Cowie I, Barreto C, Amaral A, Smit N, Monteiro J, Mau R, Amaral J, Fernandes L (2009) Marine & coastal habitat mapping in Timor-Leste (North Coast)—Final Report. Ministry of Agriculture and Fisheries, Government of Timor-Leste, Dili

Bouma GA, Kobryn HT (2004) Change in vegetation cover in East Timor, 1989–1999. Nat Res Forum 28:1–63

Chappell J, Veeh HH (1978) Late Quaternary tectonic movements and sea-level changes at Timor and Atauro Island. Geol Soc Am Bull 89:356–368

Clough BF, Tan DT, Phuong DX, Buu DC (2000) Canopy leaf area index and litter fall in stands of the mangrove *Rhizophora apiculata* of different ages in the Mekong delta, Vietnam. Aq Bot 66:311–320

Cowie I (2006) A survey of flora and vegetation of the proposed Jaco-Tutuala-Lore National Park, Timor-Leste (East Timor). Report to Birdlife International. Northern Territory Herbarium, Palmerston

Durand E (2002) East Timor: a country at the crossroads of Asia and the Pacific. Silkwork Books, Chaing Mai

FAO (2007) The world's mangroves, 1980–2005. FAO Forestry Paper 153. FAO, Rome

Hall R (2012) Sundaland and Wallacea: geology, plate tectonics and paleogeography. In: Gower DJ, Johnson KG, Richardson JE, Rosen BR, Ruber L, Williams ST (eds) Biotic evolution and environmental change in southeast Asia. Cambridge University Press, Cambridge, pp 32–78

Hantoro WS, Nganro N, Shofiyah S, Narulita I, Sofjan J (1997) Recent climate variation signals from corals in Timor. Quat Intern 37:81–87
Jesus A de (2012) Mangrove ecosystem condition in Liquisa sub district, Timor-Leste. Depik 1:136–143
Kirono D (2010) Climate change in Timor-Leste—a brief overview on future climate change projections. Final report to the Department of Climate Change and Energy Efficiency, Canberra
Macauley J (2003) Timor-Leste: newest and poorest of Asian nations. Geography 88:40–46
MacKinnon J, Beudels RC, Robinson AH (1982) National conservation plan for Indonesia. V.4. Nusa Tenggara. FAO-FO-INS/78/ Field report. FAO, Rome
Mayr E (1944) The birds of Timor and Sumba. Bull Am Mus Nat Hist 83:126–194
McWilliam AR (2001) Prospects for the Sacred Grove: valuing *lulic* forests on Timor. Asia Pac J Anthrop 2:89–113
McWilliam AR (2003) Timorese seascapes: perspectives on customary marine tenures in East Timor. Asia Pac J Anthrop 3:6–32
O'Connor S, Ono R, Clarkson C (2011) Pelagic fishing at 42,000 years before the present and the maritime skills of modern humans. Science 334:1117–1121
Penny D (2012) China and Southeast Asia. In: Metcalfe SE, Nash, DJ (eds) Quaternary environmental change in the tropics. Wiley-Blackwell, Chicester, pp. 207–235
Record S, Charney ND, Zakaaria RM, Ellison AM (2013) Projecting global mangrove species and community distributions under climate change. Ecosphere 4(3): art34 http://dx.doi.org/10.1890/ES12-00296.1, Accessed March 2013
Reid DG, Dyal P, Williams ST (2010) Global diversification of mangrove fauna: a molecular phylogeny of *Littoraria* (Gastropoda: Littorinidae). Mol Phyto E vol 55:185–201
Richardson JE, Costion CM, Muellner AN (2012) The Malesian floristic interchange: plant migration patterns across Wallace's Line. In: Gower DJ, Johnson KG, Richardson JE, Rosen BR, Ruber L, Williams ST (eds) Biotic evolution and environmental change in Southeast Asia. Cambridge University Press, Cambridge, pp 138–165
Saenger P, Snedaker SC (1993) Pantropical trends in mangrove above-ground biomass and annual litterfall. Oecologia 96:293–299
Sandlund OT, Bryceson I, Carvalho D de, Rio N, Silva J da, Silva MI (2001) Assessing environmental needs and priorities in East Timor. Final Report UNEP, Dili and Norwegian Inst Nature Res, Trondheim
Trainor CE (2005) Waterbirds and coastal seabirds of Timor-Leste (East Timor): status and distribution from surveys in Aug 2002–Dec 2004. Forktail 21:61–78
Trainor CE, Coates BJ, Bishop KD (2007) The birds of Timor-Leste. BirdLife International and Dove Publications, Indonesia
Wever L (2008) Assessing management challenges and options in the coastal zone of Timor-Leste. Griffith centre for coastal management research report No. 86. Griffith University Gold Coast, Southport
Wilkie ML, Fortuna S, Souksavat O (2003) Status and trends in mangrove area extent worldwide. Forest resources assessment working paper No. 63. FAO, Rome
Wyatt ASJ (2004) Preliminary survey of the nearshore coastal marine environment of the south coast of East Timor: a baseline for assessing the impacts of a developing nation. BE Thesis, Univ Western Australia, Perth

Vulnerability of Mangroves to Climate Change

Joanna C. Ellison

Contents

Abstract Mangroves are valuable coastal resources in the Asia-Pacific region, providing protection against the impacts of wind and wave energy, construction wood, and promotion of better water quality. Fishery products derived from mangroves provide local communities with a large proportion of their daily protein intake. However, mangroves are among the most impacted ecosystems of all, with many countries in the region having lost 50 % of their mangroves in the last 20 years, degraded and converted to other uses. Mangrove ecosystems are also vulnerable to climate change impacts, particularly sea-level rise. Inter-tidal mangrove forests occur on low gradient sedimentary shorelines, where if inundation period increases then forest health, productivity and recruitment are affected. This may be exacerbated by the wind and wave impacts of extreme storms. Increased air and sea temperatures along with enhanced atmospheric carbon dioxide may alter processes in mangroves of respiration, photosynthesis and forest productivity. Changes in rainfall and humidity affect processes of sediment inputs, groundwater and salinity, and result in changed productivity and sediment elevation. The response of mangrove habitats in different coastal

J. C. Ellison (✉)
School of Geography and Environmental Studies, University of Tasmania,
Locked Bag 1376, Launceston, Tasmania 7250, Australia
e-mail: joanna.ellison@utas.edu.au

I. Faridah-Hanum et al. (eds.), *Mangrove Ecosystems of Asia*,
DOI 10.1007/978-1-4614-8582-7_10,

locations to climate change impacts is subject to factors of coastal behaviour, such as mangrove community composition, site tidal range, salinity regime, sedimentology and shore profile. Vulnerability assessment of climate change impacts can provide site-specific synthesis of these different factors, and allow appropriate adaptation actions to be prioritised.

1 Introduction

The greatest areas of the world's mangrove forests are established on tropical low energy, low gradient shorelines, in inter-tidal habitats such as deltas and estuaries. Tree species have special adaptations for a wet, saline environment such as aerial roots, salt regulation strategies and vivipary (Fig. 1). At their locations between marine systems and terrestrial freshwater systems, mangrove ecosystems play an integral role to benefit both. They provide protection to both terrestrial and estuarine systems from high energy marine processes, preventing erosion and providing people substantial protection from tropical storms. Mangrove forests and their substrates filter the suspended sediment discharged by rivers, which protects coral reefs and seagrass beds offshore from turbidity. Mangroves have traditionally provided a number of products for people living adjacent to them (Spalding et al. 2010), such as fuelwood and timber, and compounds that affect living tissue which are of use for tanning and medicinal products (Ewel et al. 1998).

An important value of mangroves to people is the habitat they provide for economically and ecologically important fish genera (Robertson and Duke 1990; Kimani et al. 1996; Baran and Hambrey 1999; Mumby et al. 2004; Chitaro et al. 2005). Such crustacean and fish species during juvenile growth benefit from the sheltered conditions found in mangroves, as well as abundant food and reduced predation relative to conditions offshore (Mumby et al. 2004), increasing the adult fish biomass through provision of refuge from predators, and the provision of plentiful food that increases the survivorship of juveniles. Mangroves therefore positively influence the community structure of fish in offshore waters (Ley and McIvor 2002).

2 Present Status

Despite these values of mangroves, many mangrove areas have been lost or degraded (Valiela et al. 2001; FAO 2003; Giri et al. 2011), and consequently coastal people are becoming deprived of resources upon which they have traditionally depended. Worldwide the mangrove area has fallen from 198,000 km^2 estimated for 1980, to at most 150,000 km^2 by 2000 (FAO 2003), shown by Giri et al. (2011) to be as low as 137,760 km^2. Other impacts include the clearing of mangroves for aquaculture facilities, coastal development and agriculture, along with mangrove habitat degradation by overharvest of trees for timber, and pollution in both solid waste and issues with water quality.

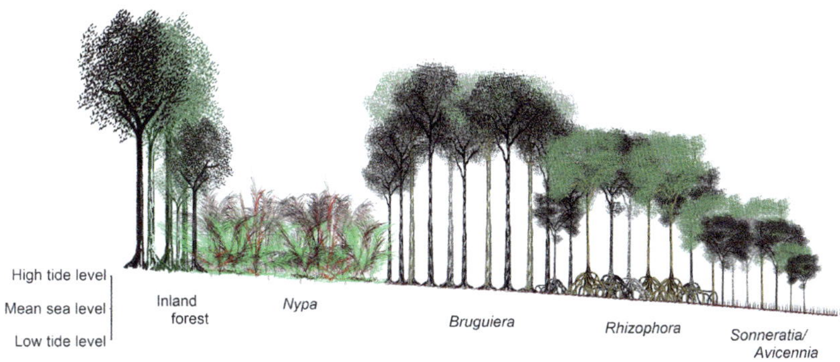

Fig. 1 Typical mangrove zonation of the Asia-Pacific region

Table 1 Mangrove species and areas in Southeast Asian countries. (Sources: Valiela et al. 2001; FAO 2003; Spalding et al. 2010; Giri et al. 2011)

ASEAN Country	Mangrove species	Mangrove area, km^2	Year of estimate	Mangrove area, km^2	Year of estimate
Brunei Darussalam	29	70	1983	173	c. 2000
Cambodia	5			728	1997
Indonesia	45	45,421	1980's	31,893	c. 2000
Laos	n/a	0	n/a	0	n/a
Malaysia	36	7,300	1980	7,097	c. 2000
Myanmar	24	5,171	1965	5,029	c. 2000
Philippines	30	4,500	1920	2,565	c. 2000
Singapore	31	18	1983	5	c. 2000
Thailand	36	3,724	1961	2,484	c. 2000
Viet Nam	29	4,000	1945	1,056	c. 2000
Total	51			51,030	

South East Asian countries have mangrove forests of among the highest biodiversity in the world, with a total area in the region of 51,030 km^2 (Table 1). The World's mangrove biodiversity is at its greatest in the Indonesia/ Malaya/ Philippines Archipelago (Fig. 2), this area having 36–47 of the 70 mangrove species recorded in this region (Duke et al. 1998; Polidoro et al. 2010). Southern New Guinea (including West Papua) is the location of the most diverse centre of the eastern group of mangrove species (Duke et al. 1998). The region however has among the highest rates of mangrove loss in the world, losing 628 km^2 per year in recent decades. As the area data collected for these assessments was before the 2004 Asian Tsunami, the cause has been attributed to human impacts (Valiela et al. 2001; Manhas et al. 2006; Duke et al. 2007).

The greatest cause of this loss in the South East Asian region has been conversion of mangrove intertidal areas to mariculture ponds (such as shrimps) (Valiela et al. 2001; Armitage 2002). Pond culture is responsible for 50 % of mangrove loss in the Philippines, and 50–80 % in Southeast Asia (Wolanski et al. 2000). There is further indirect damage from this to coastal values, such as discharge of nutrient rich waters which cause eutrophication, associated depletion of natural stocks of fish and

Fig. 2 Mangrove global distributions, with indicative graph of low latitude mangrove species diversity

crustaceans, and accumulation of toxins at the facility that cause it to be unusable after a few years, leading abandonment as the site becomes degraded (Wolanski et al. 2000), then conversion of further mangrove forest elsewhere. As well as direct clearing of mangroves for aquaculture facilities or other types of coastal development, widespread mangrove degradation has occurred along with increasingly populated coastal zones, with 26 % of mangrove forests degraded by unsustainable exploitation for production timber and fuelwood (Valiela et al. 2001).

Of special concern in the South East Asian region are two mangrove species listed on the IUCN Red List as Critically Endangered (Polidoro et al. 2010), which is the category of highest likelihood of future extinction. The first is *Sonneratia griffithii,* now rare across its distribution in South East Asia and India, where Ong (2003) has noted over the past 60 years a loss of 80 % of mangrove area. Owing to mangrove habitat clearance for coastal development, rice farming and shrimp aquaculture the species is reported in many areas be locally extinct. The second very rare species is *Bruguiera hainesii* also recently listed as Critically Endangered, only recorded from scattered locations in Myanmar, Thailand, Malaysia and Indonesia (Kress et al. 2003).

Camptostemon philippinense has been recently listed by the IUCN as Endangered, with at most 1,200 trees remaining due to habitat clearance for aquaculture and wood resources (Polidoro et al. 2010). Also listed as Endangered is *Heritiera globosa*, now only found in western Borneo in Indonesia, where its mostly riverine habitats have been destroyed by logging and conversion to oil palm and wood products commercial plantations (Polidoro et al. 2010).

Reduction in mangrove area, increasing pressures on rarer species and other threats make mangroves more vulnerable to climate change impacts (Ellison 2012). Loss of the coastal protection function of mangroves along with their valued resources may have significant economic, social and environmental consequences for the future sustainability of coastal communities.

3 Mangrove Change Impacts on Mangroves

Global climate change projections that affect the mangrove habitat are changes in temperature and rainfall, ultraviolet light, atmospheric carbon dioxide levels and changes in sea level. Research investigating the impacts of these changes on mangroves is reviewed below.

3.1 Temperature Rise

Unlike corals, mangroves do not have an upper temperature limiting factor that is within the current climate change projections, where the highest degree of warming projected by 2090–2099 relative to 1980–1999 is 6.4 °C (IPCC 2007). Thermally altered locations such as power station effluents has allowed research on mangrove tolerance of higher temperatures (Canoy 1975; Banus 1983), with results showing tolerance of the projected temperature increases (IPCC 2007).

Mangrove photosynthesis at higher latitudes becomes limited by prevalence of lower temperatures (Steinke and Naidoo 1991), causing mangrove distributions to become less diverse and then absent with increasing latitude, by the occurrence of frost. Approaching this limit mangroves become dwarfed in tree size and reduce in species diversity. With increased temperatures mangrove species are expanding their ranges, such as invading saltmarsh habitats further south into higher latitudes than they previously occurred on the east Australian coast (Rogers et al. 2006).

Biochemical processes in plants and soils are changed by increases in temperatures of water and air, with both respiration and photosynthetic carbon gain being affected (Lovelock and Ellison 2007). High midday leaf temperatures in warm conditions bring raised vapour pressure deficits between mangrove leaves and air, which causes stomatal closure (Clough and Sim 1989; Cheeseman 1994; Cheeseman et al. 1997). Decreased humidity along with rainfall when combined with raised temperatures may reduce productivity in lower latitudes in higher temperature parts of the day (Lovelock and Ellison 2007). By contrast, increased primary production could be expected to occur at higher latitudes as raised temperatures enhance the growing season.

Effects of increased temperature on primary production may however be varied by other environmental factors that influence photosynthetic rates and stomatal aperture behaviour, such as nutrient availability, humidity and rainfall. Temperature is more important than increasing atmospheric CO_2 in affecting productivity of mangroves at the ecosystem level (Luo et al. 2010); however, when temperatures and ambient CO_2 both increase then significant changes in primary net productivity have been found to occur in mangroves.

3.2 Increased Atmospheric CO_2

Atmospheric CO_2 concentrations as shown from Mauna Loa have increased from 316 ppm in 1959 to 394.5 in March 2013 (Tans 2013) and are projected to increase

40–110 % by 2030 to reach between 730 and 1,020 ppm by the end of this century (IPCC 2007). As well as causing global warming, increased CO_2 being a key reactant in photosynthesis directly affects plant growth and development and influences respiration, which may affect ecological and physiological processes in plants. As a result, plant primary production is sensitive to change in concentrations of atmospheric CO_2 (Drake et al. 1999). Mangroves have a C_3 pathway of carbon fixation in photosynthesis (Clough et al. 1982), and these plants in increased atmospheric CO_2 show increased productivity (Warrick et al. 1987) as well as with more efficient water use.

In other higher plants, doubled atmospheric CO_2 concentrations often enhance photosynthesis and growth (Drake et al. 1997), however this depends on other interacting environmental factors. Experimental work in mangrove *Rhizophora mangle* seedlings grown in doubled levels of CO_2 demonstrated significantly increased total stem length and branching activity as well as total leaf area and biomass, relative to seedlings grown in unenhanced CO_2 (Farnsworth et al. 1996). Increased CO_2 effects on growth may only occur however where high salinity is not impacting mangroves (Ball et al. 1997). Using a biogeochemical process model Luo et al. (2010) found that increasing atmospheric CO_2 concentration differentially affected net primary productivity of different mangrove species, with the greatest stimulating effects at sites where mangroves suffer the greater soil salinity stress relative to a site with lower salinity.

Elevated atmospheric CO_2 conditions that may occur in the future (IPCC 2007) are likely to increase mangrove primary production, although with variability over the range of mangrove environments (Lovelock and Ellison 2007). Further increased atmospheric CO_2 concentrations may mitigate the likely negative effects of reduced rainfall and humidity, and also alter species dominance in mangrove communities as well as bring mangrove colonisation of adjacent currently more freshwater habitats.

3.3 Precipitation Changes

Climate change projections of future precipitation changes have shown some differences between the models used, in that it may become wetter or drier for areas of the world where mangroves occur (IPCC 2007). Mangrove distributions through the world (Fig. 2) demonstrate that more productive and diverse forests of taller stature grow on coasts with higher rainfall, while on coastlines of lower rainfall mangroves are of lower diversity, height and biomass, and are of narrower margins (Duke et al. 1998; Kumara et al. 2010). Species diversity in mangroves of Australia is greatest in estuaries with moderated salinity regimes including both rainfall and catchment runoff. On drier coasts, increasing salt tolerance occurs at the expense of growth (Duke et al. 1998), resulting in lower height of the canopy. For example, at the saline, arid site of Lake MacLeod on the west coast of Australia tree canopy heights of *Avicennia marina* are 2.5–4.0 m (Ellison and Simmonds 2003), compared with wetter sites on the east coast of Australia where mangrove canopies are commonly 10 m or higher at similar latitudes (Mackey 1993).

In wetter conditions, mangrove substrates are leached of salt by rainfall and by outflow through the habitats of fresh water river discharge and groundwater outflow, also bringing nutrients. In drier conditions, high concentrations of salt result from evaporation from the mangrove substrate at low tide, in some cases causing saline flats to occur towards high tide levels. Salinity stress to mangroves has been shown at in the Gulf of Carpentaria by Conacher et al. (1996), where a sequence of drier than normal wet seasons caused higher soil salinities, resulting in mortality of *Avicennia marina*.

Two physiological adaptations enable mangrove survival in saline water, salt excretion in species such as *Aegiceras* and *Aegialitis*, and salt exclusion in species such as *Laguncularia* and *Rhizophora* (Scholander et al. 1962). Species that excrete salts eliminate salt through salt glands in the leaves, allowing them to continue photosynthesis while utilizing ocean water in transpiration. Salt excluders also can cease or diminish transpiration and photosynthesis when exposed to saline water.

Increased salinity in the mangrove habitat that can lead to salt stress may result from several factors related to climate change: reduced rainfall, more seasonal rainfall, reduced humidity and sea-level rise. Reduced rainfall is expected to cause reduction in mangrove diversity, photosynthesis, and growth rates along with substrate subsidence (Smith and Duke 1987; Rogers et al. 2005, 2006; Whelan et al. 2005). Reduced humidity is expected to cause reduced productivity and species diversity (Clough and Sim 1989; Cheeseman et al. 1991; Cheeseman 1994; Ball et al. 1997).

The mangrove species *Rhizophora mangle* seedlings were grown under different salinities (Stern and Voight 1959), with results showing that survival, growth and height all declined in increasing salt concentrations. Gas exchange characteristics in mangroves showed decreased photosynthetic capacity with increased salinity (Ball and Farquhar 1984a). *Aegiceras corniculatum* was shown to be the more sensitive relative to *Avicennia marina*, while the gas exchange characteristics of the latter during increased salinity showed decrease in evaporation rate and stomatal conductance, along with CO_2 assimilation rate and intercellular CO_2 concentration (Ball and Farquhar 1984b).

Increased evaporation along with decreased rainfall may reduce the extent of intertidal mangrove areas, with loss of the landward mangrove zone to saline flats. Raised salinity also may decrease mangrove productivity to cause reduced growth, with varied effects on species. Such reduced precipitation and humidity along with rise in mean sea level may also cause changed competitive interactions between different mangrove species present in a community.

Reduced salinity and exposure to sulphate may occur in mangroves as a result of increased rainfall (Snedaker 1995), along with increased supply of nutrients. This would lead to increased sedimentation rates, along with other effects of enhanced groundwater and less saline habitats resulting in an increase in mangrove productivity and diversity (Smith and Duke 1987; Krauss et al. 2003; Whelan et al. 2005; Rogers et al. 2006). With increased rainfall the area of mangrove may increase in some areas, as previously unvegetated areas of the landward fringe become colonized, along with increased growth and diversity of adjacent mangrove zones. The effects of increased rainfall on mangroves are therefore likely to be beneficial.

3.4 UV Changes

Loss of stratospheric ozone due to human impacts has caused increases in the lower troposphere of ultraviolet-B (UV-B) radiation, which can be damaging to proteins and nucleotides in plant tissues. In the Southern Hemisphere since the 1970s, ground level UV-B radiation has increased by approximately 6 % at mid-latitudes and up to 130 % over Antarctica (Madronich et al. 1998).

Mangrove leaves contain a suite of pigments that absorb UV-B radiation, likely as a result of their evolution in tropical latitudes where UV-B radiation levels are high (Lovelock et al. 1992; Krause et al. 2003; Lovelock and Ellison 2007). Effects on mangroves include altered morphology and some reduction in photosynthetic rates (Caldwell et al. 2003). Increased UV-B radiation is likely to have a significant effect on subtidal primary producers, the effects on intertidal plants expected to be less significant (Day and Neale 2002). Moorthy and Kathiresan (1997) grew *R. apiculata* seedlings under UV light levels associated with stratospheric ozone deletion finding an increase in net photosynthesis at lower levels but a decrease at higher levels.

Impacts of UV-B radiation are most likely to more affect plants where shade leaves become exposed to full solar radiation (Krause et al. 2003), which may occur as a result of wind or storm damage.

3.5 Storms and High Energy Wave Impacts

Climate change projections warn of increased strength of tropical hurricanes and cyclones, combined with larger storm surges and extreme waves (Nicholls et al. 2007). While tsunami events are not linked with climate change, their wave effects can be similar to storm surges hence insight can be gained from them into impacts of large storms on mangroves. Mangroves have been shown to reduce the energy of waves during storm and tsunami events, and provide resilience to substrate erosion through their root mats (Massel et al. 1999; Mazda et al. 2002; Danielsen et al. 2005; Dahdouh-Guebas et al. 2005; Katharesan and Rajendran 2005; Vermaat and Thampanya 2006).

The dense foliage of mangroves combined with friction effects of aerial roots provides some facility in reducing wave power, including reducing damage from extreme events such as tsunami waves (Mazda et al. 2007). It has been shown in a *Rhizophora* forest that at high tide levels there is a 50 % decrease in wave energy within 150 m of the seaward edge (Brinkman et al. 2007). The degree of protection obviously is subject to the size of storm waves, and the density of the forest (Yanagisawa et al. 2009), but with a tsunami wave of over 6 m mangroves are mostly destroyed.

Assessments after the 2004 Asian tsunami showed that human deaths and loss of property in some coastal villages were reduced by the shielding function of coastal vegetation (Katharesan and Rajendran 2005; Dahdouh-Guebas et al. 2005; Walters et al. 2008). The amount of reduction of wave energy in tsunamis and cyclonic storms is influenced by the tree density of the mangrove forest and types and heights of aerial root systems.

Following cyclone damage to mangroves can be complete defoliation over the narrow area of cyclone paths (Jaffar 1993). In southern Florida, Craighead and Gilbert (1962) described hurricane impacts on mangroves of the species *Avicennia germinans*, *Rhizophora mangle*, *Conocarpus erectus* and *Laguncularia racemosa*. Widespread mortality was recorded to have occurred, of 25–75 % across large areas and some locations as high as 90 %. This was not due to defoliation or tree damage, as some trees later sprouted new leaves, rather it was caused by sediment deposits of over 10 cm depths burying roots leading to oxygen deficiency.

Storm impacts can deposit marine sediments into mangrove seaward margins to either cause mortality or build shore parallel sand ridges or chenier ridges. Sediment erosion and deposition can however misrepresent the amount of elevation change because of subsurface processes being altered as a result of the storm (Cahoon 2006).

Wave action has increasing impact with higher water levels such as surges, because reduction in friction with increased water depth causes wave energy increase. Increased inundation occurs with sea level rise.

3.6 Sea Level Rise Impacts

Sea level rise associated with warming temperatures as caused by climate change is projected globally to reach 0.18–0.59 m by 2099, involving rates of rise of 1.5–9.7 mm a^{-1}. However, relative sea level trends at different coastlines may be significantly different from global mean sea level rates owing to local or regional uplift/ subsidence (Meehl et al. 2007).

Changes in sea level at a coastline can result from variation in the volume of ocean water or adjustment movement of the land, continental shelf or ocean floor (Fig. 3), and a coastal location will experience relative sea-level change or stability owing to a combination of these factors. Some coastal areas experience long-term relative sea level rise owing to tectonic subsidence, sediment compaction or fluid extraction (Syvitski et al. 2009; Nicholls and Cazanave 2010), to which global recent and projected sea-level rise mainly caused by expanding oceans and ice melt is adding. Sites with deltaic subsidence such as the Ganges-Brahmaputra delta experience higher rates of relative sea level rise and so are more exposed. Such trends in relative sea level are measured by long-term tide gauges (Bindoff et al. 2007); however, most mangrove coastlines in the developing world lack such records.

Mangroves occur between mean sea levels and the high tide elevations (Fig. 1), making the system particularly vulnerable to habitat changes with sea level rise. Across the intertidal slope, mangrove species have micro-elevation habitat preferences controlled by frequency of inundation and salinity, leading to shore- parallel species zones. Figure 1 shows a simplified mangrove zonation typical of the South East Asian region, with different species each favouring an elevation range from mean sea-level up to high tide levels, where the mangroves change to landward communities.

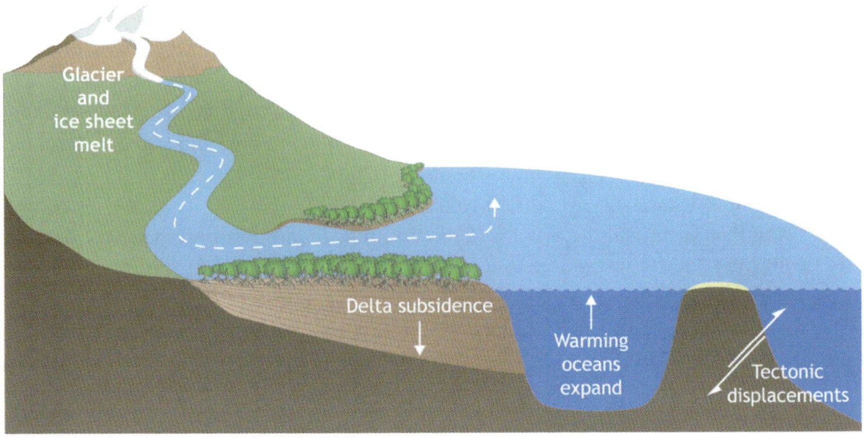

Fig. 3 Causes of relative sea-level change on mangrove coastlines

Table 2 Elevations of mangrove zones in the Ajkwa/Tipoeka mangrove estuaries, West Papua, Indonesia. (Source: Ellison 1993)

Position	Mangrove zone	Elevation relative to mean sea level (m) (error ± 0.2)
Seaward edge fringe	*Avicennia/Sonneratia*	− 0.35 to + 0.15
Seaward major zone	*Rhizophora* dominant	+ 0.15 to + 1.1
Landward major zone	*Bruguiera* dominant	+ 1.1 to + 1.6
Mangrove landward margin	Mixed mangrove/Freshwater forest	Above + 1.6

Demonstrating this sea-level control of mangrove distributions, in extensive estuarine mangroves of SW Papua in Indonesia, elevations of mangrove zones (Fig. 1) were accurately surveyed to show consistent zone elevations across a 1.9 m range within the mesotidal intertidal range of 3.5 m (Table 2). This sea-level control of mangrove locations as shown by topographic survey was also found in other studies (Boto and Bunt 1981; Wolanski et al. 1992; Ellison 1993).

Tidal ranges will influence on mangrove vulnerability to sea level rise, with greater exposure occurring in areas of narrower tidal range relative to those of wider tidal range (Lovelock and Ellison 2007; Ellison 2012). Geomorphological setting and associated sediment budgets also influence vulnerability to sea level rise (Ellison 2009). Mangroves exist in active coastal settings, influenced by tidal movements, wave action, and catchment runoff including floods and storm action. These are all direct influences on the sediment balance of the mangrove area, as well as factors such as sediment supply. This sediment budget of a mangrove area is summarised in Fig. 4, where net sedimentation is a balance of volumes of sediment entering or leaving the mangrove environment, influencing whether the substrate erodes or accretes.

Surface elevation tables along with a sediment surface horizon marked with an exotic layer such as feldspar clay enables substrate accretion to be distinguished from both shallow subsidence and elevation change (Cahoon et al. 2002). Across

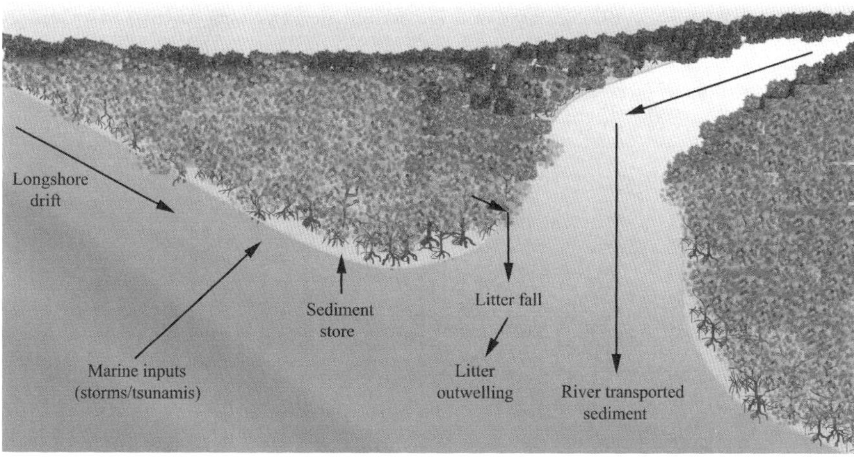

Fig. 4 Sediment budget of a mangrove swamp. (Source: Ellison 2009)

a variation of sites, Cahoon et al. (2006) found that vertical accretion averaged 5 mm a^{-1}, elevation change 1 mm a^{-1}, and shallow subsidence 4 mm a^{-1}. These results over the last few years show similarity to long-term net rates of mangrove sediment accretion shown by stratigraphic radiocarbon dates (Ellison 2009). Net sediment accretion allows the mangrove substrate to keep level with sea-level rise, so as not to alter the inundation preferences of trees.

In conditions where relative sea-level rise exceeds net sediment accretion rates, mangroves retreat landward. This is demonstrated for the last few thousand years from mangrove swamps of SW Papua as shown in Fig. 5, which is a pollen stratigraphic diagram typical of the coastline (Ellison 2005). At lower depths a landward *Bruguiera* zone was present at the core site for several thousand years, changing around 3,000 years ago to a more seaward *Rhizophora* zone. Net sediment accretion rates since that time were 0.5–0.6 mm a^{-1}, while the rate of sea-level rise was slightly more rapid at 0.67 mm a^{-1}, resulting in gradual landward migration.

Spatial analysis of mangrove area change over the last few decades can show loss of mangroves from the seaward edge, which is a key sign of sea level rise impacts (Gilman et al. 2008). In Cameroon mangroves, Ellison and Zouh (2012) used spatial analysis to show that over two-thirds of the shoreline edge of mangroves had suffered dieback at rates of up to 3 m a^{-1} over the period 1975–2007, and a mangrove island offshore suffered 89 % loss. Global averages of sea level rise in 1961–2003 were 1.8 (1.3–2.3) mm a^{-1} (IPCC 2007). The net sedimentation rate under mangroves was shown by one core to be 2.5 mm a^{-1}.

In conditions of sedimentation surplus, mangroves colonise seaward either into bays especially offshore of river mouths or over reef flats. Panapitukkul et al. (1998) demonstrated mangrove progradation of about 38 m a^{-1} at Pak Phanang Bay in SE Thailand with high rates of river delivery of sediment. Such mangrove areas with high sedimentation rates will be more resilient during rising sea levels.

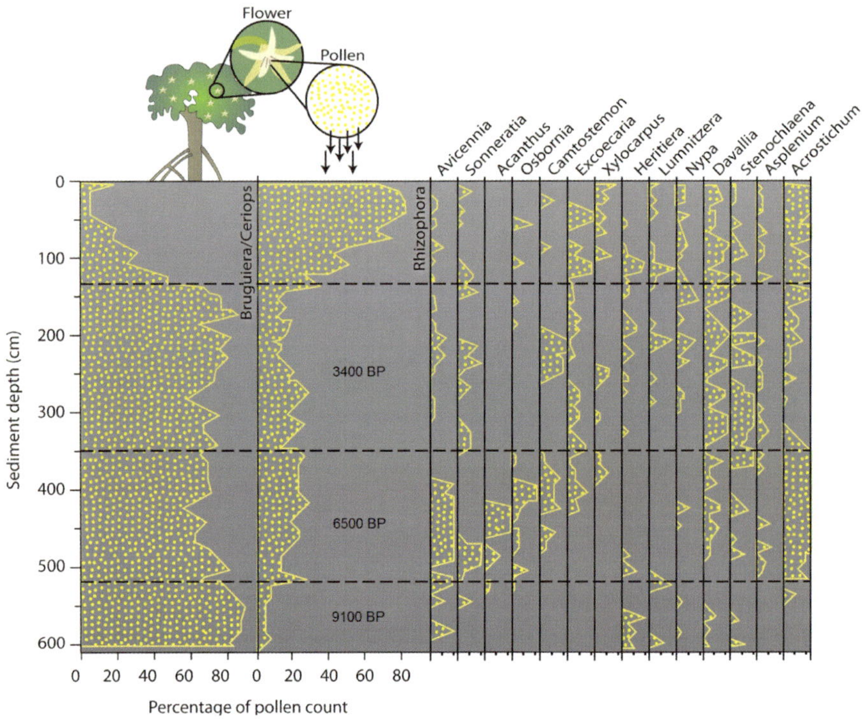

Fig. 5 Late Holocene vegetation changes in the Tipoeka Estuary, West Papua, and Indonesia. (Source: C. Collier from Ellison 2005)

4 Vulnerability Assessments

Current knowledge therefore indicates that climate change effects on mangroves are significant, may already be occurring, and may continue to be a threat even after atmospheric CO_2 emissions are decreased or stabilized. Even the best-case projection scenarios indicate that climate change will continue for centuries to come (IPCC 2007). Therefore it is essential that mangrove managers adopt actions that increase the resilience of mangroves to climate change impacts. To achieve this, management of mangroves can develop appropriate strategies that allow climate change adaptation.

In collaboration with a range of stakeholders and local communities and with support from the UNEP Global Environment Facility, WWF recently conducted trials in three countries to strengthen and build the capacity of local mangrove managers to assess mangrove vulnerability to climate change impacts, and interpret results to identify appropriate adaptation. These trials allowed the development of a generalised methodology for vulnerability assessment and adaptation prioritisation (Ellison 2012).

Table 3 Summary of the methodology components of a mangrove vulnerability assessment. (Source: Ellison 2012)

Number	Methodology	Approach
1	Initial review of existing information	Desk-top computer searches, inquiries of stakeholders
2	Forest assessment of mangroves	Transect-based permanent plots, and rapid condition assessment methods
3	Recent spatial changes of mangroves	Air photograph and satellite image analysis of change by GIS
4	Ground surface elevations in and behind the mangroves	High technology survey or water level correlation
5	Relative sea level trends	Tide gauge analysis, or stratigraphy, radiocarbon dating and pollen analysis
6	Adjacent ecosystem resilience	Coral reef and seagrass monitoring standard methods
7	Downscaled climate (rainfall) modelling	Assess available projections
8	Compilation of local community knowledge	Facilitated workshops Structured questionnaire surveys
9	Synthesis of results	Identify specific vulnerabilities to select adaption priorities

The methods of vulnerability assessment allow identification of climate change impacts already occurring, and components vulnerable to future impacts. It identifies any non-climate stresses prevalent which may exacerbate the impacts caused by climate change, or limit the system's ability to absorb change. The methodologies are listed in Table 3.

Vulnerability to climate change impacts can be demonstrated through the assessment process by several results. Results that show increased vulnerability are:

1. Mangroves in poor condition such as suffering from unsustainable exploitation will have reduced resilience to climate change perturbation, particularly mangrove species that are already threatened or close to extinction (Polidoro et al. 2010).
2. Coastal areas that are subsiding will experience higher rates of relative sea level rise relative to coastlines that are tectonically stable. Such subsidence may be due to tectonic movement or deltaic compaction/ isostatic adjustment/ fluid extraction.
3. Mangroves that occur in a micro-tidal area will be more perturbed by sea-level rise than mangroves that occur in a macro-tidal area. For example, a 30 cm sea-level rise would totally relocate the intertidal zone upslope in a 30 cm tidal range such as the Caribbean, but only move the intertidal zone by 20 % in a 3 m tidal range.
4. Low sedimentation rates under mangroves, as sediment accretion allows mangrove substrate to "keep up" with sea level rise and so maintain their inundation frequencies.
5. Mangrove zones that occupy a smaller elevation bracket in the tidal spectrum will be more perturbed by sea-level rise than mangrove zones that occupy a wider elevation bracket.

During the vulnerability assessment climate change impacts on mangroves as reviewed above can be identified by a number of these methodologies:

1. Spatial change over time across the mangrove area. This would be demonstrated by retreat along with tree mortality at the seaward edge as shown by Ellison and Zouh (2012), and mangrove recruitment at the inland landward edge.
2. Significant change in the timing of flowering and fruiting of trees, which may alter ecological connections with pollination and dispersal species, for example.
3. Reduced health, reduced growth, loss of reproduction and mortality of mangrove trees.
4. Lack of recruitment (seedling growth) under mangrove species of that same species. Recruitment under mangroves of a species that normally occurs lower in the tidal spectrum.

Results from the different components of the vulnerability assessment (Table 3) can be used to identify adaptation options that increase resilience of the mangrove area to climate change impacts (Ellison 2012). The most critical components to the vulnerability assessment are the exposure factors of site-specific sea level rise relative to sediment supply, along with the sensitivity factors of forest assessment by permanent plots, analysis of recent spatial change and sedimentation rates. These identify the "tipping point" factors for the mangrove ecosystem with respect to climate change impacts. Local community knowledge is an indicator of social vulnerability rather than a direct measure of ecosystem vulnerability.

5 Conclusions

Assessments of vulnerability of many ecosystems to climate change impacts focus on climatic modelling of temperature and rainfall changes to evaluate exposure (Dixon et al. 2003; Donner et al. 2005; Vos et al. 2008; Klausmeyer and Shaw, 2009), whereas for mangrove ecosystems the key exposure factor is projected sea-level rise. The effects of projected temperature increase and increased atmospheric CO_2 are likely to be mostly beneficial to mangroves, increasing mangrove productivity and biodiversity (Nicholls et al. 2007; Gilman et al. 2008) particularly at higher latitudes. The benefits of CO_2 increase are subject to the limiting factors of nutrient availability, humidity and salinity (Ball et al. 1997). Changes in rainfall and humidity changes may affect mangroves more, with reduced rainfall having the impact of decreasing productivity and biodiversity and causing relative subsidence. Increased rainfall is likely to be beneficial, causing productivity and sedimentation along with enhanced groundwater and less saline habitats. Of all climate change impacts, the impacts of relative sea-level rise are thought to be the most detrimental to mangrove ecosystems (Gilman et al. 2008; Krauss et al. 2010), shown first by mortality of mangroves at the seaward edge and landward migration, but such impacts are dependent on sediment supply and human modifications/ barriers to migration.

Storms may also cause a local "tipping point" of change as the mangrove area becomes further under stress from rising sea-level, when a storm causing mortality that could normally be recovered from, but with sea-level-rise stress this may be hindered. Increased waves and wind may cause changes in forest composition and extent (Semeniuk 1994). Storms may also increase vulnerability to rising sea level through erosion and substrate subsidence (Cahoon 2006).

The response of mangrove habitats to climate change in different coastal locations may be subject to a range of factors of coastal behaviour, such as shore profile, sedimentology, tidal range and salinity regime, and the mangrove species composition. Different impacts of climate change reviewed in this chapter may cause stress on mangrove species such as increasing inundation, salinity and reduction in fresh water availability. Mangrove species differ in their tolerance to physiological stresses (Ball 1988), and their relative performance in a stressful environment will be ultimately determined by the trade-off between strategies required to maximise acquisition of limiting resources and those necessary to tolerate or avoid factors that disrupt metabolism (McKee 1995). This may be different according to the relative balance of climate change stresses in each situation.

Detailed site-based mangrove vulnerability assessments can identify aspects of mangrove areas that are already under stress, from climate change and other factors (Ellison 2012). Reducing threats to mangroves from other factors such as unsustainable use can increase their resilience to future climate change. Such assessment allows identification of specific factors of vulnerability in different mangrove areas, and allows prioritisation of adaptation actions that will reduce that identified vulnerability.

Acknowledgments The research in this review was funded by the UNEP Global Environmental Facility project "Coastal Resilience to Climate Change: Developing a Generalizable Method for Assessing Vulnerability and Adaptation of Mangroves" awarded to the World Wildlife Fund (WWF).

References

Armitage D (2002) Socio-institutional dynamics and the political ecology of mangrove forest conservation in Central Sulawesi, Indonesia. Global Environ Change 12:203–217

Ball MC, Farquhar GD (1984a) Photosynthetic and stomatal responses of two mangrove species, *Aegiceras corniculatum* and *Avicennia marina*, to long term salinity and humidity conditions. Plant Physiol 74:1–6

Ball MC, Farquhar GD (1984b) Photosynthetic and stomatal responses of grey mangrove, *Avicennia marina*, to transient salinity conditions. Plant Physiol 74:7–11

Ball MC (1988) Salinity tolerance in the mangroves *Aegiceras corniculatum* and *Avicennia marina*. I. Water use in relation to growth, carbon partitioning, and salt balance. Aust J Plant Physiol 15:447–464

Ball MC, Cocharane MJ, Rawson HM (1997) Growth and water use of the mangroves *Rhizophora apiculata* and *R. stylosa* in response to salinity and humidity under ambient and elevated concentrations of atmospheric CO_2. Plant Cell Environ 20:1158–1166

Banus MD (1983) The effects of thermal pollution on red mangrove seedlings, small trees and mangrove reforestation. In: Ogden JC (ed) Coral reefs, seagrass beds and mangroves: Their interaction in coastal zones of the Caribbean. UNESCO Reports in Marine Science 23. UNESCO, Paris, pp 114–127

Baran E, Hambrey J (1999) Mangrove conservation and coastal management in Southeast Asia, What impact on fishery resources? Mar Pollut Bull 37:431–440

Boto KG, Bunt JS (1981) Tidal export of particulate organic matter from a Northern Australian mangrove system. Estuar Coast Shelf Sci 13:247–255

Bindoff NL, Willebrand J, Artale V, Cazenave A, Gregory J, Gulev S K, Hanawa K, Le QC, Levitus S, Nojiri Y, Shum CK, Talley LD, Unnikrishnan A (2007) Observations: oceanic climate change and sea level. In: Solomon S, Qin D, Manning M, Chen Z, Marquis M, Averyt KB, Tignor M, Miller HL (eds) Climate change 2007: the physical science basis. Contribution of working group I to the fourth assessment report of the intergovernmental panel on climate change. Cambridge University Press, Cambridge and New York, pp 385–482

Brinkman RM, Massel SR, Ridd PV, Furukawa K (2007) Surface wave attenuation in mangrove forests. Proc 13th Austral Coastal Ocean Engineer Conf 2:941–949

Cahoon DR, Lynch JC, Perez BC, Segura B, Holland R, Stelly C, Stephenson G, Hensel P (2002) High precision measurement of wetland sediment elevation: II. The Rod Surface Elevation Table. J Sediment Res 72:734–739

Cahoon DR (2006) A review of major storm impacts on coastal wetland elevations. Estuar Coast 29:889–898

Cahoon D, Hensel P, Spencer T, Reed DJ, McKee KL, Saintilan N (2006) Coastal wetland vulnerability to relative sea-level rise: Wetland elevation trends and process controls. In: Verhoeven JTA, Beltman B, Bobbink R, Whigham DF (eds) Wetlands and natural resource management. Ecological Studies, Vol. 190. Springer, Berlin, pp 271–292

Canoy MJ (1975) Diversity and stability in a Puerto Rican *Rhizophora mangle* L. forest. In: Walsh GE, Snedaker SC, Teas HJ (eds) Proc Internat Sympos Biol Manag Mangr. University of Florida, Gainesville, pp 344–356

Caldwell MM, Ballaré CL, Bornman JF, Flint SD, Björn LO, Teramura AH, Kulandaivelu G, Tevini M (2003) Terrestrial ecosystems, increased solar ultraviolet radiation and interactions with other climatic change factors. Photochem Photobiol Sci 2:29–38

Cheeseman JM, Clough BF, Carter DR, Lovelock CE, Eong OJ, Sim RG (1991) The analysis of photosynthetic performance in leaves under field conditions, a case study using *Bruguiera* mangroves. Photosyn Res 29:11–22

Cheeseman JM (1994) Depressions of photosynthesis in mangrove canopies. In: Baker NR, Bowyer JR (eds) Photoinhibition of photosynthesis, from molecular mechanisms to the field. Bios, Oxford, pp 377–389

Cheeseman JM, Herendeen LB, Cheeseman AT, Clough BF (1997) Photosynthesis and photoprotection in mangroves under field conditions. Plant Cell Environ 20:579–588

Chitaro P, Usseglio P, Sale PF (2005) Variation in fish density, assemblage composition and relative rates of predation among mangrove, seagrass and coral reef habitats. Environment Biol Fishes 72:175–187

Clough BF, Andrews TJ, Cowan IR (1982) Physiological processes in mangroves. In: Clough BF (ed) Mangrove ecosystems in Australia. Australian National University, Canberra, pp 193–210

Clough BF, Sim RG (1989) Changes in gas exchange characteristics and water use efficiency of mangroves in response to salinity and vapour pressure deficit. Oecologia 79:38–44

Conacher CA, O'Brien C, Horrocks JL, Kenyon RK (1996) Litter production and accumulation in stressed mangrove communities in the Embley River estuary, North-eastern Gulf of Carpentaria, Australia. Marine Freshw Res 47:737–743

Craighead FC, Gilbert VC (1962) The effects of Hurricane Donna on the vegetation of Southern Florida. Q J Fla Acad Sci 25:1–28

Dahdouh-Guebas F, Jayatissa LP, Di Nitto D, Bosire JO, Lo Seen D, Koedam N (2005) How effective were mangroves as a defence against the recent tsunami? Curr Biol 15:R443–447

Danielsen F, Soerensen M, Olwig M, Selvam V, Parish F, Burgess N, Hiraishi T, Karunagaran V, Rasmussen M, Hansen L, Quarto A, Nyoman S (2005) The Asian tsunami: a protective role for coastal vegetation. Science 310:643

Day TA, Neale PJ (2002) Effects of UV-B radiation on terrestrial and aquatic primary producers. Annu Rev Ecol Syst 33:371–396

Dixon RK, Smith J, Guill S (2003) Life on the edge: vulnerability and adaptation of African ecosystems to global climate change. Mitig Adapt Strat Gl Ch 8:93–113

Donner SD, Skirvin WJ, Little CM, Oppenheimer M, Hoegh-Guldberg O (2005) Global assessment of coral bleaching and required rates of adaptation under climate change. Glob Change Biol 11:2251–2265

Drake BG, Gonzalez-Meler MA, Long SP (1997) More efficient plants, a consequence of rising atmospheric CO_2? Annu Rev Plant Physiol Plant Mol Biol 48:607–737
Drake BG, Azcon-Bieto J, Berry J, Bunce J, Dijkstra P, Farrar J, Gifford RM, Gonzalez-Meler MA, Koch G, Lambers H, Siedow J, Wullschleger S (1999) Does elevated atmospheric CO_2 inhibit mitochondrial respiration in green plants? Plant Cell Environ 22:649–657
Duke NC, Ball MC, Ellison JC (1998) Factors influencing biodiversity and distributional gradients in mangroves. Global Ecol Biogeog Let 7:27–47
Duke NC, Meynecke JO, Dittmann S, Ellison AM, Anger K, Berger U, Cannicci S, Diele K, Ewel KC, Field CD, Koedam N, Lee SY, Marchand C, Nordhaus I, Dahdouh-Guebas F (2007) A world without mangroves. Science 317:41
Ellison JC (1993) Mangrove retreat with rising sea-level, Bermuda. Estuar Coast Shelf Sci 37:75–87
Ellison JC, Simmonds S (2003) Structure and productivity of inland mangrove stands at Lake MacLeod, Western Australia. J R Soc West Aust 86:21–26
Ellison JC (2005) Holocene palynology and sea-level change in two estuaries in Southern Irian Jaya. Palaeogeogr Palaeoclimatol Palaeoecol 220:291–309
Ellison JC (2009) Geomorphology and sedimentology of mangrove swamps. In: Wolanski E, Cahoon D, Perillo GME (eds) Coastal wetlands: an ecosystem integrated approach. Elsevier Science, Amsterdam, pp 564–591
Ellison JC, Zouh I (2012) Vulnerability to climate change of mangroves: assessment from Cameroon, Central Africa. Biology 1:617–638. doi:10.3390/biology1030617
Ellison JC (2012) Climate change vulnerability assessment and adaptation planning for mangrove systems. WWF US, Washington DC
Ewel KC, Twilley RR, Ong JE (1998) Different kinds of mangrove forests provide different goods and services. Glob Ecol Biogeogr Lett 7:83–94
FAO (2003) Status and trends in mangrove area extent worldwide. In: Wilkie ML, Fortuna S (eds) Forest resources assessment working paper no. 63. Forest Resources Division, FAO, Rome
Farnsworth EJ, Ellison AM, Gong WK (1996) Elevated CO_2 alters anatomy, physiology, growth, and reproduction of red mangrove (*Rhizophora mangle* L.). Oecologia 108:599–609
Gilman E, Ellison JC, Duke NC, Field C, Fortuna S (2008) Threats to mangroves from climate change and adaptation options: a review. Aquat Bot 89:237–250
Giri C, Ochieng E, Tieszen LL, Zhu Z, Singh A, Loveland T, Masek J, Duke N (2011) Status and distribution of mangrove forests of the world using earth observation satellite data. Glob Ecol Biogeogr 20:154–159
Intergovernmental Panel on Climate Change (2007) Climate change 2007: the physical basis. Advice for policy makers. Contribution of working group I to the fourth assessment report of the intergovernmental panel on climate change. IPCC, Geneva
Jaffar M (1993) Country report on mangrove ecosystem in the Republic of Fiji. In: Nakamura T (ed) Proceedings seminar and workshop on integrated research on mangrove ecosystems in Pacific islands region II. Japan International Association for Mangroves, Tokyo, pp 107–135
Katharesan K, Rajendran N (2005) Coastal mangrove forests mitigates tsunami. Estuar Coast Shelf Sci 65:601–606
Kimani EN, Mwatha GK, Wakwabi EO, Ntiba JM, Okoth BK (1996) Fishes of a shallow tropical mangrove estuary, Gazi, Kenya. Mar Freshw Res 47:857–868
Klausmeyer KR, Shaw MR (2009) Climate change, habitat loss, protected areas and the climate adaptation potential of species in Mediterranean ecosystems worldwide. PLoS One 4(7):e6392. doi:10.1371/journal.pone.0006392
Krause GH, Galle A, Gademann R, Winter K (2003) Capacity of protection against ultraviolet radiation in sun and shade leaves of tropical forest plants. Funct Plant Biol 30:533–542
Krauss KW, Allen JA, Cahoon DR (2003) Differential rates of vertical accretion and elevation change among aerial roots types in Micronesian mangrove forests. Estuar Coast Shelf Sci 56:251–259
Krauss KW, Cahoon DR, Allen JA, Ewel KC, Lynch JC, Cormier N (2010) Surface elevation change and susceptibility of different mangrove zones to sea-level rise on Pacific high islands of Micronesia. Ecosystems 13:129–143
Kress WJ, DeFilipps RA, Farr E, Kyi DYY (2003) A checklist of the trees, shrubs, herbs, and climbers of Myanmar. Smithsonian Institution. Contrib US Natl Herb 45:1–590

Kumara MP, Jayatissa LP, Krauss KW, Phillips DH, Huxham M (2010) High mangrove density enhances surface accretion, surface elevation change, and tree survival in coastal areas susceptible to sea-level rise. Oecologia 164:545–553

Ley J, McIvor CC (2002) Linkages between estuarine and reef fish assemblages: Enhancement by the presence of well developed mangrove shorelines. In: Porter JW, Porter KG (eds) The Everglades, Florida Bay, and coral reefs of the Florida Keys. An ecosystem sourcebook. CRC Press, Boca Raton, pp 539–562

Lovelock CE, Clough BF, Woodrow IE (1992) Distribution and accumulation of ultraviolet-radiation-absorbing compounds in leaves of tropical mangroves. Planta 188:143–154

Lovelock CE, Ellison JC (2007) Vulnerability of mangroves and tidal wetlands of the Great Barrier Reef to climate change. In: Johnson JE, Marshall PA (eds) Climate change and the Great Barrier Reef: a vulnerability assessment. Great Barrier Reef Marine Park Authority and Australian Greenhouse Office, Townsville, pp. 237–269

Luo Z, Sun OJ, Wang E, Ren H, Xu H (2010) Modeling productivity in mangrove forests as impacted by effective soil water availability and its sensitivity to climate change using Biome-BGC. Ecosystems 13:949–965. doi:10.1007/s10021–010-9365-y

Mackey AP (1993) Biomass of the mangrove *Avicennia marina* (Forsk.) Vierh. near Brisbane, South-eastern Queensland. Aust J Mar Freshw Res 44:721–725

Madronich S, McKenzie RL, Bjorn LO, Caldwell MM (1998) Changes in biologically active ultraviolet radiation reaching the earth's surface. J Photochem Photobiol B Biol 46:5–19

Manhas RK, Negi JDS, Kumar R, Chauhan R (2006) Temporal assessment of growing stock, biomass and carbon stock of Indian forests. Clim Change 74:191–221

Massel SR, Furukawa K, Brinkman RM (1999) Surface wave propagation in mangrove forests. Fluid Dyn Res 24:219–249

Mazda Y, Magi M, Nanao H, Kogo M, Miyagi T, Kanazawa N, Kobashi D (2002) Coastal erosion due to long term human impact on mangrove forests. Wetlands Ecol Manag 10:1–9

Mazda Y, Wolanski E, Ridd PV (2007) The role of physical processes in mangrove environments: manual for preservation and utilisation of mangrove ecosystems. Terrapub, Tokyo

McKee KL (1995) Interspecific variation in growth, biomass partitioning, and defensive characteristics of Neotropical mangrove seedlings: response to light and nutrient availability. Am J Bot 82:299–307

Meehl GA, Stocker TF, Collins W, Friedlingstein P, Gaye AT, Gregory JM, Kitoh A, Knutti R, Murphy JM, Noda A, Raper SCB, Watterson IG, Weaver AJ, Zhao Z-C (2007) Global climate projections. In: Solomon S, Qin W, Manning M, Chen Z, Marquis M, Averyt KB, Tignor M, Miller HL (eds) Climate change 2007: the physical science basis. Contribution of working group I to the fourth assessment report of the intergovernmental panel on climate change. Cambridge University Press, Cambridge, pp 747–846

Moorthy P, Kathiresan K (1997) Influence of untraviolet-B radiation on photosynthetic and biochemical characteristics of a mangrove *Rhizophora apiculata*. Photosynthetica 34:465–471

Mumby PJ, Edwards AJ, Arias-González JE, Lindeman KC, Blackwell PG, Gall A, Gorczynska MI, Harborne AR, Pescod CL, Renken H, Wabnitz CCC, Llewellyn G (2004) Mangroves enhance the biomass of coral reef fish communities in the Caribbean. Nature 427:533–536

Nicholls RJ, Wong PP, Burkett VR, Codignotto JO, Hay JE, McLean RF, Ragoonaden S, Woodroffe CD (2007) Coastal systems and low-lying areas. In: Parry ML, Canziani OF, Palutikof JP, van der Linden PJ, Hanson CE (eds) Climate change 2007: impacts, adaptation and vulnerability. Contribution of working group II to the fourth assessment report of the intergovernmental panel on climate change. Cambridge University Press, Cambridge, pp 315–356

Nicholls RJ, Cazanave A (2010) Sea-level rise and its impact on coastal zones. Science 328: 1517–1520

Ong JE (2003) Plants of the Merbok mangrove, Kedah, Malaysia and the urgent need for their conservation. Folia Malaysiana 4:1–18

Panapitukkul N, Duarte CM, Thampanya U, Kheowvongsri P, Srichai N, Geerz-Hansen O, Terrados J, Boromthanarath S (1998) Mangrove colonisation: mangrove progression over the growing Pak Phanang (SE Thailand) mud flat. Estuar Coast Shelf Sci 47:51–61

Polidoro BA, Carpenter KE, Collins L, Duke NC, Ellison AM, Ellison JC, Farnsworth EJ, Fernando ES, Kathiresan K, Koedam NE, Livingstone SR, Miyagi T, Moore GE, Nam VN, Primavera JH, Salmo III SG, Sanciangco JC, Sukardjo S, Wang Y, Yong JWH (2010) The loss of species: mangrove extinction risk and geographic areas of global concern. PLoS ONE 5(4):e10095. doi:10.1371/journal.pone.0010095

Robertson AI, Duke NC (1990) Recruitment, growth and residence times of fishes in a tropical Australian mangrove system. Estuar Coast Shelf Sci 31:725–745

Rogers K, Saintilan N, Cahoon DR (2005) Surface elevation dynamics in a regenerating mangrove forest at Homebush Bay, Australia. Wetl Ecol Manag 13:587–598

Rogers K, Saintilan N, Heijnis H (2006) Mangrove encroachment of salt marsh in Western Port Bay, Victoria, the role of sedimentation, subsidence, and sea level rise. Estuaries 28:551–559

Scholander PF, Hammel HT, Hemmingsen E, Garey W (1962) Salt balance in mangroves. Plant Physiol 37:722–729

Semeniuk V (1994) Predicting the effect of sea-level rise on mangroves in northwestern Australia. J Coast Res 10:1050–1076

Smith III TJ, Duke NC (1987) Physical determinants of inter-estuary variation in mangrove species richness around the tropical coastline of Australia. J Biogeogr 14:9–19

Snedaker SC (1995) Mangroves and climate change in the Florida and Caribbean region: scenarios and hypotheses. Hydrobiologia 295:43–49

Spalding M, Kainuma M, Collins L (2010) World Atlas of Mangroves. Earthscan, London

Steinke TD, Naidoo Y (1991) Respiration and net photosynthesis of cotyledons during establishment and early growth of propagules of the mangrove, *Avicennia marina*, at three temperatures. S Afr J Bot 57:171–174

Stern WL, Voight GK (1959) Effect of salt concentration on growth of red mangrove in culture. Bot Gaz 131:36–39

Syvitski JPM, Kettner AJ, Overeem I, Hutton EWH, Hannon MT, Brakenridge GR, Day J, Vorosmarty C, Saito Y, Giosan L, Nicholls RJ (2009) Sinking deltas due to human activities. Nature Geosci 2:681–687

Tans P (2013). Trends in atmospheric carbon dioxide—Mauna Loa. NOAA/ESRL. www.esrl.noaa.gov/gmd/ccgg/trends. Accessed 21 Feb 2013

Valiela I, Bowen JL, York JK (2001) Mangrove forests, one of the world's threatened major tropical environments. BioScience 51:807–815

Vermaat JE, Thampanya U (2006) Mangroves mitigate tsunami damage: a further response. Estuar Coast Shelf Sci 69:1–3

Vos CC, Berry P, Opdam P, Baveco H, Nijhof B, O'Hanley J, Bell C, Kuipers H (2008) Adapting landscapes to climate change: examples of climate-proof ecosystem networks and priority adaptation zones. J Appl Ecol 4:1722–1731

Walters BB, Ronnback P, Kovacs JM, Crona B, Hussain SA, Badola R, Primavera JH, Barbier E, Dahdouh-Guebas F (2008) Ethnobiology, socio-economics and management of mangrove forests: a review. Aquat Bot 89:220–236

Warrick RA, Gifford RM, Parry ML (1987) CO_2 climate change and agriculture. In: Bolin B, Doos BR, Jager J, Warrick RA (eds) The greenhouse effect, climate change, and ecosystems. Wiley, Chichester, pp 393–473

Whelan KRT, Smith III TJ, Cahoon DR, Lynch JC, Anderson GH (2005) Groundwater control of mangrove surface elevation, shrink and swell varies with soil depth. Estuaries 28:833–843

Wolanski E, Mazda Y, Ridd P (1992) Mangrove hydrodynamics. In: Robertson AI, Alongi DM (eds) Tropical mangrove ecosystems. American Geophysical Union, Washington DC, pp 43–62

Wolanski E, Spagnol S, Thomas S, Moore K, Alongi D, Trott L, Davidson A (2000) Modelling and visualising the fate of shrimp pond effluent in a mangrove-fringed tidal creek. Estuar Coast Shelf Sci 50:85–97

Yanagisawa H, Koshimura S, Goto K, Miyagi T, Imamura F, Ruangrassamee A, Tanavud C (2009) The reduction effects of mangrove forest on a tsunami based on field surveys at Pakarang Cape, Thailand and numerical analysis. Estuar Coast Shelf Sci 81:27–37

Impacts of Climate Change on Asian Mangrove Forests

Asish Kumar Parida, Vivekanand Tiwari and Bhavanath Jha

Contents

Abstract Mangroves are woody trees and shrubs which thrive in the inhospitable zone between land and sea along the tropical and subtropical coasts of the globe. Mangroves have made significant contributions to the economical status of the coastal communities of tropical regions for centuries, affording a large number of goods and services such as wood and timber production, salt production, support for commercial and subsistence fisheries, protection of shoreline from cyclones and typhoons and controlling coastal erosion. It has been estimated that the total mangrove area of the world was 137,760 km^2 in 2000, and Asia occupies the largest mangrove covering area in the world. The mangrove forests of Asia are currently threatened by many human activities. Besides the overexploitation of mangrove ecosystems by human activities, climate changes pose serious impacts on Asian mangrove forests. The recent studies that have focused on various climate change components affecting the mangrove forests will be discussed in this chapter with special emphasis on Asian mangrove forests. The changes in the temperature, atmospheric CO_2 concentration, precipitation, storms, ocean circulation patterns, hydrology (flows of tidal and fresh

A. K. Parida (✉) · V. Tiwari · B. Jha
Discipline of Marine Biotechnology and Ecology,
CSIR- Central Salt and Marine Chemicals Research Institute (CSIR-CSMCRI),
Gijubhai Badheka Marg, 364002 Bhavnagar, Gujarat, India
e-mail: asishparida@csmcri.org

I. Faridah-Hanum et al. (eds.), *Mangrove Ecosystems of Asia*,
DOI 10.1007/978-1-4614-8582-7_11,

water within the mangrove ecosystem), sea level rise and anthropogenic activities are the major climate change components affecting the mangrove ecosystems of Asia. Damage of mangrove ecosystems caused by climate change will significantly affect the socio-economic lives of coastal communities of Asia. Climate change driven loss of the mangrove ecosystem also results in higher risk to human safety due to the loss of protection from coastal hazards such as flooding, soil erosion, storm waves and tsunami. Therefore, the immediate attention of biologists and ecologists is needed to protect the mangroves of the Asian coast lines as well as the world which are vulnerable to climate-change driven damage.

1 Introduction

The mangroves are an assortment of woody trees and shrubs which flourish in the inhospitable zone between land and sea along the tropical and subtropical coastlines of the globe (Parida et al. 2005). The mangroves are the salt tolerant plants that grow in varying concentrations of salinity based on their exposure to the tidal flow at the time of high tide and low salinity in the monsoon. The mangrove forest is one of the most productive terrestrial ecosystems (Kathiresan and Bingham 2001). It is habitat and breeding ground for several marine and terrestrial species. The mangrove forests act as buffer between sea and land to protect coastal areas from different natural geological events (Alongi 2008; Barbier et al. 2008). The mangrove forests are distributed in an area of about 15.2 million hectares in 123 countries and different territories of the world. Asia occupies the highest mangrove covering area (37 %), followed by Africa (21 %), North and Central America (15 %), South America (12.6 %) and the remaining 12.4 % are in Australia, Papua New Guinea, New Zealand and the South Pacific Islands (FAO 2007; Sandilyan and Kathiresan 2012). Based on the satellite data, Giri et al. (2011) have reported that total area covered by mangrove forests in 2000 was 137,760 km^2 distributed in 118 countries. They have also reported that Asia has the largest area of mangroves (42 % of total mangrove forest) followed by Africa (20 %), North and Central America (15 %), Oceania (12 %) and South America (11 %). The global distributions of mangrove forests are shown in Fig. 1. Although, the global distribution of mangrove forests varies in different reports but all the reports have stated that the largest area of mangrove forests is occupied by Asia (Fig. 1). The mangrove forests of the tropical regions of Asia, distributed along the Arabian Sea, Bay of Bengal and the Gulf of Thailand, are enriched with diversified species of mangroves (Gopal 2013). Among the Asian countries, Indonesia alone inhabits 54 % of the total mangrove forest of Asia (Fig. 2). The *Sundarban* mangrove forest which is distributed in two countries, India and Bangladesh, is the single largest mangrove forest of the world, covering about 10,000 ha in the delta region of the river Ganga and Brahmaputra (Gopal and Chauhan 2006; Gopal 2013). Ellison et al. (1999) have reported that the richest species composition of mangroves is found in Indo-Malayan mangrove forests. The coastal region of China is covered by mangrove forests of 21,000 ha (Chen et al. 2009; Gopal 2013).

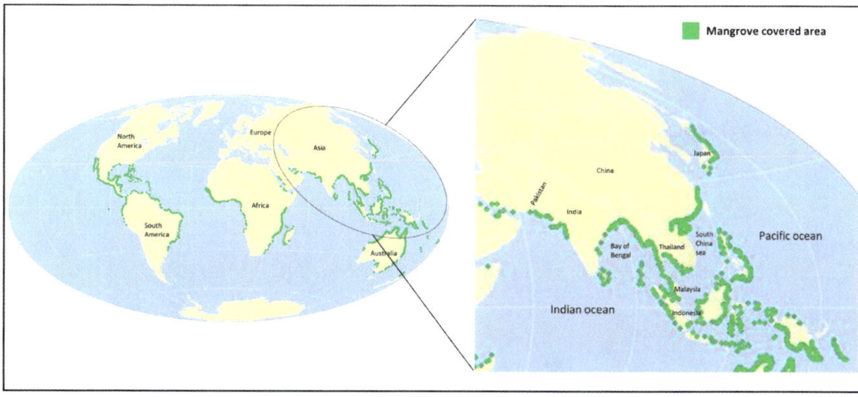

Fig. 1 Global distribution of mangrove forests. Distributions of mangroves in Asian countries are shown in inset. (Source: Giri et. al 2011)

Fig. 2 Distribution of mangrove forests (shown as % of total Asian mangrove area) in different Asian counties (Source: Giri et al. 2011)

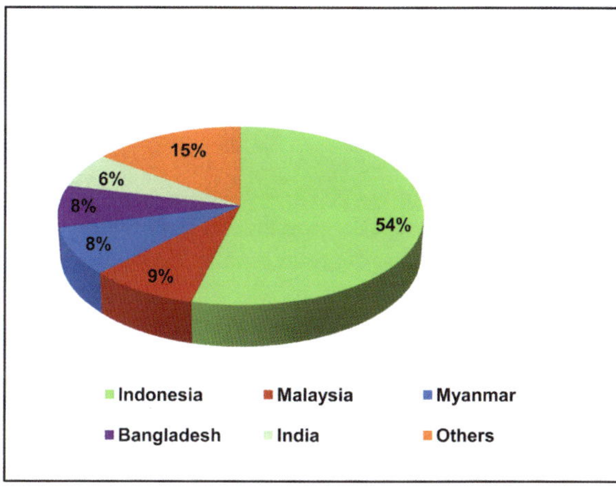

The mangroves play a crucial role in mitigating the global warming by reducing the CO_2 load from the environment. About 22.8 million metric tons of carbon (11 % of the total input of terrestrial carbon into the ocean) is sequestered by the mangroves and associated soil microorganisms (Jennerrjahn and Ittekot 2002; Giri et al. 2011). The mangroves and associated soil microorganisms provide more than 10 % of essential organic carbon to the global oceans (Dittmar et al. 2006). These data revealed that the carbon sequestration by the mangrove ecosystem is 50 times higher than any of the tropical forest ecosystems (Sandilyan and Kathiresan 2012). Traditionally mangroves are used for several years for the well being of humans residing in the coastal region. The mangrove woods are used as firewood and for construction of materials. Due to the presence of high energy and strength in the wood, it is used for timber and charcoal production. The mangroves are also used as the source of potential medicines. Some of the mangrove species such as *Bruguiera gymnorrhiza*, *Rhizophora mucronata*, *Acanthus ilicifolius*, *Arthrocnemum indicum*

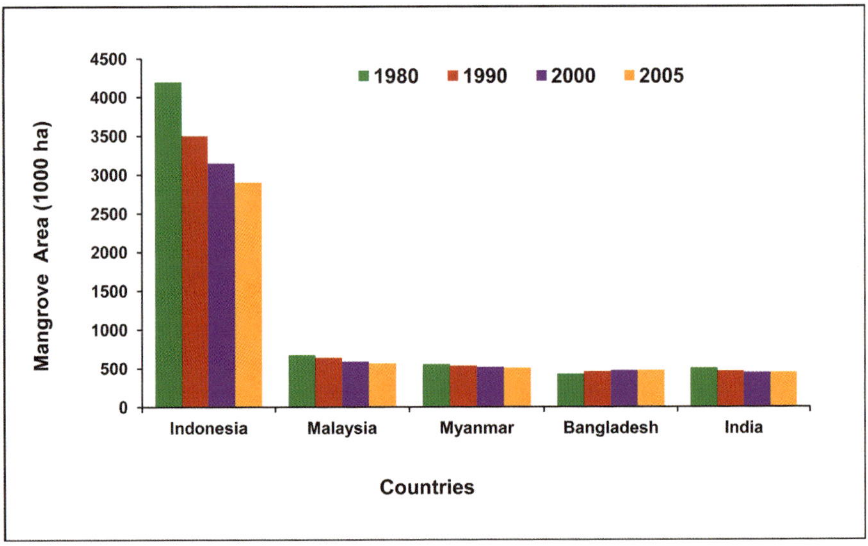

Fig. 3 Status and trends of mangrove area in major Asiatic countries from 1980–2005. (Source: FAO 2007)

and *Sesuvium portulacastrum* have been reported to be used in the treatment of diarrhoea, blood pressure, asthma, leprosy, rheumatism, snake bites, ulcer, bacterial and viral infections (Sandilyan and Kathiresan 2012). The haemorrhage is reported to be cured by the fermented juice of *Sonneratia caseolaris* , whereas the half-ripened fruit is used in the treatment of cough (Sandilyan and Thiyagesan 2010; Sandilyan and Kathiresan 2012). Several fungal species associated with the mangroves are the sources of several hormones, vitamins, enzymes and antibiotics (Sharma 2009). The root system of mangroves is so dense like a web and protects fertile sediments from washing away by the rivers to the ocean in the estuarine area. Mangrove forests act as feeding ground and nurseries for seafood such as crabs, clams, oysters, fish and shrimps. The mangrove forests are the primary nursery area for important fishes and crustaceans (Mumby et al. 2004; Sandilyan and Kathiresan 2012). Mangrove forests support 80 % of the global fish catching (Farley et al. 2010; Sandilyan and Thiyagesan 2010; Sandilyan and Kathiresan 2012). It has been reported that a mangrove-rich area supports higher catch of fishes and capitulates 70 times higher fisheries income than a mangrove-sparse area (Kathiresan and Rajendran 2002). The fishermen of Pichavaram mangrove forest in Tamil Nadu state of India harvest 208 tons of prawn, 19 tons of fish and 9 tons of crab in a year (Selvam et al. 2002). It has been estimated that the mangrove forests are the source of the products with annual values equivalent to 200,000–900,000 USD per hectare (Wells et al. 2006).

In spite of its high economic, medicinal and ecological importance, mangrove forests are being destroyed by several anthropogenic activities and also by some of the natural calamities. As shown in Fig. 3, almost 35 % of the total mangroves were lost during the period from 1980–2000 (Giri et al. 2011; MA 2005) and if continued to be lost with this speed, all the mangrove forests will disappear in the coming

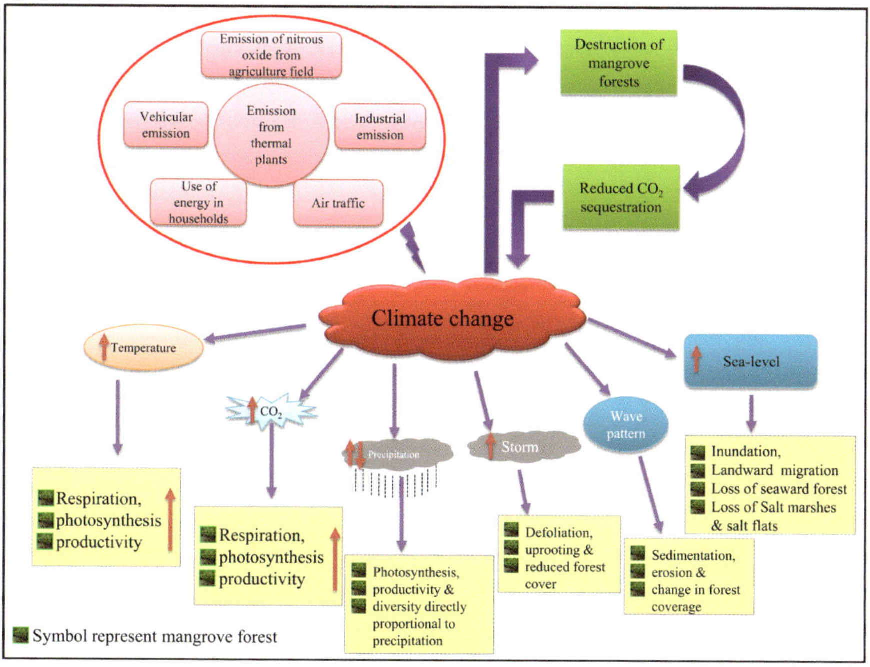

Fig. 4 Schematic representation of various climate change components and predicted effects on the mangrove ecosystem. The *upward* and *downward arrows* respectively indicate the increase and decrease in various climatic, physiographic and physiological factors

100 years (Duke et al. 2007). The climate change is a foremost threat to mangrove ecosystems of the world. Loarie et al. (2009) have predicted that the flat and flooded landscapes such as salt marshes and mangrove forests are likely to face the most rapid rate of climate change. It has been reported that the change occurs as a result of global warming and it is associated with increases in temperature, salinity and atmospheric CO_2 concentration which subsequently affect the structure, function and distribution of the mangrove vegetation (Field 1995; Das et al. 2002). The effects of various climate change factors affecting the mangrove ecosystem are discussed in the following headings.

2 Human Threats to Mangrove Ecosystems of Asia

The major threats to the mangrove ecosystem created by humans include overexploitation of mangrove forest resources by local people residing in coastal areas, destruction of forests for development of agriculture land, salt extraction, aquaculture, urban development and diversion of freshwater for irrigation (UNEP 1994). Spalding et al. (1997) reported that the global coverage of mangroves was

181,000 km^2. However, the Food and Agriculture organization recently reported that this figure may be reduced to 150,000 km^2 and many mangrove species are on the verge of extinction as a result of human activities (Polidoro et al. 2010). The human impacts on mangroves have increased severely in recent years with 50–80 % losses of mangrove forest cover in many countries as compared to that which existed 50 years ago (Macintosh et al. 2011). About 75 % of the mangrove area has been lost in the Philippines compared to the mangroves which existed in the 1950s (Primavera 2000). The reduced availability of the freshwater due to diversion or utilization of river waters may also have a crucial role in declination of mangroves (Gopal 2013). The development of shrimp aquaculture ponds accounts for the loss of 20–50 % of the mangrove ecosystem worldwide (Franks and Falconer 1999). It has been projected that the mangroves are supposed to decline by another 25 % by 2025 in developing countries (McLeod and Salm 2006). In some countries of Asia, such as Indonesia which has the world's largest intact mangroves, the higher rate of loss (about 90 %) has been projected in two provinces such as Java and Sumatra (Bengen and Dutton 2003). One-fourth of the total Indonesian mangroves lost in the near past is due to shrimp farming whereas the remaining three-fourths are due to the development of rice fields, over-exploitation of resources, infrastructure development and lack of action towards regeneration of degraded mangroves (Giesen et al. 2006). It has been reported that about 4,011 biological species exist in Indian mangrove forests alone (Sandilyan and Kathiresan 2012), but due to lack of knowledge, awareness and appropriate management strategies this has resulted in huge losses of the ecosystem. The world's largest mangrove forest in Sundarban is the preferred habitat of about 300 tree and herb species and about 425 wildlife species including the Royal Bengal Tiger (Haq 2010). A 60 % area of this forest is shared by Bangladesh and the remaining 40 % by India, covering about one million hectares in the deltaic region of the rivers Ganga, Brahmaputra, and Meghna (Erwin 2009). It has been reported that the large extents of the Sundarban mangrove forests have been destroyed due to conversion of these areas into paddy fields and shrimp farming (Erwin 2009). The construction of dams, barrages and embankments on rivers for diverting upstream water for different human requirements and to control flood has reduced the freshwater inflow to the Sundarban mangrove forest and seriously affecting its biodiversity (Erwin 2009). Several animal species are already extinct from the Sundarbans whereas 11 mangrove species are highly prone to extinction (Polirado et al. 2010). Some of these mangrove plant species, like *Aegiceras floridum*, *Camptostemon philippinensis*, and *Heritiera globsa,* are prevalent to Asia. The mangrove species *Rhizophora annamalayana* is confined to a particular region of Pichavaram mangrove forests of Southern India (Kathiresan 1999). Approximately 55 % of the mangrove areas were destroyed in Thailand in 35 years time as compared to the period before 1961 (Charuppat and Charuppat 1997). In contrast to the above reports, FAO 2003 have estimated the loss of 22 % in the period of 1973–2000. In spite of these losses, the Royal Forest Department has initiated an appreciable effort to re-plant mangroves in abandoned shrimp farms, and this initiative caused an increase in forest area during the last decade (Giesen et al. 2006). Large extents of Chinese mangrove forest were destroyed for development of agriculture and aquaculture lands (Li and Lee 1997).

Use of herbicides and bombardment during the war destroyed 149,123 ha of mangrove forests until 1971 in the Mekong delta of southern Vietnam (Hong 2003), and the destruction were also continued in the later years for development of aquaculture fields. By the year 1999, about 62 % of the degraded mangrove areas were re-planted (Hong 2003). These developments in the coastal areas lead to eutrophication of the estuarine and coastal water bodies resulting in turbid water, reduced dissolved oxygen, imbalance in nutrient ratios and their cycling (Diaz and Rosenberg 2008). The toxic algal bloom is responsible for the loss of coastal habitats (Rabalais and Gilbert 2009). Higher sedimentation has also posed a threat of increased rate of mangrove forest degradation (Alongi 2002; Victor et al. 2006; Thampanya et al. 2006; Wolanski 2007).

3 Climate Change Components Affecting Asian Mangroves

The climate is a major factor that determines not only the spatial distribution of the world's plants and animals, but also the endeavours of human beings such as agriculture and forestry (Preston et al. 2006). Therefore, the climate change is expected to have significant global outcomes. The cause of this climatic change lies in the dependence of humans upon fossil fuels as the primary source of energy. Up till now, human consumption of fossil fuels has grown in step with the global population and economy, and the unintended side-effects of their combustion have been a significant change in the composition of the Earth's atmosphere (Preston et al. 2006). The atmosphere has several components, many of which are naturally-occurring gases referred to as 'greenhouse gases', due to their capability to trap heat. The primary greenhouse gas is water vapour and other important greenhouse gases are carbon dioxide (CO_2), methane (CH_4), and nitrous oxide (N_2O). The surface of the earth is warmed by the energy released from the sun passes through the atmosphere. Although most of this heat is simply radiated back into space, some is trapped by greenhouse gases (Preston et al. 2006). This causes a warming effect on the atmosphere and ultimately keeps the planet at an average annual temperature of approximately 15 °C, and without this process the global average surface temperature would be closer to − 18 °C (Preston et al. 2006). The flows of greenhouse gases to the atmosphere have increased their concentration and subsequently the natural greenhouse effects have been magnified due to the combustion of fossil fuels and clearing of lands for many centuries. It has been reported that CO_2 levels have increased by approximately 36 % as compared to their concentrations prior to the industrial revolution (Preston et al. 2006). At the end of 2005, the average atmospheric CO_2 concentration was 379 parts per million (ppm) which is higher than at any point over at least the past 650,000 years (Siegenthaler et al. 2005). In the meantime, other greenhouse gases such as N_2O and CH_4 have also increased by 17 % and 151 %, respectively (Spahni et al. 2005). The consequence of all these changes to the atmosphere is the warming of the earth.

In developing countries of Asia, urbanisation and development of infrastructures imposed the destruction of various forest ecosystems. Also, increasing living standard, industrialisation, and uncontrolled emission from vehicles leads to continuous increases in the greenhouse effect, ultimately creating environmental imbalance. Increased emission of greenhouse gas and other factors results in climate changes which are characterised as rise in sea level and change of its wave pattern, increase in atmospheric temperature, increase in atmospheric CO_2 concentration, changes in wind velocity and their patterns (Fig. 4). Mangrove ecosystems are facing challenges to survive under these adverse climatic conditions. All the above factors of climate change do not act in isolation; instead they affect the mangroves ecosystem synergistically. Various climate change factors are predicted to affect the mangrove ecosystem by affecting the dispersal of propagules, gene flow, biomass allocation, productivity, forest cover, forest growth and sediment retention (Alongi 2008, Fig. 4). The plant morphology, photosynthesis and respiration of the mangroves are also predicted to be affected by the climate change factors (Alongi 2008). Apart from the sea-level rise, the major visible changes observed in climates are the rise of temperature and irregularities in precipitation. It has been expected that 1–13 % of mangrove forests will be lost due to climate change in the Asia/Pacific region in the near future and it might be higher in an individual nation (Preston et al. 2006). The effects of the climate change are not uniform in various regions of Asia and it mainly depends on the geographical conditions of the local area. The accurate data regarding the climate change has not yet been reported for the Indian sub-continent, but it has been predicted that there would not be much difference from the global trend. The sea-level rise and the erosion of the coast have been observed in Indian western coast (Singh 2003).

3.1 *Impacts of Alterations in Temperature*

By the end of the 21st century the temperature rise due to global warming and climate change in tropical Asia will be greater than the mean expected global rise in temperature. This has been proposed based on the MMD-A1B models of IPCC (Christensen et al. 2007). This rise in temperature is also not even uniform around Asia. The expected rises in different parts of Asia are 3.3 °C in South Asia, 3.7 °C in central Asia and 2.5 °C in southeast Asia. The rise in sea-water temperature does not have much of an adverse impact on mangroves, but the atmospheric temperature affects the mangroves significantly. Some of the mangrove species show reduced leaf formation at temperatures higher than 25 °C (Saenger and Moverly 1985). Mangrove root structures and seedlings are affected by temperatures higher than 35 °C (UNESCO 1992). The photosynthesis in almost all mangrove species is inhibited at a leaf temperature above 38 °C (Clough et al. 1982; Andrews et al. 1984). Thus, it is obvious that the increase in atmospheric temperature will adversely affect the mangrove ecosystem. It has been stated that with increasing temperature, mangroves will move pole-ward (UNEP 1994; Field 1995; Ellison 2005), but it is not possible for all species. Asia has been divided into four regions, the arid and semi-arid region including northern areas of India and Pakistan and western China; the temperate region including eastern China, the Tibetan Plateau, and the Korean Peninsula; north tropical Asia includes central and southern India, Sri Lanka, Bhutan, Bangladesh, Myanmar, Vietnam, Laos, Cambodia, and Thailand; and south tropical Asia includes the Maldives, Philippines, Malaysia, Papua New Guinea, Indonesia and East Timor (Preston et al. 2006). On the basis of the results from different climate models it has been reported that the temperatures throughout the Asia/Pacific region are projected to increase over the 21st century (Preston et al. 2006). As indicated by climate models, the arid and semi-arid areas of Western China of Central Asia, show higher rates of annual temperature rise, which is estimated to be at 2030 and 2070, 1.2 and 3.2 °C, respectively, followed by 0.9 and 2.4 °C in temperate over Asia, and 0.8 and 2.1 °C in north and south tropical Asia respectively (Preston et al. 2006). Ellison (2010) has reported that the mangroves will be able to deal with the temperature stress posed due to increase in climatic temperature. However, Gilman et al. (2008), based on the report of Field (1995) and Ellison (2000), have proposed that the increase in surface temperature will affect the mangroves in four different ways: (i) changes in the species diversity or composition, (ii) variation in flowering and fruiting time (plant phenology), (iii) increase in productivity of mangroves at temperatures below the upper threshold, and (iv) migration of mangroves towards higher latitudes under favourable physiographic conditions and sufficient supply of propagules.

Mitra et al. (2009) have reported that the surface water temperature in the Sundarban mangrove forest is rising at the rate of 0.5 °C per decade over the past three decades, which is eight times greater than the global warming rate of 0.06 °C per decade. This alarming rate of temperature rise in the Sundarban mangrove forest makes it one of the worst climate change hotspots of the globe (Haq 2010). This makes a clear challenge for the survival of flora and fauna in this forest.

3.2 Impacts of Alteration in CO_2 Concentration

Combustion of fossil fuels contributes significantly to the climate change by emission of CO_2. The atmospheric CO_2 concentration has increased from 280 ppm to 370 ppm during the period from 1880 to 2000 (IPCC 2001). This increased level of CO_2 is absorbed by the ocean water and affects the oceanic carbon balance. The photosynthesis of the mangrove plants increases with the rise in atmospheric CO_2 concentration and consequently the growth rate also increases (UNEP 1994). However, this rise in atmospheric CO_2 concentration indirectly affects the mangrove forests. Mangrove forests depend on the coral reefs for protection from the wave action. But, increased atmospheric temperature and CO_2 concentration lead to mass bleaching and weaken the growth of coral reefs (Hoegh-Guldberg 1999). Farnsworth et al. (1996) have reported that the seedlings of the mangrove *Rhizophora mangle* show higher growth rate at CO_2 concentrations of 700 ppm than that at 350 ppm. It has also been reported that the height of *R. mangle* is not increased by an increase in CO_2 concentration;

instead, the branching is promoted at high CO_2 concentration (Farnsworth et al. 1996). The earlier flowering was observed in *R. mangle* plants grown under elevated CO_2 concentration. The productivity of this mangrove also decreased in later years. Farnsworth et al. (1996) have observed the lower nitrogen level in leaves of plants grown under higher CO_2 concentration and suggested that the decrease in productivity of the mangrove might be due to the nutrient deficiency. Conversely, the impact of increasing CO_2 concentration due to climate change on mangroves is not similar in all species, because it acts synergistically in combination with salinity, temperature, availability of nutrients and hydrological actions (Field 1995).

3.3 Impacts of Alteration on Precipitation

Climate change affects the annual precipitation rate. The amount and timing of precipitation are also affected by climate change. Increased emission of greenhouse gases results in rise of atmospheric temperature that leads to higher evaporation from the ocean. Higher evaporation may result in increased precipitation. The global mean annual precipitation is expected to increase by 25 % in 2050 due to the increasing global warming (IPCC 2001; Gilman et al. 2008). The changes in precipitation are not uniform throughout the globe. The annual precipitation may increase or decrease based on the geographical architecture of that particular region (Walsh and Ryan 2000; IPCC 2001) and may be induced due to other hydrological factors. It has been reported that typhoons or tropical storms have resulted in heavy precipitation which lead to a decrease in salinity of the estuarine area, increase in total suspended matters, siltation, decrease in water transparency and temporary eutrophication in the Wenchang/Wenjiao Estuary of China (Herbeck et al. 2011; Krumme et al. 2012). The Asiatic and western pacific region is the world's most typhoon affected area with an average of 26.9 ± 4.3 typhoons per year (Wang et al. 2010). These changes in precipitation may pose intense effects on both the growth and coverage of mangroves (Field 1995; Snedaker 1995). The growth, productivity and survival of seedlings of mangroves decreased under conditions of reduced annual precipitation. The salinity of mangrove forests increases under the condition of low water availability and it affects the changes in species composition of the mangrove forests favouring more salt tolerant species (Ellison 2000, 2004). The diversified mangrove-covered areas decrease and the landward zone covered by mangroves turned into unvegetated hypersaline flats as a consequence of reduced annual precipitation (Snedaker 1995). An increase in annual precipitation results in higher growth rate in some species and increases the area of mangrove forests with diversified species (Field 1995). A further increase in annual precipitation may create a flooding situation and this leads to migration of some of the mangrove species towards the land. This may also increase the competition between mangroves and salt marshes (Harty 2004). The precipitation is not evenly distributed all over the Asiatic countries. The monsoon rain is found to be increased in South and East Asia whereas it is decreased in the north-western part of India (Bhaskaran and Mitchell 1998). This observation is consistent with the prediction of Preston et al. (2006) regarding the monsoon variation in Asia. The effect of climate change on the precipitation has been studied (Ramanathan et al. 2005; Chung and Ramanathan 2003). It was observed that the presence of aerosol particles which include incomplete combustion product and dust in the atmosphere may reduce the rain fall significantly in South Asia (Ramanathan et al 2005; Chung and Ramanathan 2003). Preston et al. (2006) also reported that the presence of aerosols may results in reduced monsoon rain in India. It has been expected that the further climate change may create the threat of higher precipitation in the monsoon and there may be great unpredictability of the annual precipitation (Wang et al. 2001). The mangroves growing in the estuarine regions are most affected by the precipitation. Higher precipitation leads to more runoff and decrease in the salinity.

Mangroves show increased productivity and growth under conditions of less salinity and increased nutrient availability (Slob 2012). The recent assessment report of IPCC expected a significant rise in the precipitation in northern and central Asia and a decline in some parts of southern Asia (Solomon et al. 2007). Lower precipitation and increased evaporation result in an increase in salinity. Due to the increased salinity the mangrove plants accumulate salts in their tissues and thus the water availability to the plants is reduced and finally productivity of the mangroves is reduced (Field 1995). Increased salinity increases the sulphate content of the sea water and, in this condition, the anaerobic decomposition of the mangrove peats occurs which make them more vulnerable to sea level rise (Snedaker 1995). The increased salinity also results in reduced seedling growth and survival, increase in the competition between the mangrove species, declination in diversity and conversion of mangrove areas to hyperasaline flats (Field 1995; Duke et al. 1998). Mangroves are also predicted to be migrated towards the salt marshes and freshwater wetlands under reduced precipitation conditions (Saintilan and Wilton 2001; Rogers et al. 2005). Inversely, higher precipitation increases the supply of sediments and nutrients and decreases the salinity and sulphate content of the water. In these conditions, productivity and diversity of the mangroves are found to be higher all around the globe (Ellison 2000; Gilman et al. 2008). In addition, an increase in peat production is observed which enhances the resilience of mangroves towards the sea-level rise (Snedaker 1995).

3.4 Impacts of Alteration in Wind Intensities, Storms and Wave Pattern

Mangroves are affected by an additional threat caused by increase in frequency and/or intensities of tropical storms due to climate change (Trenberth 2005). There is no report of changes in frequency or areas of storm formation, but it is predicted that the wind intensities will probably increase by 5–10 % due to the climate change (IPCC 2001). High wind velocity and storms may cause physical damage to the mangrove forests leading to mass mortality. The hurricanes cause peat collapse of the mangrove forests and ultimately reduce the recovery and regeneration ability of the mangrove species as those are physically damaged by the hurricane

(Cahoon et al. 2003). Due to the differences in their recovery and regeneration rates, landward shifting of some of the mangrove species is observed (Roth 1997). Storms may also cause flooding in combination with sea-level rise in the mangrove forests. Photosynthesis and water conductivity of the mangroves are decreased due to inundation (Naidoo 1983). It has also been reported that the mangrove trees die due to the reduced availability of oxygen caused by inundation of lenticels of the aerial root (Ellison 2004). One of the best examples of a devastating natural catastrophe is the tsunami on 26th December 2004 in the Indian Ocean region with seismic magnitude $M_W = 9.0$. It was the largest earthquake of the last 40 years. In this tsunami, the seawater turned into waves of 30 m height. Several habitats including mangroves, coral reefs, seagrass beds and other coastal vegetations were destroyed along with lots of people and infrastructures (Alongi 2008). The maximum damage to the mangrove forests by the tsunami was reported in the South Andaman islands of India (3,825–10,200 ha) followed by Aceh province, Sumatra, Indonesia (300–750 ha) and the Andaman coast of Thailand (306 ha), whereas the damage or loss was minimal in the North Andman islands (India) and Sri Lanka (UNEP 2005; Chang et al. 2006; Iverson and Prasad 2007; Alongi 2008). The uprooting of a large number of mangrove tress takes place due to massive soil erosion caused by inundation (UNEP 2005). In South Andaman of the Andaman and Nicobar Islands, 30–80 % of trees of *Rhizophora* spp. growing towards the sea died due to continuous inundation, but some of the mangroves like *Avicennia marina* and *Sonneratia alba* growing behind the *Rhizophora* spp. were not affected by the tsunami (Dam Roy and Krishnan 2005). Several reports of post tsunami impact on South-eastern India, the Andaman Islands, and Sri Lanka suggests that the mangroves provide shield against calamity like tsunamis (Alongi 2008). It has been proposed that the intensity of tsunami waves weakens significantly by passing through the mangrove forests and also depends upon the density of the mangrove forest (Hiraishi and Harada 2003). The mangroves are more susceptible to frequent hurricanes hitting due to its location in the coastal region. The areas with frequent hurricanes have lower canopy size, smaller diameter and the forests are found to be less complex (Pool et al. 1977). Loss of mangroves in frequent hurricanes are marked by the stripping of the leaves, break down of branches and uprooting of the mangrove plants. Hurricanes may deposit voluminous sediments which block the lenticels and suffocate the pneumatophores resulting in the death of mangrove (Slob 2012). The IPCC has projected that due to changes in the climatic conditions, the intensities of tropical cyclonic wind and precipitation are supposed to increase in some parts of the globe during the 21st century (Solomon et al. 2007; Gilman et al. 2008). The height of the storms surges are also expected to increase concomitantly with the increase in the recurrence of strong winds and low pressure conditions (Solomon et al. 2007; Gilman et al. 2008). Bangladesh is one of the most vulnerable countries to climate change (Climate change and Bangladesh Department of Environment, Government of People's Republic of Bangladesh. Climate Change Cell, Dhaka, 2007). There is considerable evidence of rise in the intensity or frequency of heat waves, tropical cyclones, storm surges, high rainfall, flood, land erosion, tornadoes, drought, salinity intrusion, etc. that severely affects the mangrove ecosystem of this country (Hossain et al. 2012). The Sundarban, a

World Heritage Site, is severely affected by climate change, crucially from increasing salinity and extreme weather events like tropical cyclones. About one-third of the Sundarban forest was destroyed by recent cyclone Sidr (Arefin 2011).

Climate change also affects the hydrological patterns of oceanic water bodies. Variation in salinity and heat changes the circulation pattern of ocean waves (Bindoff et al. 2007; Gilman et al. 2008). It has been reported that the distribution of propagules and genetic structure of the mangroves are affected by changes in the wave circulation pattern (Lovelock and Ellison 2007; Gilman et al. 2008). However, the IPCC (2001) have reported that there is no evidence of changes in the ocean circulation pattern due to the climate change.

3.5 Impacts of Rising Sea Level

The major impacts of the climate change are rising in sea level due to glaciers and Antarctic ice sheets melting as a consequence of global warming and increase in atmospheric temperature (Ragoonaden 2006). Gilman et al. (2008) have predicted that sea level rise due to increasing global warming can be the utmost threat to the existing mangroves. Mangroves can cope with a sea-level rise up to 9–10 cm/100 years (Singh 2003). Approximately 260 km of the Andaman and Nicobar Islands of the Indian coast are inhabited with mangroves and have limited possibility to cope with the sea level rise. The impact of sea level rise on mangrove distribution and species composition in this area may be devastating when the rate of sea level rise exceeds more than 10 cm/100 years (Singh 2003). The Fourth Assessment Report of Intergovernmental Panel on Climate Change (IPCC) has anticipated that the global rise in sea level from the 1990 to the 2090s will be 18–59 cm and an average rate of 1.8–5.8 mm per annum (Rahmstorf 2010). The sea-level rise causes serious threats to the coastal ecosystems due to excess flooding, erosion of the sediments, and inhibition of different nutrient cycling and gaseous exchange. The rise in sea level will result in a vertical increase in water column, and limitations of landward margins may create a situation of water logging, eventually causing death of mangroves and associated flora (Jagtap et al. 2003). In some of the Southeast Asian countries, the impact of sea level rise will have a serious issue on the mangrove community (Aksorakaoe and Paphavasit 1993). The World Wildlife Fund for Nature Conservation (WWF) has estimated that about 7,500 ha of mangrove forest area of Sundarban in Bangladesh are at risk of submergence in the near future (Arefin 2011). So far, the non-climate related anthropogenic activities pose a higher threat to mangroves as compared to the sea level rise. The anthropogenic stressors are the major cause of the global loss of mangroves and it is estimated to be 1–2 % per year (Gilman et al. 2008). During the last quarter century 35–86 % of the mangroves have been lost due to anthropogenic activities (Valiela et al. 2001; Gilman et al. 2008; Di Nitto et al. 2013). However, it has been predicted that sea-level rise may cause a significant percentage of future losses of mangroves. Studies of mangrove susceptibility to change in relative sea-level, mainly from the western Pacific and wider Caribbean regions, have suggested that most of the mangrove sites will not be adjusted with the current rates of relative sea-level rise (Gilman et al. 2008). There is a requirement of long-term studies in order to conclude whether the effects of sea level rise are long-term trends or cyclical short-term patterns, and whether this is a global or regional phenomenon. Gilman et al. (2008) proposed that the relative sea-level rise could be a significant cause of future losses in regional mangrove areas and it is estimated to be 10–20 %.

A vast area of the world's largest mangrove forest, Sundarban distributed in India and Bangladesh, is vulnerable to continuous rise of the sea-level (Sarwar 2005; MOEF 2010). It has been predicted that the climate change will cause an annual temperature rise of 0.4 °C in Bangladesh, resulting in increased intensity and recurrence of cyclonic storms (Haq 2010). The expected rise of sea level is 4 mm per year. These climate change factors will cause an increase in salinity and decrease flow of fresh water in the Sundarbans (Haq 2010). The combined flows of fresh water from the Ganges and the salty sea water from the Bay of Bengal supports a balanced growth of flora and fauna in the Sundarbans. However, the ecological balance of this forest is now being threatened due to increased siltation as a result of decreased downstream flow of rivers running through and around it (Haq 2010). The density of vegetation growth and canopy closures decreases from east to west due to the increase in salinity intrusion of the Sundarbans from east to west. The adverse effects of increased salinity of the Sundarbans ecosystem resulted in tops dying disease of Sundari trees (Haq 2010). The epidemic of the top-dying disease caused a large scale destruction of Sundari trees in the Sundarbans. The disease was first detected during 1930 and large-scale effects have been observed since 1980. The top-dying disease caused a destruction of about 34,287 cum of Sundari trees per year as reported in the survey of Forest Department of Bangladesh from 1994–1996 (Haq 2010).

3.5.1 Adaptation Strategies of Mangroves for Survival Under Rising Sea Level

Mangrove plants have developed adaptations like aerial rooting systems with pneumatophores for respiration, stilt rooting systems for support of plants, root knees and plank roots to maintain higher gaseous space at the time of water-logging and better a physiological defense system to survive in muddy, uneven, and saline conditions (Tomlinson 1986; Hogarth 1999; Baskin and Baskin 2001; Parida and Jha 2005). The most important adaptation of mangroves towards the sea level rise is the ability to grow and expand upward and landward (McLeod and Salm 2006). Due to their viviparous nature, they can expand easily on the area covered by sediment soil. The low-lying islands lack river inflow and, where carbonate settling occurs by corals, the expansion of mangrove forest is limited due to the unavailability of sediments and they pose higher threats to sea-level rise (Gopal 2013; McLeod and Salm 2006). Krauss et al. (2008) proposed that the response of mangroves towards the increase in sea-level rise are of three types: (i) the position of the mangroves is found stable at the same location, when the sea-level did not change relative to mangrove surface, (ii) seaward movement of the mangroves are observed when the sea-level decreased, and (iii) landward movement of the mangroves is observed when the sea-level increased relative to the mangrove surface. The mangroves which migrated towards the sea died back due to the stresses like weakening of the rooting system and uprooting, increased salinity and increase in frequency and depth of inundation caused by sea-level rise (Ellison 2006; Lewis 2005; Gilman et al. 2008). The survival of the mangrove species in the new habitats depends on their ability to cope with the rate of increasing relative sea-level (Lovelock and Ellison 2007; Di Nitto et al. 2008; Gilman et al. 2008), slope of the new habitat where the migration takes place and the presence of man-made hindrances (Gilman et al. 2008). The Gujarat state of India has a large extent of tidal mudflat areas which are suitable for the fast recovery of hardy mangrove species like *Avecennia* sp. but not for the genus like *Rhizophora, Ceriops, Sonneratia* and *Aegiceras* (Singh 2003). Gilman et al. (2008) have put forwarded that the four salient factors determine the pliability of mangroves to the relative sea-level rise. These factors are (i) the rate of relative sea-level rise in comparison to the mangrove sediment surface (Gilman et al. 2007b, 2008), (ii) the differences in rate of sedimentation and colonization of different mangrove species to the new habitats (McKee et al. 2007; Lovelock and Ellison 2007; Gilman et al. 2008), (iii) the slope of the new habitats in relation to current habitats and other barriers (Gilman et al. 2007a) and (iv) the cumulative effects of all the stresses.

3.5.2 Species Specific Responses to Sea-level Rise

The different species of mangroves have adapted to changes in environmental factors such as flooding, water logging and high salinity by some structural modifications of wood, bark and leaves (Yáñez-Espinosa and Flores 2011). During flooding, the vessel density of the vascular system increased in *Bruguiera gymnorrhiza* (Xiao et al. 2010), whereas the vessel diameter and fibre wall thickness reduced (Xiao et al. 2009; 2010). Yáñez-Espinosa et al. (2004) have reported that to avoid cavitation damage at the time of high flooding, numerous vessels are required and this might be the reason for increase in vessel density. Reduction in fiber wall thickness and vessel diameter might be an adaptation to keep the equilibrium between growth, conductivity, safety and mechanical strength (Xiao et al. 2009). Adaptations to modify bark anatomy are characterized as the formation of hypertrophied lenticels in immersed stems (Mielke et al. 2005), highly developed aerenchyma in the bark and formation of adventitious roots in mangrove associates (Yáñez-Espinosa et al. 2008; Yáñez-Espinosa and Flores 2011).

4 Conclusion and Future Perspectives

Mangroves are an important component of the coastal ecosystem. The mangrove forests have high economical values at the ecosystem as well as at the component levels (Hogarth 1999; Parida 2010). The mangroves play a crucial role in the economy of coastal communities of tropics offering a large number of goods and services such as wood and timber production, salt production, supporting fisheries, controlling the coastal erosion and protecting the shoreline from cyclones and storms (Parida et al. 2005). Mangrove ecosystems act as a refuge, feeding ground and nursery for many species of cyanobacteria, red algae, manglicolous fungi, invertebrates, fish and shrimp (Hogarth 1999). The mangroves are currently threatened to a large extent by various anthropogenic activities. The pressures of urbanization and industrialization near coastal regions without sufficient implementation of the legislation have caused the destruction of vast areas of mangroves in many developing countries of Asia (Sandilyan and Kathiresan 2012). Mangrove forests are also facing threats from global warming and subsequent climate change. The major climate change factors affecting mangrove ecosystems are change in temperature, increased atmospheric CO_2 concentration, precipitation, storms, ocean circulation patterns, changes in sea level, and anthropogenic activities. Global emissions of greenhouse gases are expected to raise the world temperature and the sea level at a significant rate. The mangrove forests would be more susceptible to climatic changes and resultant sea level rise because of their unique location at the inhospitable zone between land and sea. Due to the global warming and sea level rise, the landward migration of the mangroves may occur. The inadequacy of the landward margin would cause vertical rise of seawater resulting in water logging and ultimately destroying the mangroves and other associated flora and fauna.

The increasing population, overexploitation of the forest materials by traditional users, increasing agriculture and aquaculture, urbanization and industrial development and increase in frequency of natural disasters like cyclone, tsunami and hydrological changes leads to significant loss of the mangrove forests in the coastal regions of Asian countries. This will result in imbalance of the local environment, reduced CO_2 sequestration and loss of important animal and plant species inhabiting the mangrove ecosystem. Therefore, there is a crucial requirement to conserve this vital ecosystem for global interests, otherwise, the ecosystem services and other utilities provided by the mangroves will be reduced or lost perpetually (Duke et al. 2007; Sandilyan and Kathiresan 2012). A proper management strategy is needed for sustainability of the Asian mangrove forests. The predicted impacts of different climate change factors on the mangrove ecosystem and strategies to alleviate the effects of climate change have been summarized in Table 1. To mitigate the effects of climate change and to protect the precious mangrove ecosystem, there is an urgent need of management strategies. The existing mangroves should be strictly protected from encroachment and cutting. The expansion of mangrove cover by enhancing the regenerating potential and planting of vulnerable and threatened mangroves species in the intertidal areas are some of the essential management options. The adaptability of the mangrove species should be studied and the species with less adapting ability to climate change should be facilitated to regenerate in new areas (Singh 2003). The consistent monitoring of the climate changes and their effects on mangroves should be recorded to get continuous inputs for necessary management strategies. The response of mangroves to climate change will differ from location to location and hence an area specific plan on the basis of inputs of continuous recording of changes

Table 1 Predicted impacts of various climate change factors on mangrove ecosystem and strategies to alleviate the effects of climate change

Climatic factors	Likely impacts	Alleviation strategies	Reference
Temperature rise	Change in species diversity and composition, variation in flowering and fruiting time	Plantation of mangroves will regulate local climate by cloud formation, prevent over-evaporation and dissipation of heat to surrounding	Farley et al. 2010
Increase CO_2 concentration	Increase productivity up to certain extent, promote branching, reduced nutrient availability and productivity from subsequent years	Plantation of mangrove promote fixing of atmospheric CO_2 and thus regulate the global warming	Donato et al. 2011
Change in precipitation	Reduced precipitation will cause increase in salinity, reduce sediment supply, reduce productivity and finally death of certain species. Increased precipitation will result in increased productivity, increased sediment supply and higher diversity	Connect mangrove forest with river, backwater and creeks. This will facilitate continuous water supply during reduced precipitation and drainage during heavy precipitation	McLeod and Salm 2006
Wind intensity	Physical damage leading to mass mortality, peat collapse resulting in reduced recovery and regeneration ability of the mangrove species	Plantation of mangrove species with dense vegetation in front of the sea to reduce the velocity and intensity of wind and storm	Sandilyan 2007; Sandilyan and Thiyagesan 2010; Farley et al. 2010
Altered ocean wave circulation pattern	Propagule distribution and gene flow will be affected due to changes in wave pattern	Management of proper distribution of sediments and propagules	Lovelock and Ellison 2007
Sea-level rise	Lead to inundation, flooding and landward migration of mangrove	Develop salt flats and salt marshes behind the mangrove forest to promote the landward migration; Removal of anthropogenic barriers like roads, buildings, etc., which prevent the landward migration; Facilitate drainage by connecting with river and creek which will reduce the inundation	McLeod and Salm 2006

should be made for implementation (Singh 2003; Kaushik and Khalid 2011). Some of the important management strategies proposed by McLeod and Salm (2006) for the protection of mangrove ecosystems are: (i) to allow the peat building by continuous supply of freshwater and sediments, (ii) facilitation of landward migration of mangroves during flooding resulting from sea-level rise or high precipitation, (iii) management of proper distribution of sediments and propagules, and (iv) the factors that promote speedy recovery and regeneration of mangroves should be improved. Approximately 50 % of the total Indian mangrove forest covers have been protected jointly by the Government of India and the state government by constituting a dozen Marine Protected Areas (MPAs) under the Wildlife Protection Act, 1972. Apart from this, three mangrove supported areas—Sundarban of West Bengal, the Great Nicobar Island of Andaman and Nicobar and the Gulf of Mannar of Tamil Nadu—have been declared as Marine Biosphere Reserves (Singh 2003). Several management strategies for recovery of mangrove forests have resulted in significant increases of forest areas in India. The remote sensing study of mangrove forests depicted that the total cover area has increased from 4,244 km^2 in 1991 to 4,871 km^2 in 1999 (FSI 1999). In Gujarat state alone, the mangrove forest covers have increased by two-fold of its area reported in the last two decades (Singh 2003).

Acknowledgements Financial support from the Council of Scientific and Industrial Research (CSIR), New Delhi, India (project no. BSC 0109) is gratefully acknowledged. VT is thankful to CSIR, New Delhi for financial support in the form of a Senior Research Fellowship.

References

Aksorakoe S, Paphavasit N (1993) Effects of sea level rise on the mangrove community in Thailand. Malays J Trop Geogr 24:29–34

Alongi DM (2002) Present state and future of the world's mangrove forests. Environ Conserv 29:331–349

Alongi DM (2008) Mangrove forests: resilience, protection from tsunamis, and responses to global climate change. Estuar Coast Shelf Sci 76:1–13

Andrews TJ, Clough BF, Muller GJ (1984) Photosynthetic gas exchange and carbon isotope ratios of some mangroves in North Queensland. In: Teas HJ (ed) Physiology and management of mangroves (Tasks for vegetation science) vol 9. Dr. W. Junk, The Hague

Arefin ABMS (2011) Climate change impact in Bangladesh with reference to forest and biodiversity. In: Panda G, Pandia BP (eds.) Community action in a climate change regime: managing biodiversity, regional centre for development cooperation. Bhubaneswar, pp 4–14

Barbier EB, Koch EW, Silliman BR, Hacker SD, Wolanski E, Primavera J, Granek EF, Polasky S, Aswani S, Cramer LA, Stoms DM, Kennedy CJ, Bael D, Kappel CV, Perillo GME, Reed DJ (2008) Coastal ecosystem-based management with nonlinear ecological functions and values. Science 319:321–323

Baskin CC, Baskin JM (2001) Seeds: ecology, biogeography and evolution of dormancy and germination. Academic Press, London

Bengen DG, Dutton IM (2003) Interactions between mangroves and fisheries in Indonesia. In: Northcote TG, Hartman GF (eds) Fishes and forestry—worldwide watershed interactions and management. Blackwell Scientific, Oxford

Bhaskaran B, Mitchell JFB (1998) Simulated changes in Southeast Asian monsoon precipitation resulting from anthropogenic emissions. Int J Climatol 18:1455–1462
Bindoff NL, Willebrand J, Artale V, Cazenave A, Gregory J, Gulev S, Hanawa K, Le QC, Levitus S, Nojiri Y, Shum C, Talley L, Unnikrishnan A (2007) Observations: oceanic climate change and sea level. In: Solomon S, Qin D, Manning M, Chen Z, Marquis M, Averyt K, Tignor M, Miller H (eds) Climate change 2007: the physical science basis. Contribution of working group I to the fourth assessment report of the intergovernmental panel on climate change. Cambridge University Press, Cambridge
Cahoon DR, Hensel P, Rybczyk J, McKee KL, Proffitt CE, Perez BC (2003) Mass tree mortality leads to mangrove peat collapse at Bay Islands, Honduras after Hurricane Mitch. J Ecol 91:1093–1105
Chang SE, Adams BJ, Alder J, Berke PR, Chuenpagdee R, Ghosh S, Wabnitz C (2006) Coastal ecosystems and tsunami protection after the December 2004 Indian Ocean tsunami. Earthq Spectra 22:863–887
Charuppat T, Charuppat J (1997) The use of Landsat-5^{TM} satellite images for tracing the changes of mangrove forest areas of Thailand. Royal Forestry Department, Bangkok
Chen L, Wang W, Zhang Y, Lin G (2009) Recent progresses in mangrove conservation, restoration and research in China. J Plant Ecol 2:45–54
Christensen JH, Hewitson B, Busuioc A, Chen A, Gao X, Held I, Jones R, Rupa Kumar K, Kwon W-T, Laprise R, Magaa Rueda V, Mearns L, Menéndez CG, Räisänen J, Rinke A, Sarr A, Whetton P (2007) Regional climate projections. In: Solomon S, Qin D, Manning M, Chen Z, Marquis M, Averyt KB, Tignor M, Miller HL (eds) Climate change 2007: the physical science basis. Contribution of working group I to the fourth assessment report of the intergovernmental panel on climate change. Cambridge University Press, Cambridge
Chung CE, Ramanathan V (2003) South Asian haze forcing: remote impacts with implications to ENSO and AO. J Climate 16:1791–1806
Clough BF, Andrews TJ, Cowan IR (1982) Primary productivity of mangroves. In: Clough BF (ed) Mangrove ecosystems in Australia- structure function and management. AIMS with ANU Press, Canberra
Dam Roy S, Krishnan P (2005) Mangrove stands of Andamans vis-á-via tsunami. Curr Sci 89:1800–1804
Das AB, Parida A, Basak UC, Das P (2002) Studies on pigments, proteins and photosynthetic rates in some mangroves and mangrove associates from Bhitarkanika, Orissa. Mar Biol 141:415–422
Di Nitto D, Dahdouh-Guebas F, Kairo JG, Decleir H, Koedam N (2008) Digital terrain modelling to investigate the effects of sea level rise on mangrove propagule establishment. Mar Ecol Prog Ser 356:175–188
Di Nitto D, Neukermans G, Koedam N, Defever H, Pattyn F, Kairo JG, Dahdouh-Gueb F (2013) Mangroves facing climate change: landward migration potential in response to projected scenarios of sea level rise. Biogeosciences Discuss 10:3523–3558
Diaz RJ, Rosenberg R (2008) Spreading dead zones and consequences for marine ecosystems. Science 321:926–929
Dittmar T, Hertkorn N, Kattner G, Lara RJ (2006) Mangroves, a major source of dissolved organic carbon to the oceans. Glob Biogeochem Cycles 20. doi:10.1029/2005GB002570
Donato DC, Kauffman JB, Murdiyarso D, Kurnianto S, Stidham M, Kanninen M (2011) Mangroves among the most carbon-rich forests in the tropics. Nat Geosci 4:293–297
Duke NC, Ball MC, Ellison JC (1998) Factors influencing biodiversity and distributional gradients in mangroves. Glob Ecol Biogeogr 7:27–47
Duke NC, Meynecke JO, Dittmann S, Ellison AM, Anger K, Berger U, Cannicci S, Diele K, Ewel KC, Field CD, Koedam N, Lee SY, Marchand C, Nordhaus I, Dahdouh-Guebas F (2007) A world without mangroves? Science 17:41–42
Ellison AM, Farnsworth EJ, Merkt RE (1999) Origins of mangrove ecosystems and the mangrove biodiversity anomaly. Glob Ecol Biogeogr 8:95–115

Ellison J (2000) How South Pacific mangroves may respond to predicted climate change and sea level rise. In: Gillespie A, Burns W (eds) Climate change in the South Pacific: impacts and responses in Australia, New Zealand, and small islands states. Kluwer, Dordrecht
Ellison JC (2004) Vulnerability of Fiji's mangroves and associated coral reefs to climate change. Review for the World Wildlife Fund. University of Tasmania, Launceston
Ellison JC (2005) Impacts on mangrove ecosystems. The great greenhouse gamble: a conference on the impacts of climate change on biodiversity and natural resource management. Conference proceedings, Sydney
Ellison J (2006) Mangrove paleoenvironmental response to climate change. In: Gilman E (ed) Proceedings of the symposium on mangrove responses to relative sea-level rise and other climate change effects. Society of Wetland Scientists 2006 Conference, Cairns, 9–14 July 2006. Western Pacific Regional Fishery Management Council and United Nations Environment Programme Regional Seas Programme, Honolulu and Nairobi
Ellison JC (2010) Proceedings of the ASEAN conference on biodiversity, Singapore, 21–23 October 2009. ASEAN Centre for Biodiversity, Laguna, Philippines
Erwin KL (2009) Wetlands and global climate change: the role of wetland restoration in a changing world. Wetl Ecol Manage 17:71–84
FAO (2003) Status and trends in mangrove area extent worldwide. In: Wilkie ML, Fortuna S (eds) Forest resources assessment working paper no. 63. Forest Resources Division, 10 FAO, Rome
FAO (2007) The world's mangrove 1980–2005. FAO forestry paper 153, Rome
Farley J, Batker D et al (2010) Conserving mangrove ecosystems in the Philippines: transcending disciplinary and industrial borders. Environ Manage 45:39–51
Farnsworth EJ, Ellison AM, Gong WK (1996) Elevated CO_2 alters anatomy, physiology, growth, and reproduction of red mangrove (*Rhizophora mangle* L.). Oecologia 108:599–609
Field CD (1995) Impacts of expected climate change on mangroves. Hydrobiologia 295:75–81
Franks T, Falconer R (1999) Developing procedures for the sustainable use of mangrove systems. Agr Water Manage 40:59–64
FSI (1999) State of forest report, 1999. Forest Survey of India, Ministry of Environment and Forest, Dehradun
Giesen W, Wulffraat S, Zieren M, Scholten L (2006) Mangrove guidebook for southeast Asia. FAO and Wetlands International. Dharmasarn, Bangkok
Gilman E, Ellison J, Coleman R (2007a) Assessment of mangrove response to projected relative sea-level rise and recent historical reconstruction of shoreline position. Environ Monit Assess 124:112–134
Gilman E, Ellison J, Sauni Jr I, Tuaumu S (2007b) Trends in surface elevations of American Samoa mangroves. Wetl Ecol Manage 15:391–404
Gilman EL, Ellison J, Duke NC, Field C (2008) Threats to mangroves from climate change and adaptation options. Aquat Bot 89:237–250
Giri C, Ochieng E, Tieszen LL, Zhu Z, Singh A, Loveland T, Masek J, Duke N (2011) Status and distribution of mangrove forests of the world using earth observation satellite data. Glob Ecol Biogeogr 20:154–159
Gopal B (2013) Future of wetlands in tropical and subtropical Asia, especially in the face of climate change. Aquat Sci 75:39–61
Gopal B, Chauhan M (2006) Biodiversity and its conservation in the Sundarban mangrove ecosystem. Aquat Sci 68:338–354
Haq SA (2010) Impact of climate change on "Sundarbans", the largest mangrove forest: ways forward. 18th Commonwealth Forestry Conference 2010, Edinburgh
Harty C (2004) Planning strategies for mangrove and saltmarsh changes in Southeast Australia. Coast Manage 32:405–415
Herbeck LS, Unger D, Krumme U, Liu SM, Jennerjahn TC (2011) Typhoon-induced precipitation impact on nutrient and suspended matter dynamics of a tropical estuary affected by human activities in Hainan, China. Estuar Coast Shelf Sci 93:375–388

Hiraishi T, Harada K (2003) Greenbelt tsunami prevention in South Pacific region. Report of the port and airport research institute 42:1–23. http://eqtap.edm.bosai.go.jp/useful_outputs/report/hiraishi/data/papers/greenbelt.pdf, Accessed 13 April 2013
Hoegh-Guldberg O (1999) Climate change, coral bleaching and the future of the world's coral reefs. Mar Freshwater Res 50:839–866
Hogarth PJ (1999) The biology of mangroves. Oxford University Press, New York
Hong PN (2003) Mangrove forests in Vietnam—present status and challenges. Presented at the international symposium on conservation and wise use of mangroves in Southeast Asia, Brunei, Darussalam, 6–8 October 2003
Hossain MA, Reza MI, Rahman S, Kayes I (2012) Climate change and its impacts on the livelihoods of the vulnerable people in the southwestern coastal zone in Bangladesh. In: Filho WL (ed) Climate change and the sustainable use of water resources. Springer, Berlin, pp 237–259
IPCC (Intergovernmental Panel on Climate Change) (2001) Climate change 2001: the scientific basis. In: Houghton JT et al (ed) Contribution of working group I to the third assessment report of the intergovernmental panel on climate change. Cambridge University Press, Cambridge
Iverson LR, Prasad AM (2007) Using landscape analysis to assess and model tsunami damage in Aceh province, Sumatra. Landscape Ecol 22:323–331
Jagtap TG, Ansari ZA, Coutinho FB, Charulata S, Gaidhane DM (2003) Impact of climate change on mangrove forests along the southern west coast: a case study from Kasargod, Kerala, India. In: Ravindranath NH, Sharma S, Garg A, Bhattacharya S, Murthy I (eds) Proc workshop vulnerability and adaptations due to climate change on Indian agriculture, forestry and natural ecosystem. NATCOM Publication, New Delhi, pp 136–149
Jennerrjahn TC, Ittekot V (2002) Relevance of mangroves for the production and deposition of organic matter along tropical continental margins. Naturwissenschaften 89:23–30
Kathiresan K (1999) Rhizophora x annamalayana Kathir. (Rhizophoraceae), a new nothospecies from Pichavaram mangrove forest in south-eastern India. Environ Ecol 17:500–501
Kathiresan K, Bingham BJ (2001) Biology of mangroves and mangrove ecosystems. Adv Mar Biol 40:81–251
Kathiresan K, Rajendran N (2002) Fishery resources and economic gain in three mangrove areas on the south-east coast of India. Fish Manage Ecol 9:277–283
Kaushik G, Khalid MA (2011) Climate change impact on forestry in India. In: Lichtfouse E (ed) Alternative farming systems, biotechnology, drought stress and ecological fertilisation. Springer, Dordrecht, pp 319–344
Krauss KW, Lovelock CE, McKee KL, López-Hoffman L, Ewe SML, Sousa WP (2008) Environmental drivers in mangrove establishment and early development: a review. Aquat Bot 89:105–127
Krumme U, Herbeck LS, Wang T (2012) Tide- and rainfall-induced variations of physical and chemical parameters in a mangrove-depleted estuary of East Hainan (South China Sea). Mar Environ Res 82:28–39
Lewis III RR (2005) Ecological engineering for successful management and restoration of mangrove forests. Ecol Eng 24:403–418
Li MS, Lee SY (1997) Mangroves of China: a brief review. Forest Ecol Manage 96:241–259
Loarie SR, Duffy PB, Hamilton H, Asner GP, Field CB, Ackerly DD (2009) The velocity of climate change. Nature 462:1052–1055
Lovelock CE, Ellison JC (2007) Vulnerability of mangroves and tidal wetlands of the Great Barrier Reef to climate change. In: Johnson JE, Marshall PA (eds) Climate change and the Great Barrier Reef: a vulnerability assessment. Great Barrier Reef Marine Park Authority and Australian Greenhouse Office, Australia
MA (Millennium Assessment) (2005) Mille ecosystems and human well-being. Synthesis Island Press, Washington
Macintosh DJ, Epps MM, Abrenilla O (2011) Ecosystem approaches to coastal resources management: the case for investing in mangrove ecosystems. In: Food for all investing in food security in

Asia and the Pacific—issues, innovations, and practices. Asian Development Bank, Philippines, pp 140–155

McKee KL, Cahoon DR, Feller I (2007) Caribbean mangroves adjust to rising sea level through biotic controls on change in soil elevation. Glob Ecol Biogeogr 16:545–556

McLeod E, Salm RV (2006) Managing mangroves for resilience to climate change. IUCN, Gland

Mielke MS, Matos EM, Couto VB, Almeida AF de, Gomes FP, Oliveira Mangabeira PA (2005) Some photosynthetic and growth responses of *Annona glabra* L. Seedlings to soil flooding. Acta Bot Brasilica 19:905–911

Mitra A, Gangopadhyay V, Dube A, Schmidt AC, Banerjee K (2009) Observed changes in water mass properties in the Indian Sundarbans (northwestern Bay of Bengal) during 1980–2007. Curr Sci India 97:1445–1452

MOEF (Ministry of Environment and Forests) (2010) Climate change and India: a 494 assessment—a sectoral and regional analysis for 2030s, INCCA report 2. Ministry of Environment and Forests, New Delhi

Mumby PJ, Edwards AJ, Arisa-Gonza'Lez JE et al (2004) Mangrove enhance the biomass of coral reef fish communities in the Caribbean. Nature 427:533–536

Naidoo G (1983) Effects of flooding on leaf water potential and stomatal resistance in Bruguiera gymnorrhiza (L.) Lam. New Phytol 93:369–376

Parida AK, Jha B (2010) Salt tolerance mechanisms in mangroves: a review. Trees 24:199–217

Parida AK, Mittra B, Das AB, Das TK, Mohanty P (2005) High salinity reduces the content of a highly abundant 23-kDa protein of the mangrove *Bruguiera parviflora*. Planta 221:135–140

Polidoro BA et al (2010) The loss of species: mangrove extinction risk and geographic areas of global concern. PLoS One 5:1–10

Pool DJ, Snedaker SC, Lugo AE (1977) Structure of mangrove forests in Florida, Puerto Rico, Mexico, and Costa Rica. Biotropica 9:195–212

Preston BL, Suppiah R, Macadam I, Bathols J (2006) Climate change in the Asia/Pacific region: a consultancy report prepared for the climate change and development roundtable. Climate change impacts and risk. Commonwealth Scientific and Industrial Research Organisation (CSIRO), Australia

Primavera J (2000) Philippine mangroves: status, threats and sustainable development. International workshop Asia–Pacific cooperation, research for conservation of mangroves, Okinawa, Japan, 26–30 Mar 2000

Rabalais NN, Gilbert D (2009) Distribution and consequences of hypoxia. In: Urban E, Sundby B, Malanotte-Rizzoli P, Melillo JM (eds) Watersheds, bays and bounded seas: the science and management of semi-enclosed marine systems. Island Press, Washington DC

Ragoonaden S (2006) Sea level activities and changes on the islands of the Western Indian Ocean. WIOJMS 5:179–194

Rahmstorf S (2010) A new view on sea level rise. Nature reports on climate change (4). http://dx.doi.org/10.1038/climate.2010.29. (http://www.nature.com/climate/2010/1004/full/climate.2010.29.html?-session¼user_pref:42F948C2164af3635AUmUp340782). Accessed 20 Mar 2013

Ramanathan V, Chung C, Kim D, Betttge T, Buja L, Kiehl JT, Washington WM, Fu Q, Sikka DR, Wild M (2005) Atmospheric brown clouds: impacts on South Asian climate and hydrological cycle. PNAS. USA 102:5326–5333

Rogers K, Saintilan N, Heijnis H (2005) Mangrove encroachment of salt marsh in Western Port Bay, Victoria: the role of sedimentation, subsidence, and sea level rise. Estuaries 28:551–559

Roth LC (1997) Implications of periodic hurricane disturbance for the sustainable management of Caribbean mangroves. In: Kjerfve B, Lacerda LD, Diop EHS (eds) Mangrove ecosystem studies in Latin America and Africa. UNESCO, Paris

Saenger P, Moverly J (1985) Vegetative phenology of mangroves along the Queensland coastline. Proc Ecol Soc Aust 13:257–65

Saintilan N, Wilton K (2001) Changes in the distribution of mangroves and saltmarshes in Jervis Bay, Australia. Wetl Ecol Manage 9:409–420

Sandilyan S (2007) Mangrove—the evergreen emerald forest. Eco News 13:21
Sandilyan S, Kathiresan K (2012) Mangrove conservation: a global perspective. Biodivers Conserv 21:3523–3542
Sandilyan S, Thiyagesan K (2010) Mangroves—the oceanic woodland. Sci India 13:11–14
Sarwar MGM (2005) Impacts of sea level rise on the coastal zone of Bangladesh. Master's thesis, Lund University
Selvam V, Gnanappazham L, Navamuniyammal M, Ravichandran KK, Karunagaran VM (2002) Atlas of mangrove wetlands of India. Part 1. M.S. Swaminathan Research Foundation, Chennai
Sharma ND (2009) Mycological foray: a culture missing in our country. Curr Sci 97:1708–1709
Siegenthaler U, Stocker TF, Monnin E, Luthi D, Schwander J, Stauffer B, Raynaud D, Barnola JM, Fisher H, Masson-Delmotte V, Jouzel J (2005) Stable carbon cycle-climate relationship during the late Pleistocene. Science 310:1313–1317
Singh HS (2003) Potential impact of climate change on mangroves in India. The XII World Forestry Congress, Québec City, Canada, 21–28 September 2003
Slob JH (2012) Mangroves and climate change: past and future. Master's thesis, Utrecht University
Snedaker S (1995) Mangroves and climate change in the Florida and Caribbean region: scenarios and hypotheses. Hydrobiologia 295:43–49
Solomon S, Qin D, Manning M, Chen Z, Marquis M, Averyt KB, Tignor M, Miller HL(eds)Cchange (2007) the physical science basis. In: Contribution of working group I to the fourth assessment report of the intergovernmental panel on climate change. Cambridge University Press, Cambridge
Spahni R, Chappellaz J, Stocker TF, Loulergue L, Hausammann G, Kawamura K, Fluckiger J, Schwander J, Raynaud D, Masson-Delmotte V, Jouzel J (2005) Atmospheric methane and nitrous oxide of the late Pleistocene from Antarctic ice cores. Science 310:1317–1321
Spalding MD, Blasco F, Field CD (eds) (1997) World mangrove atlas. International society for mangrove ecosystems (ISME), Okinawa
Thampanya U, Vermaat JE, Sinsakul S, Panapitukkul N (2006) Coastal erosion and mangrove progradation of Southern Thailand. Estuar Coast Shelf Sci 68:75–85
Tomlinson PB (1986) The botany of mangroves. Cambridge University Press, Cambridge
Trenberth K (2005) Uncertainty in hurricanes and global warming. Science 308:1753–1754
UNEP (United Nations Environment Programme) (1994) Assessment and monitoring of climatic change impacts on mangrove ecosystems. UNEP Regional Seas Reports and Studies. Report no. 154
UNEP (2005) After the tsunami: rapid environmental assessment. UNEP, Geneva
UNESCO (1992) Coastal systems studies and sustainable development. Proceedings of the COMAR interregional scientific conference, UNESCO, Paris, 21–25 May 1991
Valiela I, Bowen JL, York JK (2001) Mangrove forests: one of the world's threatened major tropical environments. BioScience 51:807–815
Victor S, Neth L, Golbuu Y, Wolanski E, Richmond RH (2006) Sedimentation in mangroves and coral reefs in a wet tropical island, Pohnpei, Micronesia. Estuar Coast Shelf Sci 66:409–416
Walsh KJE, Ryan BF (2000) Tropical cyclone intensity increase near Australia as a result of climate change. J Climate 13:3029–3036
Wang B, Wu R, Lau K-M (2001) Interannual variability of the Asian summer monsoon: contrasts between the Indian and the western North Pacific-East Asian Monsoon. J Climate 14:4073–4090
Wang B, Yang Y, Ding Q-H, Murakami H, Huang F (2010) Climate control of the global tropical storm days (1968–2008). Geophys Res Lett 37:11
Wells S, Ravilous C, Corcoran E (2006) In the front line: shoreline protection and other ecosystem services from mangroves and coral reefs. United Nations Environment Programme World Conservation Monitoring Centre, Cambridge
Wolanski E (2007) Estuarine ecohydrology. Elsevier, Amsterdam
Xiao Y, Zuliang J, Wang M, Lin G, Wang W (2009) Leaf and stem anatomical responses to periodical waterlogging in simulated tidal floods in mangrove *Avicennia marina* Seedlings. Aquat Bot 91:231–237

Xiao Y, Wang W, Chen L (2010) Stem anatomical variations in seedlings of the mangrove *Bruguiera gymnorrhiza* grown under periodical waterlogging. Flora—Morphol Distrib Funct Ecol Plants 205:499–505

Yáñez-Espinosa L, Flores J (2011) A review of sea-level rise effect on mangrove forest species: anatomical and morphological modifications. In: Casalegno S (ed) Global warming impacts—case studies on the economy, human health, and on urban and natural environments. InTech, Rijeka

Yáñez-Espinosa L, Terrazas L, Angeles G (2008) The effect of prolonged flooding on the bark of mangrove. Trees-Struct Funct 22:77–86

Yáñez-Espinosa L, Terrazas L, López-Mata L, Valdez-Hernández JI (2004) Wood variation in *Laguncularia racemosa* and its effect on fibre quality. Wood Sci Technol 38:217–226

Quantification of Soil Organic Carbon Storage and Turnover in Two Mangrove Forests Using Dual Carbon Isotopic Measurements

Lianlian Yuan, Jinping Zhang, Chengde Shen, Hai Ren, Hongxiao Liu and Kexin Liu

Contents

Abstract Mangrove ecosystems are highly productive and play an important role in tropical and global coastal carbon (C) budgets. However, soil organic carbon (SOC) storage and turnover in mangrove forests are still poorly understood. Based on C and C isotopic measurements of soil cores from two natural mangrove forests in Southern China, SOC density was 674.41 Mg ha^{-1} in one forest (site 1, a *Aegiceras*

H. Ren (✉) · L. Yuan · J. Zhang · H. Liu
South China Botanical Garden, Chinese Academy of Sciences,
510650 Guangzhou, P. R China
e-mail: renhai@scib.ac.cn

J. Zhang · H. Liu
Graduate University of Chinese Academy of Sciences, 100039 Beijing, P. R China

C. Shen
Key Laboratory of Isotope Geochronology and Geochemistry,
Guangzhou Institute of Geochemistry, Chinese Academy of Sciences,
510640 Guangzhou, P. R China

K. Liu
State Key Laboratory of Nuclear Physics and Technology,
Peking University, 100871 Beijing, P. R China

I. Faridah-Hanum et al. (eds.), *Mangrove Ecosystems of Asia*,
DOI 10.1007/978-1-4614-8582-7_12,

corniculatum-dominated high tidal flat) and 372.83 Mg ha^{-1} in the other forest (site 2, a *Bruguiera gymnorrhiza* + *Kandelia candel*-dominated middle tidal flat). SOC storage in the upper 100 cm in mangrove forests in China was estimated to be 13.65–24.68 Tg. SOC δ^{13}X values in the two mangrove forests ranged from −29.36 to −25.90 ‰. SOC $\delta^{13}C$ was enriched with depth at 20–70 cm at site 1 (which is similar to the trend in most terrestrial ecosystems) but not at site 2, probably because the latter but not the former forest experienced frequent tidal flushing of the surface soil. SOC $\delta^{13}C$ enrichment with depth at site 1 was not due to depletion of $\delta^{13}C$ of atmospheric CO_2 by fossil fuel emissions or to the difference between root and leaf^{13}C, but possibly resulted from preferential microbial decomposition; this inference was supported by the Rayleigh distillation model, which also indicated that C was mainly from the parent *A. corniculatum* forest. C and stable C isotopic measurements indicated that tidal fluctuation greatly impacted SOC deposition in these mangrove forests; the high tidal flat (site 1) and the middle tidal flat (site 2) showed terrestrial and oceanic deposition characteristics, respectively. ^{14}C from the testing of thermonuclear weapons had penetrated to 50–60 cm in the two forests. SOC turnover time varied with depth. The surface SOC turnover time at both sites was about 0.5 year, because most surface SOC consisted of easily decomposed litter. SOC turnover time at 20–60 cm at site 1 and at 25–50 cm at site 2 was 4.44–26.04 year. Abundant roots apparently accounted for the short SOC turnover times at these middle soil layers.

1 Introduction

Located at the intertidal zone, the mangrove ecosystem is characterized by high primary productivity, rapid organic matter deposition, and low CO_2 and CH_4 efflux. The mangrove ecosystem, therefore, is an important carbon (C) sink for atmospheric CO_2 (Bouillon et al. 2008; Choi and Wang 2004; Gonneea et al. 2004; Tamooh et al. 2008). Although understanding the dynamics of soil organic carbon (SOC) in mangrove ecosystems will help us to better constrain global oceanic-C budgets (Bouillon et al. 2008, 2003; Kristensen et al. 2008), mangrove ecosystems have been ignored in most global C budgets (Chmura et al. 2003). Chmura et al. (2003) estimated that the top 100 cm of global mangrove forest soils contains about 10,000 Tg of SOC. SOC storage and dynamics in mangrove ecosystems deserve our attention.

SOC density in mangrove forests varies greatly at different sites. SOC densities at 0–100 cm depth have been estimated to be 57.3 Mg ha^{-1} in Japan, 288.65 Mg ha^{-1} in Hong Kong, and 592 Mg ha^{-1} in Australia (Khan et al. 2007; Matsui 1998; Zhang et al. 2007). SOC density at 0–100 cm has not been reported for mangrove forests in China.

C isotopes are widely used to reconstruct ecological processes and to trace ecological activity (Bouillon and Bottcher 2006; West et al. 2006), and the use of C isotopes has greatly improved our understanding of the C cycle in below-ground ecosystems (Staddon 2004). In general, the stable isotope ratio of C ($\delta^{13}C$) can be used to determine the source and decomposition of C. Thus, SOC $\delta^{13}C$ and other indicators have been used to determine the sources (mangrove-derived, oceanic, or terrestrial) of organic C in coastal wetlands (Bouillon et al. 2003; Gonneea et al. 2004). However, there have been few studies of global mangrove organic matter $\delta^{13}C$ (Muzuka and Shunula 2006). The radioactive isotope ratio of C ($\Delta^{14}C$), in contrast, can be used to determine the age and turnover of C. Using an SOC $\Delta^{14}C$ model, Choi and Wang (2004) found that turnover times of SOC at soil depths of 0–10 cm in low, middle, and high tidal flats of Florida were 16–31 years, 18–57 years, and 10–38 years, respectively. SOC turnover rates in mangrove forests, and especially in subsurface soil layers, have seldom been studied.

In this study, we estimated SOC storage and turnover in two natural mangrove forests in Southern China based on analyses of C and C isotopes. The specific objectives were to determine (1) SOC storage in the two mangrove forests, (2) the controlling mechanism of vertical variation of SOC $\delta^{13}C$, and (3) SOC turnover rates in the mangrove forests.

2 Materials and Methods

2.1 Sites

The two sites were located on the west coast of the Leizhou Peninsula, about 70 km northwest of Zhanjiang, and belong to the Zhanjiang Mangrove National Nature Reserve, which is the largest mangrove reserve in China. Soil cores were excavated in rarely disturbed, near-pristine natural mangrove forests (Fig. 1), which had an area of approximately 270 ha and was more than 80 years old. Site 1 was located at the high tidal flat and was dominated by a pioneer *Aegiceras corniculatum* community. Site 2 was located at the middle tidal flat and was dominated by a middle-to-late-stage *Bruguiera gymnorrhiza* and *Kandelia candel* community. Therefore, the two sites had quite different plant communities. Table 1 summarizes the vegetation indexes and characteristics of the top 20 cm of soil.

For both sites, the mean annual temperature is 22.9 °C, and the average temperature in the coldest month (January) is 15.5 °C. The annual average precipitation is 1,711 mm, with the rainfall typically occurring from April to October. The average annual relative humidity is 80 %. The tidal regime is diurnal, and the average tidal amplitude is 2.52 m, with a maximum of 6.25 m (He et al. 2007). According to Ren et al. (2008), the average bulk density of the typical mangrove acid soil is 2.6×10^3 kg m^{-3}.

2.2 Soil Sampling

During ebb-tide in June 2006, one soil core (11 cm in diameter and 100 cm in length) was excavated from each site. The cores were taken adjacent to the mangrove trees. Visible characteristics of the soil are provided in Table 2. The cores were divided into segments based on depth and the characteristics of soil and roots. The sampling

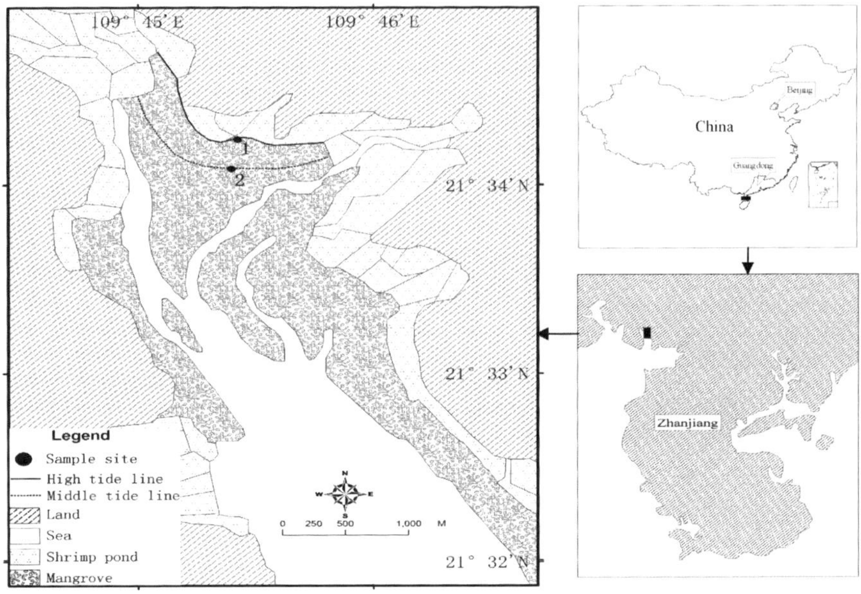

Fig. 1 Locations of soil cores at the two sites in mangrove forests of Zhanjiang, Southern China. The numbers represent: *1 Aegiceras corniculatum* site, *2 Bruguiera gymnorrhiza* + *Kandelia candel* site

Table 1 Vegetation indexes and attributes of the top 20 cm of soil in two mangrove forests (site 1 and 2) in Southern China

Site	Dominant plant species	Vegetation index Average height (cm)	Soil attribute Average DBH (cm)	pH	Salinity (%)	Total N (%)	Total P (%)
1	*Aegiceras corniculatum*	225	2.30	6.03	2.53	0.23	0.05
2	*Bruguiera ymnorrhiza*	220	5.60	6.10	2.75	0.18	0.06
	Kandelia candel	270	3.70				

interval was 2 cm for the top 20 cm depth, 5 cm for the 20–50 cm depth, 10 cm for the 50–80 cm depth, and 20 cm for the 80–100 cm depth. These soil segments were transported to the Carbon Isotope Laboratory of the Guangzhou Institute of Geochemistry, where they were frozen at − 26 °C for later analyses.

2.3 Carbon Isotope Analyses

The frozen soil segments were thawed to room temperature. After visible roots and fragmentary stones were removed, a 20 to 35-g subsample of each segment was freeze-dried under vacuum for 48 h, ground using a mortar and pestle, and then passed

Table 2 Visible characteristics of soil cores from two mangrove forests in Southern China

Site	Depth (cm)	Characteristics
1. *Aegiceras corniculatum*	0–20	Black sludge with high organic matter content, mixed with some red solid, granular structure, very loose, abundant plant debris and roots, full of crustacean and molluscan macrofauna
	20–60	Grayish black, half sludge, visible roots (root amount decreasing with depth), less plant debris
	60–100	Gray, high sand content, tight, no rootlets
2. *Bruguiera gymnorrhiza* + *Kandelia candel*	0–25	Black sludge, loose, abundant plant roots and macrofauna, occasionally brown solid
	25–50	Grayish black, half sludge, large roots
	50–100	Gray, higher sand content, tight

through a 1-mm screen to remove rootlets and coarse sand. The subsamples were boiled in 2 M HCL for 10 min to remove carbonate and then rinsed repeatedly with distilled water until a neutral pH was reached. Water was removed by oven-drying at 90 °C for 24 h.

Subsamples for ^{12}C and ^{14}C analysis were then loaded into sealed, evacuated quartz tubes and combusted with CuO at 860 °C for 2 h. The CO_2 thus generated was cryogenically purified with dry-ice and liquid nitrogen. Stable carbon isotope ratios in one portion of the generated CO_2 were determined using a Finnigan MAT-251 mass spectrometer with a precision of 0.2 ‰ at the State Key Laboratory of Loess and Quaternary Geology, CAS. Results were reported in δ-notation relative to V-PDB as: $\delta^{13}C = ((^{13}C/^{12}C)_{sample}/(^{13}C/^{12}C)_{standard} - 1) \times 1000$. The other portion of the generated CO_2 was catalytically reduced to graphite AMS targets using the method of Vogel et al. (1987). Radioactive carbon isotope ratios were determined in the generated graphite using AMS at the Institute of Heavy Ion Physics of Peking University. The error in ^{14}C is less than 1.5 ‰ of modern for all samples. Results were recorded as: $\Delta^{14}C = ((^{14}C/^{12}C)_{sample}/(^{14}C/^{12}C)_{standard} - 1) \times 1000$.

SOC content was calculated from the quantity of CO_2 generated from the subsamples. SOC density was calculated by multiplying SOC content by the bulk density and thickness of soil layer. SOC storage was calculated by multiplying SOC density by the distribution area.

2.4 Bomb $\Delta^{14}C$ Based SOC Turnover Rate

Bomb ^{14}C produced by atmospheric thermonuclear weapons testing is widely used to trace short-term SOC turnover at decade scales (Chen et al. 2002; Cherkinsky and Brovkin 1993; Telles et al. 2003; Townsend et al. 1995; Trumbore 1996). The equations used to calculate the turnover rate are:

$$\frac{I(1955)}{I_0} = \frac{m}{m+\lambda} \quad (1)$$

$$I(t) = I(t-1) - (m+\lambda) \cdot I(t-1) + m \cdot I_0(t) \quad (2)$$

where I (1955) is the SOC ^{14}C radioactivity in 1955, I (t) is the SOC ^{14}C radioactivity in the year of sampling ($t > 1955$), I ($t-1$) is the SOC ^{14}C radioactivity in year ($t-1$), I_0 (t) is the ^{14}C radioactivity in the atmosphere in year t, I_0 is the ^{14}C radioactivity of the modern carbon standard, λ is the ^{14}C decay constant (1/8267 year^{-1}), and m is the SOC turnover rate (yr^{-1}).

Equation (1) describes the SOC ^{14}C radioactivity of a stable and closed section of soil. Equation (2) is a mathematical expression for the dynamics of ^{14}C radioactivity of SOC that is exchanging carbon with atmospheric CO_2. The SOC ^{14}C radioactivity in year t is determined by that in year ($t-1$), the ^{14}C loss attributable to natural decay of ^{14}C and SOC decomposition, and the incorporation of ^{14}C from atmosphere in year t through formation of new organic matter.

During numerical simulation, one value of m was selected first, and then I (1955) was calculated by Eq. (1). The values for I (1955) and m were then put into Eq. (2), and the SOC ^{14}C radioactivity in the sampling year (2006) was obtained by iterative calculation. When two possible m values were obtained (Townsend et al. 1995), one was selected by considering the m values of the adjacent layers.

Values of I_0 (t) in 1955–1996, 1997–2003, and 2004–2006 were obtained from Hua and Barbetti (2004), Levin and Kromer (2004), and Levin et al. (2008), respectively.

For samples with $\Delta^{14}C$ less than zero, the influence of "bomb^{14}C" is negligible because of the low SOC turnover rates, although these layers might have been slightly affected by "bomb^{14}C". We assume that these layers contain no fresh organic matter. These soil layers were therefore considered stable and virtually closed. Based on Eq. (1), the SOC turnover rate of these layers can be calculated as:

$$m = -\lambda(\frac{1000}{\Delta^{14}C} + 1) \quad (3)$$

3 Results

3.1 *Aegiceras Corniculatum Site (Site 1)*

SOC content at site 1 increased from 5.39 % at 0–2 cm to 6.37 % at 6–8 cm, generally decreased at 8–70 cm, and then increased at 70–100 cm. The lowest SOC content was 0.84 % at 60–70 cm (Fig. 2a). The calculated SOC density to a depth of 100 cm at site 1 was 674.41 Mg ha^{-1}.

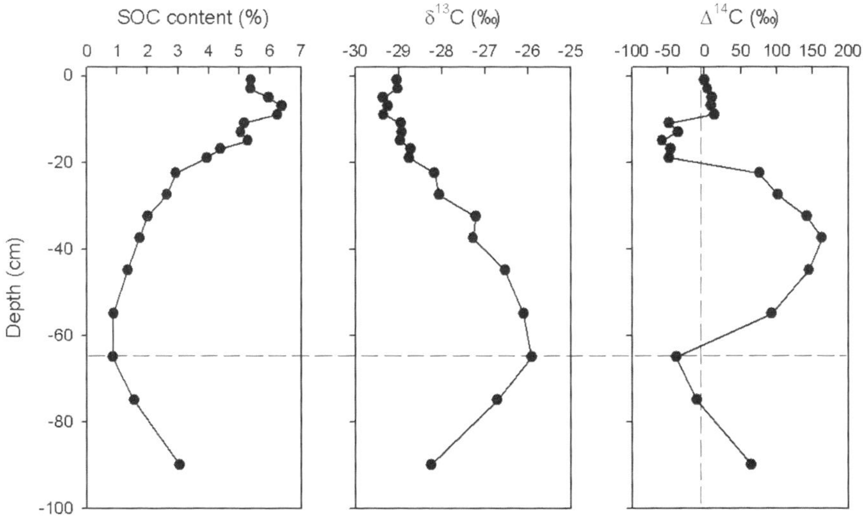

Fig. 2 Vertical patterns in the soil core from the *A. corniculatum* site. **a** SOC content. **b** SOC $\delta^{113}C$. **c** SOC $\delta^{114}C$

SOC $\delta^{13}C$ values were relatively constant in the upper 20 cm layer, became gradually less negative at 20–70 cm, and then turned more negative at depths of 70–100 cm. SOC $\delta^{13}C$ values throughout the soil core ranged from -29.36 to -25.92 ‰ (Fig. 2b).

The curve describing the vertical distribution of SOC $\Delta^{14}C$ values could be divided into five sections. Those at 0–10 cm, 20–60 cm, and 80–100 cm were greater than zero, indicating that the soil in these sections contained modern carbon polluted by atmospheric thermonuclear weapons testing. Those at 10–20 cm and 60–80 cm were less than zero. The highest and lowest values of SOC $\Delta^{14}C$ throughout the soil core were 162.90 ‰ and -58.05 ‰, respectively (Fig. 2c).

At site 1, the curves describing SOC content, $\delta^{13}C$, and $\Delta^{14}C$ all changed direction at 60–70 cm (Fig. 2a, 2b, 2c).

The SOC turnover rate at site 1 could be divided into four sections (Table 3). The high SOC turnover rate at 0–10 cm ranged from 1.9729 to 1.9781 year^{-1}. The very low turnover rate at 10–20 cm ranged from 0.0020–0.0032 year^{-1}. The turnover rate at 20–60 cm ranged from 0.0384–0.2251 year^{-1}, and that at 60–100 cm ranged from 0.0030–0.0116 year^{-1}.

3.2 *Bruguiera gymnorrhiza + Kandelia candel (Site 2)*

SOC content at site 2 was 1.67–2.24 % at 0–50 cm. The lowest SOC content was 0.53 % at 50–60 cm. SOC content tended to increase at 60–100 cm (Fig. 3a). SOC density to a depth of 0–100 cm at site 2 was 372.83 Mg ha^{-1}.

Table 3 SOC turnover rate at the *A. corniculatum* site and the *B. gymnorrhiza* + *K. candel* site

Depth (cm)	Turnover rate (yr^1)	
	A. corniculatum	*B. gymnorrhiza* + *K. candel*
0–2	1.9781	1.9704
2–4	1.9767	1.9733
4–6	1.9742	0.0068
6–8	1.9749	0.0089
8–10	1.9729	1.9657
10–12	0.0024	1.9712
12–14	0.0032	1.9698
14–16	0.0020	1.9710
16–18	0.0025	1.9688
18–20	0.0024	1.9636
20–25	0.2251	1.9382
25–30	0.1202	0.1784
30–35	0.0674	0.0485
35–40	0.0384	0.0384
40–50	0.0645	0.0384
50–60	0.1401	0.0016
60–70	0.0030	0.0036
70–80	0.0116	0.0036
80–100	0.0077	0.0018

At site 2, SOC $\delta^{13}C$ fluctuated slightly in the top 20 cm and was relatively stable at 20–100 cm. The mean SOC $\delta^{13}C$ for 0–100 cm was – 26.40 ‰ (Fig. 3b).

The curve describing the vertical distribution of SOC $\Delta^{14}C$ values at site 2 could be divided into six sections. Soil layers at 0–4 cm, 8–50 cm, and 70–80 cm had positive SOC $\Delta^{14}C$ values; the other soil layers had negative SOC $\Delta^{14}C$ values. The highest and lowest values of SOC $\Delta^{14}C$ throughout the soil core were 182.10 ‰ and – 69.36 ‰, respectively (Fig. 3c).

The curves describing SOC content and SOC $\Delta^{14}C$ at site 2 tended to change direction at 50–60 cm (Fig. 3a, 3c).

The SOC turnover rate at site 2 could be divided into four sections (Table 3). SOC turnover rate was 1.9382–1.9733 $year^{-1}$ at 0–4 cm and 8–25 cm, 0.0068–0.0089 $year^{-1}$ at 4–8 cm, 0.0384–0.1784 $year^{-1}$ at 25–50 cm, and 0.0016–0.0036 $year^{-1}$ at 50–100 cm.

4 Discussion

4.1 SOC Storage in Mangrove Forests

Chmura et al. (2003) estimated that SOC storage in the upper 50 cm of global mangrove forests was 5,000 ± 400 Tg. Because the sites in the current study were located in large mangrove forests in the largest mangrove reserve in China, data from our soil cores were useful for making coarse estimates of SOC storage in mangrove forests in

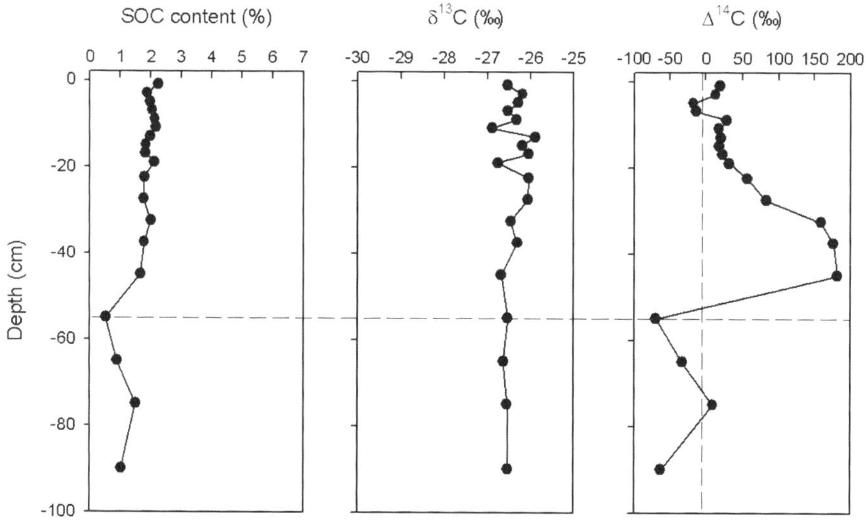

Fig. 3 Vertical patterns in the soil core from the *B. gymnorrhiza* + *K. candel* site. **a** SOC content. **b** SOC δ^{113}C. **c** SOC δ^{114}C

China. Given that SOC density at the *A. corniculatum* site was 674.41 Mg ha^{-1} and that at the *B. gymnorrhiza* + *K. candel* site was 372.83 Mg ha^{-1}, and that the area of mangrove forests in China is about 366 km^2 (Adeel and Pomeroy 2002; Spalding et al. 1997), we estimated that the upper 100 cm of soil in mangrove forests in China stored about 13.65–24.68 Tg of SOC.

Because mangrove forests are located at the intertidal zone, tidal action greatly affects SOC deposition (Middleton and McKee 2001; Tam and Wong 1998). SOC density at the *B. gymnorrhiza* + *K. candel* site (located at the middle tidal flat) was 55.28 % of that at the *A. corniculatum* site (located at the high tidal flat). This difference was largely due to differences in SOC content in the top 30 cm of soil (see Fig. 2a, 3a) resulting from tidal effects. The middle tidal flat was flooded regularly, and the high tidal flat was flooded only occasionally (Chen et al. 2007; Choi and Wang 2004; Middleton and McKee 2001; Tam and Wong 1998). At the middle tidal flat, the mangrove-derived organic matter was deposited over a wide area, and the organic matter on the surface was subject to tidal erosion. At the high tidal flat, however, the mangrove-derived organic matter was not spread so widely and was not subject to substantial erosion.

4.2 *Vertical Patterns of SOC δ^{13}C in Mangrove Forests*

The change in SOC δ^{13}C values with depth differed greatly between the two sites in that SOC δ^{13}C values changed substantially with depth at the *A. corniculatum* site but not at the *B. gymnorrhiza* + *K. candel* site (Fig. 2b, 3b). SOC δ^{13}C at 0–20 cm at the

A. corniculatum site was relatively constant, reflecting the high productivity and rapid deposition of organic matter of mangrove forests (Gonneea et al. 2004; Jennerjahn and Ittekkot 2002; Woodroffe 1992). SOC $\delta^{13}C$ was enriched at 20–70 cm at the *A. corniculatum* site. This result agreed with results obtained in terrestrial ecosystems. Numerous studies have indicated that, with the exception of soils in peatlands, SOC $\delta^{13}C$ tended to become enriched with soil depth. As discussed in the next three paragraphs, several factors might explain this enrichment (Balesdent et al. 1993; Boutton 1996; Ehleringer et al. 2000; Wynn et al. 2006).

First, because the burning of fossil fuels since the start of the industrial revolution has released CO_2 with low $\delta^{13}C$ values (approximately -27 ‰) into the atmosphere, the $\delta^{13}C$ of atmospheric CO_2 has been depleted by up to 1.3 ‰ during the past 200 years. With the mixing of C derived from biomass to soil during depleting $\delta^{13}C$ of atmospheric CO_2, the $\delta^{13}C$ values of the younger SOC at the surface should be more negative than that of the older SOC at the deeper soil layers. Thus the enrichment of SOC $\delta^{13}C$ with depth will be expected. But the enrichment with depth in this study did not result from the burning of fossil fuels. SOC $\Delta^{14}C$ values at 20–60 cm were more than zero (Fig. 2c), indicating that the SOC at this layer was formed after atmospheric thermonuclear weapons testing, which occurred much later than the industrial revolution.

Second, roots generally have $\delta^{13}C$ values 1–3 ‰ higher than those of other plant tissues (Vonfischer and Tieszen 1995). Thus, deeper soil C, which comes from roots, is expected to have higher $\delta^{13}C$ values than surface C, which comes mainly from dead leaves and stems. However, the surface litter decays rapidly and little is preserved in mangrove forests, and most subsurface soil C is derived from roots (Gleason and Ewel 2002; Jennerjahn and Ittekkot 2002; Middleton and McKee 2001). Therefore, it seems unlikely that the $\delta^{13}C$ difference between roots and leaves can explain the enrichment of SOC $\delta^{13}C$ with depth at these middle soil layers.

Third, many studies have indicated that fractionation by microorganisms results in SOC $\delta^{13}C$ enrichment with depth (Agren et al. 1996; Powers and Schlesinger 2002; Santruckova et al. 2000; Schweizer et al. 1999). Microorganisms preferentially metabolize molecules containing the lighter ^{12}C, so that the remaining SOC tends to be ^{13}C enriched. This process is quantified in the Rayleigh distillation model (Ehleringer et al. 2000), which fits the data as we will discuss in the next section.

From the above discussion, we conclude that SOC $\delta^{13}C$ enrichment at 20–70 cm at the *A. corniculatum* site did not result from the burning of fossil fuels and probably did not result from differences in $\delta^{13}C$ in roots vs. leaves. We suspect that the enrichment resulted from microbial fractionation.

Compared to the SOC $\delta^{13}C$ values at the *A. corniculatum* site, the values at the *B. gymnorrhiza* + *K. candel* site, which was located at the middle tidal flat, were relatively stable. We suspect that this results from substantial tidal effects at the middle tidal flat. The changing tides at the middle tidal flat frequently mix the organic matter derived from the mangrove forest with that derived from the ocean. In addition, the strong tides at this location tend to spread organic matter derived from the mangrove forest over a wide area, resulting in less organic matter on the surface. As a result, the top 30 cm SOC $\delta^{13}C$ values at the middle tidal flat were much higher than those at the high tidal flat.

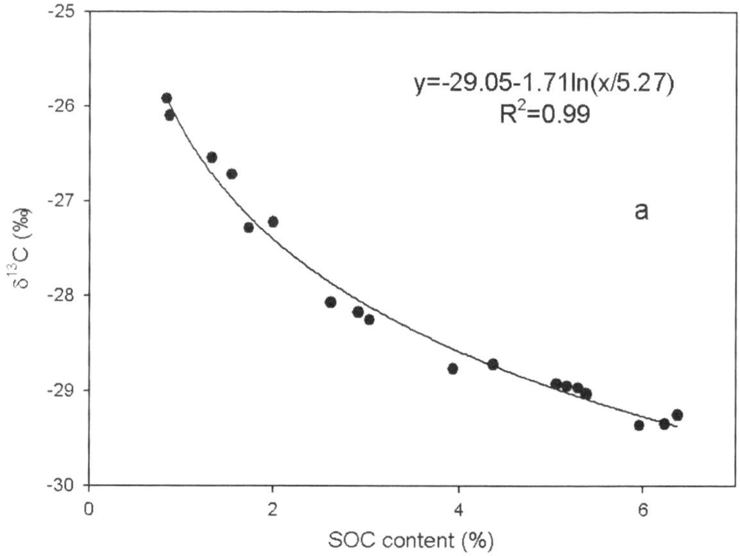

Fig. 4 Relationship between SOC content and SOC $\delta^{13}C$ in the soil core from the *A. corniculatum* site. The curve and equation are based on the Rayleigh distillation equation. (Rayleigh 1896)

4.3 Relationship Between SOC Content and SOC $\delta^{13}C$ in Mangrove Forests

The Rayleigh distillation equation (Rayleigh 1896) relating logarithmic SOC content and SOC $\delta^{13}C$ has been used in many studies to determine the initial values of SOC content and SOC $\delta^{13}C$ (Balesdent et al. 1993; Balesdent and Mariotti 1996; Powers and Schlesinger 2002; Schweizer et al. 1999; Wynn 2007; Wynn et al. 2006). The relationship between SOC content and SOC $\delta^{13}C$ at the *A. corniculatum* site was well described by the Rayleigh distillation model (Fig. 4). According to the equation generated by this model, the initial SOC content was 5.27 % and the initial SOC $\delta^{13}C$ value was − 29.05 ‰ ($R^2 = 0.99$). From Fig. 2a and 2b, we can see that the average values of SOC content and SOC $\delta^{13}C$ in the upper 20 cm at the *A. corniculatum* site were 5.31 % and − 29.03 ‰, respectively. Both of these values were very close to the initial values predicted by the Rayleigh distillation model. This suggests that the main source of C in the SOC at this site was organic matter from the parent mangrove forest. In general, SOC in mangrove forests contains C from several sources, including the parent mangrove forest, oceanic sea grass and phytoplankton, and terrestrial inputs (Bouillon et al. 2008, 2003; Gonneea et al. 2004). Oceanic organic matter tends to be deposited in deeper water, away from shore, whereas mangrove-derived organic matter tends to be deposited along the land margin (Machiwa 2000). Because the *A. corniculatum* site was located at the high tidal flat, it seems likely that little oceanic organic matter had deposited there and that organic matter from the mangrove forest was the main source of C.

4.4 SOC Turnover in Mangrove Forests

SOC turnover times differed markedly at different soil depths in the mangrove forests. Vertical patterns of SOC turnover times in both sites could be divided into three sections, although there were some discontinuities in the patterns. Turnover times of less than 1 year were common at the upper layers, although some upper layers had turnover times of several hundred years. The middle layers had turnover times of several years to decades, and the bottom layers had turnover times of hundreds of years.

Turnover time of the surface SOC was about 0.5 years because most of the surface SOC consisted of easily decomposed litter. Middleton and McKee (2001) also found that the surface litter decomposed rapidly in mangrove forests, and Trumbore et al. (1995) calculated a turnover time of less than 1 year for surface SOC.

The discontinuities in SOC turnover times at the upper layers (whereby adjacent layers had very different turnover times) were also found in other studies of mangrove forest decomposition (see Fujimoto et al. 1999). We suspect that these unusual layers were deposited by erosion of inland soil, with greater deposition at the high tidal flat than at the middle tidal flat.

With the exception of the layers with unusually long turnover times, SOC in the upper 60 cm at the *A. corniculatum* site and in the upper 50 cm at the *B. gymnorrhiza* + *K. candel* site was modern C. The bottom depth of modern C was usually less than 25 cm (Bol et al. 1999; Wang et al. 2005). In the present study, bomb ^{14}C was found as deep as 50–60 cm. We suspect that the modern C detected in this study was mainly derived from mangrove roots; mangrove roots, including pneumatophore and buttress-like roots, are powerful and penetrate deep into soil. Many studies have reported that root biomass in mangrove forests is very high, that mangrove roots decompose relatively slower, and that deep SOC in mangrove forests was derived mainly from roots (Fujimoto et al. 1999; Gleason and Ewel 2002). Root biomass could account for half of the total biomass in mangrove forests (Briggs 1977), and fine-root biomass might account for 66 % of the total root biomass (Komiyama et al. 1987). Tamooh et al. (2008) found that fine roots were distributed mainly at 0–60 cm and decreased with depth in mangrove forests. This agrees with the root and modern C distributions in this study (see Table 2, Fig. 2c, 3c).

SOC turnover time at 20–60 cm at the *A. corniculatum* site and at 25–50 cm at the *B. gymnorrhiza* + *K. candel* site was 4.44–26.04 year. In general, SOC turnover is much slower deeper in the soil than near the surface. However, Fontaine et al. (2007) found that the fresh supply of organic C could provide an energy source for microorganisms at the subsurface layers. Consequently, turnover of SOC at the subsurface layers could be much faster than before. In this study, apparent mangrove roots were working the fresh supply of organic C, and shortened the SOC turnover time at these middle soil layers.

SOC turnover times at 60–100 cm at the *A. corniculatum* site and at 50–100 cm at the *B. gymnorrhiza* + *K. candel* site was 86.21–625.00 year. These slow turnover times probably resulted from reduced input of fresh organic matter, the resistance of old SOC to decomposition, and reduced oxygen for aerobic decomposition.

5 Conclusions

The results of this paper, which are based on measurement of C and C isotopes in soil cores, indicate that mangrove forests store large quantities of C. SOC density was 372.83 Mg ha^{-1} in one forest (located at a high tidal flat) and 674.41 Mg ha^{-1} in the other forest (located at a middle tidal flat). Based on these densities, SOC storage in mangrove forests in China was estimated to range from 13.65 to 24.68 Tg. SOC at the high tidal flat evidently resulted from terrestrial deposition while SOC at the middle tidal flat evidently resulted from oceanic deposition.

SOC was $\delta^{13}C$ enriched with depth at the middle soil layers of the high tidal flat, possibly due to preferential microbial decomposition. This inference was supported by the Rayleigh distillation model, which also indicated that the carbon in SOC was mainly from the parent mangrove forest at the high tidal flat.

At both the high and middle tidal flats, turnover time was about 0.5 years for surface SOC and was 5.61–26.04 years for SOC in the middle soil layers. Mangrove roots probably accounted for the short turnover time at the middle soil layers. Further quantification of root dynamics is required to increase our understanding of C turnover in the below-ground mangrove ecosystem.

Acknowledgements We thank Professor Li Zhi'an, Senior Engineer Zhang Qianmei, Mr Zou Bi (the South China Botanical Garden), and Director Xu Fanghong of the Administration of Zhanjiang Mangrove National Natural Reserve for their help in field sampling. We also thank Doctor Ding Ping of the Guangzhou Institute of Geochemistry, CAS for help and assistance in field sampling and laboratory operation. Special thanks for help in improving the manuscript to Professors Shen Weijun and Jian Shuguang, and Master Candidate Li Pingheng (the South China Botanical Garden). This research was sponsored by funds from the Chinese Academy of Sciences (No. KSCX2-SW-132) and The Ministry of Science and Technology (No. 2007BAC28B04).

References

Adeel Z, Pomeroy R (2002) Assessment and management of mangrove ecosystems in developing countries. Trees 16:235–238

Agren GI, Bosatta E, Balesdent J (1996) Isotope discrimination during decomposition of organic matter: a theoretical analysis. Soil Sci Society Amer J 60:1121–1126

Balesdent J, Mariotti A (1996) Measurement of soil organic matter turnover using 13C natural abundance. In: Boutton TW, Yamasaki S (eds) Mass spectrometry of soils. Marcel Dekker, New York, pp 83–111

Balesdent J, Girardin C, Mariotti A (1993) Site related delta 13C of tree leaves and soil organic matter in a temperate forest. Ecology 74:1713–1721

Bol RA, Harkness DD, Huang Y, Howard DM (1999) The influence of soil processes on carbon isotope distribution and turnover in the British uplands. Euro J Soil Sci 50:41–51

Bouillon S, Bottcher ME (2006) Stable isotopes in biogeosciences—preface. Org Geochem 37:1197–1199

Bouillon S, Dahdouh-Guebas F, Rao A, Koedam N, Dehairs F (2003) Sources of organic carbon in mangrove sediments: variability and possible ecological implications. Hydrobiologia 495:33–39

Bouillon S, Connolly RM, Lee SY (2008) Organic matter exchange and cycling in mangrove ecosystems: recent insights from stable isotope studies. J Sea Res 59:44–58

Boutton TW (1996) Stable carbon isotope ratios of soil organic matter and their use as indicators of vegetation and climate change. In: Boutton TW, Yamasaki S (eds) Mass spectrometry of soil. Marcel Dekker, New York, pp 311–339

Briggs SV (1977) Estimates of biomass in a temperate mangrove community. Aus J Ecol 2:369–373

Chen QQ, Sun YM, Shen CD, Peng SL, Yi WX, Li Z, Jiang MT (2002) Organic matter turnover rates and CO_2 flux from organic matter decomposition of mountain soil profiles in the subtropical area. Catena 49:217–229

Chen QQ, Gu HQ, Zhou JZ, Meng Y, Hu KL (2007) Trends of soil organic matter turnover in the salt marsh of the Yangtze River estuary. J Geograph Sci 10:101–113

Cherkinsky AE, Brovkin VA (1993) Dynamics of radiocarbon in soils. Radiocarbon 35:363–367

Chmura GL, Anisfeld SC, Cahoon DR, Lynch JC (2003) Global carbon sequestration in tidal, saline wetland soils. Global Biogeochem Cy 17:1–22

Choi YH, Wang Y (2004) Dynamics of carbon sequestration in a coastal wetland using radiocarbon measurements. Global Biogeochem Cy 18:1–12

Ehleringer JR, Buchmann N, Flanagan LB (2000) Carbon isotope ratios in belowground carbon cycle processes. Ecol Appl 10:412–422

Fontaine S, Barot S, Barre P, Bdioui N, Mary B, Rumpel C (2007) Stability of organic carbon in deep soil layers controlled by fresh carbon supply. Nature 450:277–288

Fujimoto K, Imaya A, Tabuchi R, Kuramoto S, Utsugi H, Murofushi T (1999) Belowground carbon storage of Micronesian mangrove forests. Ecol Res 14:409–413

Gleason SM, Ewel KC (2002) Organic matter dynamics on the forest floor of a Micronesian mangrove forest: an investigation of species composition shifts. Biotropica 34:190–198

Gonneea ME, Paytan A, Herrera-Silveira JA (2004) Tracing organic matter sources and carbon burial in mangrove sediments over the past 160 years. Estuar Coastal Shelf Sci 61:211–227

He BY, Lai TH, Fan HQ, Wang WQ, Zheng HL (2007) Comparison of flooding-tolerance in four mangrove species in a diurnal tidal zone in the Beibu Gulf. Estuar Coastal Shelf Sci 74:254–262

Hua Q, Barbetti M (2004) Review of tropospheric bomb C-14 data for carbon cycle modeling and age calibration purposes. Radiocarbon 46:1273–1298

Jennerjahn TC, Ittekkot V (2002) Relevance of mangroves for the production and deposition of organic matter along tropical continental margins. Naturwissenschaften 89:23–30

Khan MNI, Suwa R, Hagihara A (2007) Carbon and nitrogen pools in a mangrove stand of Kandelia obovata (S., L.) Yong: vertical distribution in the soil-vegetation system. Wetlands Ecol Manag 15:141–153

Komiyama A, Ogino K, Aksornkoae S, Sabhasri S (1987) Root biomass of a mangrove forest in southern Thailand. 1. Estimation by the trench method and the zonal structure of root biomass. J Trop Ecol 3:97–108

Kristensen E, Bouillon S, Dittmar T, Marchand C (2008) Organic carbon dynamics in mangrove ecosystems: a review. Aquat Bot 89:201–219

Levin I, Kromer B (2004) The tropospheric (CO_2)-C-14 level in mid-latitudes of the Northern Hemisphere (1959–2003). Radiocarbon 46:1261–1272

Levin I, Hammer S, Kromer B, Meinhardt F (2008) Radiocarbon observations in atmospheric CO_2: determining fossil fuel CO_2 over Europe using Jungfraujoch observations as background. Sci Total Env 391:211–216

Machiwa JF (2000) C-13 signatures of flora, macrofauna and sediment of a mangrove forest partly affected by sewage waste. Tanzan J Sci 26:15–28

Matsui N (1998) Estimated stocks of organic carbon in mangrove roots and sediments in Hinchinbrook Channel, Australia. Mangroves Salt Marshes 2:199–204

Middleton BA, McKee KL (2001) Degradation of mangrove tissues and implications for peat formation in Belizean island forests. J Ecol 89:818–828

Muzuka ANN, Shunula JP (2006) Stable isotope compositions of organic carbon and nitrogen of two mangrove stands along the Tanzanian coastal zone. Estuar Coastal Shelf Sci 66:447–458

Powers JS, Schlesinger WH (2002) Geographic and vertical patterns of stable carbon isotopes in tropical rain forest soils of Costa Rica. Geoderma 109:141–160

Rayleigh JW (1896) Theoretical considerations respecting the separation of gases by diffusion and similar processes. The London, Edinburgh, and Dublin Philosophical Magazin 42:493–498

Ren H, Jian SG, Lu HF, Zhang QM, Shen WJ, Han WD, Yin ZY, Guo QF (2008) Restoration of mangrove plantations and colonisation by native species in Leizhou bay, South China. Ecol Res 23:401–407

Santruckova H, Bird MI, Lloyd J (2000) Microbial processes and carbon-isotope fractionation in tropical and temperate grassland soils. Func Ecol 14:108–114

Schweizer M, Fear J, Cadisch G (1999) Isotopic (C-13) fractionation during plant residue decomposition and its implications for soil organic matter studies. Rapid Commu Mass Spectro 13:1284–1290

Spalding MD, Blasco F, Field CD (1997) World mangrove atlas. International society for mangrove ecosystems, Okinawa, Japan, P 178

Staddon PL (2004) Carbon isotopes in functional soil ecology. Trends Ecol Evol 19:148–154

Tam NFY, Wong YS (1998) Variations of soil nutrient and organic matter content in a subtropical mangrove ecosystem. Water Air Soil Poll 103:245–261

Tamooh F, Huxham M, Karachi M, Mencuccini M, Kairo JG, Kirui B (2008) Below-ground root yield and distribution in natural and replanted mangrove forests at Gazi bay, Kenya. Forest Ecol Manag 256:1290–1297

Telles ECC, Camargo PB, Martinelli LA, Trumbore SE, Costa ES, Santos J, Higuchi N, Oliveira RC Jr (2003) Influence of soil texture on carbon dynamics and storage potential in tropical forest soils of Amazonia. Global Biogeochem Cy 17:1040

Townsend AR, Vitousek PM, Trumbore SE (1995) Soil organic matter dynamics along gradients in temperature and land use on the island of Hawaii. Ecology 76:721–733

Trumbore SE (1996) Applications of accelerator mass spectrometry to soil science. In: Boutton TW, Yamasaki S (eds) Mass spectrometry of soils. Marcel Dekker, New York, pp 311–339

Trumbore SE, Davidson EA, Camargo PB, Nepstad DC, Martinelli LA (1995) Belowground cycling of carbon in forests and pastures of Eastern Amazonia. Global Biogeochem Cy 9:515–528

Vogel JS, Nelson DE, Southon JR (1987) Carbon-14 background levels in an accelerator mass spectrometry system. Radiocarbon 29:323–333

Vonfischer JC, Tieszen LL (1995) Carbon isotope characterization of vegetation and soil organic matter in subtropical forests in Luquillo, Puerto-Rico. Biotropica 27:138–148

Wang L, Ouyang H, Zhou CP, Zhang F, Song MH, Tian YQ (2005) Soil organic matter dynamics along a vertical vegetation gradient in the Gongga Mountain on the Tibetan Plateau. J Integrative Plant Biol 47:411–420

West JB, Bowen GJ, Cerling TE, Ehleringer JR (2006) Stable isotopes as one of nature's ecological recorders. Trends Ecol Evol 21:408–414

Woodroffe C (1992) Mangrove sediments and geomorphology. In: Robertson A, Alongi D (eds) Tropical mangrove ecosystems. American Geophysical Union, Washington DC, pp 7–41

Wynn JG (2007) Carbon isotope fractionation during decomposition of organic matter in soils and paleosols: implications for paleoecological interpretations of paleosols. Palaeogeogra Palaeoclimatol Palaeoecol 251:437–448

Wynn JG, Harden JW, Fries TL (2006) Stable carbon isotope depth profiles and soil organic carbon dynamics in the lower Mississippi Basin. Geoderma 131:89–109

Zhang HB, Luo YM, Wang MH, Zhao QG, Zhang GL (2007) Soil organic carbon storage and changes with reduction in agricultural activities in Hong Kong. Geoderma 139:412–419

The Relationship Between Mangrove Deforestation and Economic Development in Thailand

Toyokazu Naito and Suphakarn Traesupap

Contents

Abstract Mangrove in Thailand has been steadily deforested from 1961 to 1996 and has been reduced to about half of the original area. The Environmental Kuznets Curve (EKC) hypothesis, however, posits that economic development eventually reverses resource degradation. This hypothesis is examined using pooled data on mangrove loss and Gross Provincial Product (GPP) from 23 provinces in Thailand in various years between 1975 and 2004. The empirical results show strong evidence of an EKC relationship between mangrove loss and GPP. In addition, since shrimp farming is considered to be one of the main causes of mangrove deforestation, the relationship between shrimp farming and mangrove loss is examined. Shrimp farming is found to significantly affect the extent of mangrove deforestation. The development of extensive and semi-intensive shrimp farming techniques quickens mangrove deforestation, but intensive shrimp farming, which developed during the 1990s, reduces mangrove loss.

T. Naito (✉)
Department of Human and Cultural Studies, Kyoto Gakuen University,
621-8555 Kameoka, Kyoto, Japan
e-mail: naito@kyotogakuen.ac.jp

S. Traesupap
Coastal Development Centre, Faculty of Fisheries,
Kasetsart University, 10900 Bangkok, Thailand

I. Faridah-Hanum et al. (eds.), *Mangrove Ecosystems of Asia*,
DOI 10.1007/978-1-4614-8582-7_13,

1 Introduction

As an economy develops, a society's wealth increases; at the same time, the environment becomes polluted and the resources are depleted. During the 1970s and 1980s, the shrimp farming industry in Thailand developed rapidly and income levels rose drastically; however, many mangrove forests were cut down to create pond habitat for shrimp farming. In the 1990s, mangrove deforestation in Thailand became a major concern in Japan, one of the main importers of shrimp products from Thailand. Will mangrove deforestation continue or will the forests recover as Thailand's economy continues to develop?

The relationship between economic development and environmental degradation has been a concern among environmental economists since the 1970s, and the discussion has been framed within the "Environmental Kuznets Curve (EKC)" hypothesis since the 1990s. The EKC hypothesis posits that environmental degradation and resource destruction increase in the early stages of economic development but eventually decline as the economy develops and per capita income increases. The origin of the EKC hypothesis is the so-called "Kuznets Curve," in which Kuznets (1955) postulated that the relationship between the extent of income inequality and the level of income can be represented in an inverted-U shape. Thus the EKC hypothesis is also called the "inverted-U" hypothesis, with environmental degradation represented on the vertical axis and the level of per capita income on the horizontal axis.

Although no study exists that focuses on the EKC relationship between mangrove deforestation and income level, there are a number of studies dealing with the EKC hypothesis and deforestation in the literature (Antle and Heidebrink 1995; Cropper and Griffiths 1994; Koop and Tole 1999; Lopez and Galinato 2005; Panayotou 1993; Shafik 1994a; Shafik and Bandyopadhyay 1992). In those studies, the existence of the EKC relationship between deforestation and income level has been demonstrated empirically. In other words, most of the studies show that if a society becomes rich, the forest that has been destroyed eventually recovers. However, the results of these studies depend on the data used (cross-country data, panel or pooled data, or a single country's data); estimation models used (equation form, explained variable used, or explanatory variables used); and estimation methods used (one point fixed model, fixed effect model, or random effect model).

Some of these studies have estimated an EKC turning point, in which economic development eventually reverses forest degradation. Those estimated EKC turning points exist between \$ 5,000 and 8,000 (1985 US\$ level) (Barbier and Burgess 2001; Bhattarai and Hammig 2001; Cropper and Griffiths 1994; Koop and Tole 1999; Lopez and Galinato 2005), income levels that are far beyond the per capita income of the countries possessing tropical forests. Therefore, there is no guarantee that current deforestation will be reduced by an increase of per capita income, as the EKC hypothesis suggests. According to one future prediction, even if the income level reaches the EKC turning point in the future, this turning point may be reached too late for forests to recover from deforestation.

In this study, we first examined the existence of the EKC relationship in the particular case of mangrove deforestation in Thailand. We used pooled data on mangrove-covered area, gross provincial product (GPP), provincial population, and provincial shrimp farming production from 23 provinces across Thailand with mangrove forests during various years between 1975 and 2004 (data for 2009 is not included for reasons explained in Sect. 4). If evidence of the EKC relationship was convincing, then we examined the possible determinants of the EKC relationship and analyzed their impact on the EKC relationship.

In past studies on the EKC relationship to deforestation, many determinants of the EKC relationship have been found (Antle and Heidebrink 1995; Barbier and Burgess 2001; Bhattarai and Hammig 2001; Cropper and Griffiths 1994; Lopez and Galinato 2005; Panayotou 1993; Panayotou and Sungsuwan 1994; Shafik 1994b), for example, population growth rate, population density, price of products (wood, fuel, and other substitutes), structural factors (agricultural production, agricultural products export, technological change, and distance from markets), political factors (investment, accumulated debt, international trade, and land use), and institutional factors (economic system, political stability, political freedom, and security of ownership). Panayotou (1997) and Barbier and Burgess (2001) have particularly claimed that the industrial share is an important determinant of the EKC relationship.

In general, mangrove deforestation has been attributed to increased demand for land due to population growth and economic development in Thailand. Therefore, by using the population growth rate as a factor to express population growth and the industrial share of shrimp farming as a factor to show economic development, we can analyze the impact of those determinants on the EKC relationship.

Our initial results show strong evidence of the existence of an EKC relationship between mangrove loss and per capita income, correlating with many previous studies. This implies that mangrove forests in Thailand deteriorated during the 1970s and 1980s but would have recovered as the economy subsequently developed. The EKC turning point, which is the starting point of recovery, is at $ 5,600 (1985 US$ level). However, the average per capita income of 23 provinces in Thailand, which is calculated from the collected data, is around $ 4,000 (1985 US$ level) (and that number is based on 2004 data), which implies that the EKC turning point has not been reached yet in Thailand.

Our analysis of the impact of determinants on the EKC relationship shows that an increase in the population growth rate shifts EKC upward and accelerates mangrove deforestation. On the other hand, the results show that an increase in GPP growth rate shifts EKC downward and reduces mangrove deforestation. In other words, if the economic growth increases, then the mangrove forest is recovered. The results also show that shrimp farming significantly affects the extent of mangrove deforestation. More specifically, the development of extensive and semi-intensive shrimp farming techniques quickens mangrove deforestation, but intensive shrimp farming, which developed during the 1990s, reduces mangrove loss.

This chapter is organized as follows. Section 2 presents the background of mangrove deforestation and the development of the shrimp farming industry in Thailand. Section 3 provides empirical models, the hypothesis test, and estimation techniques. Section 4 explains the data used in this study. In Sect. 5, the results are reported. The final section discusses the results.

2 Mangrove Deforestation and Shrimp Farming

Mangrove, known as *manggi* in the Malay language, is a unique plant colony found in coastal streams and intertidal estuaries. Mangrove is mainly found in the subtropical and tropical zones north and south of the equatorial zone, approximately between 25° and 30° N. and S. in latitude (Walter 1971). Most mangroves are woody trees or shrubs that belong to the *Rhizophoraceae* family, which is characterized by salt tolerance, thick leaves, and many aerial roots. Mangrove forms an important ecosystem between land and sea area that provides vital plant and animal habitat. Moreover, mangrove plays an important role in stabilizing shorelines by protecting coastal streams and estuaries against tidal wave and soil erosion. Mangrove in Thailand inhabits 23 provinces (six provinces on the coast of the Andaman Sea and 17 provinces on the coast of the Gulf of Thailand), approximately half of the total 2,614 km coastline of Thailand (Auksronkaw 2002) (see Fig. 1).

Figure 2 shows changes in the area covered by mangrove in Thailand (23 provinces) from 1961 to 2009. Mangrove in Thailand has been steadily deforested from 1961 to 1996 and has been reduced to about half of the original area, from 3,679 km^2 in 1961 to 1,685 km^2 in 1996. In general, mangrove loss has been attributed to an increase in the demand for land as a result of population growth and economic development. The land converted from mangrove forest has been used for aquaculture (especially shrimp farming); agriculture; tin mining; salt production; urbanization; construction of houses, factories, roads, and ports; and power plants. Illegal cutting for household timber and fuel has also impacted mangrove loss (Aksornkoae and Tokrisna 2004).

After 1996, however, mangrove reforestation began to increase and reached 2,658 km^2 in 2004, which is about three-quarters of the area covered by mangrove in 1961. In other words, in the 8 years between 1996 and 2004, mangrove forests recovered approximately half the loss experienced in previous years. Unfortunately, since then deforestation resumed, and the total area of mangrove forest was reduced to 2,440 km^2 by 2009. According to our field studies in Thailand, much mangrove forest has been recently converted to the land for mining (selling soil including minerals), creating docks, creating communities and infrastructure, coastal aquaculture, and coastal tourist areas.

Figure 3 shows mangrove deforestation and its conversions to other land uses from 1991 to 1996 (the mangrove area in 1961 is set to 100 %). The five bar graphs represent four regions in Thailand and the four-region total. These four regions are the eastern region east of Bangkok (including five provinces), the central region near Bangkok (including six provinces), the southern region on the coast of the Gulf of Thailand (including six provinces), and the southern region on the coast of the Andaman Sea (including six provinces). Looking overall at the four regions, 55 % of the total mangrove area in 1961 had been deforested and about 18 % (one third of destroyed mangrove area) converted into shrimp ponds by 1996. Another 2 % of the mangrove area had been converted into new residential land and 35 % converted for other uses.

Fig. 1 Habitats of mangrove in Thailand (23 provinces). Translated Thai on the Fig. in Auksronkaw 2002 into English

The percentage of mangrove land converted into shrimp ponds is 44 % in the eastern region, 39 % in the southern region along the Gulf of Thailand, 23 % in the central region, and 2 % in the southern region along the Andaman Sea. Conversions into new residential land are very small—7 % in the eastern region, 5 % in the central region, 2 % in the southern region along the Gulf of Thailand, and almost 0 % in the southern region along the Andaman Sea. Conversions into other uses are most common in the central region near Bangkok at around 64 %, which is two times higher than the other three regions. A possible explanation might be that the demand for land for economic development is higher in the Bangkok area than in other regions.

The previous section demonstrated that deforestation in Thailand accounted for a loss of half of all mangrove forests during the 35 years from 1961 to 1996, and about one-third of the destroyed area has been converted into shrimp ponds. How has the shrimp culture industry in Thailand been developed during the same period? Figure 4 shows changes in the number of shrimp farms (line graph) and culture areas

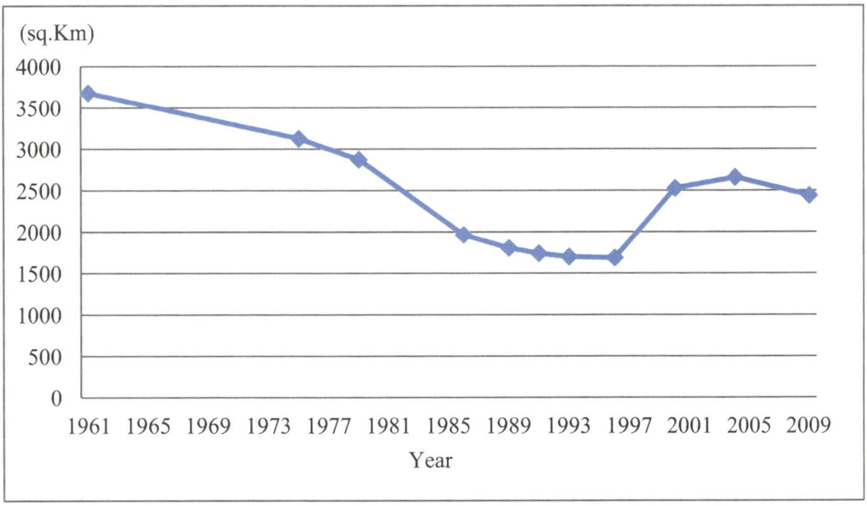

Fig. 2 Changes of mangrove area in Thailand (23 provinces), 1961–2009. (These are generated by using the data from Geo-Informatics, National Park, Wildlife and Plant Conservation)

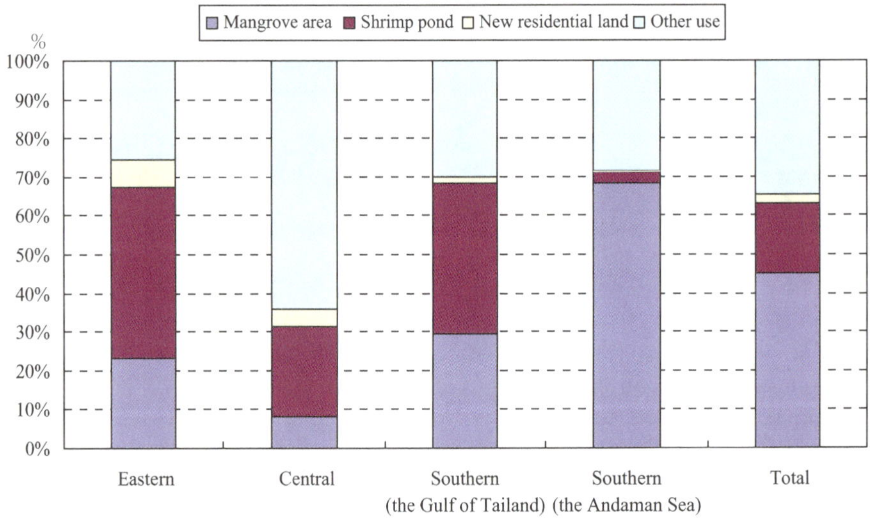

Fig. 3 Mangrove deforestation and its conversions to the other land uses, 1991–1996. The mangrove area in 1961 is set to 100 %. (Generated by using the data Geo-Informatics, National Park, Wildlife and Plant Conservation)

(area chart) in Thailand between 1972 and 2009. Both graphs clearly show a steady rise in the development of the shrimp culture industry in Thailand until 2003.

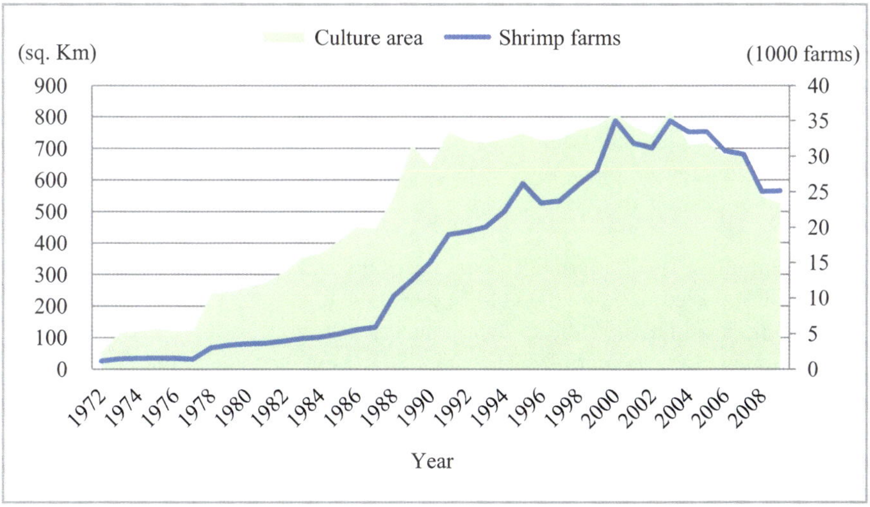

Fig. 4 Changes of the number of shrimp farms and culture area in Thailand, 1972–2009. (Generated by using the data from the Statistics of Shrimp Culture, Department of Fisheries, Ministry of Agriculture and Cooperatives)

It is helpful to examine the historical background of shrimp farming in Thailand. Here we describe the history of shrimp culture in Thailand with reference to Aksornkoae and Tokrisna (2004). In its early stages, shrimp production in Thailand was a by-product of salt production, wild shrimp strayed into the salt pans and were harvested. Subsequently, farmers stocked intentionally captured natural larvae in brackish ponds (a mixture of seawater and fresh water), then harvested them once they matured. This shrimp culture method is called "extensive shrimp farming." Most brackish water ponds were usually converted from or were part of a salt pan. During this early period, the species raised were banana shrimp (*Peneaus merguiensis*) and school shrimp (*Metapenaeus* sp.), which depended on natural larvae and fed on wild seaweed and plankton.

According to the census, extensive shrimp farming is defined as a culture method that uses only natural larvae and feeds in pond water that is derived from canals (Taya 2003). The culture ponds used are relatively large, approximately 20–30 hectares per pond. It takes between 45 and 90 days for this method to raise shrimps, so the cost is low but the productivity is also low. In the semi-intensive shrimp farming method, farmers stock the larvae from the hatcheries (less than 24,000 fries per 1 rai [= 1,600 m^2] of a pond), use artificial feed, and manage pond water by pumping from canals to improve the productivity. This culture system fits in between the extensive and intensive systems (a fuller explanation appears later).

In the late 1960s, the number of shrimp farms started to increase gradually as rising shrimp prices made shrimp farming more profitable than salt production. Shrimp ponds were constructed on the coastline and the banks of canals by clearing mangrove trees and taking advantage of the natural tide system for water exchange. The success of artificial incubation for banana shrimp and black tiger shrimp (*Peneaus monodon*) in 1973 accelerated the increase in shrimp farms. The number of shrimp farms gradually increased from 1972 to 1987, but after 1988 to 2000 they increased more rapidly due to the introduction of a new shrimp species, the black tiger shrimp, which had a higher profit rate because of its higher tolerance and better survival rate. In 1983, a multinational company from Taiwan started a joint venture with local investors, which used black tiger shrimp in an "intensive shrimp farming" system. The development of this new technology caused a rapid increase in the number of shrimp farms after 1988.

The intensive shrimp farming system uses a small pond, such as a rice field, in which shrimp is cultured by high larvae density and artificial mixed feed. To protect against disease infection, the farmers manage water quality by adding antibiotics and chemical products such as nutrients for 24 h. Moreover, the farmers settle paddle wheel machines in shrimp ponds to maintain oxygen in the water, which is consumed during the decomposition of wastes from shrimp and artificial feeds. The culture ponds are small at 0.5–1 ha per pond. According to the census, intensive shrimp farming is defined as a culture method that stocks more than 24,000 larvae per 1 rai, feeds 3–5 times per day, settles 1 paddle wheel per 1–2 rai of a pond, and takes 4–5 months for growth (Taya 2003).

In 1991, the Thai government enforced a new law (Cabinet Resolution) that prohibited the conversion of conserved and fertile mangrove areas into shrimp ponds, which slowed down the increase in shrimp farms temporarily during 1992 and 1993. Under this law, it became impossible to construct new shrimp ponds in the coastal area. However, investment in intensive shrimp farming continued to steadily increase since it did not require large shrimp ponds, which caused a rapid increase in the total number of shrimp farms from 1988 to 2003 (although there was a temporary decline because of the Asian economic crisis in 1997). This increase depended on the success of black tiger shrimp hatcheries; of course, at the same time, it was supported by the development of the shrimp feed industry, the technique of intensive shrimp farming, and the high shrimp price in the international market.

After 2003, the number of shrimp farms decreased sharply. According to our field studies in Thailand, this was due to falling shrimp prices in the international market, possibly caused in part by an increase in shrimp production in other countries (such as Indonesia, Vietnam, and India). Indeed, even in the United States, the largest importer of shrimp from Thailand, shrimp imports from Ecuador have increased gradually. Moreover, Thai shrimp farmers are confronted with the problems of infectious disease and lack of parental shrimp with the result that many farmers have gone out of business.

The area chart in Fig. 4 shows changes in the shrimp culture area in Thailand from 1972 to 2009. The shrimp culture area, as well as the number of shrimp farms, showed an upward trend from 1972 to 1991. During the period between 1992 and 2003, however, the size of the shrimp culture area became stable. This was due to the Cabinet Resolution prohibiting the conversion of conserved and fertile mangrove areas into shrimp ponds; it also reflected the shift from extensive to intensive shrimp farming techniques. Intensive shrimp farming did not require the large shrimp ponds; therefore, this shift in the shrimp culture method gradually decreased mangrove deforestation.

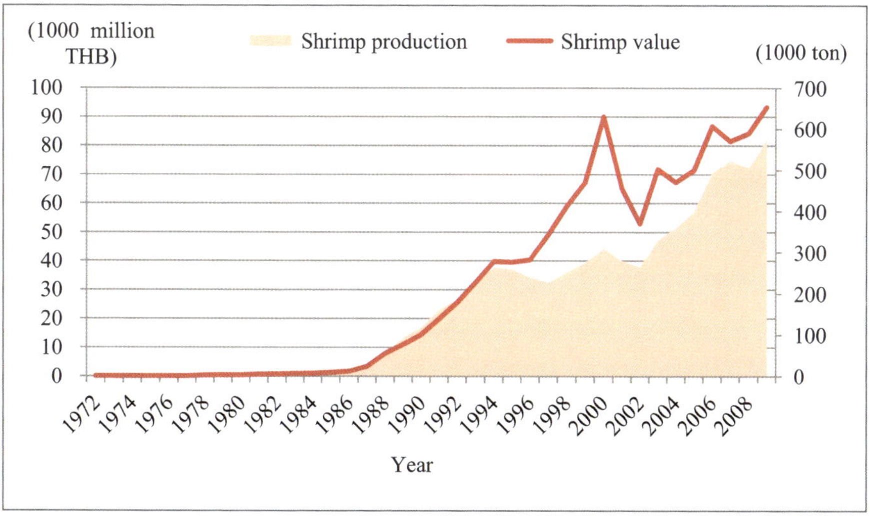

Fig. 5 Changes of shrimp production and value in Thailand, 1972–2009. (Generated by using the data from Statistics of Shrimp Culture, Department of Fisheries, Ministry of Agriculture and Cooperatives)

The expansion of the culture area stabilized because of not only the shift to intensive shrimp farming, but also the development of a new culture system, called a "closed shrimp farming system." In this system, once the water is poured into the shrimp ponds, it needs almost no replacement. The farmers monitor the water condition and maintain a constant salt concentration. The water in the pond is imported from the sea by truck and is carefully inspected before being poured into the pond. This system rapidly spread in the central region near Bangkok from around 1996, so that shrimp ponds have been constructed not only in the coastal areas but also in agricultural land in inland areas. In November 1998, however, the National Environment Board banned shrimp farming in freshwater areas (particularly in the central region) out of concern about land chlorination and environmental degradation, since too many agricultural lands were converted into ponds.

After 2003, the shrimp culture area also decreased sharply in cooperation with the number of shrimp farms. Afterwards many shrimp farmers went out of business, so many shrimp ponds are left. These shrimp ponds were, however, converted into agricultural land. On this reclaimed land, farmers are producing agricultural products with high salinity tolerance, for example, rice, coconut, papaya, dragon fruit, and banana (only one kind of banana).

Figure 5 shows changes in shrimp production (line graph) and shrimp value (area chart) in Thailand from 1972 to 2009. Both graphs show an upward trend during the whole period, although the number of shrimp farms and area in Fig. 4 show a recent downward trend after 2003. Shrimp production and value increased gradually until 1987, then experienced a rapid increase from 1988 to 2000, which is partially attributable to the impact of intensive black tiger shrimp farming beginning in 1983. The rapid growth of production is also due to the stability of the international market price of shrimp and Japanese shrimp demand (Traesupap et al. 1999).

Shrimp production slowed down twice (from 1995 to 1997 and from 2001 to 2002), due to widespread shrimp infectious disease caused by the deterioration of water quality and nourishment. After 2002, however, shrimp disease problems were solved and shrimp production returned to the upward trend. Shrimp production increased steadily over the period from 2003 until 2009, largely through developments in shrimp farming techniques and introduction of educational programs for farmers by the Thai government.

Shrimp value also has rapidly increased since the advent of Black Tiger intensive shrimp farming beginning in 1983. Shrimp value increased sharply in 2000, given high market prices and high international demand. This occurred because of widespread shrimp infectious disease in the previous year. Shrimp value also increased from 2003 to 2009 although there were fluctuations. This is because farmers began to produce shrimp according to a schedule to keep a good market price as well as the progress of the productivity of shrimp farming. Recently, the organic shrimp farming is very popular, since its price is 20 % higher than regular shrimp. They are mostly exported to advanced countries.

3 Empirical Model and Hypothesis Test

To analyze the EKC relationship between mangrove deforestation and income level, the empirical mode in this study uses the quadratic reduced form, which has been used for empirical EKC studies in general. Although a log-quadratic model has been utilized in many empirical studies of the EKC relationship to deforestation, we could not use it because the mangrove deforestation index included some minus values.

The quadratic model includes per capita gross provincial production (GPP) and its square term to test the EKC hypothesis. Previous EKC studies for deforestation examined the reason why the EKC relationship exists and showed important determinants of EKC, such as population growth rate, population density, price of products, structural factors, political factors, and institutional factors. Hence this study adds the important determinants of mangrove deforestation—increasing population pressure (population growth rate and population density) and industrial structural change on the economy (industrial share)—to the empirical model.

Population growth pressure indicates rising land demand, which is the main cause of deforestation, so it is always included in the empirical model. Population density data at the provincial level, however, is not available, so only the population growth rate is included as an explanatory variable in the model. Moreover, as a structural variable, the GPP share of the shrimp industry has a strong relationship to mangrove deforestation and is therefore included as one of explanatory variables in the model.

Industrial share affects the EKC relationship between mangrove deforestation and income level as follows. In the early stages of economic development, mangrove forests are intact. However, as the economy begins to change structurally, agriculture and fisheries shift to aquaculture and manufacturing. During this stage, both economic development and mangrove deforestation are underway. As the economy continues to develop and moves into the second structural change, those industries shift again to the service, informational and technological industries. In this stage, society can afford to pay attention to environmental degradation, and environmental protection laws are enforced and reforestation projects begun, which finally reduces mangrove loss.

The empirical model in this study is represented by the following equation:

$$MD_{it} = \alpha_i + \beta_1 Y_{it} + \beta_2 (Y_{it})^2 + \beta_3 \Delta Y_{it} + \beta_4 P_{it} + \beta_5 S_{it} + \beta_6 D_{it} + \varepsilon_{it}, \quad (1)$$

where, $i(= 1, \ldots, n)$ is each province (23 provinces) and t is each year. Hence, MD_{it} indicates a mangrove deforestation index for the i-th province in the t-th year. In this study, as a mangrove deforestation index, both 'annual mangrove deforestation' and 'total mangrove deforestation' (used by Shafik and Bandyopadhyay 1992) are utilized. The former is the yearly change in mangrove area and the latter is the change in mangrove area between the earliest date, 1975, and latest date.

Y_{it} is the gross provincial product (GPP) per capita for the i-th province in year t. $(Y_{it})^2$ is its square value; ΔY_{it} represents average GPP growth rate for the i-th province in year t; P_{it} is population growth rate for the i-th province in year t; S_{it} is shrimp value share for the i-th province in year t; and D_{it} is a dummy for the shock from the Asian economic crisis in 1997 and 1998 (1 in the crisis year, otherwise zero). Moreover, α_i is an intercept term that reflects technical innovation and cultural and social structure in the i-th province, β's are coefficients for each variable, and ε's are disturbance terms.

As a pooled regression, the empirical model can be estimated using three different estimation techniques: the one-point fixed model, the fixed-effect model, and the random-effect model. In the preliminary estimation, the no-effect model was significantly rejected in favor of the fixed-effect model using the F test, and the random-effect model was significantly rejected in favor of the fixed-effect model using the Hausman test. Therefore, in this estimation, the fixed-effect model is employed (see Chap. 13 in Greene 2003).

The fixed-effect model is also referred to as the least squares dummy variable (LSDV) model, which has a cross-section group-specific constant term in the estimation model. Since the pooled data used in this study was collected from 23 different provinces, there should be included some province-specific historical and structural differences. Therefore, the use of the fixed effect model is a reasonable choice here.

If the pooled data used in this model has different sizes according to each province size, the existence of a heteroscedasticity problem can be suspected. Hence, in the preliminary estimation, we also conducted the Wald test, after which the null hypothesis of homescedasticity was significantly rejected as predicted (there exists heteroscedasticity). To correct this problem, the weighted least square (WLS) approach is utilized in general. This technique transforms the variance of observation to give larger weight to observation with small variance. In this estimation, the two-stage feasible generalized least square (FGLS) technique is employed since only the cross section weight is considered.

On the other hand, since the pooled data used includes time-series data, the autocorrelation (AR) problem should be examined by using the Durbin-Watson (DW) statistic. If the null hypothesis (no autocorrelation) is rejected, the AR term should be included in the model to correct the autocorrelation. In that case, the iterated feasible generalized least square (iterated FGLS) approach is utilized for the estimation; however, care should be taken not to lose too many degrees of freedom in the estimation. While the effects of all combinations of the explanatory variables on mangrove deforestation are estimated by using the estimation model (1), at the same time, the EKC relationship between mangrove deforestation and income level is determined. To show the existence of the EKC hypothesis, the null hypothesis of both zero coefficients of β_1 and β_2 ($\beta_1 = \beta_2 = 0$) should be rejected and the alternative hypothesis ($\beta_1 > 0$ and $\beta_2 < 0$) accepted. If this alternative hypothesis is satisfied and demonstrates the evidence for the existence of EKC, it is useful to estimate an EKC turning point, which indicates the income level at which mangrove deforestation begins to decline. The EKC turning point can be calculated by dividing estimated coefficient, β_1 by $-2\beta_2$.

The sign for the estimated coefficient of the GPP growth rate, β_3 is unpredictable since it is situational. For example, if mangrove only plays an input role on the production process in Thailand, the increase in the GPP growth rate accelerates mangrove deforestation. However, if the GPP grows and the technology develops and mangrove is no longer necessary for inputs, then the increase on the GPP growth rate reduces mangrove loss. The sign for the estimated coefficient of population growth rate, β_4, is expected to be positive, because the rising pressure on population growth causes the increase in land demand, which quickens mangrove deforestation.

It is also difficult to predict the sign for the estimated coefficient of the industrial share on shrimp farming, β_5. In general, the development of the shrimp farming industry accelerates mangrove deforestation, but that is the case for extensive shrimp farming. As mentioned in Sect. 2, shrimp farming has shifted from extensive to intensive shrimp farming techniques, which reduces mangrove deforestation. Therefore, the expected sign for the shrimp industrial share depends on the difference in shrimp value between the two culture techniques. Finally, the sign for the estimated coefficient of dummy variable, β_6, is expected as minus. An economic shock always decreases the production level in an economy, thereby working to reduce environmental degradation.

4 Data

This analysis uses pooled data on mangrove areas that combine cross sections on 23 provinces in Thailand and a time series of 10 years (1961, 75, 79, 86, 89, 91, 93, 96, 2000, and 2004). The data on mangrove areas, except for 1961 data, are very accurate since they are derived from a Landsat-5^{TM} satellite. Unfortunately, for 1961, data on six provinces are missing and data for the eastern and central areas are also extremely small values compared to 1975. Therefore, we suspect measurement errors or differences in methodology in 1961, and have removed the 1961 data from the analysis.

Data was collected for 2009. However, since the mangrove area in 2009 was smaller than in 2004, the relationship between mangrove deforestation and economic development is shown by not an inverted-U shape but an N shape (See Fig. 2). In this study, we examine the existence of the EKC relationship in case of the inverted-U shape only, which means it is difficult to test the EKC hypothesis including the data in 2009. If the 2009 data is to be included, a different empirical model pertaining to N-shape hypothesis might be necessary. Hence, the data from 2009 is excluded from this study.

From the data on mangrove areas, we created two types of indexes, a "total mangrove deforestation" index and an "annual mangrove deforestation" index, which indicate the level of mangrove deforestation (the terms were first used by Shafik and Bandyopadhyay in 1992). The total mangrove deforestation index is calculated by dividing the difference between the mangrove area in the base year of 1975 ($M_{i,75}$) and the one in year t ($M_{i,\,t}$) by the one in 1975 ($M_{i,75}$), and multiplying by 100:

$$ML_{i,t} = \frac{(M_{i,75} - M_{i,t}) \times 100}{M_{i,75}}, \quad (i = 1,\ 2, \ldots,\ 22 \text{ provinces except Bangkok}).$$

The annual mangrove deforestation index is simply calculated by extracting the mangrove area in year t-1 ($M_{i,\,t-1}$) from the one in year t ($M_{i,\,t}$);

$$ML_{i,t} = M_{i,t} - M_{i,t-1}, \quad (i = 1,\ 2, \ldots,\ 23 \text{ provinces}).$$

The former index is the change rate of mangrove areas between the base year of 1975 and other years (the unit is %), which is available for 22 provinces except Bangkok since its data in 1975 is zero. The latter index is the change of mangrove areas between years (the unit is square km), which is available for all 23 provinces. Therefore, each index has eight data points in the time series, so that the available observation numbers are 176 for the total mangrove deforestation index and 184 for the annual mangrove deforestation index.

The gross provincial product (GPP) expresses the provincial level of gross domestic product (GDP). Because the early studies of EKC used the 1985 US$ basis, we did the same, by converting the nominal GPP into the real GPP by using a GDP deflator with the 1985 US$ basis (= 100). This allows us to compare our results with the early studies and possibly avoid some autocorrelation problems in the time series data. In our analysis, the GPP per capita is used, so each GPP is divided by the population in each province. Moreover, the GPP growth rate is not the per capita level but the per province level.

As mentioned in Sect. 2, the data for Thailand's total number of shrimp farms, culture area, shrimp production, and shrimp value are available from 1972 to 2002. Data on the provincial level, however, are limited for the same period. Since the provincial data for mangrove area, GPP, and population are available from 1975 to 2004, we utilized the data for shrimp production in 1976 and 2002 as proxies for 1975 and 2004, respectively. The industrial share of shrimp farming to total GPP is calculated by dividing the shrimp production in each province by the GPP in each province (1985 US$ level). There are many zero levels of shrimp production and value in the early years of the data; in those cases, however, the zeros were kept intact and used for the estimation.

Finally, the sources of the data are listed in Table A.1 in the Appendix. The statistics for all data are also shown in Table A.2 in the Appendix. We can see there is a wide variance in the data from the table. The size of the data indeed depends on the situation in each province. The average of GPP data is 49,880 Thailand Baht, which is converted into US$ 1,833 (1985 US$ level).

5 Empirical Results

Table 1 shows the estimation results of the regression Eq. (1) in the case of model I, which uses total shrimp industrial share. The left-hand column presents the estimation results using the total mangrove deforestation index as an explained variable (case 1); the right-hand column presents results based on the annual mangrove deforestation index as an explained variable (case 2). In the preliminary estimation for case 1, Durbin-Watson (DW) statistics shows 1.350 ($< d_l = 1.57$), which indicates the existence of a positive autocorrelation problem. We added the AR (1) term in the fixed effect model to correct the problem; however, we could not estimate the model because of the singular variance-covariance matrix. Therefore, we used the no effect model with the AR(1) and AR(2) terms by using iterated feasible weighted least square (FGLS) in model I.

In case 1, the number of cross sections in the data is 22, because the Bangkok province is excluded as mentioned in the previous section. The time series data includes eight years, from 1979 to 2004, so that the number of total observations is 176, but the degrees of freedom reduced to 132 by correcting autocorrelation. The adjusted R^2 is 0.916, which indicates the high explanatory power of the model. The F-value is also very high at 179.9. After correcting autocorrelation, the DW statistics are 1.887 ($> d_u = 1.78$), in which the null hypothesis of no autocorrelation ($\rho = 0$) was not rejected.

In case 1, the estimated coefficient for the GPP per capita has the expected positive sign ($\beta_1 > 0$) and is statistically significant at the 1 % confidence level. The estimated coefficient for the GPP per capita squared has the expected negative sign ($\beta_2 < 0$) and is statistically significant at the 5 % confidence level. These results strongly suggest the existence of the EKC hypothesis. The estimated coefficient for the GPP growth rate shows negative sign and is statistically significant at the 10 % confidence level, which means that the increase of GPP growth reduces mangrove deforestation. Moreover, the estimated coefficient for the population growth rate has the expected positive sign, but it is not satisfied with the 10 % level of significance (it is 11 %

Table 1 Estimates for model I

Explanatory variables	Case 1 (total deforestation index)	Case 2 (annual deforestation index)
Constants	– 20.65	– 6.502
	(23.75)	(5.062)
GPP per capita	0.481	0.172
	(0.179)***	(0.083)**
GPP per capita squared	– 0.0010	– 0.00049
	(0.0005)**	(0.00021)**
GPP growth rate	– 0.069	– 0.087
	(0.036)*	(0.055)
Population growth rate	0.304	1.535
	(0.189)	(0.411)***
Industry share of:	0.040	– 0.435
Shrimp farming	(0.189)	(0.460)
Dummy	– 9.050	– 29.93
For Asian economic crisis	(2.249)***	(7.699)***
Adjusted R^2	0.916	0.410
DW	1.887	2.130
F-value (P-value)	179.9 (0.000)	5.538 (0.000)
The number of cross-sections	22	23
The number of time series	8 (1979–2004)	8 (1979–2004)
Observations (with AR)	176 (132)	184
EKC turning point (1985 US$)	$ 8451	$ 6505
Estimation model	No effect + AR + WLS	Fixed effect + WLS

Standard errors are in parentheses
EKC turning points are calculated by using international exchange rate (US$ 1 = 27.21THB)
*,**, and *** are statistically significant at 10, 5, and 1 % significance level

significance, however). This weakly suggests that the rising of the population growth rate accelerates mangrove deforestation.

On the other hand, in the preliminary estimation of case 2, using the DW test, the null hypothesis is not rejected (DW statistics are $2.130 < 4\text{-}d_u = 2.22$), making it unnecessary to correct autocorrelation. Hence, in case 2 the fixed effect model is estimated by two-step FGLS. The number of cross sections in the data is 23, all of which are provinces possessing mangrove. The number of observations is 184, and all were gathered from a time series of 8 years, from 1979 to 2004. However, the adjusted R^2 is 0.410, which indicates a lower explanatory power of the model than in case 1, and the F-value reduces from 179.9 to 5.538.

In the estimation results in case 2, the estimated coefficients for the GPP per capita and the GPP per capita squared have the expected signs ($\beta_1 > 0$ and $\beta_2 < 0$) and are both statistically significant at the 5 % confidence level. These results, like the case 1 results, strongly suggest the existence of the EKC hypothesis. The estimated coefficient for the GPP growth rate also shows a negative sign but is not statistically significant. Hence, in case 2 we cannot reach a conclusion about the relationship between GPP growth rate and the EKC. The estimated coefficient for the population growth rate has the expected positive sign, as in case 1, and is statistically significant at the 1 % confidence level. This strongly suggests that the increase in the population growth rate accelerates mangrove deforestation.

The estimated coefficients for the industrial share of shrimp farming have a positive sign in case 1 and a negative sign in case 2, neither of which are statistically significant. These signs depend on how many shares there are based on data from extensive or intensive shrimp farming, as mentioned in Sect. 3. The industrial share from extensive farming accelerates mangrove deforestation but the industrial share from intensive farming reduces mangrove destruction. Since the industrial share results here are derived from a combination of these two farming methods, we cannot conclude anything statistically. Therefore, we analyze model II by dividing all industrial shares into categories of extensive or intensive shrimp farms.

Table 2 presents the estimation results for model II, which includes two parts of the industrial share of the extensive and intensive shrimp farms (the semi-intensive shrimp farm is included in the share of the extensive shrimp farm since the size of their shrimp ponds is very similar). In the same way as model I, the total mangrove deforestation index is used as an explained variable in case 1 and the annual mangrove deforestation index is employed as an explained variable in case 2. Since the data for the extensive and intensive shrimp farming are only available from 1987 to 2002, we used the data for 1987 and 2002 as proxies for 1986 and 2004, respectively. Therefore, the number of observations is 154 in case 1 and 161 in case 2.

In case 1 of model II, the preliminary estimation shows DW statistics of 1.368 ($< d_l = 1.57$), which indicates positive autocorrelation, so we corrected the problem by adding an AR(1) term in the model (after correction, DW statistics are 2.053). Hence, in case 1 the fixed effect model is estimated by using iterated FGLS with an AR(1) term. The estimation results in case 1 show that the adjusted R^2 is 0.930, which indicates a high explanatory power of the model; the F-value is also very high at 60.58. The estimated coefficient for the GPP per capita and the GPP squared both have the expected signs and both are statistically significant at the 1 % confidence level. Hence, the results in case 1 also suggest the existence of the EKC hypothesis in model II. The estimated coefficient for the GPP growth rate shows a negative sign like model I and is statistically significant at the 5 % confidence level. However, the estimated coefficient for the population growth rate has a negative sign, which is the inverse sign in model I and is not statistically significant.

In case 2, there is no autocorrelation problem in the preliminary estimation; therefore, we estimated the fixed effect model by using the two-step FGLS. The time series data includes 7 years between 1986 and 2004 and the number of observations is 161. The adjusted R^2 is 0.439, which indicates a lower explanatory power of the model than in case 1 and the F-value is also smaller than in case 1. The estimated coefficient for the GPP per capita and the GPP squared both have the expected signs as well as in case 2 in model I and are statistically significant at the 5 and 10 % confidence levels, respectively. Hence, the EKC hypothesis is satisfied in case 2. The estimated coefficient for the GPP growth rate shows a negative sign and is statistically significant at the 10 % confidence level (it is not statistically significant in case 2 in model II). The estimated coefficient for the population growth rate has the expected positive sign and is statistically significant at the 1 % confidence level.

Table 2 Estimates for model II

Explanatory variables	Case 1 (total deforestation index)	Case 2 (annual deforestation index)
Constants	− 101.4	− 4.099
	(77.463)	(4.370)
GPP per capita	0.932	0.105
	(0.163)***	(0.061)*
GPP per capita squared	− 0.0022	− 0.0003
	(0.0005)***	(0.0002)**
GPP growth rate	− 0.099	− 0.119
	(0.045)**	(0.059)**
Population growth rate	− 0.134	1.573
	(0.102)	(0.388)***
Industry share of:		
Extensive and semi-intensive	− 3.239	8.724
Shrimp farming	(1.755)*	(2.249)***
Intensive shrimp farming	0.363	− 0.795
	(0.070)***	(0.429)*
Dummy	− 9.576	− 28.77
For Asian economic crisis	(1.680)***	(7.200)***
Adjusted R^2	0.930	0.439
DW	2.053	2.096
F-value (P-value)	60.58 (0.000)	5.325 (0.000)
The number of cross-section	22	23
The number of time series	7 (1986–2004)	7 (1986–2004)
Observations (with AR)	154(132)	161
EKC turning point (1985 US$)	$ 7690	$ 5615
Estimation model	Fixed effect + AR + WLS	Fixed effect + WLS

EKC turning points are calculated by using international exchange rate (US$1 = 27.21THB)
Standard errors are in parentheses
*,**, and *** are statistically significant at 10, 5, and 1 % significance level

The estimated coefficient for the industrial shares in case 1 has the opposite sign from the one in case 2, but they are both statistically significant. In case 1, the estimated coefficients for the industrial share of both the extensive and semi-intensive shrimp farming and the intensive shrimp farming do not have the expected signs, but they are statistically significant at the 10 % and 1 % confidence levels, respectively. In case 2, however, both industrial shares have the expected signs; the former is statistically significant at the 1 % confidence level and the latter at the 10 % confidence level. Therefore, the results strongly suggest that extensive and semi-intensive shrimp farming accelerate mangrove deforestation and intensive shrimp farming reduces mangrove destruction in case 2.

6 Discussion

First of all, we examined the existence of the EKC hypothesis and an EKC turning point based on the estimation results. In both models I and II, the estimated coefficients for the GPP per capita and the GPP per capita squared were satisfied with the expected signs. Also, they were statistically significant at the 1 % or 5 % confidence levels except the GPP per capita term in model II, which was significant at the 10 % level. Therefore, we can conclude that our results provide strong evidence of the existence of an EKC relationship between mangrove deforestation and income level in Thailand. That is, mangrove deforestation in Thailand increases as the income level rises, but the forests begin to recover once income reaches a threshold level.

From the estimated coefficients of β_1 and β_2 in model I, the EKC turning points are calculated as $ 8,451 in case 1 and $ 6,505 in case 2 (1985 US$ base). In model II, the EKC turning points are computed as $ 7,690 in case 1 and $ 5,615 in case 2 (1985 US$ base). The EKC turning points in case 1 (the total mangrove deforestation index is used) in both models I and II are very similar to the results of a study by Lopez and Galinato (2005), in which turning points were calculated between $ 7,000 and $ 8,000 in the case of deforestation in Brazil, Malaysia, and the Philippines.

On the other hand, the EKC turning points in case 2 (the annual mangrove deforestation index is used) in both models I and II are very close to those in the study by Barbier and Burgess (2001), which were computed as $ 6,182 in the case of deforestation in Asia. The fact that our estimated EKC turning points are very close to previous estimates strengthens the evidence of the existence of the EKC hypothesis. There is a difference of about $ 2,000 in the EKC turning points of the total mangrove deforestation index and the annual mangrove deforestation index. This is because the former index recovers more slowly than the latter index (the shape of the EKC in the former index is flatter than the latter index).

Although the minimum EKC turning point calculated was $ 5,615 (in case 2 in model II), it is impossible for mangrove loss to recover if the turning point is far from the present GPP per capita in Thailand. If we calculate the GPP per capita in 23 provinces in Thailand from the collected data, it was about $ 4,000 (1985 US$ base) in 2004. Hence, the EKC turning point that is the starting point for mangrove loss recovery has not yet been reached in Thailand. Based on the collected data, however, the annual mangrove deforestation indexes show minus values in 22 provinces out of 23 provinces in 2000, which indicates recovery of mangrove loss in Thailand has already begun.

Next, we examined the effects of the shrimp farming industry on mangrove deforestation in Thailand. When the total industrial share of shrimp farming as a whole was used in model I, we did not get any useful results at all. However, when we included two divided industrial shares by shrimp culture technology in model II, we did get useful results. In model II, the estimation results in case 2 are stable and robust compared to the ones in case 1, because the former results did not change much between models I and II but the latter results did. Therefore the estimation results in case 2 are more reliable than the ones in case 1. This might be the case because the autocorrelation was corrected at the expense of losing many degrees of freedom in case 1, which was not necessary in case 2.

Hence, we examined the relationship between shrimp farming and mangrove deforestation based on case 2 only. From the estimation results, it was confirmed that the development of extensive and semi-intensive shrimp farming techniques accelerated mangrove deforestation (shifted EKC upward) and the development of intensive shrimp farming reduced mangrove loss. As stated in Sect. 2, many mangrove forests were cut down to create ponds for shrimp farming in the early stages of extensive shrimp farming; however, it was no longer necessary to clear forest in the 1980s, when intensive shrimp farming started to develop. The results of this study provide evidence that the development of technology in shrimp culture contributes to the reduction of mangrove deforestation.

In addition, the results of the factor analysis for mangrove deforestation clearly demonstrate that the rise of the population growth rate accelerated mangrove deforestation by shifting the EKC upward. This result supports the viewpoint that the fundamental cause of mangrove deforestation is increased demand for land due to population growth. It is also clear that as the GPP growth rate increases, mangrove loss is reduced and the EKC shifts downward. The faster the GPP growth, the higher the mangrove loss recovery.

Also worthy of mention are the estimation results of the dummy variable used for expressing the effect of the Asian economic crisis in 1997 and 1998 on mangrove deforestation. The estimated coefficients for the dummy variable have the expected negative signs and are statistically significant at the 1 % confidence level in all cases in both models. These results strongly suggest that the Asian economic crisis slowed down the economy in Thailand, which reduced mangrove loss. In the same way that Moomaw and Unruh (1997) demonstrated the relationship between the EKC hypothesis and the Oil Crisis in 1979, these results demonstrate that the Asian economic crisis in the 1990s had the effect of stabilizing mangrove deforestation.

While the existence of an EKC relationship between mangrove deforestation and income level is indicated, some caution should be used in interpreting the empirical results. It is impossible to generalize the EKC hypothesis from these results, as pointed out by Arrow et al. (1995) : “Economic growth is not a panacea for environmental quality.” In this study, we examined mangrove deforestation in Thailand, where economic development had been far ahead of other developing Asian countries. Hence, the EKC hypothesis confirmed in this study does not necessarily fit in Indonesia and Vietnam, which are facing the same problem of mangrove deforestation. Moreover, although mangrove trees recover relatively easily, it may take hundreds of years for primeval forests to recover, and it may be impossible for fishery resources to recover.

This study remains incomplete due mainly to lack of data. We need annual data on mangrove area and pre-1975 data for more precise analysis; we also need data on population density for each province, which is always used in EKC studies as the causing factor of deforestation. Moreover, to examine the causing factor of the EKC, we should not only include industrial share as an explanatory variable in the model, but also international trade values, political factors (policies for land use and investment for reforestation projects), and institutional factors (ownership and corruption). Indeed, the Thai government enforced a new law (Cabinet Resolution) that prohibited the conversion of mangrove areas into shrimp ponds in 1991 and 1998 and began the major project of reforestation. Another factor that deserves attention is the tendency of companies in Thailand to bribe government officials to break environmental protection laws. These analyses are left for future research.

Finally, it is important to mention about the mangrove area data for 2009. The data for total mangrove area in 2009 provides us with a new problem to solve for the EKC hypothesis. The mangrove was deforested again in 2009, although the mangrove area should be recovered more than the one in 2004 according to the EKC hypothesis. There are two points of view about the relationship between mangrove deforestation and economic development. One is that the N-shaped EKC hypothesis is a valid and accurate conception of the conditions in Thailand which are a result of the time lag between developed and undeveloped areas in the country. As mangrove in developed areas (big cities) recover, one in undeveloped areas is deforested. The other view is that the situation is best illustrated by a inversed U-shaped hypothesis. In this view, the recovery of mangrove area in 2001 and 2004 is explained by technological developments in shrimp farming, which means that the pressure of mangrove deforestation is still continuing even in 2009. For analysis of these two views, we have to wait for the next data of mangrove area in 2013 since the satellite data is issued an official announcement once every 4 years.

Acknowledgments This research was supported by a 2012 Kyoto Gakuen University Research Promotion Grant. We greatly appreciate their support. This chapter was written by developing an existing working paper (Naito and Traesupap 2006) with new data for 2009. We are deeply grateful to Ms. Nareerat Veerapong, a graduate student of Kasetsart University, for help in collecting this new data.

Appendix

Table A.1 List of data sources

Data	Source
Mangrove area (23 provinces)	Geo-Informatics, National Park, Wildlife and Plant Conservation
GPP (23 provinces)	Office of the National Economic and Social Development Board, Office of the Prime Minister
GDP deflator	Economic and Financial Statistics, Bank of Thailand
International exchange rate	Bank of Thailand
Population (23 provinces)	Registration Division, Local Administration Department, Ministry of Interior
Shrimp value (23 provinces)	Statistics of Shrimp Culture, Department of Fisheries, Ministry of Agriculture and Cooperatives

Table A.2 Summary statistics for data

Explanatory variables	Annual deforestation index	Total deforestation index	GPP per capita	GPP per capita squared	GPP growth rate	Population growth rate	Industrial share of shrimp farming
Mean	2.548	35.79	49.88	5182	18.18	6.794	4.861
Median	1.189	37.83	30.74	944.9	16.13	5.095	1.919
Maximum	142.7	100.0	335.1	112308	147.0	34.80	49.72
Minimum	−181.3	−411.6	8.150	66.40	−42.87	−4.769	0
Standard deviation	33.11	61.46	52.05	12943	24.56	6.576	7.793
Observations number	184	176	184	184	184	184	184
Cross-section number	23	22	23	23	23	23	23

References

Aksornkoae S, Tokrisna R (2004) Overview of shrimp farming and mangrove loss in Thailand. In: Barbier EB, Sathirathai S (eds) Shrimp farming and mangrove loss in Thailand. Edward Elgar, London, pp 37–51

Antle JM, Heidebrink G (1995) Environment and development: theory and international evidence. Econ Dev Cult Chang 43(3):603–625

Arrow K, Bolin B, Costanza R, Dasgupta P, Folke C, Holling CS, Jansson B, Levin S, Maler K, Perrings C, Pimentel D (1995) Economic growth, carrying capacity, and the environment. Sci 268(5210):520–521

Auksronkaw S (2002) Population and coastal resources. Matee Senior Research Fund. The Thailand Research Fund. College of Population Studies, Chulalongkorn University

Barbier EB, Burgess JC (2001) The economics of tropical deforestation. J Econ Surveys 15(3):423–421

Bhattarai M, Hammig M (2001) Institutions and the environmental Kuznets curve for deforestation: a cross country analysis for Latin America, Africa and Asia. World Dev 29(6):995–1010

Cropper M, Griffiths C (1994) The interaction of population growth and environmental quality. Am Econ Rev 84(2):250–254

Greene WG (2003) Econometric analysis, 5th edn. Prentice Hall, pp 283–303

Koop G, Tole L (1999) Is there an environmental Kuznets curve for deforestation? J Dev Econ 58(1):231–244

Kuznets S (1955) Economic growth and income inequality. Am Econ Rev 45(1):1–28

Lopez R, Galinato G (2005) Deforestation and forest-induced carbon dioxide emissions in tropical countries: how do governance and trade openness affect the forest-income relationship? J Env Dev 14(1):73–100

Moomaw WR, Unruh GC (1997) Are environmental Kuznets curves misleading us? The case of CO_2 emissions. Env Dev Econ 2(4):451–463

Naito T, Traesupap S (2006). Is shrimp farming in Thailand ecologically sustainable? J Fac Econ Kyoto Gakuen Univ (working paper) 16(1):55–75

Panayotou T (1993) Empirical tests and policy analysis of environmental degradation at different stages of economic development. Working Paper WP238. Technology and Employment Programme, ILO

Panayotou T (1997) Demystifying the environmental Kuznets curve: turning a black box into a policy tool. Env Dev Econ 2(4):465–484

Panayotou T, Sungsuwan S (1994) An econometric analysis of the causes of tropical deforestation: the case of Northeast Thailand. In: Brown K, Pearce DW (eds) The causes of tropical deforestation. University College London Press, London

Shafik N (1994a) Economic development and environmental quality: an econometric analysis. Oxf Econ Pap 46:757–773

Shafik N (1994b) Macroeconomic causes of deforestation: barking up the wrong tree? In: Brown K, Pearce DW (eds) The causes of tropical deforestation. University College London Press, London

Shafik N, Bandyopadhyay S (1992) Economic growth and environmental quality: time series and cross-country evidence. Background paper for the world development report 1992, WPS904. The World Bank, Washington, DC

Taya K (2003) Shrimp farming and trade in Asia (The book title is translated Japanese into English by authors.), Ch. 4. Seizando (written in Japanese)

Traesupap S, Matsuda Y, Shima H (1999) An econometric estimation of Japanese shrimp supply and demand during the 1990s. Aquacul Econ Manag 3(3):215–221

Walter H (1971) Ecology of tropical and subtropical vegetation. Oliver & Boyd, Edinburgh

Remote Sensing Technology: Recent Advancements for Mangrove Ecosystems

Mohd Nazip Suratman

Contents

Abstract Mangroves are considered to be a unique component of coastal zones in the tropical and sub-tropical regions. They not only play critical roles in fulfilling important socio-economic benefits to coastal communities, but also in ensuring sustainability of coastal ecosystems. Despite the important aspects of their roles, habitat destruction continues to be a major threat due their conversion into other types of land uses. For the last two decades, the use of remote sensing techniques for a variety of mangrove analyses have been reported by many authors. These applications are essential for mangroves as they often involve an extensive area because more often accessibility and larger topographic maps are not available for detailed mapping. General conclusions are that the newer generations of data (i.e., very high resolution [VHR], hyper spectral and synthetic aperture radar [SAR]) and techniques have improved the accuracy in characterizations of mangroves over the traditional remote sensing systems. This chapter reviews recent advancements in remote sensing for

M. N. Suratman (✉)
Faculty of Applied Sciences and Centre for Biodiversity and Sustainable Development, Universiti Teknologi MARA, 40450 Shah Alam, Malaysia
e-mail: nazip@salam.uitm.edu.my

I. Faridah-Hanum et al. (eds.), *Mangrove Ecosystems of Asia*,
DOI 10.1007/978-1-4614-8582-7_14,

mangroves with a particular emphasis on data, techniques and their applications. Results of experiment and investigation for mangrove assessments are reported. Several new approaches and novel applications of recent techniques are also highlighted. In addition, several constraints of the use of remote sensing applications for mangrove monitoring are identified. Finally, future opportunities of data and techniques for mangrove assessments are described.

1 Introduction

Mangroves represent transitional ecosystems where the ocean, land, and freshwater meet. They are easily recognized as they are located at tideland mud or sand flats inundated daily with sea water (Suratman 2008). Similar to the tropical rainforests, mangrove forests continue to be degraded at rapid rates through different types of human activities (Saenger et al. 1983; Hamilton and Snedaker 1984). However, in contrast to the tropical rainforests, mangroves are lacking a high level of research and conservation efforts, which result in limited understandings of these ecosystems. To address these issues, this paper begins by providing an overview on historical perspectives of remote sensing technology, followed by a general description on its applications. A short overview of mangrove ecosystems is included, followed by a detailed description on recent advances in remote sensing data and techniques of mangrove characterization in terms of mangrove productivity, mapping and monitoring, species discrimination, estimations of biomass, carbon sequestration, leaf area index (LAI) and tree heights.

2 An Overview of Remote Sensing Technologies

Remote sensing has been used for many decades. During the First World War, aerial reconnaissance was among the early applications over wide areas, it allowed the positions of the opposing armies to be monitored more safely than a ground-based survey (USGS 2000). In addition, aerial photographs also allowed for accurately mapping and updating military maps in a rapid manner with strategic positions. Today, remote sensing is widely utilised for many applications including monitoring mangrove ecosystems. In comparison with traditional aerial photography, medium-resolution satellite imagery has the following advantages: (1) the frequency of data collection, (2) global availability of remote sensing data, (3) data are suitable for digital analysis and classification, and (4) data are gathered at relatively low cost (Wilkie and Finn 1996).

Remote sensing also has many advantages over ground-based surveys in that large land areas can be surveyed at one time, and areas of land or sea can be included that are otherwise inaccessible (Keiner and Yah 1998; Guidon and Edmonds 2002). Remote sensing can reduce cost and improve efficiency of forest inventories

Table 1 Characteristics of remote sensing systems. (Sources: Wilkie and Finn 1996; Pitt et al. 1997; Lillesand and Kiefer 2000)

Sensor	Platform	Spatial resolution	Image swath	Characteristics	Weather requirements	Potential uses
Landsat MSS[a]	Satellite	79 m	185 km	Visible + NIR[g] (4 bands)	Cloud-free	Mapping stratification
Landsat TM[b]	Satellite	30 m	185 km	Visible + NIR & MIR[h] (6 bands) + thermal IR[i]	Cloud-free	Mapping stratification
SPOT-HRV[c]	Satellite	10 m 20 m	65 km	Panchromatic, visible, NIR (3 bands)	Cloud-free	Mapping stratification
IRS[d]	Satellite	5.8 m 23 m 70 m	148 km	Visible, NIR, (4 bands)	Cloud-free	Mapping stratification
RADARSAT-1	Satellite	> 8 m	0–50 km	Microwave	None	Sampling mapping stratification
CASI[e]	Aircraft	> 30 cm	150 m–5 km	Visible + NIR (1–19 bands)	Clear to light cloud cover	Stratification Sampling
Aerial photo	Aircraft	Based on scale	0–25 m	Visible + NIR	Clear to light cloud cover	Mapping stratification sampling
Video graphy	Aircraft	> 4 cm	0–25 m	Visible, NIR, MIR & thermal IR (3–8 bands)	Clear to light cloud cover	Mapping stratification
Laser LiDAR[f]	Aircraft	10–50 cm	0–1 km	Generally NIR	No rain	Sampling

[a]Multispectral Scanner
[b]Thematic Mapper
[c]Systeme Pour l'Observation de la Terre-High Resolution Visible
[d]Indian Remote Sensing
[e]Compact airborne spectrographic imager
[f]Light Detection and Ranging
[g]Near-infrared
[h]Mid-infrared
[i]Near-infrared

if remotely-sensed data are well correlated with important field measurements and are available when needed in the sampling design (Czaplewski 1999) and cover large areas (Lindgren 1985).

On the other hand, remote sensing has limitations that prevent it from totally replacing ground-based survey methods. These are partly related to spatial, spectral, and temporal resolutions of the various sensors. Also, there are problems with the all-weather capabilities (see Table 1) data analysis and data interpretations. Also, not all important information is related to the electromagnetic spectrum. In this respect, remotely-sensed data should be considered as a complementary source of information, rather than a substitute for ground-based data gathering. However, the insight that it provides into the environmental status and processes is valuable

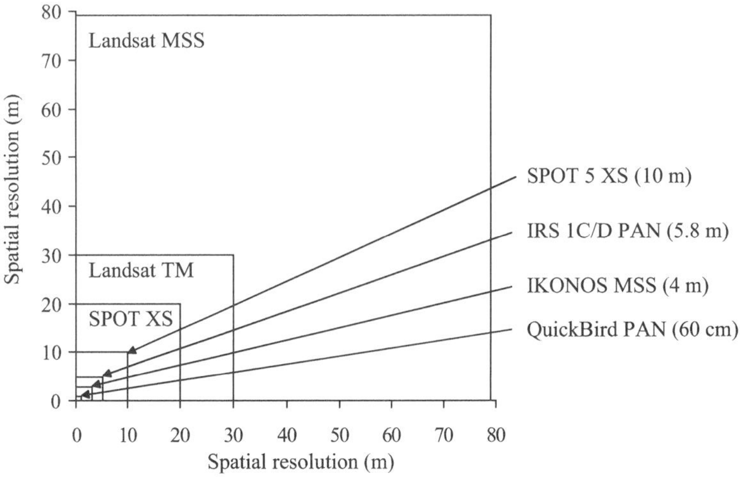

Fig. 1 Illustration of traditional and recent sensor systems with respect to their spatial resolution

3 Development of Remote Sensing

Aerial photographs have been used routinely in forestry since the 1950s and have played a key role in forest mapping and inventory systems up to the present (Aldrich 1979; Leckie and Gillis 1995). Today, other remote sensing technologies have improved capability and resolution, and are conducted using satellites or aircraft platforms and a variety of sensors.

The pixel sizes of selected operational sensor systems are compared in Fig. 1. The first earth resources technology satellite (ERTS-1 or Landsat 1), with a MSS, was launched in 1972 and had a resolution of 79 × 79 m with four spectral bands. Ten years later, it was improved with the addition of Landsat TM imagery. Landsat TM on Landsat 4 improved the resolution to 30 m and covered a wide range of the electromagnetic spectrum with seven bands, including a thermal band and two mid-infrared bands. Together, the Landsat series permit a retrospective image interpretation possible back to 1972 (IUFRO 1994).

The SPOT satellite was launched in 1986 and has a 20 m spatial resolution for multispectral and 10 m for panchromatic modes. SPOT with panchromatic, visible and near-infrared bands, is useful for vegetation studies including health assessments. By 1995, images with 5.8 m resolution were available from the IRS satellite (Lillesand and Kiefer 2000).

Canada's RADARSAT, which was launched in 1995, represents an operational space-borne active sensor technology (Table 1). In this system, the target area on the ground is scanned by microwave radiation. The reflected and back-scattered radiation then provides information about the surface, sub-surface, physical, and dielectric[1]

[1] An indication of the reflectivity and conductivity of the materials.

Table 2 A selection of various currently operational and future sensor platforms. (Source: Kätsch and Vogt 1999 [revised])

Sensor	Country	Year of launch	Type of sensor	Spatial resolution (m)	Stereo capability
RADARSAT	Canada	1995	SAR[b]	28×25 10×9	
SUNSAT[a]	South Africa	1999	MSS	15	Yes
QuickBird	USA	2001	PAN[c]/MSS	0.6/2.44	
IKONOS 2	USA	1999	PAN/MSS	1.0/4.0	Yes
SPOT 4	France	1998	PAN/MSS	10/20	Yes
SPOT 5	France	2002	PAN/MSS	2.5/5/10	Yes
Landsat TM-7	USA	1999	PAN/MSS	15/30–60	

[a]The sensor has not been active since 2001 due to technical problems
[b]Synthetic Aperture Radar
[c]Panchromatic

properties (Leckie 1998). Microwave sensors have the highly advantageous properties of operating independently of sun illumination and are usually insensitive to weather conditions or cloud cover. These characteristics are particularly suitable to monitoring phenomena in the tropic regions (Thompson et al. 1993; Toan 1995; Salas et al. 2002), although the full capability of radar has yet to be exploited (Leckie 1998) (Table 2).

The first commercial imaging satellite (IKONOS) was launched in September 1999 from the Vandenberg Air Force Base, California. This satellite provides one-metre resolution panchromatic images and four-metre multispectral images (Lidov 1999). Test images from IKONOS prove the superior quality of the new system (Baltsavias et al. 2001). Many of the newly launched sensor systems feature high spatial geometric resolutions in combination with stereo capabilities such as SUNSAT, IKONOS, SPOT 4, and SPOT 5 (Table 1). These characteristics will make images suitable for the application of traditional photogrammetric techniques to extract altimetric information, such as a digital elevation model (DEM). Evaluations using IKONOS images are still on-going for studying different topographic terrain and applications; however, recent results in mountainous areas are promising for small area mapping (Toutin and Cheng 2000).

QuickBird is now the highest-resolution commercial remote sensing satellite offering imagery with 60-cm resolution. QuickBird was launched in October 2001, and collects multispectral and panchromatic imagery simultaneously with 16.5×16.5 km swath width at nadir (Euroimage 2002).

In May 2002, SPOT 5 was launched from the Guiana Space Centre in Kourou, French Guiana. It offers enhanced capabilities compared to SPOT 4 in terms of improved resolution (up to 2.5 m) and will also be used to create coverage of five continents with digital terrain models (CNES 2002).

3.1 Applications of Remote Sensing in Developing Countries

In developing countries, the use of satellite imagery data as a component in resource inventories and information systems has been reported by many authors (e.g., Lachowski and Dietrich 1978; Wacharakitti and Morain 1978; Aldrich 1979; Lal et al. 1990; Bong 1991; Rao et al. 1991). General conclusions were that remote sensor imagery has proven to be a more authoritative source of data than was formerly possible. For example, the Philippines government believed that its evergreen rainforest cover still accounted for 57 % of the land base during the early 1970s, but a remote-sensing survey carried out by Lachowski and Dietrich (1978) in 1976 revealed that the actual amount was only 38 %. In this survey, the authors used Landsat imagery with support from ground data, and considered that the methodology was sufficiently comprehensive for the results to be characterised as accurate within 95 % accuracy. A similar example occurred in the early 1970s in Thailand, where the government believed that 48 % of the country was under forest cover, largely monsoon deciduous forest. A 1978 Landsat survey revealed that the actual cover amounted to only 25 % (Wacharakitti and Morain 1978). In India, the Department of Forestry estimated 23 % of land area as forested, but a Landsat survey estimated the amount as less than 10 % (Lal et al. 1990).

Inspired by the revealing results reported for the Philippines and Thailand, and motivated by growing evidence of forest depletion in their countries, a good number of other tropical countries have undertaken remote-sensing surveys of their tropical forest cover. In many different countries remote sensing revealed that forest cover was in fact less—often a good deal less—than was previously thought (Malingreau 1986).

Today, many developing countries are involved in the systematic monitoring of renewable resources. Given constraints of time, money, and skilled manpower, countries must evaluate effective methods to obtain reliable and timely resource data. Traditional ground methods are time-consuming and expensive for regional or national resource inventory programs. Remote sensing from aircraft and satellites has gained world-wide recognition as an efficient method to provide resource information that is often technically and economically feasible compared to ground methods (FAO 1996).

In Malaysia, aerial photography has been used effectively for several decades. The first complete coverage was obtained in 1967 at a 1:25,000 scale with black and white panchromatic film (Kamaruzaman and Mohd Rasol 1995). Landsat MSS and TM data have been used for land use surveys and the results have shown that it is possible to map various natural land cover types and man-made features, including terrain change, forest areas, soil types, dams, and urban areas (Salleh 1976; Mahmood et al. 1983). However, the resolution of the MSS was found to be unsuitable for mapping Malaysian agricultural land utilisation due to small farm sizes and irregular cropping patterns (Darus 1989). Another study conducted by the Asean Institute of Forest Management (AIFM) in 1989 showed that Landsat TM imagery could be used to detect and classify forest disturbances and provide data to update forest resource maps through the integration of remote sensing and a geographic information system (GIS) (Zahriah et al. 1989). Landsat TM has been used to detect deforestation and to identify suitable areas for tourism-related development in Langkawi Island, Malaysia (Kamaruzaman and Mohd Rasol 1995). Another study was conducted by Kamaruzaman and D'Souza (1996) to determine the applicability of SPOT-HRV in the State of Pahang, Malaysia for detecting logging activities. It was shown that physical features and forest disturbances could be detected by this image.

In India, the use of remote sensing goes back over 30 years. The first aerial photographs for forestry purposes were acquired in 1963. Applications of satellite imagery in forestry date back to 1975. The first attempt to assess forest cover in India by satellite imagery interpretation was made in 1984–1985 by the National Remote Sensing Agency (NRSA 1983). This exercise was done visually and resulted in an estimate of the forest cover for the country of 0.64 million km^2, or 19.5 % of the geographical area, in contrast to the previously recorded figures of 22.8 % (Rao et al. 1991). The years between 1980 and 1990 were dominated by satellite remote sensing for forest resources assessment, monitoring, wildlife habitat evaluation, and fire damage assessment (Kushwaha 1987). Subsequently, vegetation cover and forest type mapping was done by the Forest Survey of India that involved preparing forest cover maps of 1:250,000 for the entire country, to be repeated every two years for monitoring (Kohl and Kushwaha 1994). This project revealed that non-forest areas could generally be mapped with an accuracy of 80–95 % in flat undulating areas if the trees were in full foliage. Currently, approximately 70 % of India has been covered on a thematic map (FAO 1998).

An FAO/UNDP project helped Myanmar assess forest resources with the use of satellite data from a 1970 Landsat image. This project, which was conducted from 1981–1991, provided reliable information on forest resources for about 90 % of the area. Since 1991, the country has been conducting field forest inventories every year, covering 2 million ha using remote sensing and GIS technologies (FAO 1998).

In Sri Lanka, forest cover assessment maps using Landsat imagery were produced in 1991–1992. Indicative inventories of non-forest land and detailed periodic inventories of plantations were carried out for assessing resources. A forest resource assessment was done using 1:20,000 aerial photographs for natural forests and 1:10,000 and 1:20,000 for plantations (FAO 1998).

From 1995 to 1997, the Forest Department of Bangladesh completed an inventory program in hill and coastal forest areas, with the assistance from an international development agency. This was a unique inventory in the sense that a socio-economic survey was also conducted along with the forest inventory to understand the behavioural pattern of the users. Forest statistics were generated with continuous resource change assessments. SPOT-HRV data were used to generate signatures of different types of forest vegetation (FAO 1998).

In some parts of the world, conversion of mangrove forest into other types of land use (e.g., residential areas, airports, agricultural lands, fishponds, etc.) takes place continuously (Hartono 1994). Satellite imagery data have been used for various mangrove forest analyses and monitoring in many parts of the tropical world. For example, in Bangladesh, computer-processed Landsat data, with additional data from 1:30,000 aerial photographs, permitted two mangrove species to be distinguished with 71 % classification accuracy (Heller and Ulliman 1983). SPOT-HRV satellite data have been used for more than 10 years in a mangrove forest analysis. For mapping purposes, SPOT-HRV images have been used in Vietnam (Bong 1991) and in Guinea (Moreau and Vercesi 1989). Monitoring of mangroves has been performed with SPOT-HRV and aerial photographs in East and West Java (Hartono and Muljosukojo 1990). Based on a combination of image analyses, five classes of mangrove vegetation were identified: (1) *Lumnitzera* spp. (2) *Avicennia* spp. (3) *Rhizophora* spp. (4) mixed floristries, and (5) degraded mangrove. In this study, a confusion matrix analysis was performed and an overall 94 % classification accuracy was achieved. In other parts of the image, rice fields, villages, home state gardens, rivers, creeks, and irrigation channels were identified.

In Thailand, Landsat MSS images were used in the form of 1:1,000,000 diazochrome additive-colour composites and 1:500,000 black-and-white images of bands 4, 5, and 7 and of bands 5 and 7. Together with additional information from the field and from aerial photographs, maps made from the Landsat images were used to determine the total forest cover. Comparing this information with forest-cover data either from aerial photographs or Landsat imageries with earlier dates permitted a rough calculation of the reduction of the forest cover over large areas, at a relatively low cost (Morain and Klankamsoon 1978). Miller et al. (1978) utilised Landsat imagery covering the years 1972 through 1977 for determining the expansion of shifting cultivation in north-eastern Thailand. Additional information from 1:20,000 to 1:60,000 aerial photographs on shifting cultivation, irrigated rice, hill evergreen forest, and other forest types grouped together was also incorporated. Mapping of the different values of MSS band 7, displayed by assigning grey levels to various levels of difference in tone (scene brightness), permitted detection of shifting cultivation at one-year intervals. The difference in maps of MSS band 5 was in showing where permanent agriculture was encroaching on the forest.

In Tanzania, remote sensing technology has been applied in the production of forest cover maps and inventories of plantations and natural forests. For example, Sylvander et al. (1988) successfully utilised satellite imagery for delineation of vegetation types in Eastern Tanzania using Landsat MSS false composites at a scale of 1:250,000. Double sampling with aerial photographs for estimating the volume of Miombo woodlands was done by Temu (1981). He found that the method was effective, especially for the areas where access was poor.

This review shows that forest inventories and monitoring work in developing countries makes extensive use of remote sensing data. Area information on forest types from satellite data mapping has generally been successful, but identification of species has been difficult.

3.2 Forest Stand Parameters Estimation using Remote Sensing Data

3.2.1 Relationship Between Forest Stand Parameters and Remote Sensing Data

Information about forest conditions is essential for forest management planning. Forest management activities require reliable forecasts of the development of all constituent stands in the area being managed. Strategic decisions concerning forest policies to achieve management objectives require accurate information, including

Table 3 Correlation coefficients between forest variables and Landsat TM and SPOT-HRV spectral data

Spectral bands	Band	Spectral range (μm)	Volume (m^3/ha)[a]	Basal area (m^2/ha)[b]	Age (years)	Height (m)
TM band 1	Blue	0.45–0.52	−0.61	−0.27	−0.35[b]	−0.44[d]
TM band 2	Green	0.52–0.60	−0.72	−0.42	−0.54[b]	−0.56[d]
TM band 3	Red	0.63–0.69	−0.69	−0.47	−0.53[b]	−0.48[d]
TM band 4	Near-infrared	0.76–0.90	−0.76	−0.47	−0.45[b]	−0.54[d]
TM band 5	Mid-infrared	1.55–1.75	−0.63	−0.43	−0.62[b]	−0.62[d]
TM band 7	Mid-infrared	2.08–2.35	−0.55	−0.48	−0.59[b]	−0.53[d]
HRV band 1	Green	0.50–0.59	−0.77	−0.18	−0.67[c]	−0.18[c]
HRV band 2	Red	0.61–0.68	−0.63	−0.35	−0.40[c]	−0.35[c]
HRV band 3	Near-infrared	0.79–0.89	−0.82	−0.41	−0.42[c]	−0.41[c]

[a]Ripple et al. (1991)
[b]Brockhaus and Khorram (1992)
[c]Danson (1987)
[d]Nilsson (1997)

stand growth forecasts. In the last decade, many studies have shown that spectral radiance recorded by satellite remote sensing can be related to several forest parameters. Forest inventory studies have found that many tree and stand variables, such as wood volume, biomass, basal area, diameter, and stand age, show strong inverse relationships with red, near-, and mid-infrared bands from Landsat TM and red and near-infrared bands from SPOT (e.g., Horler and Ahern 1986; Danson 1987; Poso et al. 1987; Ripple et al. 1991; Ardö 1992; Brockhaus and Khorram 1992; Nilsson 1997). Ripple et al. (1991) argued that this was because the understory has a highly reflective shrub and herb layer. Young stands with lower wood volumes have higher radiance in all TM and HRV bands than older stands which have more shadows, thus causing the strong inverse relationships. Table 3 summarizes the correlation coefficients (r) between some forest variables and Landsat TM and SPOT-HRV spectral data from various sources.

Studies using the near-infrared band of SPOT and the near- and mid-infrared bands of Landsat TM in Douglas-fir (*Pseudotsuga menziesii* (Mirb.) Franco) forests in Oregon have found reflectance and wood volume-related parameters to be well-correlated when using data averaged at the forest-stand scale with correlation values as high as −0.89 (Ripple et al. 1991). Studies that have not involved spatial averaging of data beyond the pixel scale produce relationships between reflectance and wood volume that have much lower r values, especially at higher wood volumes (Franklin 1986; Danson 1987). For example, Franklin (1986) presented a study, which included basal areas exceeding 100 m^2/ha, that had a relation between Landsat reflectance and wood volume with correlation values between −0.38 and −0.54.

3.2.2 Estimation Methods

Image classification commonly uses statistical techniques to group pixels into various pre-defined classes, such as land-cover types, land-use classes, and vegetation types (e.g., Bolstad and Lillesand 1992; Brockhaus and Khorram 1992; Kamaruzaman and Mohd Rasol 1995). According to Leckie (1990), a discriminant analysis based on Bayesian[2] maximum likelihood is the most common algorithm used for classification analysis. In addition, he stated that ancillary data describing soil type, slope, and previous management operations, for example, are important for improving the classification accuracy.

The ability of remotely sensed data to provide information on forest variables such as wood volume, tree height, tree diameter, and tree species composition has been reported by numerous researchers (e.g., Horler and Ahern 1986; Danson 1987; Poso et al. 1987; Ripple et al. 1991; Ardö 1992; Brockhaus and Khorram 1992). Regression functions are often used to relate these variables to the satellite data (e.g., Franklin 1986; Ahern et al. 1991; Ripple et al. 1991; Ardö 1992; Brockhaus and Khorram 1992; Trotter et al. 1997). This requires that the correlation between the variables and the satellite data be sufficiently strong. The regression models used in many studies relate different stand variables to functions of spectral band, band products, band ratios, and band transformations (Jakubauskas and Price 1997; Scheer et al. 1997).

A study conducted in the boreal forest by Ardö (1992) showed that field plots established for forest planning in Sweden could be used to construct regression models that predict wood volume. The correlation value between the observed and the estimated volume was 0.83 and the standard error of estimate was 46.5 m^3/ha. Ardö concluded that there was a stronger relationship between spectral radiance and volume for compartments with small volumes than for compartments with large volumes. This agrees with Franklin (1986), who suggested that when the vegetation cover approaches 100 %, the basal area continues to increase as the stand grows older. However, the remotely-sensed signal is not affected by the increase because it is most sensitive to the degree of crown closure.

An alternative to regression technique is the k nearest neighbour (kNN) estimation method, in which forest variables are calculated as weighted means of spectrally nearby samples (Muinonen and Tokola 1990; Tomppo 1990). The method has been used operationally in the Finnish National Forest Inventory (NFI) since 1990. According to Tomppo (1990), among the advantages of this estimation method is that a vector consisting of all variables that are measured or registered in the NFI can be estimated. However, lack of or a low number of sample plots in certain forest types might lead to unreliable estimates (Moeur 1987). kNN estimates are unreliable at a pixel level, but reliable when aggregated to a community level (Tomppo 1990). For example, a study by Tokola et al. (1996) in the south of Finland with primary species of Scots pine (*Pinus sylvestris* L.), Norway spruce (*Picea abies* (L.) Karst.) and birch (*Betula* spp.) found that the standard error of estimates for wood volume on a pixel level was approximately 68–77 m^3/ha.

[2] Using the knowledge of prior events to predict future events.

Table 4 Very high resolution satellite remote sensing systems. (Source: Heumann 2011)

Remote sensing systems	No. of bands	Spectral	Platform	Spatial resolution	
				Multispectral	Panchromatic
IKONOS	4	VNIR, Pan		4 m	1 m
PRISM	1	Pan	ALOS	NA	2.5 m
Quickbird	4	VNIR, Pan		2.4 m	0.6 m
GeoEye-1	4	VNIR, Pan		1.65 m	0.41 m
WorldView-2	8	VNIR, Pan		$<$ 2 m	$<$ 0.5 m
HYPERION	220	400–2500 nm	EO-1	30 m	NA
GLAS	2	Green (532 nm)	IceSAT	70 m, 170 m	NA

4 Recent Advancements on Mangrove Remote Sensing

For the last two decades, traditional methods for mangrove monitoring such as aerial photography, SPOT and Landsat (Table 1) are dominant technology applied in mangrove ecosystems. Aerial photography is more accessible in developing nations and has been widely used for visual interpretations to map the extent of mangrove. Digital images acquired from Landsat and SPOT were used based on computational classification to detect changes between successive periods and to map individual species.

The new generations of satellite sensors are introduced not only to provide important information on mangrove ecosystems, but also to improve the techniques and accuracies obtained by the traditional approaches. In recent years, there have been rapid advances in the new types of sensors. Table 4 listed recent sensors that have the potential to improve the accuracy in classification of mangroves and species discrimination. The systems were reported to contribute in improving the estimations of mangrove biomass, canopy height and leaf area.

As shown in Table 4, the recent VHR remote sensing systems include IKONOS, Quickbird, Panchromatic Remote Sensing Instrument for Stereo Mapping (PRISM), WorldView 2, GeoEye, Geoscience Laser Altimeter System (GLAS) and HYPERION. In addition to the aforementioned passive optical satellite remote sensing system, new types of synthetic aperture radar (SAR) systems are developed. There are many reasons why radar-imagery application can be advantageous. It is an appropriate option for characterizing mangrove ecosystems under the persistent cloud cover in the tropical and sub-tropical regions. The SAR systems include Space-borne Imaging Radar (SIR-C), European Remote-Sensing Satellite (ERS-1), Japanese Earth Resource Satellite (JERS-1), RADARSAT-1, RADARSAT-2, Advanced Synthetic Aperture Radar (ASAR) and Phased Array type L-band Synthetic Aperture Radar (PALSAR).

For more than two decades, remote sensing has been widely used as a source of information for mangrove research. The number of mangrove publications using remote sensing has grown very rapidly and this is noticeable with many applications. Researchers have become increasingly aware of the potential of remote sensing to address important issues in mangroves for a long time.

4.1 Mangrove Forest Productivity

Productivity in mangroves refers to either the ecological value or the function of mangrove vegetation community. According to Komiyama et al. (2008), productivity in mangroves can be affected by their environment. Measuring productivity is not easy; however, it can be estimated by quantifying the quantity of living materials produced by mangroves over periods of time. According to Saenger et al. (1983), mangrove productivity is important because it has direct impact on the health and function of the marine food chain. In addition, higher productivity means a more diverse or larger number of organisms can be supported by a mangrove ecosystem.

Although many studies have been conducted to study the mechanisms and rate of mangrove productivity, there is lack of studies that have been carried out on productivity mapping in the mangrove. However, a few studies have demonstrated the potential of remote sensing in detecting productivity in mangroves. For example, a significant correlation was observed by Song et al. (2011) between photochemical reflectance index (PRI) and soil water salinity in their mangrove study in Galapagos Island, Ecuador. In a study of detecting quantum yield and non-photochemical quenching (NPQ) in mangrove canopy, Nichol et al. (2006) observed significant r values in the correlation of PRI vs. quantum yield. From this study, they indicated that hyperspectral has the potential to be used for photosynthesis estimations.

4.2 Mangrove Area Mapping

To study mangrove areas effectively, cost effective mapping techniques and an accurate change detection analysis are required. Traditionally, multispectral remote sensing offers many advantages and has been used to map forest areas and monitor deforestation. However, the spatial and spectral resolution of the sensors coupled with the inability to penetrate cloud limits their effectiveness.

The launch of the two latest generations of high-resolution optical data, i.e., IKONOS and QuickBird, has opened up new opportunities for mapping mangroves which are able to differentiate mangrove stands and discriminate other assemblages of species in mangroves. Identification to the species level is important in order to assess the history of mangrove growth and better understand the ecosystem functions and processes. In comparison between the two sensors, Wang et al. (2004) found that IKONOS imagery captured a more detailed spectral reflectance. In their study, the maximum likelihood classification (MLC) from IKONOS has resulted in better discrimination of mangrove species. In Pambala, Sri Lanka, Dahdouh-Guebas et al. (2005) used IKONOS to study the effectiveness of mangroves as a defense against tsunami. In this study, they applied various image composites and later classified the imagery using both unsupervised and supervised classifications. The resulting classified images were then compared with on-screen digitized images and ground-truth information. They reported that this method has successfully distinguished *Rhizophora apiculata* and *R. mucronata* . In another study in Punta Galeta, Panama,

Wang et al. (2008) tested a neural-network approach to discriminate mangroves at species level using IKONOS. They found that the clustering-based neural network classifier (CBNN) and MLC methods have improved classification accuracy when compared with the use of spectral bands only.

In Danshui River, Taiwan, Lee and Yeh (2009) calculated the normalized differences vegetation index (NDVI) from QuickBird images to create vegetation mask. Using an MLC, they obtained high classification accuracy for mangroves vs. non-mangroves discrimination. In the Texas Gulf Coast region, Everitt el al. (2008) tested two classification methods, i.e., ISODATA vs. MLC, to classify a black mangrove community and found that MLC produced more acceptable results.

4.3 Monitoring and Change Detection

Change detection enables the evaluation of mangrove changes over periods of time due to anthropogenic or natural phenomena. It is an effective way to measure and visualize changes, and therefore better understand the dynamics of mangrove ecosystems. Increase or decrease in mangrove distribution and condition can be estimated using change detection analysis. A study on detection and quantification of land use changes caused by deforestation were conducted by Rodriguez and Feller (2004) in Two Cays Archipelago, Belize. A combination of aerial photo and IKONOS were used as primary datasets. By applying three methods, i.e., principal component analysis (PCA), NDVI calculation and intensity, hue and saturation (IHS) transformation, they successfully distinguished seven main land cover classes (black and red mangroves, five non-mangrove classes and mixed forest).

Several studies were conducted using radar data to examine the relationships between mangrove canopy and backscattering response of a SAR system. For example, in tropical Australia, Lucas et al. (2002) used RADARSAT-1 to study changes in mangroves and they found that an increase in backscatter signal was due to the variability in mangrove stand structures. They found that the changes in mangrove structures from homogenous to heterogeneous stages as a result of successive growth has resulted in an increase in the scattering of L and C bands. Consistent results were obtained when the study was repeated using JERS-1 SAR and AIRSAR L-band in West Alligator River (Lucas et al. 2007). The two studies suggested that the mangrove change detection using radar data would be most effective if mangroves are bordering with non-mangrove areas and where mangroves are different in structure.

According to Kuenzer et al. (2011), an integration of SAR data and hyperspectral imagery will contribute to increase mapping accuracies. For example, in North Queensland, Australia, Held et al. (2003) integrated AIRSAR data and hyperspectral imagery derived from CASI to map mangrove ecosystems in the Daintree River. An improvement of overall classification of about 3 % was obtained in classifying species communities as compared to the accuracy from individual sensors. They suggested that the integration of such sensors allow subtle and long-term monitoring of changes in mangroves.

Recent techniques to improve the accuracy of change detection have been developed. For example, a new classification technique known an object based image analysis (OBIA) was introduced. This technique, also known as segmentation, involves grouping contiguous pixels into objects based on image properties such as tree crowns, tree species, tree ages, etc. Several have applied OBIA to detect changes in mangroves between successive periods. For example, Conchedda et al. (2008) used the OBIA technique to show the current extent of mangrove resources and subsequently map its changes from two images. In this study, a user's accuracy of about 97 % was achieved for classifying mangroves. However, the OBIA approach resulted in a lower overall change accuracy (66 %) as compared to the traditional method (76 %). They suggested that, using the OBIA technique, the segmentation of temporal images, changes between images, cannot be separated due to small size of the objects.

4.4 Species Discrimination

Given the constraints in both spectral and spatial, traditional systems of remote sensing have been unable to discriminate mangrove species with required confidence. However, a newer generation of sensors has demonstrated the possibility of mangrove discrimination which includes very high resolution and hyperspectral imagery. The VHR imagery such as IKONOS and QuickBird are capable of reducing the effects of mixed-pixels. In addition, the high spatial resolution provides sufficient details for the analysis of canopy structure.

In order to improve classification accuracy, several new classification techniques have been introduced to be applied with VHR imagery for discrimination of mangrove species. These include a fusion technique of post classification by Vaiphasa et al. (2005), followed by fuzzy classification (Neukermans et al. 2008), neural networks (Wang et al. 2008) and support machine vectors (Huang et al. 2009). Using fuzzy classification technique, Neukermans et al. (2008) used QuickBird data to discriminate four mangrove species with an overall classification accuracy of 72 %. They suggested that the spectral-only information was insufficient for individual species discrimination. This was agreeable with Wang et al. (2004) in their study in the Caribbean Coast of Panama. Using IKONOS and QuickBird imagery, they applied the MLC technique to discriminate three mangrove species and obtained an overall classification accuracy of 75 %. They recorded as low as 55 % user's accuracy for some individual species as a result of using spectral data only.

The distributions of mangroves are strongly related to ecological condition which can be used for mapping the species at a certain level of discrimination. For example, in Pak Phanang District of Thailand, Vaiphasa et al. (2005) found that the soil pH was closely associated with certain mangrove species. Therefore, using a Bayesian probability model, the soil data were integrated with ASTER imagery in a post-classification stage. This step has resulted in spectral distinction of mangrove species in the study area.

Using VHR multispectral IKONOS imagery, Kanniah et al. (2007) characterized mangrove species in a study area in Malaysia using three methods. These were MLC with individual spectral bands, MLC in combination with texture information and maximum distance classifier. The results showed that combining all spectral bands has produced highest overall classification accuracy (i.e. 81.8 %). The linear spectral unmixing applied on the image has produced output images that showed proportion maps of *Rhizophora apiculata*, *R. mucronata* and other species.

4.5 Tree Biomass Estimates

The need for understanding the roles of mangroves in the climate change scenario has generated interest in understanding the carbon-stocking ability of mangrove. The ground work could be the most accurate way in determining the biomass of mangroves; however, due to an inaccessibility of mangrove areas, the remote sensing approach seemed to be a practical way for estimating the mangrove biomass.

Various approaches of biomass estimations from high resolution imagery were assessed for mangroves. For example, in the Agua Brava Lagoon System of Nayrit, Mexico, Kovacs et al. (2004) used 1-m multispectral IKONOS imagery for estimating biomass of mangrove species based on an *in situ* measurement of leaf area index (LAI). From the statistical test, there were significant relationships between LAI and NDVI and a simple ratio (SR) of multispectral IKONOS imagery. They produced an LAI map and classified the values into four categories which include red mangroves, healthy white mangroves, poor condition white mangroves and dead mangroves.

In French Guiana, Proisy et al. (2007) estimated the total above-ground biomass using two combined techniques, a PCA of Fourier spectra and Fourier transformation, with 1-m panchromatic and 4-m NIR data from IKONOS. Using regression models, they found significant relationships between the mangrove tree stages (i.e., pioneer, mature and dead) and the principal component of a Fourier spectra. The best regression model was an estimation of total and trunk biomass of mangroves from panchromatic data which gave an R^2 value of 0.90 and standard error of estimates of 16.9 %.

Many studies have tested the potential of SAR sensors to estimate mangrove biomass from canopy characteristics (i.e., Proisy et al. 2000, 2002; Li et al. 2007; Lucas et al. 2007). Results obtained using L-band SAR indicated that biomass can be estimated up to 100–140 Mg ha^{-1}; however, a decrease in the L-band backscattering coefficient has resulted in complication in an estimation of mangrove stand biomass greater than 200 Mg ha^{-1}(Lucas 2007). For mangrove forest estimations, Proisy et al. (2002) suggested that low-frequency measurements are the best with maximum sensitivity occurring at cross-polarization mode for P- and L-bands. At the biomass level of 70 t DM ha^{-1}, saturation values occurred for C-band. For L- and P-bands, saturation values occurred at 140 t DM ha^{-1} and 160 t DM ha^{-1}, respectively. The correlation between backscattering coefficient and total biomass decreased at 250 t DM ha^{-1}.

In South China, Li et al. (2007) tested a regression model to estimate mangrove wetland biomass using Radasat-1 images. In this study, the SAR data (C-band,

HH) and NDVI were used as predictor variables for mangrove biomass. The results indicated that SAR data was better than NDVI in which it explained 45 % of biomass variance.

Biomass has also been estimated using polarimetric synthetic aperture radar (PolSAR). Research into the analysis of PolSAR data continues to reveal biomass estimation techniques. For example, Mougin et al. (1999) estimated total above-ground biomass of various mangrove forest stages. Strong relationships were found between biomass and radar data with ($R^2 = 0.94$) where P-HV showed the greatest sensitivity to total biomass. While the backscattering coefficient of P-HV saturated at 160 DM ha^{-1}, the total biomass was accurately estimated up to 240 t DM ha^{-1}. They suggested that the use of polarization ratios at different frequencies has provided useful information about penetration capability of radar wave to the mangrove canopy.

Hamdan et al. (2011) conducted a study to estimate above-ground biomass of tropical rainforest in Malaysia from L-band ALOS PALSAR data. They found a strong correlation between aboveground biomass and radar backscattering coefficient in HV polarisation from ALOS PALSAR images. From this study, above-ground biomass was estimated to be from 25.9 ± 10.9 to 569.3 ± 10.9 t ha^{-1} which covered all types of standing forests. Based on these estimations, a spatial map that showed spatial pattern of above-ground biomass for the study area was produced. Despite its limitations, the use of L-band SAR could provide an alternative for rapid assessment of carbon stocks for the study area.

4.6 Carbon Sequestration

Mangroves play an important role in the carbon cycle by removing CO_2 from the atmosphere and storing it as carbon in plant materials and soils. This process is called carbon sequestration (Suratman 2008). In the coastal zones, mangroves may potentially sequester the largest storage of carbon as about half of mass in trees is carbon. Therefore, it is important to understand the distribution of carbon storage within mangroves spatially and temporally. In a small island of Indonesia, Wicaksono et al. (2011) used two image processing techniques, i.e., vegetation index and linear spectral unmixing (ULC), to estimate carbon sequestration using Landsat ETM+ images. The images were selected as they are relatively low cost, widely available with large coverage, and therefore provide a cost effective way in mapping carbon stocks. From this study, they produced a spatial map of mangrove carbon sequestration estimates for the study area.

4.7 Leaf Area Index

Leaf area index (LAI) is a dimensionless number that refers to a single-side leaf area per unit ground. It is an important biophysical parameter for determining net primary production and assessing photosynthesis, transpiration and forest health.

Field measurements of LAI have been used to predict growth and yield and to monitor changes in canopy structure due to climate change (Gholz et al. 1991). Therefore, estimation of LAI in mangroves is important to understand the ecological processes and to predict mangrove ecosystem responses.

Despite its importance, little research has been conducted on the estimation of LAI using remote sensing in mangroves as compared to terrestrial forests. For example, in 2004, Kovacs et al., who conducted a mangrove study in Agua Brava Lagoon, Mexico observed significant correlation between field-measured LAI of white and red mangroves using simple ratio (SR) and NDVI from IKONOS imagery. The R^2 values recorded by SR and NDVI were 0.71 and 0.73, respectively, whereas the standard errors of estimate values were 0.63 and 0.65, respectively.

In the second study in the Agua Brava Lagoon System of Nayrit, Mexico conducted by Kovacs et al. (2009), they found consistent results between QuickBird's vegetation index and IKONOS. However, they observed stronger relationships between SAR data (C-band) and LAI (i.e., $R^2 = 0.82$).

In Turks and Caicos Islands, USA, Green et al. (1998) used CASI data and SPOT XS to assess their suitability for mangrove mapping. A supervised classification of bands derived from PCA and band ratios has classified mangroves vs. non-mangrove habitats with high accuracy (96 %). The nine mangrove habitats were discriminated with an overall classification accuracy of 85 %. Comparison between CASI data and SPOT XS indicated that bands 6 and 7 from the former showed the best relationship and prediction for mangrove LAI and canopy closure.

4.8 Tree and Canopy Heights

Significant relationships between radar data and structural parameters including tree height have been reported (Held et al. 2003). Canopy height has also been shown to be strongly related to mangrove species (Smith and Whelan 2006). Recent emerging airborne light detection and ranging (LiDAR) has recently been used to extract surface information. The system can be used to measure both vertical and horizontal forest structure. According to Hyppa et al. (2004), it is capable of measuring objects on the earth surface with a horizontal resolution of several meters and centimeters vertical accuracy.

In the United States, the globally available Shuttle Radar Topography Mission (STRM) digital surface model is to be used to determine the roles of mangroves as carbon sinks by development of the relationship between tree height and biomass. Simard et al. (2006) demonstrated that this dataset provides the potential to produce reasonable estimates of canopy heights in mangroves. However, both air- and space-borne LiDARs are reported to characterize vertical canopy structure better than the STRM digital surface model (Simard et al. 2006, 2008). In terms of accuracy, this approach is effective when applied to mature and tall mangroves because of small relative error. However, due to unavailability of species information, generalised allometric functions were used to estimate the standing biomass of mangrove from canopy height.

5 Conclusions and the Way Forward

The objective of this chapter was to review remote sensing technology and its recent application on mangrove ecosystems in recent years. It can be summarized that there is a growing interest in this subject and there has been a rapid advancement of the science of remote sensing for mangroves. However, it is difficult to make a detailed comparison to all reviewed studies as each research has different sets of objectives and focuses. While the approaches used for data analyses depend on a variety of factors, some consistencies in the trend of research findings exist. In recent years, VHR, hyperspectral and SAR data have demonstrated to be very valuable in improving accuracies from traditional remote sensing approaches. In terms of applications, the progress has evolved from the realm of pure research to that of world-wide day-to-day application. Nevertheless, there is an opportunity for the newer remote sensing data to be used in complement with traditional systems.

From the review, the majority of studies have focused on mapping and monitoring at the local levels which therefore offers a limited scope. Efforts in monitoring at regional and global scales are very important in order to provide more comprehensive overview on current extent, dynamics and changes of mangroves over a period of time. Furthermore, in response to effects of climate change, more research should focus on assessment methods to develop standard indicators of mangrove change. However, despite these limitations, many opportunities remain in promoting advancement, technology, application and science of remote sensing for mangroves. This can be achieved through the formation of a strong global mangrove research network which actively provides mutual support and is working towards achieving a common goal, especially in protecting the unique mangrove ecosystems.

References

Aldrich RC (1979) Remote sensing of wild land resource: a state-of-art review. USDA For. Serv., Rocky Mountain For. and Range Exp. Sta., Fort Collins, CO, Gen Tech Rep RM-71, p 56

Ardö J (1992) Volume quantification of coniferous forest compartments using spectral radiance recorded by Landsat Thematic Mapper. Int J Remote Sens 13(9):779–786

Baltsavias E, Pateraki M, Zhang L (2001) Radiometric and geometric evaluation of IKONOS geo images and their use for 3D building modelling. http://www.photogrammetry.ethz.ch/general/persons/jana/isprs/tutmapup/ISPRS_tutorial_Baltsavias_hannover.pdf. Accessed 20 May 2013

Bolstad PV, Lillesand TM (1992) Improved classification of forest vegetation in northern Wisconsin through a rule-based combination of soils, terrain, and Landsat Thematic Mapper data. For Sci 38:5–20

Bong PP (1991) Application of remote sensing techniques to study the changes of mangrove forest at Canau Peninsula, South Vietnam, In: Application of remote sensing in Asia and Oceania—Environ Change Monitoring. pp 282–285

Brockhaus JA, Khorram S (1992) Comparison of SPOT and Landsat-TM data for forest inventories. Int J Rem Sen 13(16):3035–3043

CNES (2002) Centre national d'etudes spatiales: Dossier SPOT 5. http://www.cnes.fr. Accessed 20 Apr 2013

Conchedda G, Durieux L, Mayaux P (2008) An object based method for mapping and change analysis in mangrove ecosystems. ISPRS J Photogram Rem Sens 63:578–589
Czaplewski RL (1999) Multistage remote sensing: toward an annual national inventory. J For 97(12):44–48
Dahdouh-Guebas F, Verheyden A, Kairo JG, Jayatissa LP, Koedam N (2006) Capacity building in tropical coastal resource monitoring in developing countries: a re-appreciation of the oldest remote sensing method. Inter J Sustain Dev World Ecol 13:62–76
Danson FM (1987) Preliminary evaluation of the relationships between SPOT-1 HRV data and forest stand parameters. Int J Rem Sens 8(10):1571–1575
Darus A (1989) Application of Landsat MSS data in land use mapping: A Malaysian experience. In: Proceedings of the 10th Asian Conference on Remote Sensing, Malaysia, Kuala Lumpur. 23–29 Nov 1989. pp 81–88
Euroimage (2002) QuickBird: the world's highest resolution commercial satellite. http://www.eurimage.com/Products/product_pdf/qb.pdf. Accessed 15 April 2103
Everitt JH, Yang C, Sriharan S, Judd FW (2008) Using high resolution satellite imagery to map black mangrove on the Texas Gulf Coast. J Coast Res 24:1582–1586
FAO (1996) Forest resources assessment 1990: survey of tropical forest cover and study of change processes. FAO For. Pap. 130, Rome
FAO (1998) The status of the forest resources assessment in the South-Asia sub-region and the country capacity building needs. In Proceedings of GCP/RAS/162/JPN Reg. Workshop, FAO, India
Franklin J (1986) Thematic Mapper analysis of coniferous forest structure and composition. Int J Rem Sens 7(10):1287–1301
Gholz HL, Vogel SA, Cropper WP, McKelvey K, Ewel KC, Teskey RO, Curran PJ (1991) Dynamics of canopy structure and light interception in *Pinus elliottii* stands, North Florida. Ecol Monograp 6:33–51
Green EP, Clark CD, Mumby PJ, Edwards AJ, Ellis AC (1998) Remote sensing techniques for mangrove mapping. Int J Remote Sens 19:935–956
Guidon B, Edmonds CM (2002) Large-area land-cover mapping through scene-based classification compositing. Photogram Eng Remote Sens 68(6):589–596
Hamdan O, Khali Aziz H, Abd Rahman K (2011) Remotely sensed L-Band SAR data for tropical forest biomass estimation. J Trop For Sci 23(3):318–327
Hamilton LS, Snedaker SC (Eds) (1984) Handbook for mangrove area management. IUCN/UNESCO/UNEP, Honolulu
Hartono (1994) The use of SPOT image for mangrove inventory in Cimanuk delta West Java, Indonesia. Indon J Geog 26(68):11–26
Hartono, Muljosukojo B (1990) Monitoring disappearance of mangrove by remote sensing. Indones J of Geogr 21(61):15–32
Held A, Ticehurst C, Lymburner L, Williams N (2003) High resolution mapping of tropical mangrove ecosystems using hyperspectral and radar remote sensing. Int J Rem Sens 24:2739–2759
Heller RC, Ulliman JJ (1983) Forest resource assessments. Man Remote Sens (2nd ed) 2:2229–2324
Heumann BW (2011) Satellite remote sensing of mangrove forests: recent advances and future opportunities. Prog Phy Geograp 35(1):87–108
Horler DNH, Ahern FJ (1986) Forestry information content of thematic mapper data. Int J Rem Sens 7(3):405–428
Huang X, Zhang LP, Wang L (2009) Evaluation of morphological texture features for mangrove forest mapping and species discrimination using multispectral IKONOS imagery. IEEE Geosci Rem Sens Letters 6:393–397
Hyyppa J, Hyyppa H, Litkey P, Yu X, Haggren R, Petri R, Pyysalo U, Pitkanen J, Maltamo M (2004) Algorithms and methods of airborne laser scanning for forest measurements. In: Proceedings of the ISPRS working group VIII/2 on laser scanners for forest and landscape assessment, Freiburg, Germany. XXXVI (8/W2)
IUFRO (1994) International guideline for forest monitoring. IUFRO World Series. vol 5

Jakubauskas ME, Price KP (1997) Empirical relationships between structural and spectral factors of Yellowstone lodgepole pine forests. Photogram Eng Rem Sens 63(12):1375–1381

Kätsch C, Vogt H (1999) Remote sensing from space—present and future applications in forestry, nature conservation, and landscape management. Southern Afr For J 185:14–27

Kanniah KD, Wai NS, Shin A, Rasib AW (2007) Per-pixel and sub-pixel classifications of high-resolution satellite data for mangrove species mapping. App GIS 3(8):1–22

Kamaruzaman J, Mohd Rasol RF (1995) Satellite remote sensing of deforestation in the Sungai Buloh forest reserve, Peninsular Malaysia. Int J Rem Sens 16(11):1981–1997

Kamaruzaman J, D'Souza G (1996) Quantifying disturbed hill dipterocarp forest lands in Ulu Tembeling, Malaysia with HRV/SPOT Images. Photogram Rem Sens 51:39–48

Keiner LE, Yah XH (1998) A neural network model for estimating sea surface chlorophyll and sediments from thematic mapper imagery. Rem Sens Env 66(2):153–165

Kohl M, Kushwaha SPS (1994) A four-phase sampling method for assessing standing volume using Landsat-TM data, aerial photography and field assessments. Commonwealth For Rev 73(1):35–53

Komiyama A, Ong JE, Poungparn S (2008) Allometry, biomass, and productivity of mangrove forests: a review. Aquatic Bot 89:128–137

Kovacs JM, Flores-Verdugo F, Wang JF, Aspden LP (2004) Estimating leaf area index of a degraded mangrove forest using high spatial resolution satellite data. Aquatic Bot 80:13–22

Kovacs JM, King JML, Santiago FF de, Flores-Verdugo F (2009) Evaluating the condition of a mangrove forest of the Mexican Pacific based on an estimated leaf area index mapping approach. Environ Monit Assess 157:137–149

Kuenzer C, Bluemel A, Gebhardt S, Vo Quoc T, Dech S (2011) Remote sensing of mangrove ecosystems: a review. Rem Sens 3:878–928

Kushwaha SPS (1987) Remote sensing of shifting cultivation on North-Eastern India, In: Application of remote sensing in Asia and Oceania—Environmental Change Monitoring. pp 61–77

Lachowski HM, Dietrich DL (1978) Forest inventory of the Philippines using satellite imagery. General Electric Company, Beltsville, p 116

Lal JB, Singh J, Gulati AK, Projapati RC (1990) Deforestation study in Kodagu district of Kartanaka using Landsat MSS data. Ind For 116(6):487–493

Leckie DG (1990) Advances in remote sensing technologies in forest surveys based on satellite imagery. Can J For Res 20(4):464–483

Leckie DG (1998) Forestry applications using imaging radar. In: Henderson FM, Lewis AJ (eds) Man of remote sensing, 3rd ed., vol. 2. John Wiley and Sons, NY, pp 435–519

Leckie DG, Gillis MD (1995) Forest inventory in Canada with emphasis on map production. For Chron 71(1):74–88

Lee TM, Yeh HC (2009) Applying remote sensing techniques to monitor shifting wetland vegetation: a case study of Danshui River estuary mangrove communities. Taiwan Ecol Eng 35:487–496

Li X, Yeh AGO, Wang S, Liu K, Liu X, Qian J et al (2007) Regression and analytical models for estimating mangrove wetland biomass in South China using RADARSAT images. Int J Rem Sens 28:5567–5582

Lidov L (1999) IKONOS satellite launches into space: News releases. http://www.spaceimaging .com/newsroom/releases/1999/inorbit.htm. Accessed 10 Apr 2013

Lillesand TM, Kiefer RW (2000) Remote sensing and image interpretation, 4th Ed. John Wiley and Sons, NY, p 724

Lindgren DT (1985) Land use planning and remote sensing. Martunus Nijhoff Pub. Dordrech, Netherlands, p 176

Lucas RM, Ellison JC, Mitchell A, Donnelly B, Finlayson M, Milne AK (2002) Use of stereo aerial photography for quantifying changes in the extent and height of mangroves in tropical Australia. Wetlands Ecol Manage 10:161–175

Lucas RM, Mitchell AL, Rosenqvist A, Proisy C, Melius A, Ticehurst C (2007) The potential of L-band SAR for quantifying mangrove characteristics and change: case studies from the tropics. Aquatic Conserv 17:245–264

Mahmood NN, Bruneau M, Lemen H (1983) A pilot study on the use of satellite remote sensing data for agroecological mapping in Peninsular Malaysia. MARDI Special Report 2:p 31
Malingreau JP (1986) Global vegetation dynamics: satellite observation over Asia. Int J Rem Sens 7:1121–1146
Miller LD, Nualchawee K, Tom C (1978) Analysis of the dynamics of shifting cultivation in the tropical forests of northern Thailand using landscape modelling and classification of Landsat Imagery. NASA Tech. Memo 79545, Greenbelt
Moeur M (1987) Nearest neighbour inference for correlated multivariate attributes. In: Proceedings of IUFRO Conference on Forest Growth Modelling and Prediction, Minneapolis, MN. 23–27 Aug 1987. USDA For Serv Gen Tech Rep NC-120. pp 716–723
Morain SA, Klankamsoon B (1978) Forest mapping and inventory techniques through visual analysis of landsat imagery: examples from Thailand. In: Proceedings 12th Int Sym for Remote Sensing Environ. Ann Arbour, MI, pp 417–426
Moreau N, Vercesi L (1989) Cartographie des mangrove de guinea a l'aide du satellite SPOT 1, in photo-interpretation No. 1989–1 fascicule 4, Janvier Fervier 1989 (as cited by Hartono 1994)
Mougin E, Proisy C, Marty G, Fromard F, Puig H, Betoulle JL, Rudant JP (1999) Multifrequency and multipolarization radar backscattering from mangrove forests. IEEE Trans Geosci Rem Sens 37:94–102
Muinonen E, Tokola T (1990) An application of remote sensing for communal forest inventory. In: Proceedings of the Usability of Remote Sensing for Forest Inventory and Planning, SNS/IUFRO Workshop Umeå, Sweden. 26–28 Feb 1990. pp 35–42
Neukermans G, Dahdouh-Guebas F, Kairo JG, Koedam N (2008) Mangrove species and stand mapping in Gazi bay (Kenya) using QuickBird satellite imagery. J Spat Sci 53:75–86
Nichol CJ, Rascher U, Matsubara S, Osmond B (2006) Assessing photosynthetic efficiency in an experimental mangrove canopy using remote sensing and chlorophyll fluorescence. Trees-Stru Func 20:9–15
Nilsson M (1997) Estimation of wood volume using satellite spectral data: a simulation study. PhD dissertation, Swedish Univ of Agric Sci, Sweden, p 54
NRSA (1983) Nation-wide mapping of forest and non-forest areas using Landsat false colour composites for the periods 1972–1975 and 1980–1982. Project Rep, Nath Remote Sensing Ag Hyderabad, India
Pitt DG, Wagner RG, Hall RJ, King DJ, Leckie DG, Runesson U (1997) Use of remote sensing for forest vegetation management: a problem analysis. For Chron 73(4):459–479
Poso S, Paananen R, Samila M (1987) Forest inventory by compartments using satellite imagery. Silva Finnica 21:69–94
Proisy C, Mougin E, Fromard F, Karam MA (2000) Interpretation of polarimetric radar signatures of mangrove forests. Rem Sens Env 71:56–66
Proisy C, Mougin E, Fromard F, Trichon V, Karam MA (2002) On the influence of canopy structure on the radar backscattering of mangrove forests. Int J Rem Sens 23:4197–4210
Proisy C, Couteron P, Fromard F (2007) Predicting and mapping mangrove biomass from canopy grain analysis using Fourier-based tectural ordination of IKONOS images. Remote Sens Environ 109:379–392
Rao, PPN, Jayaraman V, Chandrasekhar MG (1991) Remote sensing applications to natural resources development in India, In: Application of remote sensing in Asia and Oceania—Environ Change Monitoring. pp 131–138
Ripple WJ, Wang S, Isaacson DL, Paine DP (1991) A preliminary comparison of Landsat TM and SPOT-1 HRV multispectral data for estimating coniferous forest volume. Int J Rem Sens 12(9):1971–1977
Rodriguez W, Feller IC (2004) Mangrove landscape characterization and change in Twin Cays, Belize using aerial photography and IKONOS satellite data. Atoll Res Bull 513:1–22
Saenger P, Hegerl EJ, Davie JDS (1983). Global status of mangrove ecosystems. Commission on Ecology Papers No.3. IUCN, Switzerland, p. 88

Salas WA, Ducey MJ, Rignot E, Skole D (2002) Assessment of JERS-1 SAR for monitoring secondary vegetation in Amazonia: II. Spatial, temporal, and radiometric considerations for operational monitoring. Int J Rem Sens 10(23):1381–1399
Salleh (1976) Preliminary interpretation of ERTS imagery of Peninsular Malaysia. Malaysian Forester 39:13–16
Scheer LA, Acka A, Feldkotter C (1997) Efficient growing stock estimation from satellite data employing two-phased sampling with regression. Geo Inform Sys 10(3):22–25
Simard M, Zhang KQ, Rivera-Monroy VH, Ross MS, Ruiz PL, Castañeda-Moya E, Twilley RR, Rodriguez E (2006) Mapping height and biomass of mangrove forests in everglades national park with SRTM elevation data. Photogramm Eng Rem Sens 72:299–311
Simard M, Rivera-Monroy VH, Mancera-Pineda JE, Castaneda-Moya E, Twilley RR (2008) A systematic method for 3D mapping of mangrove forests based on Shuttle Radar Topography Mission elevation data, ICEsat/GLAS waveforms and field data: application to Cienaga Grande de Santa Marta, Colombia. Rem Sens Environ 112:2131–2144
Smith TJ III, Whelan KRT (2006) Development of allometric relations for three mangrove species in South Florida for use in the Greater Everglades Ecosystem restoration. Wetlands Ecol Manag 14:409–419
Song C, White B, Heumann BW (2011) Hyperspectral remote sensing of salinity stress on red (*Rhizophora mangle*) and white (*Laguncularia racemosa*) mangroves on Galapagos Islands. Rem Sens Letters 2(3):221–230
Suratman MN (2008) Carbon sequestration potential of mangroves in Southeast Asia. In: Bravo F, Jandl R, LeMay V, Gadow von K (eds) Managing forest ecosystems: the challenge of climate change. Springer, Netherlands, pp 297–315
Sylvander R, Bystrom M, Hogberg P (1988) The use of satellite imagery for delineation of vegetation types in eastern Tanzania. Section of Forest Mensuration and Management Rep., No. 17
Temu AB (1981) Double sampling with aerial photographs in estimating wood volumes in miombo woodlands. Div of For Rec No. 22. Morogoro.
Thompson MD, MacDonald BC, Jefferies WC (1993) Progress in operational forest mapping using airborne radar in tropical regions, In: Proceedings of Int Symp Operationalization of Remote Sensing. pp 163–175
Toan LT (1995) Assessment of ERS-1 SAR data for forest studies in Southeast Asia. Earth Obs Q 48:16–20
Tokola T, Pitkänen J, Partinen A, Muinonen E (1996) Point accuracy of a non-parametric method in estimation of forest characteristics with different satellite materials. Int J Remote Sen 17(12):2333–2351
Tomppo E (1990) Designing a satellite image-aided national forest survey in Finland. In: Proceedings of the Usability of Remote Sensing for Forest Inventory and Planning, SNS/IUFRO Workshop in Umeå, Sweden. 26–28 Feb 1990. pp 43–47
Toutin T, Cheng P (2000) Demystification of IKONOS. Earth Observation Magazine. http://www.eomonline.com/Common/Archives/July00/toutin.htm. Accessed 10 April 2013
Trotter CM, Dymond JR, Goulding CJ (1997) Estimation of timber volume in a coniferous plantation forest using Landsat TM. Int J Rem Sens 18(10):2209–2223
USGS (2000) Satellite image of environmental change: remote sensing before Landsat 1. http://edc.usgs.gov/earthshots/slow/Help-GardenCity/remotesensing. Accessed 19 April 2013
Vaiphasa C, Ongsomwang S, Vaiphasa T, Skidmore AK (2005) Tropical mangrove species discrimination using hyperspectral data: a laboratory study. Estuar Coastal Shelf Sci 65(1–2):371–379
Wacharakitti S, Morain SA (1978) Procedures for land-use analysis in developing countries: Example for Southeast Asia. In: Proceedings of the 12th International Symposium of Remote Sensing of the Environ, Manila, the Philippines. Environ Resour Inst of Michigan, Ann Arbour, MI. Vol 1. pp 587–595
Wang L, Sousa WP, Gong P, Biging GS (2004) Comparison of IKONOS and QuickBird imagery for mapping mangrove species on the Caribbean coast of Panama. Rem Sens Env 91:432–440
Wang L, Silvan-Cardenas JL, Sousa WP (2008) Neural network classification of mangrove species from multi-seasonal IKONOS imagery. Photogram Eng Rem Sens 74:921–927

Wicaksono P, Danoedoro P, Hartono H, Nehren U, Ribbe L (2011) Preliminary work of mangrove ecosystem carbon stock mapping in small island using remote sensing: above and below carbon stock mapping on medium satellite resolution. In: Proceedings SPIE 8174, Remote Sensing for Agriculture, Ecosystems, and Hydrology XIII, 81741B
Wilkie DS, Finn JT (1996) Remote sensing imagery for natural resources monitoring. Columbia Univ. Press, NY, p 295
Zahriah A, Cheah H, Layden D (1989) Forest mapping and change detection using satellite data. In: Proceedings of the 10th Asian conference of remote sensing (ACRS), Malaysia

Management Strategies for Sustainable Exploitation of Aquatic Resources of the Sundarbans Mangrove, Bangladesh

M. Enamul Hoq

Contents

Abstract The Sundarbans forms an impenetrable saltwater swamp of tidal estuaries and creeks, and is the largest mangrove forest in the world, covering about 10,000 sq km in Bangladesh and West Bengal, India. This transboundary ecosystem is extremely important both ecologically and economically as a nursery and breeding area for key fisheries including those of the Bay of Bengal. The site is notable for a long history of scientific management and wise use of its wetland resources with protected areas established along the southern periphery of Bangladesh. But a long-term ecological change is taking place in the Sundarbans, due to the eastward migration of the Ganges river. Forest cover, species diversity and ecosystem function have declined despite several forest policies, laws and management plans enacted to protect them. The effectiveness of these regulations is limited due to poor implementation. The current management situation includes a moratorium on wood extraction. For fishing, recreation and non-wood forest products exploitation is regulated through permits, fees, and forest patrols. Extraction is prohibited in the wildlife sanctuaries.

M. E. Hoq (✉)
Bangladesh Fisheries Research Institute, Mymensingh,
Mymensingh 2201, Bangladesh
e-mail: hoq_me@yahoo.com

I. Faridah-Hanum et al. (eds.), *Mangrove Ecosystems of Asia*,
DOI 10.1007/978-1-4614-8582-7_15,

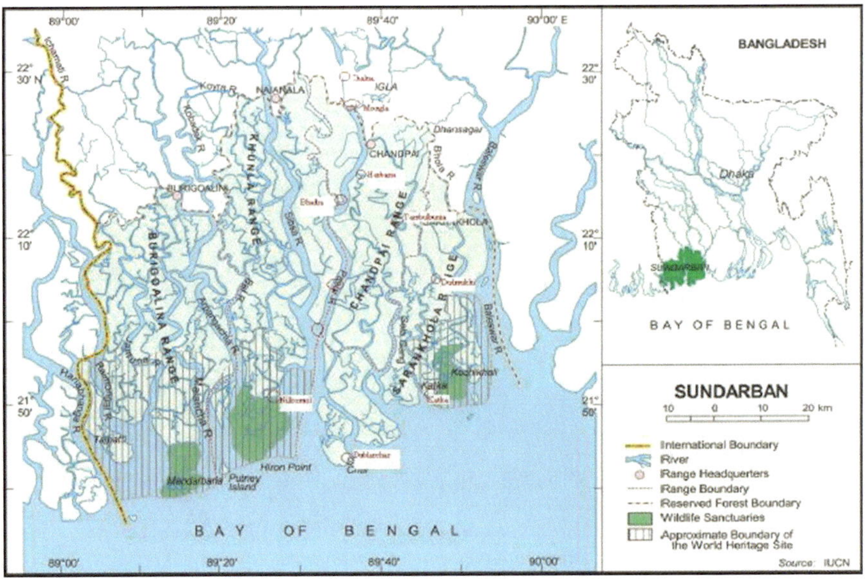

Fig. 1 The Sundarbans with the World Heritage Site

Rivers and canals identified as key fish breeding grounds have been restricted as well. In the past 15 years, land use in the Sundarbans impact zone has been affected by a significant transformation from rice-based farming systems to shrimp aquaculture, with numerous adverse social and environmental effects, including increased pressure from unsustainable extraction of resources from the Sundarbans. Currently, no monitoring and evaluation framework is being implemented to ensure that resource conditions and uses are within sustainable limits. Dialogues are underway between Bangladesh and India with a view toward enhanced collaboration in the management of this important world heritage site.

1 Introduction

The Sundarbans is the single largest tract of natural mangrove forest in the world, consisting of a total area of 601,700 ha which represents 4.07 % of the land mass and 40 % of total forest area of Bangladesh. The mangrove ecosystems of the Sundarbans support around 334 species of flora and 375 species of fauna and also supply food and support livelihood to 3.5 million local people in addition to precious wood and non-wood forest products. The forest was declared a Ramsar site by the Convention of Wetlands of International Importance in 1992, and in 1997, three wildlife sanctuaries encompassing a total of 139,698 ha was declared a World Heritage Site by UNESCO (Fig. 1). Although most of the country's forests have been an intimate interspersion of human habitations and cultivation, there are neither settlements inside the Sundarbans Reserve Forest (SRF) nor has the forest land been subjected to encroachment since their gazetting as a reserve forest. However, the traditional dependency of the local population from peripheral areas for their livelihood including fishing has been an important aspect of the Sundarbans mangrove. The SRF also represents the country's largest single carbon asset pool to market international carbon trade. Wood and forest products harvested from the SRF have in the past been a major resource, but currently a logging ban is in place (until 2015). The Sundarbans ecosystem is an important nursery and breeding area for key fisheries of coastal and marine waters of the Bay of Bengal. Despite continued degradation, the Sundarbans contributes 3 % to the country's gross domestic product out of 5 % contribution of the country's forestry sector (Roy and Alam 2012). In recent years, population pressure, economic development, and unsustainable management practices have resulted in the rapid depletion and degradation of the Sundarban's resources, threatening its biodiversity and the livelihood of the local community.

In recent times, concerns have been voiced by fishermen over the apparent declining stocks and productivity of fisheries in and around the Sundarbans. Although there is inadequate monitoring of fish stocks, fishermen have noted that they are spending more time and efforts to capture fewer and smaller fish. The current Integrated Forests Management Plan (IFMP) for the SRF, developed in 1997 for the period 1998–2010, the Conservation Management Plan (CMP) for the period 1997–2002, and the Forest Resource Master Plan (FRMP) that was developed in 1993 to cover the periods 1993–2012 do not provide sufficient guidance to address the present issues, the Department of Forest (FD) presently faces in managing the Sundarbans. Climate change, food security, recreation and tourism, co-management, biodiversity conservation, and carbon financing are the issues for which specific directions need to be incorporated into the Sundarbans' management policies. The present paper presents an in-depth analysis of the management measures, strategies and policy directions of the Sundarbans with special focus on aquatic resources of this unique mangrove ecosystem.

2 Legal Status of the Sundarbans Reserved Forest

There is a long and varied history of the legal status of the SRF recorded as far back as the Mughal empire period (1203–1538) when the whole forest area was leased to a local king (Hossain and Ahmed 1994). Records on reclamation, forest clearing and settlement stem from the late eighteenth century, and the first management legislation was the Charter of Indian Forests and the Forest Act 1876, according to which the Sundarbans was declared a Reserved Forest by the Government of British-India. Subsequently, systematic forest management became official structure, and Heining (1892) in his working plan recorded important events as the legal background to what eventually became the Forest Act of 1927 which makes provisions for Reserve Forests

Table 1 Legal changes in the management of the Sundarbans (1876–2005). (Roy and Alam 2012)

Periods	Legal initiatives	Major objectives	Major outcomes
Pre-British rule (before 1757)	No management	Resource extraction	Resource extraction
British Rule (1757–1947)	Charts of India Forest, 1855	Conservation idea generation Controlling the resources	Awareness with importance realization and first regulations regarding felling trees for revenues
Indian Forest Act 1894	Declaration of 'Reserve Forest', 1875–76 under Indian Forest Act, 1894	Introduction of formal forest policy to be administered Targeting benefits with commercial management of wood and non-wood forest products for public at large and for the local people under regulations and rights	Resource extraction
Pakistan Rule (1947–1971)	Forest Policy of Pakistan, 1955 Revised Forest Policy of Pakistan, 1962	Classification of the Sundarbans on the basis of its utility and objectives Acceleration of timber harvesting Speed up regeneration for increased harvesting Ignorance of the principle of sustainable forest use and rights of local people	Over exploitation of forest resources from the Sundarbans Protection of wildlife and habitats Realization of overuse Ecological degradation
Bangladesh Rule (1971-present)	National Forest Policy, 1979	Qualitative improvement based on modern trend and technology for extraction and utilization of forest resources Coastal mangrove plantation	There were inconsistencies as conservation leaves little incentive to expand forest-based industries and becomes detrimental to forest health by increasing degradation through illegal harvesting Inappropriate land tenure agreement caused illegal felling of mangrove trees and encroachment of the land
	Revised National Forest Policy, 1994	Multi-dimensional uses of its resources including water and fish Keeping the bio-environment intact and consideration of global warming and climate change for its existence Use of appropriate extraction technology Identification of protected areas Ensuring participation of local people	Sustainable management

and their legal status (FAO 1998). Table 1 highlights the chronological changes of policies in the management of the Sundarbans.

The first Forest Policy for Sundarbans was promulgated in 1894 under the Government of British-India that provided the foundation for all future acts and rules which underpinned the administrative basis of the SRF. The principal policy directives and legislations involved in integrated management of the SRF are:

I. The Forest Act, 1927 and its amendments
II. The Protection and Conservation of Fish Act, 1950
III. The Wildlife Ordinance 1973 and The Wildlife (Preservation) Act, 1974
IV. The Protection and Conservation of Fish (Amendment) Ordinance, 1982
V. The Marine Fisheries Ordinance, 1983
VI. The Brick Burning Act, 1991
VII. The National Forest Policy of Bangladesh, 1994
VIII. The National Environment Act, 1995
IX. The Environment Conservation Rules, 1997
X. The National Conservation Strategy, 2005

3 History of the Sundarbans' Management Regulations and Policies

In Southeast Asia, scientific management of any mangrove forest was first initiated in the Sundarbans mangrove in the 1870s when a Forest Management Division under the Government of British-India was established exclusively for the management of mangrove forests of the Sundarbans. Specific management prescriptions were gradually formulated for the sustainable exploitation of the wood and non-wood forest products. These prescriptions were primarily structured with a view to ensuring restocking of the harvested area through natural forest regeneration. The first 10 years of a forest management plan was prepared in 1893 for the SRF. Other mangrove forests in Asia have subsequently come under scientific management. Matang forest in Peninsular Malaysia and Irrawaddy delta mangrove forest in British-Burma (presently Myanmar) were brought under scientific management in 1920 while mangrove formations in Ca Man Peninsula in the Mekong delta of Vietnam were brought under scientific management in 1930. Also the first management plan for mangrove forests in Indonesia was prepared for Segara Anakah in 1930. In 1995, Myanmar has promulgated a new Forest Policy focusing on the protection of nature and sustainability of natural resources, with participation of local people for biodiversity conservation. Other countries like India, Pakistan and Thailand have been practicing scientific management of mangrove forests for several decades.

In practice, the responsibility of mangrove management at the national level has been assigned on a sectoral basis to executing agencies of the government, for example, Forestry, Fishery or Agriculture departments. Mangrove forests are very important to the livelihoods of mangrove dwellers, those who live close to mangroves and to the economy of the countries. To maintain ecological balance, all countries having mangroves have set management plans for using the mangrove resources on a sustained yield basis. The concept of sustainable use involves sustainable harvest of wood and non-wood products while at the same time maintaining the ecosystem in as natural or close to its original state, as possible. The formulation and implementation of adequate legislation and policies concerning management of mangrove ecosystems is up to the government, but stakeholder participation must be arranged at both the management planning and implementation stages. In most of the countries, the major stakeholders are forestry, fisheries, tourism, agriculture, mining and industry sectors with, in many cases, representation from both the public and private bodies. Some countries have established protected areas (PAs) with aquatic resources and many have identified priorities for protection and conservation in national environmental action plans and national conservation strategies.

Historically, the SRF has always been managed as a contiguous block of mangrove forest with no permanent human habitation inside. Afterwards, management plans aimed to assure the sustainable harvesting of forest products and maintained its coastal zone in a way that meets the need of the local community. In the SRF, the first forest policy was declared in 1894 to administrate the forest for the overall benefit of the forest dependent population, and the interest of the local community got high priority. The main objective of the policy was to extract available resources which eventually caused degradation of the forest. To avoid degradation, the 'Lloyd Plan' and the 'Working Plan' were introduced during the period of 1904–08 as a basis of its founding administration but failed to reverse or reduce the degradation. After the British colonial period, during the Pakistan era (1947–71), forest policy was enacted for huge extraction of resources where rights of the local community were denied (Kabir and Hossain 2008). After the independence in 1971, Bangladesh adopted its first National Forest Policy in 1994, but again failed to address issues such as sustainability, community participation and their livelihood. Consequently, due to lack of defined ownership and established rights, huge dependent populations of the Sundarbans have not been able to be a part of the strategies and activities aimed at conserving the forest and using its resources sustainably.

The fisheries status and management in the SRF was first scientifically studied in 1994 through a FAO assisted project (BGD/84/056), from which fisheries structure of the Sundarbans was understood; although Rabanal (1984), under another FAO assisted project (FAO:TCP/BGD/2309), observed some basic fisheries of the SRF. In-depth study on fisheries of the Sundarbans was conducted by Chantarasri (1994), where description and maximum sustainable yield (MSY) of major fisheries resources of the Sundarbans were estimated and some management issues were addressed. The data produced by that study were very comprehensive based on one year of work and after that no long-term data collection was done, which could provide more information and ensure the sustainable exploitation of the aquatic resources. In the Sundarbans, the major problem for fisheries management is the absence of fishery experts with overall responsibility of fisheries management in the Sundarbans mangrove.

In order to ensure sustainable yield management of mangroves for coastal fisheries, the mangroves are kept for providing nutrients, breeding and nursery grounds as permanent habitat. Mangrove forest management is basically based on the science and skills of geology, pedology, climatology, hydrology, ecology, silviculture, forest technology and economics—in the exploitation and conservation of both wood and non-wood resources. About 40 % of the global mangrove forests are in Asia.

Many Asian countries have established many tools that can provide better baseline information which include remote sensing, aerial tracking and environmental impact assessment studies. Still now the Sundarbans is in a traditional management system with absence of such modern techniques.

4 Fisheries Management Options and Strategies

Under the FD, the SRF has two administrative divisions, namely, Sundarbans West Forest Division and the Sundarbans East Forest Division, with two Range Offices under each division. Administratively the whole SRF has four Range Offices and for better management the SRF has been divided into 55 ecological compartments. Basically, FD has been maintaining fisheries resource management in the Sundarbans in an integrated process of information gathering, planning, consultation and implementation with enforcement of the certain rules and regulations. A key aspect of fisheries management in SRF is implementation of some regulations with a view to earn certain yearly revenue income. Presently there is inadequate or no monitoring of fish stocks in the SRF, and fisheries resources has not been inventoried for a decade. Under the present management system of the FD, the following two types of fisheries based on area and season are covered by the fisheries licensing and revenue collection system:

I. Inshore fisheries: Production zone of the SRF waterways (except Wildlife Sanctuaries) and operates in all seasons except closures time period.
II. Off-shore fisheries: The winter fishery in the coast of the Bay of Bengal (i.e. Dubla Island) and fishing in estuaries and marine zone of the Sundarbans.

The yearly fisheries production data from the Sundarbans as recorded by the FD Range Offices showed sharp fluctuations from year to year. Some reduction in production figures is noticed, particularly in case of white fish, marketable shrimp, undersized shrimp and hilsa (*Tenulosa ilisha*). However, the assumption of fishermen in the Sundarbans is around a 50 % decrease in fish catch during the last 10 years and the reduction is somewhat less in the lower part and more in the upper part of the SRF. However, in general, the perception of the fishers' community, FD and local people reveals that production of fisheries resources in the Sundarbans have been decreasing over the years. In recent years, fishing with poison (organophosphate agro-chemicals) in the canals of the Sundarbans has increased and is recognized to be highly detrimental for the aquatic resources of the SRF. Moreover, there is an increasing trend in number of fishers as well as fishing efforts observed (Fig. 2).

From fisheries production of the SRF, total revenue earned during 2010–11 was around 37 million Taka (Fig. 3). It was also observed that average earnings from fisheries resources within Taka was 5.0–10.0 million during 1999–2000 to 2006–2007 and a remarkable increase of revenue from fishery production was observed in 2007–2008. Among the total fish catch from the SRF, white fish comprised the major portion (53 %) and was followed by dried fish (22 %), shell (18 %), crab (6 %) and shrimp constituted the least proportion (about 1 %) (Akhter 2012).

Fig. 2 Increase in fishers number during 2000–2010. (Forest Department 2010)

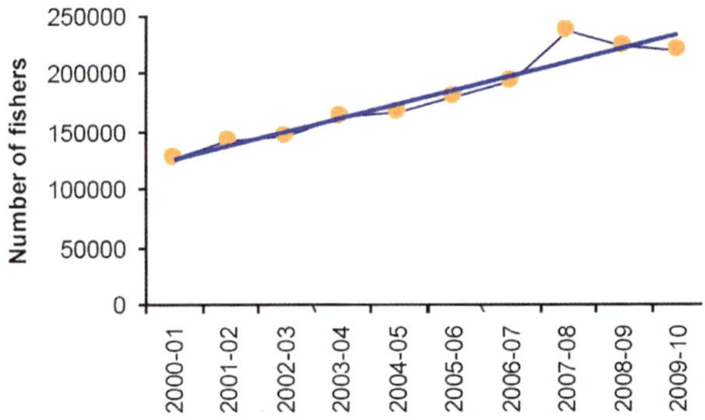

4.1 Major Components of Fisheries Management System for the Sundarbans

The Sundarbans mangrove has a distinct management history. In view of maintaining the ecological balance of mangrove resources and its sustainable utilization, the Government of Bangladesh has formulated and implemented different management policies and action plans (Fig. 4). The management of fisheries resources in the SRF from a technical point of view was first started in 1989 with the closing of 18 water canals within the SRF to accelerate fish breeding activity. A closed season for fish catch and wildlife sanctuary regulations were introduced during the last decade.

The fisheries policy and management execution practices in Bangladesh has some experience with implementation of various development projects which focus mainly on the process of development, the involvement of stakeholders and the possibilities for adapting existing policies and institutional structures. In that context, implementation of the New Management System under the Sundarbans Biodiversity Conservation Project (SBCP 2002) was a new approach for the improvement

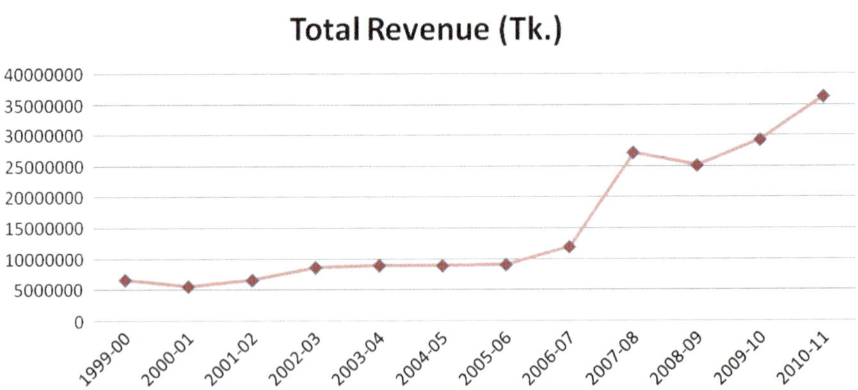

Fig. 3 Revenue from fishery resources of the Sundarbans from 1999–2000 to 2010–2011

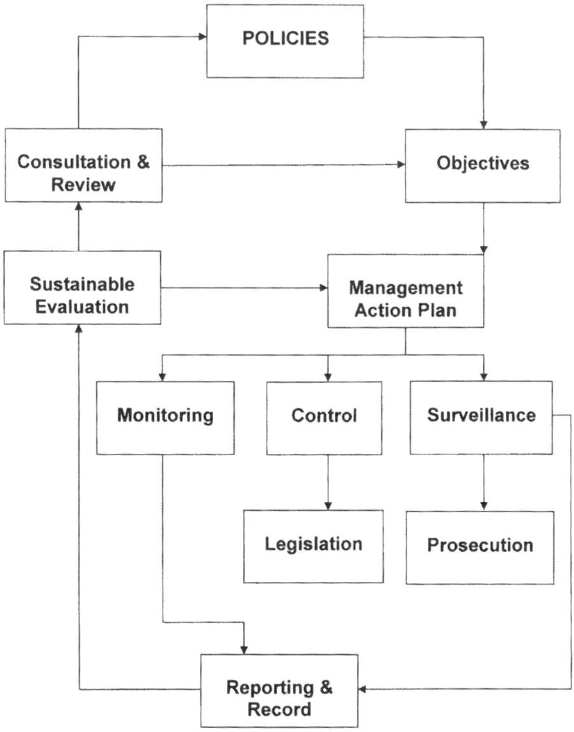

Fig. 4 Main features of fisheries management system for the Sundarbans

of management structure of the Sundarbans fishery. The SBCP was designed with the goal of securing the integrity of the environment, biodiversity and community participation of the Sundarbans. The main objective of the project was to develop a system for sustainable management and conservation of the SRF and the surrounding impact zone and near-shore marine waters. The SBCP was implemented for slightly over four years before it was suspended in September 2003 due to serious implementation delays and financial management problems. With respect to aquatic resource management, the project developed a draft fisheries management plan, a preliminary assessment of fishery stocks in the SRF and several other useful reference documents. Moreover, the Aquatic Resource Division (ARD) was established under the SBCP, and the FD staffs were trained in aquatic resources management both biologically and economically. The ARD was responsible for overall management of aquatic and fisheries resources based on sustainable utilization criteria, survey data and analysis. ARD was established with a regular work program, which consisted of aquatic biodiversity conservation, fisheries research, monitoring, control and surveillance, and fish processing and marketing. ARD also considered the aquaculture practice or mangrove silviculture practice as an alternative income source in their proposed plan and the goal for the aquaculture in Sundarbans is to reduce resource exploitation pressure in the SRF by creating a substantial number of employment and income opportunities in the impact zone (Sundarbans adjacent 17 Upazillas). After project cancellation,

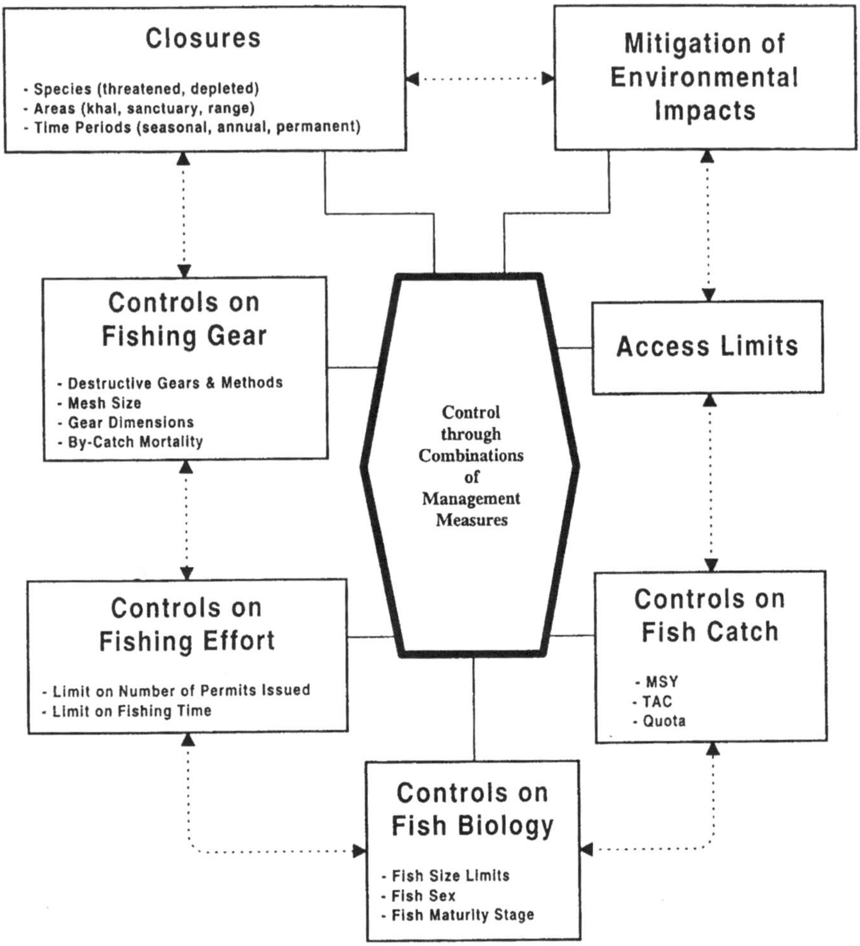

Fig. 5 Control regimes of the ARD fisheries management system. (SBCP 2002)

the ARD was dissolved and the functions were merged with the Wildlife and Nature Conservation Division of the FD. Control regimes of the ARD Fisheries Management System are presented in Fig. 5. The overall management structure of fisheries resources of the SRF is comprised of the following main features and activities:

- Establishment of management objectives and supporting policy guidance for the Sundarbans fisheries
- Developing a fisheries management action plan for each of the fisheries in the SRF
- Collection of catch data of fisheries production and biodiversity, fishing effort, and fish markets

- Developing and enacting fishing rules and regulations concerned with season closures, fishing gears, fishing effort, fish catch and access limits for fishing
- Monitoring illegal and unlicensed intruders (into the SRF and its 20 km Marine Zone), and enforcement of rules and regulations

4.2 Fish Catch Monitoring System

FD has a long history of a fish catch monitoring system in the SRF. The production monitoring system has changed since 1975. Before 1975, each Range Office-cum-Revenue Collection Station was equipped with a measuring scale, and the FD staffs weighed the catch of each fishing boat when it exited from the Sundarbans. The revenue fee was calculated based on the actual fish weight in a fishing boat. At that time, each fishing boat transported its catch out of the SRF individually. During 1975 when fish traders began entering the SRF with mechanized engine boats to collect catches from the fishermen and transport out large quantities of fish in ice it became impossible to get the actual fish catch weight. The FD then used a system of calculating boat maundage (maunds is a local measuring unit and 1 maund is approximately equals to 37.38 kg) for calculating the volume of the haul and adjusting this figure for fish weight (in maund) for fish catching/transport boats, and charging revenue fees based on a full hold. In order not to penalize individual fishermen who exited the SRF with no catch in their boats (having already sold the catch to fish transport boats inside the SRF), revenue fees were set at a low daily catch level.

At present FD does not monitor actual fish catches/production from the SRF. A main focus of the FD is on revenue collection and a set revenue fee is charged to fishermen/traders which is approximately related to a daily fish catch rate for different types of fisheries. There are two basic measures, which are required for fishing operations in the SRF. One is a boat license certificate (BLC) and the other is a fishing permit (PT).

BLC BLCs are issued for every year in July (at the beginning of the Govt. financial year) and are valid for only one year. The pre-requirement for a BLC is a certificate and attested photograph from a local public representative, i.e. Union Parisad Chairman. After having a BLC, fishermen have to pay annual fees for that BLC which is based on the 'Maundage' of their boat. Maundage (M) is calculated as follows:

$$M = L \times W \times H \times 0.356$$

where,

L Boat length (in feet)
W Maximum boat width (in feet)
H Boat height (in feet)

and for dry fish,

$$M(\text{dry fish}) = (L \times W \times H \times 2)/10$$

The BLC number is fixed on a yearly basis. Each year a pre-assessment on the production/catch of fish in the SRF is made based on the information from the fishers and observations by the FD personnel, and BLCs are issued on the basis of that assessment.

PT Every BLC holder is required a PT for fishing inside the waterways of the SRF. A PT typically has one week validity. A fishing permit specifies the type of gear and kind of fish species (group-wise) that may be caught by the permit holder, i.e., white fish fishing using gill nets and crab fishing using hook and line, etc. Payment for the PT at time of entry to forest covers only the entry fee for most of the fisheries.

After seven days of fishing, the boat returns to the same Forest Range Station where the permit was purchased. Payment of the fish catch revenue fees is done at the time of entry and is recorded on the certificate. Full payment of the catch is done at the time of exit and is recorded on the PT. The permit is surrendered to the local Forest Post Officer who gives the fishermen a certificate (CT) which shows the amount of fees paid and testifies that the fishing was carried out properly abiding all regulations. The fishermen then exit from the SRF and the permit process comes to an end. According to the FAO (1998) the permit limits are as follows:

- One BLC holder boat will get a maximum of eight times a fishing permit in a year for all gear and fishery types
- Maximum number of permits to be below 100,000 per year
- The maximum limit of permit for a BLC for a month will be restricted to three times
- Compliance is mandatory to the Bangladesh Fish Act of 1950 and its amendments
- Ensure the local community participation in fishing activities within the SRF

Revenue collection fees

- Entry fee: Tk. 7.00 (1 US$ = Tk. 78.00) per person per week for white fish and Tk. 6.00 for crabs
- Nominal catch fee is obligatory
- A total of 2.8 kg per person per day of which 10 % should be shrimp catch

Catch fee rates

- White fish (fin fish): Tk. 3.20 per kg
- Hilsa (*Tenualosa ilisha*), Sea bass (*Lates calcarifer*): Tk. 12.0 per kg
- Shrimp (small): Tk. 2.50 per kg
- Large shrimps (*Penaeus monodon, Macrobrachium rosenbergii*): Tk. 25.0 per kg, etc.

New BLC and permits can be issued in lieu of old BLC and permits that are either cancelled or surrendered but such new issuance should be within the prescribed limits of a particular Forest Station/Range. The catch/BLC/year versus number of BLC per sq km showed that in the SRF, the number of BLC has an increasing trend starting from 2004. The annual catch/BLC increased from 2006 through 2008, but decreased since 2009. A review analysis indicates that prior to overexploitation the FD could have provided BLC not more than 10 numbers per sq km per year. The first priority in issuing a new BLC will be given to those boat owners who live within a 5 km area around the Sundarbans. If the 5 km area does not fulfill the targeted BLC, then a 10 km area in the interface landscape should be considered which ensures local participation in the management of the Sundarbans. Precautionary management has to be adopted to protect the Sundarbans fisheries resources and to maintain sustainable availability of resources. Clearly, the maximum number of BLC permits has been reached and sufficient indications of overexploitation of the fishery in the SRF have been found. The suggestion to reduce the number of BLC permits below 110,000 per year is an important step towards sustainable fisheries management.

In case of a dry fishery, the Office Collector (OC) of the concerned Range office collects all the revenues at the end of the fishing season from fishing drying camps in the islands of the Sundarbans. For overstay in the SRF, except dry fishing, the fisherman is required to pay a fine of @ Tk 2.00 per person per day. If a fisherman fails to pay the fine, legal proceedings are initiated against him using the duplicate copy of the BLC and PT kept in the files of the FD. The court is informed of the infringement and issues a warrant. This legal prosecution is served by the police who attempt to apprehend the fisherman.

4.3 Controls on Fishing Effort

Prohibition on fishing in breeding season This is mainly done to facilitate and enhance fish breeding during the 1st of May to the 30th of June for the entire SRF. During January to February the entire SRF also remains closed for all types of crab fishing. Hilsa (*T. ilisha*) fishing is banned for September and October. Pangas catfish (*Pangasius pangasius*) and sea bass (*L. calcarifer*) fishing is banned on each alternate year (as per FAO 1998).

Destructive gears and methods This makes and offence to use or be in possession of a particular type of fishing gear or fishing method which has been determined to be destructive, undesirable, or otherwise illegal (i.e. poison/explosive fishing, *pata jal, behundi jal*) by the FD.

Mesh size This makes an offense to use or be in possession of a fish net which has a mesh size less than the prescribed lower limit (i.e. gill nets of less than 2 inch/5 cm bar mesh size) as fixed by the FD.

Gear dimension This makes an offense to use or be in possession of a fish net which has a mesh dimension of above or below the prescribed limit (i.e. gill net more than 100 m long) as fixed by FD.

Restriction on small mesh sized nets to reduce by-catch It is obligatory to fit special devices (i.e. turtle excluder device—TED) to fishing gears to reduce or eliminate non-targeted by-catch and other aquatic wildlife species, e.g. sea turtles, dolphins.

In the SRF, there are regular boat patrols by the FD in the rivers to check on fishing licenses and permits. But it is often inadequate for the supervision of fishing efforts in the Sundarbans. Negligence of the FD field staff in patrolling and enforcing regulations in the SRF during the overstay fishing practices and fishing in the declared closed *khals* (*canals*) and wildlife sanctuaries by fishermen severely hampered fishing effort control in the SRF. Presently, along with the FD, the Bangladesh Coast Guard, Bangladesh Navy and other law enforcing agencies play a vital role in patrolling/law and enforcement in the SRF. In near shore marine areas surveillance is done by the Coast Guard and Bangladesh Navy.

There are 16 entry stations along the north and north-eastern border of the SRF and the southern border is surrounded by the Bay of Bengal. But the main gap lies to the south-west border, where entry in no way is regulated along international boundaries, which follow the imaginary center line of rivers. This situation complicates surveillance of illegal entry and unofficial removal of forest resources. However, significant coordination among all such departments like the FD, Department of Fisheries, Coast Guard, Bangladesh Navy and other law enforcing agencies is highly desirable for successful surveillance in the Sundarbans areas.

PAs are being considered increasingly as a useful management and conservation tool in the context of disappointments with the standard management practices in a reserve forest like the Sundarbans. Although it is argued that the benefits of PAs for fisheries remain controversial, many workers have suggested that there has been some degree of success depending on the specific region and the type of fisheries in practice. No report is available so far as to the practice of PAs in the mangrove fisheries of Bangladesh, although presently the IPAC (Integrated Protected Area Co-management) project under financial support of USAID initiated such a concept in the SRF in addition to terrestrial-protected areas of Bangladesh (IRG 2010). PAs are considered to be a powerful tool for inshore fisheries management within the SRF. But it may not produce expected results in mangrove fisheries management because many marine fishes are dependent on the ecosystem of the Sundarbans for breeding and nursing. However, for the same reason, PAs in the mangroves can enhance the off-shore fisheries.

4.4 Legislations and Protections in SRF

In Bangladesh, several management practices including gear restrictions, season closer and/or banning were practiced for the development of the coastal and marine fisheries. To reduce the impacts of the Estuarine Set Bag Net (ESBN) fishery, a 30-mm cod-end mesh size was introduced on the assumption that the juveniles would escape. However, increased mesh size virtually resulted in no catch because this gear targets mainly the juveniles in the coastal waters. Banning of a particular fishing gear which involves few people but targets large masses of fish might be a useful option for management of a particular fish stock was also suggested (Islam 2003). The legislation which empowers FD to manage the inshore and off-shore fisheries in the SRF was started with the Forest Act 1878 and is still in practice with certain modifications (Table 2).

Table 2 Existing and proposed fisheries management and conservation rules in the Sundarbans Reserve Forest. (Hoq 2007)

Legislation	Summary of regulations
Indian Forest Act (1878)	Empowers the Forest Department to manage the inshore and offshore fisheries in the Sundarbans and near shore of 20 km marine waters
Hunting and Fishing Rules (1959)	A fishing permit is required to fish in the reserved or protected forests
	Royalty may be levied on fish caught in tidal waters of the reserved and protected forests
	It is illegal to use poison, explosives or fixed engine fishing gears, or to dam in the reserve and protected forests
Major Fisheries Regulations for the SRF	*Khal* Closure Regulation (1989): Closes 18 khals permanently for fishing to ensure fish breeding
	Collection & Export of Live Crab Regulation (1995): Closes the entire SRF for crab fishing from December to February to ensure crab breeding
	Closed Season Regulation (2000): Closes fishing in the entire SRF for five species (*P. pangasius, Plotosus canius, L. calcarifer, M. rosenbergii, Scylla serrata*) during 1st May to 30 June to ensure natural breeding
Wildlife Sanctuary Regulations (1999)	Fishing is permanently prohibited in the three wildlife sanctuaries of the SRF
Other Regulations for Fisheries in the SRF	It is illegal to place nets across a *khal* and thereby completely block it
	It is illegal to fix a rope transversely across a *khal*
FAO (1994) (BGD/84/056)	Introduction of closed season
	Introduction of protected zones, i.e. fish sanctuaries
	Introduction of minimum size limit of two species—30 cm for *L. calcarifer* and 10 cm for *J. argentatus*
	Restriction on number of gillnets
	Maintenance of exploitation rates for commercial species at current levels except *P. monodon* PL
	Coordination of regulatory powers of Forest Department and Department of Fisheries for life-cycle management of migratory fish stocks, i.e. *T. ilisha* and *L. calcarifer*
FAO (1998)	Closure of small *khals* (less than 30 m wide) for 12 months within 5 km radius of Forest Stations in the SRF, in alternating years
	Permanent closure of wildlife sanctuaries and any other protected areas
	Maintenance of records of permits issued and catch for individual fishermen
	Maintenance of annual harvest limit for various species, initially *T. ilisha*, all catfishes and mud crab
	Issuance of catch quota to individual fishermen based on a share of the total allowable catch (TAC)
	Restriction of shrimp fry catch from boundary rivers only
	Release of small sized fishes back to the water caught in shrimp fry collection nets
	Prohibition on harvesting of berried crabs or female crabs with egg
	Maintenance of minimum harvesting weight of crabs, 200 g for male and 120 g for female
	Enforcement of National Fish Act to ensure minimum harvesting size limits and closed seasons

Table 2 (continued)

Legislation	Summary of regulations
Sundarbans Biodiversity Conservation Project-ADB (2002)	*P. pangasius* and *L. calcarifer* harvest should be completely banned for five years Fishing of *T. ilisha* is to be closed from November to April, that of mud crab from December to February, that of *M. rosenbergii*, *Plotosus* spp. and *Mugil cephalus* in May/June Minimum size limits should be 23 cm for *T. ilisha*, 10 cm carapace width for male mud crab and 10 cm head length for male *M. rosenbergii* It should be illegal to catch or be in possession of female mud crab and female *M. rosenbergii* having eggs and live giant oysters All gear operated by fixed engine boats should be permanently prohibited. Gear having very small mesh netting which catch larvae of fish and shrimps should be permanently prohibited *L. calcarifer* fishing is restricted in the marine zone, sport fishing is restricted in the wildlife sanctuaries, long lining for white fish is restricted in Satkhira Range, and giant oyster collection is locally restricted by certain revenue collection Forest Stations Turtle exclusion devices are mandatory for set bag nets The only legal mesh size for all gill nets, lift nets, shore seines and set bag nets is 5 cm. No other mesh sizes (smaller or larger) are permitted Cast nets are not permitted for commercial fishing, but may be used for subsistence fishing by non-fishermen only (i.e. wood cutters, honey gatherers) Capture of wildlife, sea snakes, large sharks, large rays, large sawfish, and very large *L. calcarifer* and *P. pangsius* is strictly prohibited

5 Regulations and Policy Implications on the Sundarbans Fishery

5.1 Access Right by Local Community

The Sundarbans reserve forest and water areas are controlled by the Government through the FD. Traditional right of access to the Sundarbans mangrove is through membership in SIZ (Sundarbans Impact Zone) village community which is influenced by the local elite. These rights are not formally regulated, but are the birth right of the resident of the community (Roy et al. 2012). For the time being, this common property and access rights are no longer recognized in the tenure system. The privatizing of public lands, including some of the mangrove forest, evolved through the leasing system of government. Thus, multiple uses of common resources are transferred from traditional community to single use private control. Because of poverty, less education and political weakness of coastal communities, external capital has generally been used to extract resources, infrastructures or develop coastal lands. Consequently, most of the benefits of coastal development flow are away from such communities to the investors. As economic needs increased, common property values declined and the traditional rules governing access weakened to the point of ineffectiveness. Concurrently, the local elites, whose main interest had been in controlling access to the forest and its resources, shifted to rent-taking from the new resource extractors. The extractors increasingly came from urban areas far away from the mangrove forest. At the beginning of the 21st century, impoverishment of communities that depend upon the SRF is accompanied by increased exploitation, invention by outsiders and powerlessness of the local community. An overview of stakeholder matrix for the development of the SRF fisheries is presented in Table 3.

The management practices and regulations within the SRF should consider the concern of the local community, as so far, the traditional and cultural wisdom of the local people have sustained the unique ecosystem for generations. For the time being, the traditional management regime has been replaced by state monopoly and control, which has led to total commercialization of resource extraction and uses. In the beginning of the 1980s the commercial shrimp industry entered into the Bangladesh coast as well as in the SIZ. Shrimp farms were established in the areas of the Sundarbans, and at present is the major activity in the buffer zone (around 20 km of the SRF). The existing SRF management system has no policy guidelines regarding the shrimp industry and also has poor regulation and control on shrimp fry collection in the buffer zone.

The increased population with fewer alternative livelihood opportunities poses a serious threat to the Sundarbans mangrove, which is one of the main causes of climate change vulnerability of the local people and ecosystem. Moreover, dependence of local people in the forest is high (28 % of the population in the landscape are dependent on the SRF) and in future undoubtedly this dependence will increase many fold, which is likely to aggravate the existing pressure on the forest management and protection. There are more than 1.0 million people directly involved with the different resources extraction from the SRF. The pressure on the SRF for resources extraction has increased tremendously as the number of collectors has increased many-fold over recent decades, resulting in huge reduction in per capita resource collection from the SRF. With the high increase in living cost added to that scenario, the people and the community, especially that of the bottom layer in the value chains, tend to fall in the process of pauperization.

5.2 *Biodiversity and Conservation*

The collection of fisheries resources has significantly declined in the SRF. Some of the species are getting rarer or shifting their habitat. Number of harvesters, e.g., fishermen or worker for dry fish industry, increased many fold. It was estimated to be over 0.9 million fish collectors, most of which are fisher laborers. Other actors in the fish sector were estimated at more than 0.2 million, most of whom are *Farias/Beparis* (middlemen). Because of gradual displacement from agriculture due to increased salinity, a greater number of people is pouring into the landscape as resource collectors. Most SRF resource extractions are merely seasonal and consequently there is

Table 3 Key stakeholder matrix for the development and management of the Sundarbans Reserve Forests (SRF) fisheries

Stakeholder's category	Stakeholder	Responsibilities	Who are they	What they suppose to give	What they get	Impression and conclusions
Primary	Ministry of Environment & Forest	Policy & planning for the SRF	GOs	Policy, plan and services	Achievement of targets, financial assistance	Policy reflects conservation without consideration of users perception No integration with the Ministry of Fisheries & Livestock
Primary	Sundarban Forest Division and Aquatic Resource Division	Implementation of forest enacted policy & plan	GOs	Expert, service, assistance, management	Good-will	No fisheries expert Only revenue collection option
Secondary	Ministry of Fisheries & Live stock	Fisheries policy & plan making	GOs	Policy, plan and service	Achievement of targets, financial assistance	No specific policy for the SRF fisheries
Secondary	Coastal Fisheries Development Projects	Implementation of fisheries (inland and marine) development projects & making management action plan	GOs	Expert, service, assistance, training, technology	Good-will	No access inside the SRF No co-ordination with the Forest Department
Secondary	Bangladesh Navy and Coast guard	Border protection, law & order and life safety	GOs	Service (patrolling the Bay of Bengal and coast)	Good-will	Less equipped for effective surveillance
Secondary	FAO, WB, UNDP, ADB, GDC, GEF, WFC	External support for conservation, development and management	Dev. Partners	Expert, grant, assistance, higher studies, training, advice	Good-will	Less initiative for forest department capacity building
Secondary	NGOs (CARE, CARITAS, World Vision, RUPANTOR)	Local development for poverty alleviation	NGOs	Support, assistance, training, advice	Money, recognition	No activity for fish resource management
Primary	Fishermen	Fishing	User	Support	Income, skill, knowledge, training	No ownership attitude Destructive and low mesh size gear use Destruction of forest
Secondary	Dealer or trader or money investor	Trading & financing	User	Support	Income, profit	Pay lower price to fishers and high profit generation
Secondary	Shrimp farmer	Shrimp farming	User	Support	Sales volume, profits	Lower payment for resource
Secondary	Fish processing plants	Process of fish and export to foreign country	User	Support	Sales volume, profits	Lower payment for resource

high pressure on the fishery for subsistence, and per capita collection has reduced to a large extent. The major income share of the collectors is taken away by the higher level intermediaries such as the *Mahajans* or the *Aratdars* (money lenders) due to the *dadons* (private money lending) system prevailing in the coastal areas. In the coast, *dadons* and poverty operate in a vicious circle, which hinder the empowerment of the local community. Transportation/preservation cost, especially for the fishers, is very high and the time needed for the transportation/collection is also long to render the fishers more vulnerable.

The FD is mandated to conserve all aquatic resources inside and near-shore water of the Sundarbans. There is a national policy to protect the forest and provide for sustainable use of the Sundarbans resources. Also, a land use zoning framework for the World Heritage Site is in place for PAs, the buffer zone, a commercial zone for sustainable harvesting and a subsistence living area. None of the policy framework ensures protection of the Sundarbans without community participation.

The Bangladesh Government has set a target of bringing 5 % of the country under PAs by 2015 in the Millennium Development Goals (MDG). Moreover, the government is planning to conserve 17 % of its terrestrial and inland water and 10 % of coastal and marine areas potential for biodiversity and ecosystem services by 2020 under the PA network of Aichi target. At present there are 34 forest PAs covering 2,654.03 sq km in Bangladesh representing about 1.8 % of the country's landmass. However, there are about 1.6 million ha of forests in the country which are protected by restricting biotic interferences. Such forests are maintaining a status quo almost similar to the criteria set for different PAs categorized by the IUCN, but so far have not been officially declared as PAs through existing regulations. These forests are managed mainly for conserving the biodiversity through protection of ecological integrity and others are managed for providing sustainable flow of goods and services to meet the community needs. Moreover, there are 12 Ecological Critical areas (ECA) in the country. Considering the criteria set by the IUCN more forest areas already obtained PA status in Bangladesh. Necessary policies are being framed to regulate different activities inimical to conservation of biodiversity in the present ECAs.

The near-shore marine water of the Sundarbans is of outstanding importance for maintaining biodiversity. Therefore, the IFMP of the Sundarbans included 12 nautical miles in the Bay of Bengal (to the south of the Sundarbans) as a marine protected area (MPA). Such extension will add 1,603 sq km more and thereby the Sundarbans itself will account for 7,620 sq km of marine and coastal PAs. Moreover, the Department of Fisheries have declared two marine reserves (69,800 ha) in the Bay of Bengal in the year 2000. However, wildlife sanctuaries and national parks of coastal areas can also be counted as MPAs in Bangladesh. The Governance Matrix of Bangladesh PAs with categories of different PAs has been proposed depending on the management strategies as mentioned in the IUCN classifications (Table 4).

The integrated management of mangrove wood and non-wood products depends on an understanding of the ecological parameters for forest productivity (primary production) and the biological role that the primary production from the forest plays

Table 4 Matrix of the protected area category and governance (marine and coastal)

Governance types/Protected Area Category	A. Governance by Government	B. Shared Governance	C. Private Governance	D. Governed by indigenous People and local communities
Ia. Strict Nature Reserve	Nil	Nil	Nil	Nil
Ib. Wilderness	16,352.23 ha	139,699 ha	Nil	Nil
II. National Park	6,074.58 ha	Nil	Nil	Nil
III. Natural Monument	Nil	Nil	Nil	Nil
IV. Habitat/Species management	32,723 ha	Nil	Nil	Nil
V. Protected Land-scape/Seascape	69,800 ha 11,055 ha (ECA)	Nil	Nil	Nil
VI. Protected Area with Sustainable Use of Natural Resources	461,030 ha	Nil	Nil	Nil

Forests—6,55,779 ha; ECA—11,055 ha; Marine—69,800 ha; Total = 7,36,634 ha
Ib A: Nijhum Dweep
Ib B: Sundarbans East, West and South Wildlife Sanctuaries
II A: Sonarchar, Tangragiri WS
IV A: Established Coastal Plantation (30,000 ha), No extraction, Char Kukrimukri, Chandpai, Dudhmukhi and Dhangmar WS, Kuakata NP
V A: Declared as Marine Protected Area; Teknaff Peninsula and St. Martin Island (ECA)
VI A: Sundarbans Reserved forests

in the mangrove food web of aquatic resources (secondary production). An understanding of the role of key species in maintaining the equilibrium of a particular ecosystem is likewise essential.

The CBFM (Community Based Fisheries Management) project could be a good example of co-management in the inland waters of Bangladesh (Thompson et al. 2003). In order to facilitate an effective fisheries management system in the Sundarbans, a similar option should be operated in the SRF that demonstrates GO-NGO (non-government organization) partnerships can be mutually beneficial. In the SRF, fishing is banned in the sanctuaries which are 23 % of the total areas of the Sundarbans. The rest of the water area is open for fishing except the small canals where fishing is prohibited on every alternative year. This openness encourages commercial shrimp farming in the SIZ with an increased interest of local elites to treat this farming as a Blue Revolution, akin to the agricultural Green Revolution of the mid-60s. Several studies revealed that large-scale commercial shrimp aquaculture is least desirable in terms of degradation and comparative evaluation of policy scenarios from ecological and social perspectives (Knowler et al. 2009). In the context of present trend of resource extraction in the SRF, CBFM may be considered as an alternative management strategy to reduce degradation of the SRF by allowing the local community in management with defined property rights and decision-making power.

Otherwise, due to a lack of ownership and alternative livelihood options poaching, illegal tree felling, over- and unauthorized resource collection will cause continuous degradation of the Sundarbans.

6 Vision for the Future-Desired Management Condition

The present on-going World Bank and German Development Cooperation non-lending technical assistance projects, 'Sundarbans Ecosystem and Livelihood Security (SEALS) (2011–2015)' and 'Integrated Protected Area Co-management (IPAC) (2008–2013)' will provide an opportunity to address major issues in the Sundarbans ecosystem including aquatic resource and habitat management, ecology and biodiversity, alternative income generation, climate change and sea level rise, etc., that are currently affecting the SRF. The main focuses of SEALS project have two interlinked goals: (1) improved Government capacity to protect and manage the SRF and (2) sustainable resource extraction and dependence of the SRF surrounding communities on resources obtained from the Sundarbans and their exposure to sustainable reduced natural disasters.

A bi-country (Bangladesh-India) strategic action plan is essential for the conservation and sustainability of the SRF. Existing management plans and worthy proposals could be included for implementation both by India and Bangladesh for betterment of the Sundarbans. The Central Government of India allocated Rs. 30 million of the Rs. 116 million Integrated Coastal Zone Management project in 2010 to be spent by West Bengal, most of which was for the Indian part of the Sundarbans. The funds are for prevention of river erosion, construction of storm shelters, and promotion of eco-tourism and livelihood improvement in the Sundarbans. In addition, there is a central government grant of Rs. 45 million for repairing embankments at critical areas in the Sundarbans forest. Bangladesh has similar allocations of Tk. 70 million for its Sundarbans management and conservation.

The IFMP developed by the FD in 1998 included a long-term vision for the SRF. While many of the previous plans are still appropriate, there are changes needed to further emphasize or better reflect the present situation. The following strategies are proposed by the FD (Bangladesh) as the long-term vision for the management of the SRF:

- The SRF shall continue to provide subsistence resources at a level in which the sustainability of the resource is ensured, though emphasis will be on reducing dependency by surrounding communities and improving current management practices.
- Traditional users will acquire a greater awareness and shared responsibility and a share in the financial benefits as a result of co-managing the resources and will act accordingly to help conserve them.
- The FD will involve the local community in the overall management of the SRF.

- The FD will develop its capacity including infrastructure, logistics and managerial capacities and seek technical assistance from relevant agencies where appropriate in the SRF management.
- Development and efficient operation of alternative income generation in the SIZ will help depress the demand for resources currently obtained from the SRF.
- The SRF landscape will be managed to ensure that essential ecological conditions and services are maintained. The wildlife sanctuaries will be managed to provide a secure habitat and biodiversity for wildlife resources.
- Specific sites, infrastructure and routes in designated areas of the SRF will be developed and/or maintained to provide quality ecotourism.
- In order to take advantage of the increasing tourism, the FD will seek public/private partnerships consistent with the guidelines and principles established by the Government of Bangladesh to improve ecotourism services and facilities in the SRF.
- The effects anticipated to result from climate change will be recognized and adaptive management strategies developed and implemented in order to ensure the maintenance of ecosystem goods and services.
- The Sundarbans, as the largest contiguous mangrove ecosystem in the world and befitting its world heritage site designation, will become the international recognized example of collaborative management of a mangrove ecosystem. This can occur with provisions for sustainable financing for more effective management efforts in the SRF in tandem with a broad range of programmes supporting tenure rights, poverty reduction and sustainable socio-economic development in the SRF.

Acknowledgements Sincere thanks to Dr. A. K. Yousuf Haroon, National Technical Advisor, Support to BOBLME Project, BFRI, Mymensingh for his suggestions and critically reviewing the draft.

References

Akhter M (2012) Present status, harvesting, management and legislations pertinent to the Sundarbans fisheries. In: Hoq ME, Yousuf Haroon AK (eds) Sundarbans fisheries of Bangladesh: current status and potentialities, pp 1–13. SBOBLME Pub/Rep. 6. Support to Sustainable Management of the Bay of Bengal Large Marine Ecosystem (BOBLME) Project, Bangladesh Fisheries Research Institute, p 56

Chantarasri S (1994) Fisheries resources management for Sundarban reserved forest. Integrated resource development of the Sundarbans reserved forest, Bangladesh. BGD/84/056, UNDP and FAO, p 171

FAO (1998) Integrated Resource Development of the Sundarbans Reserve Forest Project–Finding and Recommendations. Final Project Report, Rome: Food and Agriculture Organization of the United Nations

FD (Forest Department) (2010) Integrated resources management plans for the Sundarbans (2010–2020). Ministry of Environment and Forests, GoB, Bangladesh, p 281

Heining RL (1892) Working plan of the Sundarbans government forests, Khulna and 24 Paragans Districts Bengal

Hoq ME (2007) An analysis of fisheries exploitation and management practices in Sundarbans mangrove ecosystem, Bangladesh. Ocean Coastal Manag 50(5–6):411–427

Hossain Z, Ahmed I (1994) Management of forest resources. In: Mangroves of the Sundarbans, volume two: Bangladesh. IUCN, Bangkok, pp 181–202

IRG (International Resources Group) (2010) Study on the conservation and management of fisheries resources of the Sundarbans (draft). USAID Contract N EPP-1-00-06-00007-00, p 128

Islam MS (2003) Perspectives of the coastal and marine fisheries of the Bay of Bengal, Bangladesh. Ocean Coast Manag 46(8):763–796

Kabir DMH, Hossain J (2008) Resuscitating the Sundarbans: customary use of biodiversity and traditional cultural practices in Bangladesh. Unnayan Onneshan—the innovators, Dhaka, p 98

Knowler D, Philcox N, Nathan S, Delamare W, Haider W (2009) Assessing prospects for shrimp culture in the Indian Sundarbans: a combined simulation modelling and choice experiment approach. Marine Policy 33:613–623

Rabanal HR (1984) Fisheries integrated development in the Sundarbans. In: Integrated development of the Sundarbans, Bangladesh—report for TA Project (FAO:TCP/BGD/2309). FAO, Rome, p 28

Roy AKD, Alam K (2012) Participatory forest management for the sustainable management of the Sundarbans mangrove forest. Amer J Env Sc 8(5):549–555

Roy AKD, Alam K, Gow J (2012) A review of the role of property rights and forest policies in the management of the Sundarbans mangrove forest in Bangladesh. Forest Policy Econ 15:46–53

SBCP (Sundarbans Biodiversity Conservation Project) (2002) Fisheries management for the Sundarbans. Bangladesh forest department, Khulna circle, Aquatic resources division, Khulna, Bangladesh, p 26

Thompson PM, Sultana P, Islam N (2003) Lessons from community based management of floodplain fisheries in Bangladesh. J Environ Managt 69(3):307–321

Economic Sustainability for Halophyte Cash Farms in Urban Environments

Paul Bierman-Lytle

Contents

Abstract To successfully establish a halophyte cash farm requires an economic model that is embedded in the goals, development structure, implementation, management and budget from the onset. Goals must identify clearly if the project intends to be research-focused or if it is to be 'sustainably profitable', or if both, how is this accomplished. This decision sets the course for its long term viability as a cash-productive model for others to replicate. Too often, great concepts and projects fail to withstand the commercial test of time by failing to establish economic sustainability into their plans. They become victim to subsidization, grants, and capital debt which results in an addiction common to research-focused projects. This chapter will attempt to provide guidelines on how to build in economic sustainability to a halophyte cash farm project in order to demonstrate its economic value to a community and thus ensure longevity in its mission. It will provide meaningful economic value, create jobs, and enhance community sustainability by reducing carbon footprint, increasing local food or product inventories, and introducing local, regional, and global edu-tourism.

P. Bierman-Lytle (✉)
Yale University USGBC LEED Technical Advisor,
Pangaeon Global Denver, Colorado, 80237 Denver,
United States of America
e-mail: p.biermanlytle@pangaeon.com

I. Faridah-Hanum et al. (eds.), *Mangrove Ecosystems of Asia*,
DOI 10.1007/978-1-4614-8582-7_16,

Fig. 1 Title page photo, Larnaca Salt Lake, Greece, thesis project. (© 2012 Michalis Piroccas)

1 Introduction

This chapter focuses on the 'economic sustainability' of an industry, in this instance, Halophyte Farming, versus the application of 'sustainability economics', 'ecological economics', 'natural capitalism', or 'ecology-based economics'. I am defining 'economic sustainability' in the most simplistic of terms: maintaining a profit/loss proforma that will allow a particular project to support itself using current market economics of supply/demand (Baumgartner 2009; Daly 1999; Hawken et al. 1999; Soerbaum 2000; Walter 2002). To describe a halophyte cash farm project one is describing a project seeking to demonstrate 'sustainability' principles, including protection, restoration, and increased valuation of natural resources, development of regional commercial products to reduce carbon footprint resulting from foreign purchases, increased local jobs, carbon sequestration, and use of non-renewable energy technologies, to mention a few. From this perspective of a new sustainability industry, however, the challenges of entering the commercial market already emerge and will be difficult to establish. Sustainability initiatives historically fly in the face of the status quo of 'doing business as usual'[1] (Fig. 1). Among the many reasons why sustainability industries suffer from mis-alignment with the status quo is the lack of community integration where the product or services have little impact on daily lifestyles, including monetary improvements: cost reductions, higher paying, meaningful jobs, and career advantages, for example. Many of these industries also address only a margin of the population, and therefore, are limited to niche markets.

How can a halophyte cash farm project penetrate the commercial status quo, given most will not know what 'halophyte'[2] means or, if and when they do, how it will change their lives to the extent they should 'purchase' it?

Fortunately, there are examples of innovative 'sustainability' products and services that have solved this challenge.

First, let's look at what a halophyte cash farm is and what its intended mission includes.

---

[1] USGBC (2012).

[2] Salt tolerant plant species, commonly referred to as 'mangroves' Brown (1993).

2 Halophyte Cash Farm

A unique project is under development by UNESCO, entitled: ***Environmentally friendly farm for cash crop halophytes and biodiversity conservation*** (Böer et al. 2012). The proposed project is "designed to explore the scope and lay conditions for the production of cash-crop halophytes utilizing seawater as part of the strategy for developing techniques for biosaline agriculture".[3] The project recognizes the commercial challenges: "The acceptance of halophyte farm systems still meets great reservations among land-owners, land-users, and decision makers, even though the rise of population, the need to feed the growing population, and wide-spread soil- and ground-water salinization urgently requires solutions. There are means and technologies now to make use of saline waters. A functioning pilot-system needs to be developed, tested, scientifically documented, and demonstrated in an understandable way, so farmers and land-users can understand and replicate halophyte production systems profitably."[4] Replicate halophyte production systems profitably! This is the challenge and this is also the opportunity.

What are the societal/community benefits?

2.1 Aims

1. Enhance food and feed security.
2. Explore new resources for biofuel.
3. Enhance the generation of jobs and income based on halophyte R & D.
4. Reduce waste and enhance rational utilization of national water-, waste-, applicable for farms.
5. Identify and develop methods for the enhancement of sustainable human living in hyper-arid hot desert regions, based on salt-water utilization.
6. Contribute to UN MDGs, UN conventions, and national plans related to desertification, energy, food security, and biodiversity conservation.
7. Raise awareness and demonstrate feasibility of profitable halophyte production agro-systems."[5]

Cash crop products include: fodder (seeds and leaves), as well as charcoal, wood-chips, and biomass for fish-feed and honey. Seagrass farm cash products include materials such as tresses, carpets, insulation material, and livestock and fish fodder. Landscaping species will also be a dominant cash crop product.

---

[3] Böer et al. (2012) and Choukr-Allah (1996).

[4] IBID.

[5] UNESCO proposal, ***Environmentally friendly farm for cash crop halophytes and biodiversity conservation,*** pp 9–10.

The halophyte cash farm should also have the following components:

1. Seacoast locations for floating mangroves and coastal sabhka; farm should extend 'into the sea'
2. Be located near a community for food waste, compost, greywater, blackwater, animal waste
3. Easily accessible for staff and visitors
4. Animal husbandry will include approximately 77 livestock and 109 poultry to provide meat products and animal waste
5. Infrastructure technologies will be based on renewable energy, water conservation, and ecological engineering solutions, including potable water production, electricity, biofuels, sewage treatment (black and greywater), and solid waste conversion and recycling
6. Buildings will comply with ESTIDAMA[6] Pearl rating and USGBC LEED[7] green building certifications
7. Tourism will be a key component for promoting public education on halophyte sciences and benefits

Economic analysis of the proposed halophyte cash farm has not been determined to this date; however, estimates for development range as follows:

Phase 1: International advisory committee meeting = $ 150,000 USD
Phase 2: Master plan = $ 250,000–500,000 USD
Phase 3: Site selection and farm development = Cost not defined
Phase 4: Inauguration and operations = cost not defined[8]

3 Components of Economic Sustainability

Given the program description of the proposed halophyte cash farm and biodiversity conservation project, an identification of key components that will constitute economic sustainability for the project over many years is required to fully understand the scope of master planning and project preparation. The key components of economic sustainability can be summarized as:

a. Land
b. Money to build
c. Money to operate/maintain
d. Money to sustain/grow
e. Product inventory
f. People
g. Contingencies

[6] Estidama (Arabic word for 'sustainability' is a green building program required for all new buildings in the emirate of Abu Dhabi.

[7] United States Green Building Council Leadership in Energy and Environmental Design

[8] IBID, pp. 29–30.

These can be defined further:

A. Land acquisition and long-term conditions
   SOLUTION: Since land in Abu Dhabi emirate is owned by the government, establish a joint venture (JV) with government agency or with government-sponsored development entity. The JV can include terms whereby the project is granted 99-year lease terms, with a sequenced pay-back increasing relative to the financial proforma of the project. Terms of the JV can also include: budget grants as part of government education, tourism, and Plan 2030 parks and recreation, science and technology.
B. Partnerships and investors for capital development (CAPEX costs)
   a. Master planning Bierman-Lytle (2012)
      i. OTGI[9] infrastructure
      ii. Green buildings
      iii. Green transportation
      iv. Marketing/branding
   b. Design & engineering
   c. Construction
   d. Commissioning
   SOLUTION: Establish technology and product partnership agreements with vendors. In exchange for reduced product and technology costs, including materials and installation, the vendors would be offered research access to monitor their technologies and product performances, and receive term-limit exclusivity for marketing their products within the framework of the UNESCO and EAD[10] farm project. They will benefit from high profile public and media exposure as well as international tourism.
C. Partnerships and investors for operations (OPEX costs)
   a. Website
   b. Marketing/branding
   c. Product and services
      i. Harvesting
      ii. Processing
      iii. Packaging
      iv. Storage
      v. Sales & services
      vi. Distribution
   d. Nursery (see R&D)
   e. Library
   f. Food & beverage
   g. Tourism
      i. Edutourism
      ii. Agritourism

[9] Bierman-Lytle 2012 ASEEL Eco Sports Camp OTGI (Off The Grid Intelligent) infrastructure.

[10] Environmental Agency Abu Dhabi.

iii. Voluntourism
iv. Health Tourism

h. Maintenance

SOLUTION: As illustrated in other case studies below, a major effort should be focused on developing partnerships with the commercial sector that will benefit from harvesting, packaging, sales and distribution of the cash farm products. These partnerships might include regional landscaping companies, food processors, food distributors, food markets, and mariculture and livestock farmers/ranchers.

Other key partnerships to engage include the tourism industry, including local eco-adventure companies (Noukhada Adventure Company, for example), tourism operators (Big Bus Tours, for example), hotels, air and ground transportation companies (Etihad, Al Ghazal, and Tawil, for example).

Partnerships with regional and international universities will offer opportunities for student thesis projects, on-site internships, and faculty research initiatives (Masdar Institute of Science & Technology, NYU Abu Dhabi, Sorbonne, UAE University, Zayed University, to mention a few).

D. Partnerships and investors to sustain growth (R&D costs)
   i. Floating mangroves unit
   j. Land-based mangrove farm unit
   k. Seagrass unit
   l. Annual halophyte unit
   m. Salt bush and salt grass unit
   n. Indigenous fodder plants unit
   o. Animal production unit
   p. World Halophyte Garden

   SOLUTION: Both UNESCO and EAD will have access to a variety of local and international organizations, including multi-lateral and bi-lateral institutions and agencies, that can provide grants or JVs to finance specific research, as well as contribute financially to the World Halophyte Garden.

E. Species acquisition and/or development
   q. Flora
   r. Fauna

   SOLUTION: Same as above, letter 'D'.

F. Human resources development
   a. Ownership
   b. Stockholders
   c. Executive officers
   d. Management
   e. Staff
   f. Support staff

   SOLUTION: Attracting and securing high quality professionals is essential, not only to the program's success, but also in attracting and securing the key partnerships and investors mentioned above. Creating a work and living environment for executive management and senior staff is essential to entice them to relocate with their families. By providing a comprehensive program as outlined in this chapter, the farm project has a competitive advantage in global recruitment. Members of the project will be part of a leading edge, innovative program that can influence global challenges, integrating science solutions with urban partnerships, tourism, and education and youth internships.

4 Proposed Halophyte Cash Farm

To construct a master plan for the proposed halophyte cash farm project, all economic variables are taken into consideration: location to the sea, to mangrove and sabkha ecosystems, to tourism, to commercial hubs, to communities, and to public exposure and access. The proposed site for consideration is located in a prime logistic core of Abu Dhabi and surrounding communities (Fig. 2) Bierman-Lytle 2012.

As the map illustrates, the proposed site is within close proximity to (Fig. 3):

1. Residential communities: Yas Island (nationals), Al Raha, Al Bandar, Al Zeina, Khalifa City A, Masdar City, Al Bahia, and Saadiyat Island
2. Schools and Universities: Masdar Institute of Science & Technology, New York University, British & American schools, and Zayed University
3. Tourism: Yas Island (7 hotels, 10 restaurants, 6 sporting attractions), Saadiyat Island (cultural district and golf), Masdar City
4. Access: Abu Dhabi International Airport (AUH), Highways 10 and 12
5. Commercial hubs: Yas Mall, Masdar City, Khalifa City A, Al Raha Beach, Al Bandar, Al Bahia Mall
6. Nature preserves: Samilya Island, existing mangrove forests (over 800 hectare)

This location, therefore, satisfies key criteria for establishing a viable commercial (economic) sustainable future for the cash farm project. The advantages of the Abu Dhabi government to demonstrate environmental stewardship in the core of urban development is also an essential attribute of this location. While tourists and locals can enjoy the tourism attractions, they can more readily participate in tours of the cash farm project. Indeed, establishing 'package tours' with the regional hotels (seven within 5 min of the site) provides economic benefits to both the hotels and the farm.

Proximity to the international airport also provides convenient access for the halophyte experts and guests arriving from around the world. Proximity to high quality residential communities and schools provides incentive for attracting key staff and executives to relocate to the UAE.

5 Proposed Farm Layout Options

Within the context of this proposed location, the following site plan (illustration C) offers multiple choices for the facility layouts. Following the guidelines of the UN-ESCO brief, "the farm should actually extend into the sea, in order to accommodate

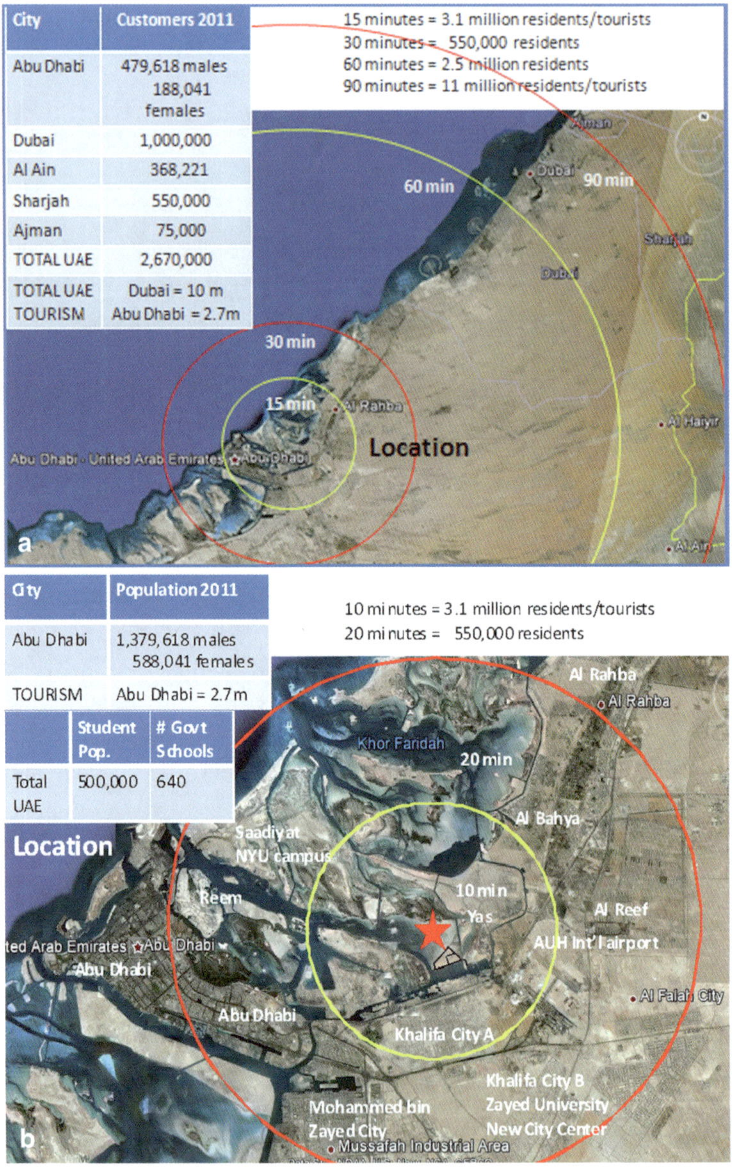

Fig. 2 Proposed site—regional—context. (© 2012 Pangaeon Middle East North Africa)

a few of the suggested modules. This corridor should have a length of ca. 5 km, and a width of ca. 1000 m."[11] As the illustration shows, the central area outlined, SITE A, represents an area of approximately 5 million m^2 or 500 ha. SITE B is approximately

[11] Böer et al. (2012).

Fig. 3 Proposed site closeup. (© 2012 Pangaeon Middle East North Africa)

460 m wide x 4,600 m long (at the longest) and 3,000 m long (at the shortest) or a total of 1,750,000 $m^2 = 175$ ha.

SITE A would comprise:

- Mangrove units
- Floating mangroves (4 units 25 × 25 m)
- Land-based mangrove units

SITE B would comprise (Fig. 4):

- Seagrass units (above high water mark) of terraced plots approximately 1,000,000 m^2 or 100 ha.

Additional units:

a. Annual halophyte unit
b. Salt bush and salt grass unit
c. Indigenous fodder plants unit

These can be located in either SITE A or SITE B as appropriate. The facilities that benefit from public access for education (schools) and tourism would be located at the lower end of SITE A:

Fig. 4 Sites A and B. (© 2012 Pangaeon Middle East North Africa)

d. Animal production unit
e. World Halophyte Garden
f. Research labs
g. Staff offices

6 Synergistic Economics

An essential component to reducing farm capital costs and increasing farm revenue is to build synergistic economies with adjacent commercial businesses . To this end, the proposed sites have an opportunity to share infrastructure resources with proposed eco sports parks. The synergies include (Figs. 5, 6, 7, 8, 9) (Bierman-Lytle 2012):

A. Equine center (animal waste generator) providing shared stables, fenced arenas, staff, fodder supply, tourism
B. Solar electric photovoltaics and/or concentrated solar for both electricity generation and potable water production (solar desalination)
C. TerraSave wastewater treatment plant for conversion of blackwater to greywater and greywater harvesting for reuse on all properties for irrigation
D. Solid waste conversion to electricity, biofuel, and other commercial by-products Bokaie 2012
E. Solar and hybrid-fuelled transportation for staff, tourists, maintenance and security

Fig. 5 Al Qurm Eco Adventure campsites Bierman-Lytle (2012). (© 2012 Eco Structures Australia)

Fig. 6 Al Qurm Eco Adventure campsites—use of solar hot water. (© 2012 Eco Structures Australia)

F. Solar-powered street and security lighting
G. Smart Grid systems integration to manage use of electricity and water consumption

Fig. 7 Cultural heritage integration. (© 2012 Exclusive Tents South Africa)

Fig. 8 Tourism activities near halophyte cash farm Bierman-Lytle (2012) ASEEL Eco Sports Park. (© 2012 Pangaeon Middle East North Africa)

Fig. 9 100% solar powered ferries: no noise, no wake and no pollution Bierman-Lytle (2012). (© 2012 Grove Boats SA)

H. ESTIDAMA and LEED green building certifications
I. Proximity to Al Qurm Eco Adventure Campsites for tourists and school/university students Bierman-Lytle 2012

7 Infrastructure

In addition to the renewable energy-fueled transportation above, the proposed farm intends to incorporate as many green technologies as possible. By creating a synergistic relationship with the Eco Sports Park facilities Bierman-Lytle 2012, these can include (Figs. 10, 11, 12, 13, 14, 15, 16, 17, 18, 19):

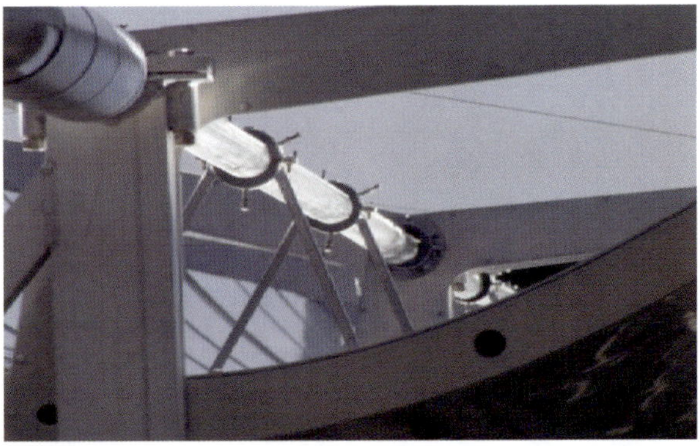

Fig. 10 Solar desalination technology. (Skyfuel, Inc.)

Fig. 11 10 MW wind electric turbine. (Vestas)

Fig. 12 Solar-electric powered pumps. (© 2012 GenPro Energy Solutions)

Fig. 13 Solar electric photovoltaics over TerraSave wastewater treatment system Neuschafer (2012). (© 2012 TerraSave)

A. Solar-electric photovoltaic production
B. Solar desalination
C. TerraSave wastewater treatment
D. Solid waste bioconversion

8 Green Buildings

The farm project will also incorporate green buildings which will save energy, water, waste, and provide healthy work environments for staff. To fit the economic model, the manmade structures should consider pre-fabricated modules. The following examples represent GREEN PRECAST affordable, yet durable and eco-friendly modules (Fig. 20) (Salvadore 2012).

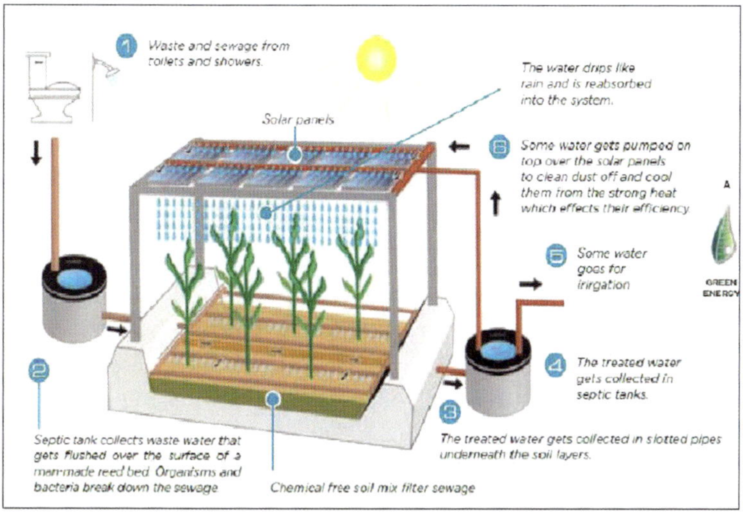

Fig. 14 TerraSave biologically engineered wastewater treatment diagram Neuschafer (2012). (© 2012 TerraSave)

Fig. 15 Solar powered street and security lighting. (© 2012 Pangaeon Middle East North Africa)

Fig. 16 Solar electric photovoltaic car park shading. (© 2012 Pangaeon Middle East North Africa)

Units can be configured as needed for offices, laboratories, workshops, residential, and storage. Precast, insulated units are 40 % cheaper than conventional construction and are fireproof, termite proof, and durable. The farm facility buildings can be erected in less than four months, providing additional time savings (Fig. 21).

Building designs are unlimited for function, cultural heritage, and climate.

Examples: Figs. 22, 23)

An example of a precast modular facility incorporating renewable energy includes a cost of approximately 2,000 AED per square meter and less than 6 months construction time.

Fig. 17 Solar desalination plant produces 500 liters potable water per day. (© 1999–2013 Middle East Desalination Research Center)

The following images demonstrate additional examples of various types of facility configurations for offices, research labs, library, and mobile laboratories on the water itself (Figs. 24, 25, 26, 27).

Given the proximity to the sea, the farm can consider optimization of mobile research units as illustrated in the two images above. This allows units to be relocated as needed.

9 Case Studies

Several halophyte cash farm projects can be noted as prototype examples, but which have not demonstrated economic sustainability to this date (Figs. 28, 29).

A. DBU Project AZ 27708: Biological wastewater treatment in land-based marine circulation systems through the integrated culture of halophytes (Neomar 2012).

ECONOMIC TARGET: SEA FISH & FOOD STOCK FOR HUMANS PLUS FEED STOCK.

The goal of this project is the biological wastewater treatment in marine recirculating aquaculture systems by an integrated culture of salt-tolerant plants (halophytes) in constructed wetlands. Although the water renewal rate of the *ocean loop* is less than 1 % of the system volume per day, there is ecological as well as economic interest in recycling the wastewater. This can be done using constructed wetlands.

Constructed wetlands are already an interesting alternative to treat domestic sewage, especially in remote areas, and are also used to treat waste water of freshwater aquaculture systems. The water treatment in constructed wetlands is an interaction of different processes; such as mechanical purification by the substrate; assimilation, conversion or removal of organic and inorganic matter by bacteria; nutrient assimilation in the scope of plant growth and adsorption or precipitation reactions on particles and roots.

The application of constructed wetlands for marine aquaculture is hardly investigated. Halophytes are an interesting supplement to the production of seafish. Many species are already used in some regions of Europe as a foodstuff and have an interesting market potential, regarding their nutritional value as well as taste.

The project works on the basic information and requirements to combine the production of halophytes in constructed wetlands with the land-based production of seafish. If the project is successful, we aim to establish a commercial pilot system.

Fig. 18 Solar electric powered vehicles for onsite transportation. (**a** © 2013 Solar Novus **b** © 2013 Solar Drive Distributors)

The long-term goal is to completely recycle the waste water of land-based fish farms and generate a marketable product at the same time. The diversification of the product portfolio will improve the profitability and sustainability of re-circulating aquaculture systems. The halophytes can be sold as food for human consumption or as mineral-rich feed plants (Figs. 28, 29, 30).

B. Sustainable Bioenergy Research Project, Abu Dhabi, UAE (Neomar 2012).

ECONOMIC TARGET: BIOFUELS FOR AEROSPACE INDUSTRY

Fig. 19 Biogas fuelled tractors. (© 2012 New Holland Agriculture South Africa)

Fig. 20 Precast modular construction saves time, labor, and material costs. (© 2012 Green Precast Systems)

Fig. 21 Precast construction featuring integrated solar electric photovoltaics as shading device. (© 2012 Pangaeon Middle East North Africa)

In January, Boeing announced that it, the Masdar Institute of Science and Technology, Etihad Airways, and Honeywell's UOP unit would establish a major research

Fig. 22 Residential 'mangrove villas' for cash farm staff. (© 2012 Eco Structures Australia)

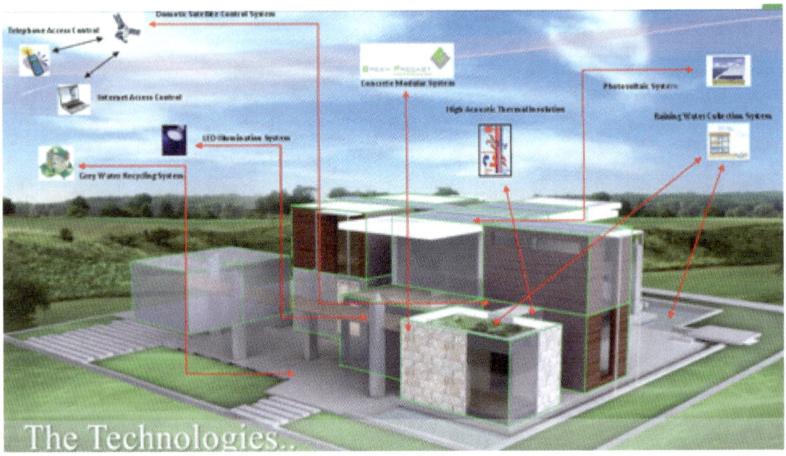

Fig. 23 Green Precast modular construction. (© 2012 Green Precast)

Fig. 24 Residential and/or office. (© 2012 Pangaeon Middle East North Africa)

institution and demonstration project in Abu Dhabi, United Arab Emirates. The Sustainable Bioenergy Research Project will use integrated saltwater agricultural systems to support the development and commercialization of biofuel sources for aviation and co-products. The Sustainable Bioenergy Research Project in Abu Dhabi is

Fig. 25 Flat, expansive roof design allows for abundant shading and surface area for solar panels. (© 2012 Pangaeon Middle East North Africa)

Fig. 26 This facility shows how the integration of the TerraSave wastewater and greywater system as an integral part of the landscaping features will complement cash farm facilities. (© 2012 Pangaeon Middle East North Africa)

taking an evolutionary approach that combines saltwater farming, mangrove forests, and the cultivation of salicornia (a species of halophytes) as potential sources for sustainable jet fuel (Fig. 30) (Boeing).

Illustration of the planned Sahara Forest Project Test- and Demonstration Centre in Aqaba, Jordan (Fig. 31)

1. Algae-facility; 2. Saltwater based greenhouses; 3. External vegetation and evaporative hedges; 4. Designed stepped protection for flash floods; 5. Facilities for research and accommodation; 6. Concentrated solar power facilities; 7. Evaporative ponds

C. SAHARA FOREST PROJECT[12]

ECONOMIC TARGETS:

1. High value crops
2. Commercial compounds from salt
3. Biomass
4. Revegetation of desert

---

[12] Sahara Forest Project (2012a, b, c), http://saharaforestproject.com/

Fig. 27 Mobile research labs will optimize cash farm area. (© 2012 New York Sun Works)

5. Potable water
6. Electricity = 324 GWh/year
7. Carbon sequestration = 50,000 t/year

LAND AREAS:

1. Revegetation = 1,500 ha
2. Solar Energy = 640 ha
3. Fodder Farm = 500 ha = 30,000 t/year
4. Salt Ponds = 480 ha
5. Infrastructure & facilities = 355 ha

Fig. 28 Artificial wetlands using halophytes. (© 2012 DBU Project AZ 27708)

Fig. 29 Sea Asparagus. (© 2012 DBU Project AZ 27708)

Fig. 30 Sustainable bioenergy research project, Abu Dhabi, UAE. (© Sustainable Bioenergy Research Project (Boeing))

Fig. 31 Sahara Forest Project. (© 2012 Sahara Forest Project)

6. Saltwater greenhouses = 300 ha = 190,000 t/year (tomatoes and melons)
7. Algae cultivation = 150 ha = 7,500 t/year bio-fuel
8. Halophyte farm = 75 ha

Total = 4,000 ha

Total electrical power requirements = 570 MW in addition to the 324 GWh/year

Total employment = approx 20,000

By establishing a saltwater value chain, the Sahara Forest Project will make electricity generation from concentrated solar power (CSP) more efficient, operate energy- and water-efficient saltwater-cooled greenhouses for growing high value crops in the desert, produce freshwater for irrigation or drinking, safely manage brine and harvest useful compounds from the resulting salt, grow biomass for energy purposes without competing with food cultivation, and revegetate desert lands. The synergies arising from integrating the technologies improve performance and economics compared to those of the individual components. In addition to its commodity outputs of food, energy and salt, the system also provides global climate benefits by sequestering CO_2 in the facility's plants and soils, and by pushing back the accelerating process of desertification through the revegetation of desert areas (Figs. 32, 33).

The large-scale Sahara Forest Project Oasis in the desert would consist of the following components and yields:

Concentratedsolar power (CSP) at a combined rating of approximately 570 MW. This power plant would yield enough power to supply all needs of the oasis and at the same time export an average of 27,000 MWh/month, or 324 GWh/year.

A combined area of **saltwater greenhouses** of 300 ha. These greenhouses would consume 20,000 m^3 of freshwater produced by the Oasis' own sun-powered thermal desalination per year, and could as an example, yield 190,000 tons of tomatoes and melons combined per year.

The oasis would have a combined area of 2,000 ha of **outdoor vegetation** and crops. Approximately 500 ha would be fodder crops grown in between evaporative

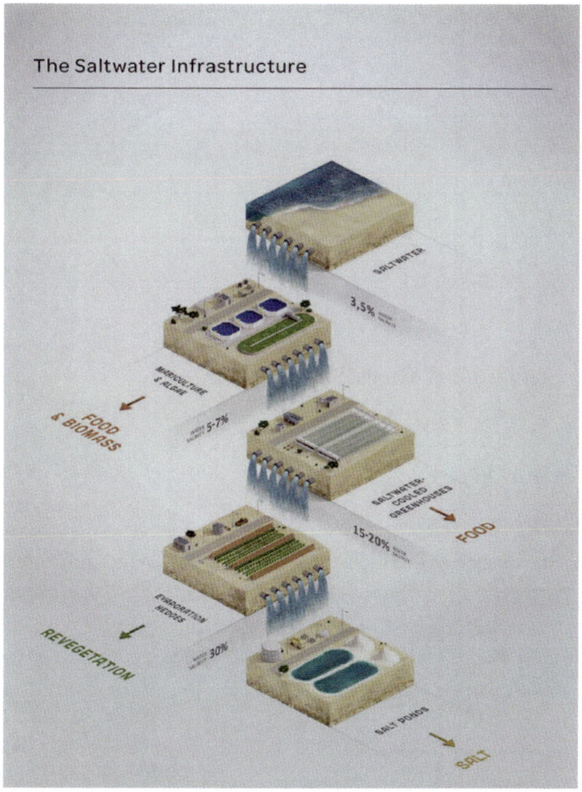

Fig. 32 Land distribution of a 4000 ha facility. (© 2012 Sahara Forest Project)

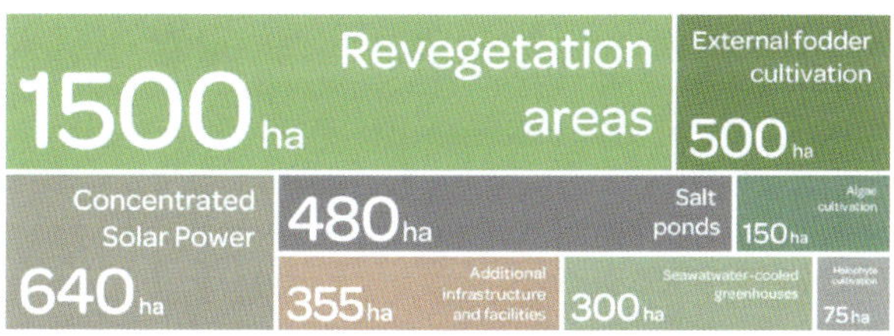

Fig. 33 Yield distribution of a 4000 ha facility. (© 2012 Sahara Forest Project)

hedges, and the remaining 1,500 ha would be free-ranging re-vegetation of the desert land. The fodder crops would have an annual yield of approximately 30,000 tons.

Furthermore, the oasis would have **algae cultivation** facilities of 150 ha, capable of extracting and yielding bio-fuel ready algae oils at the rate of 7,500 tons/year.

While the above is but one example of many possible scenarios of combinations, it illustrates how a major Sahara Forest Project Oasis can yield substantial amounts of energy, food, fuel and fodder crops—all in a closed-loop system which re-vegetates arid land and provides carbon sequestration in the magnitude of 50,000 tons/year. This particular example could employ up to 20,000 people and support up to five times as many.

Fig. 34 Mariculture. (© 2012 Sahara Forest Project)

Additional extensions (Fig. 34):

9.1 Mariculture

The farming of fish, snails, shrimp, abalone, and other aquatic animals for food can further utilize the seawater infrastructure of the Sahara Forest Project to produce additional high protein and high-value food. The mariculture ponds will be fed with fresh seawater directly from the seawater intake. This is necessary because most marine organisms cannot tolerate the higher salinities found further downstream in the SFP system. Fish or shrimp or other animals are raised in open ponds. As evaporative losses increase the salinity of a pond, the water is cycled out and refreshed with fresh seawater. Upon its removal, the 'waste' water from the pond is only mildly more saline than seawater—5 to 5.5 % salt—and is enriched in nutrients thanks to the organic wastes (droppings) of the marine animals. In this state, it is ideal for feeding into algae ponds. The system is set up as a simplified ecosystem. Algae can often tolerate mild increases in salinity, and will thrive on the nutrients in the organic wastes. By the time the water passes through the algae cultivation facility, it will have been naturally purified from all its wastes and ready for use in the greenhouse cooling systems. While providing this service—one that would otherwise have to be carried through costly filtering and purification—the algae use the nutrients they harvest, boosting yields and reducing requirements for the external supply of nutrients. To close the loop, the oil- and nutrient-rich algae can be fed back to the marine organisms, providing a high-performing feed for the farmed animals.

The cultivation of freshwater species, such as Tilapia, may also present significant opportunities to increase the value of the SFP infrastructure. As in the marine mariculture systems, the nutrient rich algae grown at the Sahara Forest Project can be used to feed the fish. In the case of a freshwater aquaculture system, this may prove to be a valuable natural pathway that can be used to transfer nutrients from the saltwater-based algae and mariculture systems to the freshwater-based agricultural system.

Fig. 35 Sahara Forest Project testing facility in Qatar, "The Local Calibration Station (LoCal)". (© 2012 Sahara Forest Project)

Fig. 36 Illustration of the Sahara Forest Project Pilot Facility in Qatar, 10,000 m^2, \$5.3 m USD. (© 2012 Sahara Forest Project)

Although no economic performance metrics are provided in this case study, the synergistic integration of multiple assets can provide a viable commercial economic model. Yet, this is still to be proven.

Key economic strategies engaged by SFP include (Figs. 35, 36, 37, 38):

1. Established and financed by Norway
2. Partnerships with major commercial entities: YARA ASA, supplier of fertilizer and QAFCO, producer of urea and ammonia.
3. Agreements with government: Aqaba Special Economic Zone Authority to provide 20 ha for stage 1 and 200 ha for future growth.
4. Establishment of two entities: SFP Foundation to conduct R&D; and SFP AS as the commercial development and distribution of products.

10 Conclusion

The economic sustainability of a halophyte cash farm requires a new business model. This model is not easy to establish because the conventional business community in the sectors of science and research, education, tourism, commercial sales, and

Fig. 37 Picture of the Sahara Forest Project Pilot Facility in Qatar under construction. (© 2012 Sahara Forest Project)

Fig. 38 Proposed Sahara Forest Project. (© 2012 Sahara Forest Project)

real estate development tend to be most comfortable when they are 100 % in control of ownership. However, this traditional approach results in a 'stove-pipe' or 'silo' mentality and places restraints on innovative business in the 21st century. As the sustainability theater has evolved, it is quite apparent to many in this field that interdependency among parts, integration and collaboration across disciplines and commerce are essential. Like an ecosystem, our modern business models must also adapt, respond, and establish synergies.

Within each component of the farm project, there exists a 'business' partner, whether it is finance, product, transportation, building, education, tourism, harvesting, packaging, distribution, sales, or media. Each of these creates opportunities for collaboration and, in fact, increases opportunities for economic sustainability for all.

In the end, therefore, the halophyte cash farm project can meet its primary mission goals if it is economically sustainable, and, in addition, it will provide a new business model for other enterprises in the 21st century.

References

Baumgartner S, Quaas MF (2009) What is sustainability economics? University of Luneburg, Working Paper Series in Economics 138

Bierman-Lytle PW (2012) Site evaluation and selection for UNESCO cash farm prototype. Powerpoint presented to Abu Dhabi Tourism Authority, Abu Dhabi

Bierman-Lytle PW (2012) Al Qurm Eco Campsites for Abu Dhabi, feasibility and site assessment. Presented to Abu Dhabi Tourism Authority, Abu Dhabi

Bierman-Lytle PW (2012) Pangaeon Parks 'ASEEL' Eco Sports Park, feasibility and site assessment. Presented to Abu Dhabi Tourism Authority, Abu Dhabi

Bierman-Lytle PW (2012) Abratour solar-powered ferry network, feasibility and route assessment. Presented to Abu Dhabi Tourism Authority, Abu Dhabi

Böer B, Al Abdessalaam TZ, Khan A, Koyro H-W, Marcum K (2012) UNESCO proposal. Environmentally friendly farm for cash crop halophytes and biodiversity conservation. UNESCO draft proposal, pp 6, 9–10, 29–30,

Bokaie M (2012) President, BORM Technologies, Conversations and discussions relative to solid waste conversion technology. Dubai

Brown L (ed) (1993) Dictionary, definition of 'halophile' (halophyte). The New Shorter Oxford English Dictionary, Clarendon Press, Oxford, p 1178

Choukr-Allah R, Malcolm CV, Hamdy A (eds) (1996) Halophytes and biosaline agriculture. Marcel Dekker, Inc., New York, pp 3–17

Daly HE (1999) Ecological economics and the ecology of economics; essays in criticism. Edward Elgar Publishers

Hawken P, Lovins A, Lovins LH (1999) Natural capitalism: creating the next industrial revolution. Little Brown & Company, New York

Neomar (2012) Integrated marine aquaculture with halophytes. DBU Project AZ 27708: Biological wastewater treatment in land-based marine circulation systems through the integrated culture of halophytes. Neomar, Denmark

Neuschafer P (2012) Founder/director, TerraSave, Conversations and discussions relative to TerraSave ecological wastewater treatment technology. Dubai

Sahara Forest Project (2012a) Greening the Desert, projects in Jordan and Qatar. http://saharaforestproject.com/. Accessed 31 July 2013

Sahara Forest Project (2012b) Qatar Folder, An information folder about The Sahara Forest Project Pilot Facility in Qatar. http://saharaforestproject.com/. Accessed 31 July 2013

Sahara Forest Project (2012c) Jordan Folder, An information folder about The Sahara Forest Project in Jordan, http://saharaforestproject.com/. Accessed 31 July 2013

Salvadore S (2012) Conversations and discussions relative to Green Precast advantages and disadvantages. MENA region director. Abu Dhabi, UAE

Soerbaum P (2000) Ecological economics. Earthscan Publications Ltd, UK

Sustainable Bioenergy Research Project (2012) Biofuels for Aerospace Industry, Masdar Institute of Science & Technology, Abu Dhabi, UAE

United States Green Building Council (2012) Advocacy: Resiliency section of USGBC. www.usgbc.org/advocacy. Accessed 31 July 2013

Walter GR (2002) Economics, ecology-based communities, and sustainability. Ecol Econ 42:81–87

Research and Development Activities Towards Sustainable Management of Mangroves in Peninsular Malaysia

Azian Mohti, Ismail Parlan and Hamdan Omar

Contents

Abstract Tsunami tragedy, in particular on 26 December 2004, has shown the importance of mangrove ecosystems in providing protection against strong waves and winds. Since then, research and development (R&D) activities have been intensified mainly to stabilize shoreline areas that could potentially be affected by tsunami in the future. In relation, a Task Force Committee of Planting Mangroves and Other Suitable Species Operation on Shorelines of the Country was formed by the Malaysian Government led by the Ministry of Natural Resources and Environment in 2005. Under this task force, a technical committee on Research and Development (JTR&D) was formed under the chairmanship of the Forest Research Institute Malaysia (FRIM). Many R&D activities were conducted by this JTR&D covering aspects of mangrove areas assessment, producing planting materials, planting in severe erosion areas, control of pests and diseases of the planted mangroves and non-mangroves species, monitoring of wave energy and carbon stock assessment. Some of the R&D activities

M. Azian (✉) · P. Ismail · O. Hamdan
Forestry and Environment Division, Forest Research Institute Malaysia (FRIM),
52109 Kepong, Selangor Darul Ehsan, Malaysia
e-mail: azyan@frim.gov.my

I. Faridah-Hanum et al. (eds.), *Mangrove Ecosystems of Asia*,
DOI 10.1007/978-1-4614-8582-7_17,

were completed, however, few are continued to obtain more concrete and long-term data. This chapter briefly elaborates outputs of the completed R&D activities. Detail of the each project can be referred in separate published materials. The outputs could significantly contribute towards sustainable management of mangroves in Peninsular Malaysia and may be applied in other mangroves elsewhere in this country.

1 Introduction

Mangrove forest in Malaysia is estimated to cover an acreage of about 566,856 ha. Malaysia is divided into two geographical regions: East Malaysia and West Malaysia. East Malaysia covers an area of mangrove approximately 467,089 ha, while West Malaysia known as Peninsular Malaysia covers an acreage of about 99,767 ha. Although this is around 1.7 % of the total land that makes up the country, Malaysian mangroves are actually the third largest in area of mangroves in the Asia-Pacific region after Indonesia and Australia.

The unique ecosystem of mangrove occurs most extensively along the protected coastal shores with muddy to sandy bottom. It is not only crucial as a breeding ground of economically important and diverse marine life (flora and fauna), but also plays an important role as a fortress to break and reduce the impact of strong waves on the shorelines, which are alternately covered and uncovered by tidal fluxes (ITTO 2002). It has been proven to be an impressive natural shoreline protector, sheltering the fragile mangrove ecosystem. The experience of the *tsunami* tragedy on 26 December 2004 and recently 11 March 2011 reported less serious destruction compared to Indonesian, Sri Lanka, Thailand and Japan, but the importance of this ecosystem in providing protection against strong waves and winds cannot be undermined. Although human life and properties located away from the epicentre of the tsunami were spared from massive destruction, the mangrove ecosystem has proven its protective roles.

In realizing the critical need to stabilize shoreline areas that could potentially be affected by future *tsunamis*, a Task Force Committee of Planting Mangroves and Other Suitable Species Operation on Shorelines of the Country was formed by the Ministry of Natural Resources and Environment, Malaysia (NRE). The National Task Force Committee was to implement the coastal rehabilitation and protection programme by planting mangrove and non-mangrove tree species. The national committee has identified two technical committees that have the credibility to conduct the project: (1) the Technical Committee on Planning and Implementation (JTPP), which is chaired by the Director General of Forestry Department Peninsular Malaysia (FDPM) and (2) the Technical Committee on Research and Development (JTR&D), which is chaired by the Director General of Forest Research Institute Malaysia (FRIM).

This chapter highlights only completed R&D activities conducted by the JTR&D. On record, the JTR&D has the tasks to identify scope of R&D activities, coordinate collaboration research between agencies, monitor implementation of R&D activities, disseminate the R&D outputs and impacts, and give technical advice to JTTP.

Table 1 Extent of mangrove forests in Peninsular Malaysia (as of 2010). (Source: Hamdan et al. 2012)

State	Extents (ha)
Johor	23,676.43
Kedah	7,841.25
Kelantan	428.95
Melaka	1,308.68
Negeri Sembilan	2,276.50
Pahang	9,039.26
Penang	43,291.97
Perak	94.02
Perlis	1,695.60
Selangor	22,530.20
Terengganu	2,925.74
Total	*115,108.60*

The JTR&D has the following R&D activities as their tasks: (1) techniques in producing planting materials that can meet national demands, (2a) innovative techniques in planting the mangroves and non-mangroves species on coastlines, (2b) monitoring of wave energy and ocean current on coastlines, (3) control of pests and diseases of the planted mangroves and non-mangroves species on coastlines, (4) monitoring and assessment on the survival of planted mangroves, and (5) other R&D projects towards sustainable management of mangroves. Some of the R&D activities were completed, however, few are continuing to obtain more concrete and long-term data. This chapter elaborates outputs of the completed R&D activities under the JTR&D.

2 Distribution and Extents of Mangroves Using Remote Sensing and GIS

A study was conducted on the use of remote sensing and GIS technology in assessing the distribution and extents of mangroves in Peninsular Malaysia. A series of satellite imagery from Landsat TM (Thematic Mapper), ETM+ (Enhanced Thematic Mapper Plus), Dan SPOT-5 (Satellite Pour l'Observation de la Terre) XS of 1989 and 2010 had been used in this study. Activities included satellite image processing and ground truth on the study site through Peninsular Malaysia (Hamdan et al. 2012). A map and extent of the mangroves including their status were successful determined as the results of this study. About 115,108.60 ha of mangrove were recorded in Peninsular Malaysia for 2010 (Table 1 and Fig. 1).

This study shows that the remote sensing and GIS technologies could be used easily and efficiently in determining mangrove areas and their status in Peninsular Malaysia. The remote sensing and GIS technologies are relatively well-known and readily available to many agencies. Therefore, the technologies can be used easily and cost-effectively by the relevant agencies in managing their mangroves for protection and sustainable utilization.

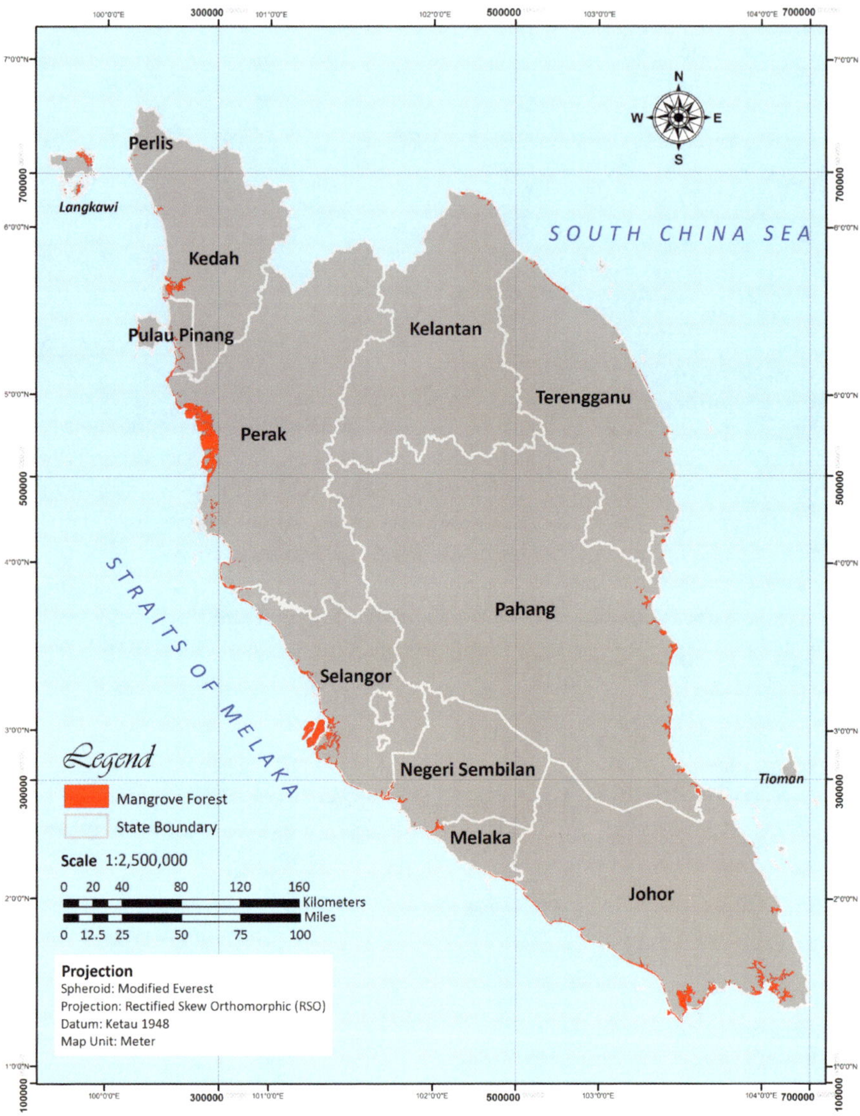

Fig. 1 Map of mangrove areas in Peninsular Malaysia for 2010

3 Earth Observation for Mangroves Changes Detection

Nearly all mangrove nations have experienced net losses in mangrove cover in recent decades as a result of human activities. Similarly, many remaining mangrove areas are no longer pristine, with most showing some level of ecosystem alteration as a result of utilization of wood or the harvesting of fish and shellfish/cockle. Spalding et al. (2010) identified six major impacts of human activities that can be translated as threats to the mangroves, *viz.* (1) conversion to other uses, (2) overharvesting, (3) overfishing, (4) pollution, (5) sedimentation and (6) alteration of flow regimes. Out of these, direct conversion to other uses is the most substantial change to the world's mangrove cover, which includes conversions to urban and industrial spaces, to aquaculture and to agriculture. Similar impacts are observed in Peninsular Malaysia from this study. In addition to these human impacts, natural processes, e.g. coastal erosion, are also the non-human impacts that destroy most of the mangroves in Peninsular Malaysia, including the tragic tsunami on 24 December 2004.

A total areal extent of approximately 111,046 ha of mangrove forests was degraded and converted during the period of 1973–2000 due to unsustainable forestry practices, illegal harvests, agriculture, aquaculture, construction of airports and harbours, industrialization and urbanization, land reclamation for coastal development, and waste disposal and pollution (Chong and Sasekumar 2002). During that time period, specific losses in five states in Malaysia to anthropogenic impacts had an alarming range between 20 and 70 % (Chan et al. 1993). The most significant losses have been in Peninsular Malaysia where large areas have been converted into coastal roads, agricultural land and housing estates (Hamdan et al. 2012).

Research (for example FAO 1997; Aizpuru et al. 2000) indicates that the extent of mangrove areas in Malaysia is decreasing, from about 700,000 ha in 1975 to 572,000 ha in 2000 due to the intensive harvesting and natural wave actions. FAO (1981, **?**) also reports that the total mangrove area worldwide has declined from 18.8 million ha in 1980 to 15.6 million ha in 2005. There has, however, been a slowdown in the rate of mangrove loss; from some 187,000 ha annually in the 1980s to 102,000 ha a year between 2000 and 2005, reflecting improved management systems and increased awareness of the value of the mangrove ecosystem. Asia has suffered the largest net loss of mangroves since 1980, with more than 1.9 million ha destroyed, mainly due to changes in land use.

There have been case studies to assess mangroves changes, for example, in Terengganu (Sulong et al. 2002) and in Selangor (Khali Aziz et al. 2009), and to identify the threats, but these are unable to represent the conditions at national level. A study (i.e. Hamdan et al. 2012) was conducted to provide the information pertaining to the threats to mangroves and to assess the changes that have taken place in recent decades in the ecosystem. Several states with large extents of mangroves in Peninsular Malaysia have been assessed to represent the current trend of mangrove changes as well as to identify the threats faced by mangroves. Three series of temporal satellite images were acquired in the years 1990, 2000 and 2010 to complete the exercise. Landsat TM, Landsat ETM+, SPOT-4 and SPOT-5 satellite images were utilized in this study. Changes were categorized into several 'from-to' classes according to factors, which are (1) land-use/land-development activities, (2) agricultural conversion, (3) aquaculture activities, (4) erosion and (5) others.

Table 2 and Fig. 2 shows the mangrove changes in the five states which have the majority of mangrove areas in Peninsular Malaysia that have been recorded for this study. It is clearly seen that within the last two decades, 25,637.89 ha of mangroves

Table 2 Changes in mangrove areas in the last two decades

State	Mangrove area (ha) 1990	Mangrove area (ha) 2000	Mangrove area (ha) 2010	Mangrove reduction (ha) 1990–2010
Johor	29,797.13	25,477.19	23,676.43	6,120.70
Kedah	9,236.24	8,322.79	7,841.25	1,394.99
Pahang	11,467.03	10,791.42	9,039.26	2,427.77
Perak	52,562.00	46,057.00	43,291.97	9,270.03
Selangor	28,954.60	24,213.14	22,530.20	6,424.40
Total	*132,017.00*	*114,861.54*	*106,379.11*	*25,637.89*

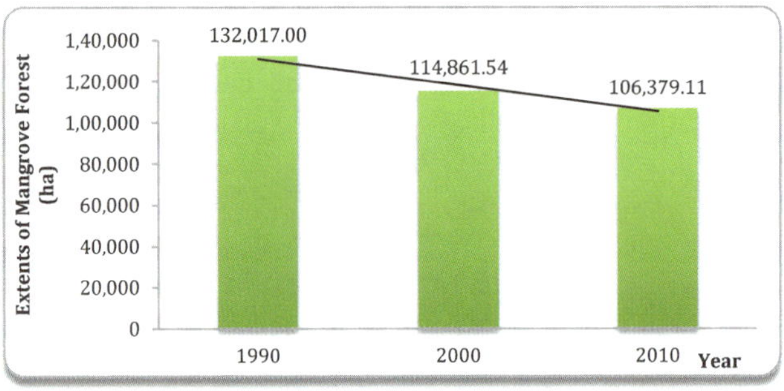

Fig. 2 Trends of mangrove changes over the last two decades (1990–2010)

have been lost. Major factors that contributed to these changes have been identified as: (1) direct conversion to other land uses, principally for aquaculture and agricultural and (2) coastal erosion. The other factors such as overharvesting and pollution affect the mangroves to a lesser degree.

4 Phenology of Mangrove Species

Studies on phenology of mangrove species were conducted in determining their flowering and fruiting behaviour. A calendar on the phenology of 25 mangrove species in Peninsular Malaysia has been established as shown in Table 3. The phenology calendar serves as a guide for researchers, foresters, students and groups who are interested in mangrove plantation establishment or in other related studies. The information on the phenology is about which species are fruiting and in which months the fruits are ready for collection.

This phenology study has recorded the flowering and fruiting behaviour of mangrove species. This information is useful for planting activities such that planting materials preparation can be in line with phenology pattern (Marzalina et al. 2011).

Table 3 Phenology pattern for mangrove species in Peninsular Malaysia. (Source: Marzalina et al. 2011)

Species	J	F	M	A	M	J	J	A	S	O	N	D
Avicennia alba							[blue]	[blue]	[blue]			
Avicennia marina						[blue]	[blue]	[blue]				
Avicennia officinalis						[blue]	[blue]					
Avicennia rumphiana						[blue]	[blue]	[blue]	[blue]			
Bruguiera cylindrica										[blue]	[blue]	
Bruguiera gymnorrhiza									[blue]	[blue]	[blue]	
Bruguiera parviflora								[blue]	[blue]	[blue]		
Callophylum inophyllum									[blue]	[blue]		
Casuarina equisetifolia			Y2	Y2								
Cerbera odollam		[blue]	[blue]									
Ceriops tagal						Y2	Y2	Y2	Y2	Y2		
Excoecaria agallocha	[blue]	[blue]					[blue]	[blue]				
Leucaena glauca	[blue]	[blue]	[blue]									
Lumnitzera littorea										[blue]	[blue]	
Lumnitzera racemosa									[blue]	[blue]		
Melaleuca cajaputi										[blue]	[blue]	
Nypa fruiticans									[blue]	[blue]		
Rhizophora apiculata									[blue]	[blue]		
Rhizophora mucronata		[blue]	[blue]									
Sonneratia alba	[red]		[blue]	[blue]	[blue]							[red]
Sonneratia caseolaris	[red]	[red]			[blue]	[blue]						[red]
Sonneratia ovata		[red]	[red]			[blue]	[blue]					
Talipariti tillaceum	[blue]	[blue]			[red]	[red]	[blue]	[blue]			[red]	[red]
Xylocarpus granatum		[blue]	[blue]	[blue]						[red]	[red]	[red]
Xylocarpus moluccensis			[red]		Y2	Y2						

Note:

[red]	Peak flowering (Year 1)
[blue]	Peak Fruting (Year 1)
Y2	Peak Fruting (Year 2)

5 Raising Planting Stocks

A study conducted on raising planting stocks found that planting stocks of mangrove species may be obtained from wildings, seeds or propagules depending on the species. As for example, *Rhizophora* spp. produce propagules which can be collected fresh from the mother trees; however, only mature and pest-free propagules should be used for planting (Marzalina et al. 2011).

Propagules can be directly planted by inserting the pointed part into the mud or the propagules can also be raised in the nursery to produce potted seedlings. Meanwhile, *Avicennia* spp. produce seeds, hence the best way of producing planting stocks is through potting the germinated seeds in the nursery for required sizes of potted seedlings. Common nursery practices could also be applied in managing the mangrove planting stock in the nursery.

6 Monitoring of Wave Energy and Ocean Current on Coastlines

Mangrove plants require certain environmental conditions to establish and grow. In general, mangrove plants are divided into two groups: (1) true or exclusive mangrove species and (2) associate mangrove species (Hamdan et al. 2012).

The Malaysian coastline, which is about 4,800 km in length, is rich in coastal resources and has an abundance of natural biodiversity. However, the National Coastal Erosion Study (NCES) has shown that about 1,390 km of the total coastline is facing erosion (Ziauddin and Siti Aishah 2010). This is due to the rapid pace of development activities in the coastal area which has resulted in a conflict between the need for immediate consumption and the need to ensure the long-term supply of these resources.

The overall erosion rate on the eastern coast was found to be 3.6 m y^{-1} and on the western coast it was estimated at 0.04–0.25 m y^{-1} (Hamdan et al. 2012). In some places the occurrence of erosion and accretion coupled with sand movement continuously takes place. The eroded sediment is transported from higher energy segments by littoral drift to lower energy segments and accumulates there.

Mangrove-dominated coastal segments exhibit less erosion while non-vegetated segments or former mangrove areas incurred substantial erosion (Thampaya et al. 2006). The dense structure of mangrove root systems possibly helps to consolidate (firm up) the coastal soil; hence, the shoreline is more resistant to erosion (Mazda et al. 1997). Furthermore, mangrove roots reduce flow and promote flocculation and sedimentation upon the soil surface, eventually allowing positive accretion. On the other hand, exposed and unconsolidated soils of non-vegetated and former mangrove land (mud-flats) are more prone to erosion. The possible and cheaper strategy in sustaining the coastal zone areas is by enhancing the total area covered by mangroves. This can be easily achieved by means of assisting the natural mangrove colonization in sheltered coastal segments through providing to or enhancing seedling fluxes on the area, protecting the seedlings from herbivory and increasing the propagules retention time with artificial shelters. If the natural colonization is not possible due to strong wave impacts, planting of suitable mangrove species with proper techniques of planting would be the best options. To facilitate decision-making for planting mangroves, demarcated maps of mangroves should be made available. The maps will provide overall information on the location and extent of estuarine, brackish water and mud-flat areas available along the coastlines.

A development of the mangrove buffer zone along coastal areas has become a research priority project between scientists in the affected coastal regions including Malaysia. The proven functions of mangroves has instigated widespread reforestation schemes to cope with the declining coastal mangroves (FAO 1994; Havanond 1995; Field 1996).

6.1 Hydraulic Parameters of the Planting Site

The National Hydraulic Research Institute of Malaysia (NAHRIM) has developed a coastal vulnerability index (CVI) for the Peninsular Malaysia coastlines (Lee and Mohd Fauzi 2010). The CVI incorporates six variables, viz. geomorphology, shoreline change rate, maximum current speed, maximum tidal range, significant wave height and sea level rise. The CVI may assist in determining the level of risk for rehabilitation activities. The assessment uses hydraulic parameters on the west coast of Peninsular Malaysia and has concluded that mangroves survive well in areas if the significant wave height is < 1 m, there is a yearly calm condition $> 30\,\%$ of the time and mean suspended sediment is $> 300\,\text{mg l}^{-1}$ (Nor Aslinda et al. 2010). Thus, if readings of the parameters go beyond these levels, it requires immediate action for improvement.

6.2 Breakwaters

Erosion and accretion are natural processes that takes place along the dynamic coastlines. Erosion along either sandy or muddy coastlines occurs as a result of a couple of factors such as wave actions that hit perpendicular to the coastlines and the swift flow of sea currents washing away the soil or sand particles. The height and frequency of waves hitting the coastlines contribute to the severity of erosion. Thus, the presence of coastal vegetation, especially mangroves, would assist to reduce soil erosion significantly. Coastal vegetation actually acts as a natural bio-shield or bio-barrier or as natural breakwater. In Peninsular Malaysia, a study of mangroves' ability to act as wave dampener was conducted on mangrove and associated species, *Rhizophora* spp., *Bruguiera* spp. and *Nypa fruticans* (Isfarita et al. 2010). The study showed that *Nypa fruticans* has the densest roots and the best species to attenuate wave followed by *Rhizophora* spp. The study conducted by Wolaski et al. (2001)

Fig. 3 L-Blocks established in study site at Sg Hj Dorani, Selangor

Fig. 4 Geotubes established in study site Sg Hj Dorani, Selangor

concluded that each mangrove species has a unique configuration of trunks, prop roots and pneumatophores that work as different drag forces, therefore resulting in different reduction rates of sea waves.

However, rehabilitating degraded coastal mangroves due to erosion through planting requires not just an improvement in the techniques of planting but also the need for the newly planted mangroves to be protected from high and strong wave impact. In areas where the wave height is more than 1 m and the yearly calm condition more than 30 % of the time (Nor Aslinda et al. 2010), a frontline breakwater is crucial in order to support the new plants to stay put.

The construction of hard and soft engineering structures such as rock armour, groyne, Labuan block, seawall and geo-textile tube (geotube) (Fig. 3) has been undertaken by the Department of Irrigation and Drainage (DID) to protect the coasts. L-Block, an engineering structure designed by the University of Malaya, has been established in Sungai Haji Dorani (Fig. 4). Therefore, a soft engineering structure such as geotube is preferable because the system creates a favourable hydrodynamic environment that may be conducive to mangrove regeneration.

7 Soil Monitoring

The study of soil monitoring was done in deciding which species of mangroves are suitable to be planted in a particular area (Wan Rasidah et al. 2010). Muddy soil is found as the best soil for mangrove planting. Mud is best characterized as soft sediment composed of a combination of organic and inorganic materials. The firmness of mud can vary from loose to hard. However, planting mangrove should be avoided in loose muddy soils because propagules or seedlings will be washed away during high tidal activities and currents.

This study concluded that different soil types have an effect on the presence and fertility of mangrove species. Suitable areas for *Rhizophora* spp. is where there is slightly structured soil which contains a lot of decaying organic matter. Meanwhile, *Avicennia* spp. are growing well in a sandy mud area with the presence of organic matter.

8 Planting in Severe Erosion Areas

This study aims to conduct research on techniques of coastal stabilization prior to planting, especially in areas classified as highly eroded with strong wave actions. Improved planting techniques need to be developed to ensure a high rate of survival of planted seedlings in a scenario of strong wave and current action, high and low tide scenario, and to reduce mortality due to barnacles, insect attacks and diseases. Planting mangroves at the coastal sites, especially facing the open sea, needs improvement upon the conventional techniques of planting in order to ensure the planted mangroves hold the ground. The plants may be uprooted and washed away. The larger sized plants with more extensive root systems promise greater and faster shore protection.

The study found that breakwater is needed in the areas where the height of the waves is more than 1 m. This is to ensure the success of the planting. Breakwater in the form of geotubes was found to be useful in a planting trial in high risk erosion areas in Selangor (Raja Barizan et al. 2010). Each of the geotubes measured 1.8 m in height, 3.7 m in width and was placed about 100 m away from the coastline of the planted site. The mud accumulated between the geotubes and the coastal line provides a space for planting mangrove species. The study found that about three years is needed in stabilising the mudflat at the back of the geotubes before the planting. This is to ensure a high rate of survival and simultaneously encourage natural regeneration in the area. This study shows that even the high-risk coastal areas could be rehabilitated; however, it required more time and sufficient budget.

9 Control of Pests and Diseases

Through years of managing mangroves, there have been few reports of any devastating effects of pests and diseases. The common incidence of pests on newly planted mangroves using propagules is crab attack. However, crab attacks are found to be localized and to overcome the crab problem, it is recommended to use potted seedlings. Barnacle infestation is also a common problem especially those planted along the shorelines. This study produced a handbook by Su Ping et al. (2010) entitled 'Pests of Planted Mangroves in Peninsular Malaysia'.

10 Earth Observation for Carbon Stock Monitoring

Mangroves located at estuarine sites have been estimated to have higher amounts of carbon (C) stored at an average of 1,074 Mg C ha^{-1}, whilst oceanic sites contained 990 Mg C ha^{-1} (Daniel et al. 2011). The two carbon pools that are critical to the mangrove forests' role as efficient and intense carbon sinks are the burial of carbon in sediments or soil organic carbon and the above-ground and below-ground living biomass (Laffoley and Grimsditch 2009; Murray et al. 2011).

It is interesting to note that although the plant biomass in the ocean and the coastal areas comprise only 0.05 % of the total plant biomass on land, it cycles a comparable amount of carbon each year (Bouillon et al. 2008; Murray et al. 2011). It has been estimated that a typical hectare of mangroves has the potential to release as much carbon as a 3–5 ha of a terrestrial tropical forest (Ong 1993; Murray et al. 2011). In a study to determine the mangrove ecosystem's capacity as a carbon source and sink, Ong (1993) had even postulated that an effective management and conservation of the mangrove ecosystem would be vital in any effort to address climate change. This may have been one of the earliest recognitions in literature, if not the first, of the importance of mangrove forests in climate change mitigation and adaptation.

However, despite being a crucial component in the global C cycle and having a potentially profound influence on climate change, the areal extent of mangrove forests has declined significantly as a result of coastal development, aquaculture expansion and over-harvesting (FAO 2007). It has been estimated that the world has seen a total loss of 35 % of mangrove area with a current estimated loss of 1–3 % per year (Murray et al. 2011; Valiela et al. 2001). This represents a rate of approximately 2–15 times faster than the loss of terrestrial tropical forests, which has an estimated loss of 0.5 % per year (Achard et al. 2002).

This is essential to form a better understanding of how the C dynamics respond to changes in time and space. It will also enable the formulation of management strategies that could incorporate the management of carbon flux to promote efficient and sustainable use of ecosystem services.

Satellite remote sensing offers the potential to methodologically quantify the carbon dynamics in a more detailed and accurate manner. The recent advancement in technology has not only improved the accuracy of remote sensing for mapping and change detection but has also led to an increased capability of estimating standing biomass (Proisy et al. 2007). It offers the ability to effectively measure and monitor large forested areas in a very consistent and robust manner (Huete 2012). Remote sensing is also a potentially effective tool because it offers a cost effective and non-intrusive method in the analysis of the characteristics of mangroves.

A study was conducted to quantify and characterize the carbon dynamics of the Matang Mangrove Forest Reserve in Malaysia by applying satellite remote sensing techniques. This mangrove forest covers an area extent of over 41,000 ha. It is the single largest mangrove tract in Malaysia and has been primarily and sustainably managed for more than a century for the production of charcoal and poles (Kamaruzaman and Dahlan 2009). Whilst an extensive range of studies has focused on the estimation of biomass and the productivity of Matang Mangroves (Putz and Chan 1986; Gong and Ong 1990; Ong et al. 1995), these studies have primarily focused on a specific date or time. There is yet to be a study that has extensively analysed the carbon flux of Matang Mangrove Forest Reserve over a period of time. This study has been designed to address the literature gap in the C dynamics of a managed mangrove forest.

The study found that the C stocks of 1991 and 2011 ranged from 1.03 to 263.65 t C ha^{-1} and 1.01 to 259.68 t C ha^{-1}, respectively (Fig. 5). Dominant C stock for both years was around 100 t C ha^{-1}. A slightly higher average in C stock during 1991 (76.62 t C ha^{-1}) as opposed to that found in 2011 (69.89 t C ha^{-1}) could have possibly been attributed to the production of charcoal and poles.

Putz and Chan (1986) found that the highest C stocks for a > 80 year-old *Rhizophora apiculata*- dominated mangrove forest was 230.0 t C ha^{-1}, and Ong et al. (1995) estimated that average C stock for 20- and 28-year-old *Rhizophora* forests was 114 t C ha^{-1} and 105.9 t C ha^{-1}, respectively. The average C values derived from 1991 and 2011 in this study are in agreement with previous research (i.e. Putz and Chan 1986, Ong et al. 1995), which was performed at Matang Mangroves.

In order to quantify the growth rate of mangroves in logged-over forests, a comparison was made between 645.3 ha of the clear-cut regime in 1991 (where C stock was relatively at 0 t C ha^{-1}) and the 20-year-old forest in the same area in 2011 (Fig. 6). In 2011, the total C accumulation in that area was 53,234 t C (ranges between 0 and 184.3 t C ha^{-1}) with an average of 136.7 t C ha^{-1}. This indicates that the mangrove trees in that particular regime have sequestered carbon at about 2,662 t C yr^{-1} with an average of 4.1 t C ha^{-1} yr^{-1}. Within the next ten years (after 2011), this regime will undergo the next rotation of a 30-year clear felling cycle that is practiced for mangrove forest production. It can be concluded that even though a 20-year mangrove is considered as a mature forest, the C stock of this planted mangrove will never be the same as natural mangrove stands such as found in VJR, which is reserved as a natural basis that can store C up to 250 t C ha^{-1}.

Considering the spatial distribution over the entire study area of about 41,540 ha in year 1991 and 40,710 ha in 2011, the total C stock in Matang Mangroves were 3.04 and 2.15 million t C in 1991 and 2011, respectively. This means that the total change of C stock within the period was 883,928 t C, with an average increased of 44,196 t C yr^{-1}. These estimates were derived from the individual pixels obtained from the image data sets of the mangrove forest. It took into account the variation of each C stock value of each pixel in the satellite images. As it did not apply the biome average method which involves the application of an estimated average C stock value to the entire areal extent, the estimates produced in this study present a more accurate estimate that reflects the profile of the forest.

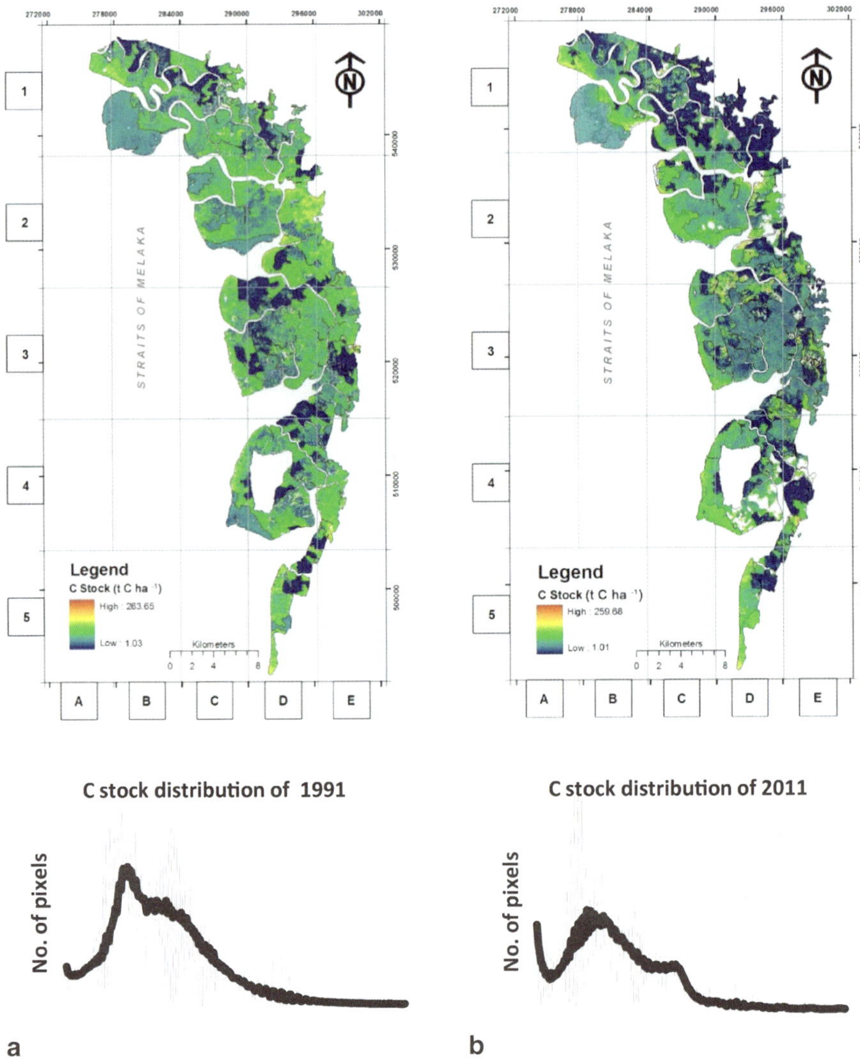

Fig. 5 Distribution of C stocks in the study area for the year 1991 (**a**) and 2011 (**b**)

Overall, the study indicated the use of vegetation indices from optical remote sensing data in the assessment of carbon stocks (and changes) on the mangrove ecosystem is still a viable technique. Multispectral optical data offer choices for C stock in mangrove forests where there is the capability to derive a number of indices that cannot be produced by other systems such as radar or lidar. This study thus suggests replication for C stock assessment on mangroves in different parts of the world.

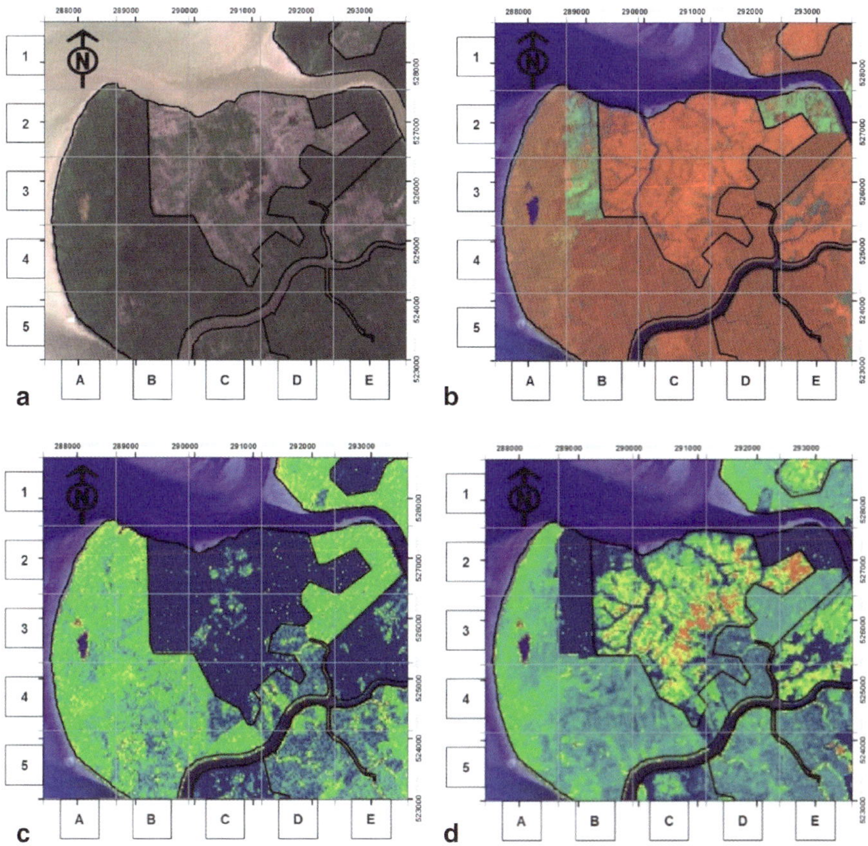

Fig. 6 Changes of C stock in logged-over areas. The cutting regime clearly appears on satellite image as depicted from the year 1991 (**a**) and the logged-over areas after 20 years as can be seen on an image from 2011 (**b**). **c** and **d** are spatial distribution of C stock for the years 1991 and 2011, respectively

11 Conclusion and Future Perspective

This chapter briefly elaborates completed R&D activities conducted under JTR&D. There is a lot of scope of studies covering aspects of mangrove area assessment, producing planting materials, planting in severe erosion areas, control of pests and diseases of the planted mangroves and non-mangroves species, monitoring of wave energy and carbon stock assessment. The completed R&D activities were published separately to be used as a guide for the stakeholders in mangrove management.

However, some R&D activities are still on-going as they require more concrete and long-term data before they could be completed. As a note, this project covered five broad components to answer sustainable management of mangroves in Peninsular Malaysia, including in non-forest reserve areas. All outputs of this project are expected to significantly contribute towards the sustainable management of mangroves and may be applied in other mangroves elsewhere in this country in which mangroves have existence as part of it's landscape ecosystem. In fact, it may be useful for other mangrove areas in the world that have similar characteristics and issues. Nonetheless, R&D activities on mangroves still need to be conducted accordingly to ensure that the management of this area is based on scientific data and findings.

References

Achard F, Eva HD, Stibig HJ, Mayaux P, Gallego J, Richards T, Malingreau JP (2002) Determination of deforestation rates of the world's humid tropical forests. Science 297:999–1002

Aizpuru M, Achard F, Blasco F (2000) Global assessment of cover change of the mangrove forests using satellite imagery at medium to high resolution. EEC Research Project No. 15017-1999-05 FIED ISP FR. Joint Research Centre, Ispra, Italy

Bouillon S, Borges AV, Castañeda-Moya E, Diele K, Dittmar T, Duke NC, Kristensen E, Lee SY, Marchand C, Middelburg JJ, Rivera-Monroy VH, Smith TJ, Twilley RR (2008) Mangrove production and carbon sinks: a revision of global budget estimates. Glob Biogeochem Cycles 22:1–12

Chan HT, Ong JE, Gong WK, Sasekumar A (1993) Socioeconomic, ecological and environmental values of mangrove ecosystems in Malaysia and their present state of conservation. In: Clough BF (ed) The economic and environmental values of Mangrove forests and their present state of conservation in the South-East/Pacific Region. Japan International Association for Mangroves, Okinawa, pp 41–82

Chong VC, Sasekumar A (2002) Coastal habitats (mangroves, coral reefs and seagrass beds) of the ASEAN region: status, utilization and management issues. Fishery Sci 68(1):566–571

Daniel CD, Kauffman JB, Murdiyarso D, Kurnianto S, Stidham M, Kanninen M (2011) Mangroves among the most carbon-rich forests in the tropics. Nature Geoscience Letters: NGEO1123

Field CD (1996) Restoration of mangrove ecosystems. International Society for Mangrove Ecosystems (ISME), Okinawa, pp 250

Food and Agriculture Organization (FAO) (1981) Tropical forest resources assessment project. Forest Resources of Tropical Asia. FAO, Rome, pp 475

Food and Agriculture Organization (FAO) (1994) Mangrove forest management guidelines. FAO Forestry paper 117. FAO, Rome, pp 319

Food and Agriculture Organization (FAO) (1997) Asia-Pacific forestry sector outlook study. Country Report—Malaysia. Working Paper Series (FAO):7/FAO Bangkok (Thailand). Regional Office for Asia and the Pacific, pp 30

Food and Agriculture Organization (FAO) (2007) The World's Mangrove 1980–2005. FAO Forestry Paper 153. FAO, Rome, pp 77

Gong WK, Ong JE (1990) Plant biomass and nutrient flux in a managed mangrove forest in Malaysia. Estur Coastal Shelf Sci 31(5):519–530

Hamdan O, Khali Aziz H, Shamsudin I, Raja Barizan RS (2012) Status of mangroves in Peninsular Malaysia. Forest Research Institute Malaysia, Kepong

Havanond S (1995) Re-afforestation of mangrove forests in Thailand. In: Khenmark C (ed) Ecology and management of mangrove restoration and regeneration in east and southeast Asia. Proceedings of the ECOTONE IV. Kasesart University, Bangkok, Thailand, pp 203–216

Huete A (2012) Vegetation indices, remote sensing and forest monitoring. Geogr Compass 6:513–532

Isfarita I, Mohd Lokman H, Sinha PC, Rosnan Y (2010) Root density analysis and wave reduction abilities of selected mangrove species of Peninsular Malaysia. In: Ahmad Jamaluddin S, Mohd

Fauzi M, Lee HL (eds) Proceedings of the National Seminar on Coastal Morphology (COSMO) 2010—The muddy coast of Malaysia, 17 June 2010, NAHRIM, pp 98–108

ITTO (2002) ITTO mangrove workplan 2002–2006. ITTO Policy Development Series No. 12, pp 6

Kamaruzaman J, Dahlan HT (2009) Managing sustainable mangrove forests in Peninsular Malaysia. J Sustain Dev 1(1):88–96

Khali Aziz H, Hamdan O, Shamsudin I, Ismail H (2009) Digital change detection of mangrove forest in Selangor using remote sensing and geographic information system (GIS). Malay Forester 72(1):61–69

Laffoley D, Grimsditch G (2009) The management of natural coastal carbon sinks. IUCN, Gland, Switzerland

Lee HL, Mohd Fauzi M (2010) Coastal vulnerable assessment for Peninsular Malaysia coastline. In: Ahmad Jamaluddin S, Mohd Fauzi M, Lee HL (eds) Proceedings of the National Seminar on Coastal Morphology (COSMO) 2010. The Muddy Coast of Malaysia, 17 June 2010. NAHRIM, pp 19–29

Marzalina M, Mohd Afendi H, Nashatul Zaimah I, Tariq Mubarak H, Ang KC, Azian M (2011) Komponen 1: Teknik penghasilan bahan tanaman spesies bakau yang mampu menampung permintaan negara. In: Aminah H, Azian M (eds) Laporan Jawatankuasa Teknikal Mengenai Penyelidikan dan Pembangunan (R&D) Program Penanaman Bakau dan Spesies-Spesies yang Sesuai di Pesisiran Pantai Negara Bagi Tahun 2009, pp 2–10

Mazda Y, Wolaski E, King B, Sase E, Ohtsuka D, Magi M (1997) Drag force due to vegetation in mangrove swamps. Mangrove Salt Marshes 1:193–199

Murray B, Pendleton L, Jenkins WA, Sifleet S (2011) Green payments for blue carbon economic incentives for protecting threatened coastal habitats. Nicholas Institute Report NI R 11-04. Duke Nicholas Institute for Environmental Policy Solutions, USA

Nor Aslinda A, Lee HL, Khairul Azuan A (2010) Laporan Institut Penyelidikan Hidraulik Kebangsaan (NAHRIM) Mengenai Lawatan Tapak Jawatankuasa Khas Penilaian Bagi Projek Inovaţif Bakau Untuk Negeri Perlis, Kedah, Pulau Pinang, Perak dan Selangor (unpublished report)

Ong JE (1993) Mangroves—a carbon source and sink. Chemosphere 27(6):1097–1107

Ong JE, Gong WK, Clough BF (1995) Structure and productivity of a 20-year-old stand of *Rhizophora apiculata* Bl. mangrove forest. J Biogeogr 22(2/3):417–424

Proisy C, Couteron P, Fromard F (2007) Predicting and mapping mangrove biomass from canopy grain analysis using Fourier-based textural ordination of IKONOS images. Remote Sens Env 109:379–392

Putz F, Chan HT (1986) Tree growth, dynamics, and productivity in a mature mangrove forest in Malaysia. Forest Eco Manag 17:211–230

Raja Barizan RS, Shamsudin I, Ismail H (2010) Unsung heroes of the mud-flats. Forest Research Institute Malaysia (FRIM), Malaysia, pp 115

Sulong I, Mohd-Lokman H, Mohd-Tarmizi K, Ismail A (2002) Mangrove mapping using landsat imagery and aerial photographs: Kemaman District, Terengganu, Malaysia. Env Dev Sustain 4:135–152

Spalding M, Kainuma M, Collins L (2010) World atlas of mangroves. Earthscan, UK, pp 319

Su Ping O, Chen S, Chong VV, Tan YS (2010) Pests of planted mangroves in Peninsular Malaysia. Forest Research Institute Malaysia (FRIM), Malaysia, pp 52

Thampanya U, Vermaat JE, Sinsakul S, Panapitukkul N (2006) Coastal erosion and mangrove progradation of Southern Thailand. Estuar Coastal Shelf Sci 68:75–85

Valiela I, Bowen JL, York JK (2001) Mangrove forests: one of the world's threatened major tropical environments. BioScience 51:807–815

Wan Rasidah K, Mohd Fakhri I, Suhaimi WC, Rozita A, Jeyanny V, Adi Fadzly AK (2010) Soil profiles of mangrove forest: case studies in Peninsular Malaysia. Forest Research Institute Malaysia (FRIM), Malaysia, p 38

Wolaski E, Mazda Y, Furukawa K, Ridd P, Kitheka J, Spagnol S, Stieglitz T (2001) Water circulation in mangroves and its implication for biodiversity. In: Wolaski E (ed) Oceanographic processes of coral reefs. CRC Press, London, pp 53–76

Ziauddin AL, Siti Aishah H. (2010) Integrated shoreline management Plan-ISMP. In: Abd Rahman AR et al (eds) Proceedings Seminar Kabangsaan Pemuliharaan Hutan Pesisiran Pantai Negera 2009: Memulihara Pesisiran Pantai Bersama Masyarakat. 1–2 December 2009, Kuching, Sarawak, pp 51–64

Climate Change Adaptation: Management Options for Mangrove Areas

Joanna C. Ellison

Contents

Abstract Mangroves are vulnerable to climate change, but this vulnerability can be reduced by adaptation planning by managers. Adaptive capacity is one of three components of vulnerability to climate change impacts, and includes both the abilities of mangroves to adapt to those impacts, and the capacity of mangrove managers to improve resilience. There are three categories of management actions: reduction of existing threats not related to climate change, direct adaptation actions and ongoing monitoring and evaluation. Actions that reduce existing threats include improvement of local management, improving legislation that facilitates mangrove protection and sustained use, establishment of strategic protected areas and rehabilitation of degraded mangrove areas. Direct adaptation actions include selection

J. C. Ellison (✉)
School of Geography and Environmental Studies, University of Tasmania,
Locked Bag 1376, Launceston, Tasmania 7250, Australia
e-mail: Joanna.Ellison@utas.edu.au

I. Faridah-Hanum et al. (eds.), *Mangrove Ecosystems of Asia*,
DOI 10.1007/978-1-4614-8582-7_18, © Springer Science+Business Media New York 2014

of "climate-smart" species in rehabilitation and choice of protected areas, management actions to promote sediment accretion in mangroves, and proactive planning for changed conditions. Ongoing monitoring actions include evaluation of mangrove extent and condition, continued monitoring of mangrove sedimentation rates, and evaluation of the success of active adaptation actions as well as actions to reduce non climate threats.

1 Introduction

Climate change is a threat to mangrove systems (Ellison 2013), and is likely to be the greatest challenge that society will face this century (Khan et al. 2012). Even the best-case scenarios of the projections available indicate that climate change will continue to be a major threat beyond 2100 (IPCC 2007). While other ecosystems such as coral reefs are most vulnerable to temperature increase, particularly in association with increased CO_2 (Hoegh-Guldberg and Bruno (2010), those effects of projected temperature increase and increased atmospheric CO_2 increase are likely to be mostly beneficial to mangroves, increasing mangrove productivity and biodiversity (Nicholls et al. 2007; Gilman et al. 2008) particularly at higher latitudes. The benefits of CO_2 increase are however subject to the limiting factors of nutrients, salinity and humidity (Ball et al. 1997). Rainfall and humidity changes have greater implications for mangroves (Ellison 2013), with reduced rainfall decreasing productivity and biodiversity and causing relative subsidence. Increased rainfall is likely to be beneficial, causing productivity and sedimentation along with enhanced groundwater and less saline habitats. Of all climate change impacts, the effects of relative sea-level rise are possibly the most detrimental to mangroves (Gilman et al. 2008; Krauss et al. 2010), shown first by mortality of mangroves at the seaward edge and migration inland, but there depending on sediment budgets and human modifications/barriers to migration.

It is therefore important that ways to increase the resilience of mangroves to climate change are developed, and increase adaptive capacity. This chapter reviews the contexts and management approaches that can assist mangrove and associated human systems deal with climate change vulnerability, and undertake appropriate adaptation.

2 Climate Change Adaptation

Climate change adaptation is defined as adjustment in economic, social or ecological systems in response to observed or expected changes in climatic conditions and their adverse impacts (Adger et al. 2005). Adaptive capacity and vulnerability have become useful concepts for analysing coupled human-environment response to climate change (Adger et al. 2007). Vulnerability is the potential to be harmed by a combination of exposure and sensitivity to stresses, and is reduced by the capacity to adapt to those stresses (Adger et al. 2007; Mertz et al. 2009).

Exposure refers to external factors, such as the type of change, its magnitude and rate, that a system or species may experience. In mangrove systems, factors of exposure to climate change include the rate of relative sea level rise and the degree of precipitation change (Ellison 2012). They also include factors concerning the setting of the mangrove system, such as the tidal range and the rate of sediment supply.

Sensitivity refers to internal characteristics of a species or system including its tolerance to changes in such factors as sea-level, temperature, rainfall, humidity, or fire. In mangroves, sensitivity is shown by a decline in forest conditions, as shown by productivity, biodiversity and recruitment (Ellison 2012). Sensitivity is also indicated by spatial change of mangroves (Gilman et al. 2007; Ellison and Zouh 2012), such as seaward edge retreat, and overall reduction in mangrove area. Sedimentation rates within the mangroves also indicate sensitivity, particularly if close to or less than the rate of relative sea level rise. Sensitivity of mangroves is also increased by reduction in resilience of adjacent systems that supply sediment or provide protection, such as coral reefs or seagrass, and adaptation actions to improve their resilience are provided by Marshall and Shuttenberg (2006), Grimsditch and Salm (2006) and Björk et al. (2008).

Adaptive capacity is the ability of a system or species to cope with or accommodate impacts of climate change with minimal disruption (Glick and Stein 2010). This can be through ecosystem or species response or through human actions that reduce vulnerability to actual or projected changes in climate. Resilience is the capability to absorb and recover from the effects of disturbance, and resistance is the ability to withstand change and continue to function. Hence, adaptation actions include those that reduce vulnerability, as well as those that enhance resilience and resistance (Adger et al. 2007).

Adaptation has been applied to the human or economic aspects of vulnerability, in consideration of human systems such as agriculture, public health, and natural hazards (Kelly and Adger 2000; Adger et al. 2005). The UNDP Adaptation Policy Frameworks addressed implementation of adaptation strategies that ensure human development in the face of climate change, including a structured approach, policies and measures (Lim et al. 2004). More recently, natural systems such as species, habitats, and ecosystems have also become assessed for their climate change vulnerability (Zhao et al. 2007; Lovelock and Ellison 2007; Nitschke and Innes 2008; Glick and Stein 2010). Vulnerability is therefore described as a combined function of three elements: exposure, sensitivity, and adaptive capacity commonly shown diagrammatically with a triangular relationship (Fig. 1).

2.1 Adaptation Approaches

Adaptive capacity can functionally be recognised as how much a system can adapt to the stresses of vulnerability as caused by exposure and sensitivity factors (Adger

Fig. 1 Vulnerability of mangroves as a combined function of exposure, sensitivity, and adaptive capacity

et al. 2007; Mertz et al. 2009). Adaptability therefore includes how much community structures, processes and practices can make adjustments in response to actual or projected changes of climate (Adger et al. 2007; Mertz et al. 2009). In the mangrove context this refers to communities, stakeholders and managers of a mangrove area, with the understanding of how mangrove systems are changing and how they may further change in the future. Adaptation processes can be anticipatory of climate change impacts, reactive after these come into effect, planned as a result of policy decisions in response to actual or anticipated changes, or an autonomous ecological change in a natural system triggered by change (Burley et al. 2012). In an ecosystem such as mangroves the adaptive capacity of the system is how much it can undergo spontaneous change, while humans can undertake all four processes, including autonomous change such as triggered by market or welfare conditions.

In mangrove systems both the ecosystem and people dependent upon their resources such as fish are best included in adaptation planning. Hence, in this context, adaptive capacity is how much the mangrove ecosystem and the people living in and around the mangroves can adjust to climate change, to moderate potential damages as this includes climate variability and extremes, deal with consequences and also notice and take advantage of any opportunities emerging as a result of climate change moderation of the ecosystem. The actions discussed in this chapter are what people managing mangrove ecosystems and stakeholders can do to increase their capacity to adapt to sea level rise impacts in particular, but also to enhance the capacity of people living nearby to assist in this process.

2.2 Stakeholder Involvement

Adaptation approaches are best participatory, involving stakeholders, who are individuals, groups, communities, organizations or government agencies that have an interest in the mangrove resource. A participatory approach to stakeholders allows understanding of their perspectives and knowledge; and it involves engaging local communities living in and adjacent to the mangrove area and including them in planned adaptation actions. Involving different stakeholders, including local communities, in the vulnerability assessment and adaptation process is an overarching principle to adaptation planning. Stakeholders usually have different exposures to climate change impacts (Khan et al. 2012), and deliberate involvement makes adaptation planning more encompassing of these different perspectives.

Stakeholder involvement can be improved by identifying and working with existing resource management structures and processes at both national and local levels. Stakeholder workshops are useful at early stages of adaptation planning as an information-sharing exercise, such as the identification of other relevant adaptation measures being undertaken.

Stakeholder contributions are enabled by ongoing communication through facilitator consultation, information exchange, workshops, meetings and sharing of reports and results. Engagement with local communities can be assisted by the delivery of materials in suitable formats, such as local-language publications. Such a focus on engagement, utilising a communications plan, should permeate all of the vulnerability assessment and adaptation approaches.

Awareness-building and education of community members improves the sustainable use of mangrove resources, helps to reduce direct human impacts, and builds capacity for climate change adaptation. This can be carried out through meetings, workshops, reconnaissance surveys of the mangroves that involve community members, and a range of other engagement activities, following the objectives shown in Table 1.

2.3 Climate Critical Mangroves

Mangrove areas with low resilience and high vulnerability are unlikely to adapt to climate change impacts without significant and immediate interventions. Vulnerability assessment can identify these particular areas as those with low productivity and biodiversity, exhibiting dieback and with low sedimentation rates, so allowing their prioritisation for adaptation actions such as special zoning and rehabilitation.

"Climate critical" species are those with restricted habitats, being less competitive species and more liable to be under represented or threatened. For example, in the SE Asian region two species of mangrove have been listed as Critically Endangered, *Sonneratia griffithii* and *Bruguiera hainesii* (Polidoro et al. 2010), primarily due to loss of habitats owing to clearance. Two further mangrove species, *Camptostemon philippinense* and *Heritiera globosa*, have been listed as Endangered, also due to

Table 1 Framework for community education regarding mangrove area climate change adaptation. (Adapted from Khan et al. 2012)

Objective	Purpose	Climate context	Content	Strategies
Convey information	Provide information	Exposure and sensitivity	Science of climate change and sea level rise	Community radio, newspapers, posters, exhibits, brochures, street plays
Build understanding	Exchange perspectives	Impact and vulnerability	Impacts on mangrove ecosystems and local people	Workshops, focus groups, community discussions, climate witness programs
Build skills	Improve capability	Capacity building	Ways to increase mangrove resilience and active adaptation planning prioritisation	Workshops, demonstration fieldtrips, use of mangrove vulnerability assessment manuals
Enable actions	Allow action planning, problem solving and monitoring	Adaptation planning	Community participation in actions to allow mangrove adaptation	Resourcing, action planning, participation, on ground works, ongoing monitoring and evaluation

the loss of mangrove areas to aquaculture and other pressures (Polidoro et al. 2010). Species such as these that are already under stress from non-climate stressors will be more sensitive to impacts brought by climate change, and reducing non-climate stressors will increase their resilience.

2.4 Reduction of Vulnerability

Vulnerability is the exposure and susceptibility to harmful stresses, combined with the ability to respond to those stresses (Adger et al. 2007; Mertz et al. 2009). Specific vulnerability of mangrove systems to climate change can be analysed (Ellison 2012) to allow adaptation strategies to be prioritised that may be applied in different ecological and socio-economic contexts.

Identification of specific site vulnerabilities such as microtidal range, low sedimentation rates or mangrove species with specific elevation bracket habitats allows the site-based prioritisation of adaptation strategies that can be adopted to reduce such vulnerabilities.

Table 2 Adaptation actions to reduce mangrove system vulnerability to climate change

Vulnerability	Adaptation priority (Sect. 3)
Tidal range small	Proactive planning for changed conditions Strategic protected areas Promotion of accretion in mangroves
Relative sea level rise high	Promotion of accretion in mangroves Proactive planning for changed conditions
Sedimentation supply rate low	Promotion of accretion in mangroves
Climate modelling shows drier conditions	Rehabilitation of degraded mangroves Monitoring
Mangrove condition, recruitment, tree growth poor and mortality high	Reduction of non-climate stressors Improved local management Improved legislation Strategic protected areas Rehabilitation of degraded mangroves Proactive planning for changed conditions
Seaward edge retreat occurring	Promotion of accretion in mangroves
Reduction in mangrove area occurring	Proactive planning for changed conditions
Elevations within mangroves tight	Promotion of accretion in mangroves Proactive planning for changed conditions
Elevations above mangroves unavailable	Proactive planning for changed conditions
Sedimentation rates in mangroves low	Promotion of accretion in mangroves
Adjacent coral reef resilience low	Reduction of non-climate stressors Rehabilitation (of reefs)
Adjacent seagrass resilience low	Reduction of non-climate stressors Rehabilitation (of seagrass)
Local community management Capacity low	Improve local management
Stakeholder involvement poor	Improved legislation Improved local management
Mangrove protection legislation weak	Improved legislation

Vulnerability assessment approaches include spatial analysis of changes in mangrove extent as indicated by comparison of past and recent aerial images, and verified by local community surveys (Ellison and Zouh 2012). Permanent plots are a well-established technique for long-term monitoring of mangroves, from which data collected can be compared, and provide a basis for monitoring mangrove community structure, biomass, growth and productivity. Mangrove condition surveys carried out in conjunction with local community participants can provide information on human impact levels or indication of natural changes. Reconstruction of the site's relative sea-level history can allow local subsidence factors to be identified to show relative sea-level rise rates, useful as most developing country locations with mangroves lack long-term tide gauges. Results from these components of vulnerability assessment contribute to identifying specific vulnerabilities, from which adaptation actions can be prioritised, as shown in Table 2. The adaptation priorities are outlined in the following sections.

2.5 Enhancement of Mangrove Resilience

Ecosystem resilience reflects the degree of perturbation that a system can absorb without changing its stability domain (Gunderson 2000) and lies in the variety of functional groups in the ecological system that provides sources for recovery. Boosting resilience is one of the more risk-averse measures of the range of climate change adaptation options that conservation management may adopt (Heller and Zavaleta 2009).

An underlying premise to increasing a system's resilience to climate change is the enhancement or protection of its natural capacity to respond to change and stress (Adger et al. 2007; Mertz et al. 2009). Most ecosystems have some capacity within their internal diversity to allow natural adaptation or adjustment, which lends greater resilience. Management actions that enhance the ecological diversity and productivity of a system provide a potential buffer against climate change impacts, and these include capacity building of local communities to better manage mangrove areas, community based monitoring of mangrove condition, and rehabilitation of degraded areas.

Mangrove ecosystems can be resilient to climate change if they are healthy, with high diversity and active sedimentation processes, and if there are inland migration areas available at suitable elevations. People living in and around mangroves can help them adapt to climate change by implementing management practices that enhance these mangrove values, and by reducing their pressure on mangrove resources.

Hence adaptation can include actions that may reduce vulnerability or increase resilience (Adger et al. 2007), and adaptive capacity is the potential ability of a system to successfully respond to climate change and variability (Adger et al. 2007). Adaptive capacity in ecological systems involves maintaining and increasing their diversity (Adger et al. 2005), and has been described as the robustness of a system to any change in resilience (Gunderson 2000). Adaptation options to increase the resilience of mangroves to climate change have been reviewed (McLeod and Salm 2006; Gilman et al. 2006a; Lovelock and Ellison 2007; Gilman et al. 2008; Gehrke et al. 2011; Waycott et al. 2011). The adaptation actions described in this chapter have been developed from those recommended in these reviews with the addition of others.

Response measures may differ between adaptation sites, because of varying socio-economic, ecological and biophysical conditions. Management strategies during adaptation to climate change are not divergent from conservation methods (Hansen et al. 2003), but there is more emphasis placed on protecting mangrove communities that have been shown to be resilient, managing for increased perturbation, and maintaining flexibility given the uncertainties of what climate change will bring.

Mangrove area adaptation planning to climate change encompasses a suite of response measures that will enhance mangrove resilience to climate change through the reduction of vulnerability. They are reviewed below as a set of targeted options that can directly reduce specific vulnerabilities. These adaptation actions have three main groups: reduction of non-climate stressors, active adaptation actions, and monitoring.

3 Adaptation Actions

3.1 Reduction of Non-climate Stressors

Reduction of non-climate stressors increases the resilience of habitats and species to the effects of climate variability and climate change (Erwin 2009; Hansen et al. 2009), and correspondingly, the vulnerability to climate change of natural resource-dependent communities is increased if their resources are degraded by overuse or if their management systems are ineffective (Adger et al. 2007).

The non-climate stressors on mangroves are substantial in many parts of the world, with human development pressure in the low-lying coastal zone leading to deteriorating water quality with pollution and other aspects of habitat degradation. Mangrove productivity has been further affected by overharvesting for timber along with other unsustainable use of resources such as fish and crabs. Mangrove habitats have been enclosed, drained and converted for aquaculture ponds, or agriculture such as rice paddies, or filled over to allow conversion to housing, tourism facilities or other coastal infrastructure. Mangrove areas worldwide have shown alarming rates of loss over the last few decades, with 188,000 km^2 in 1980 falling to 137,760 km^2 by 2000 (FAO 2003; FAO 2007; Giri et al. 2011), this due to human pressures as the assessments used pre-date the 2004 Asian tsunami (Valiela et al. 2001; Duke et al. 2007).

Unsustainable mangrove forest use is where resources are taken at a rate higher than their natural replacement. This occurs in many mangrove areas where communities are largely dependent on mangrove resources, such as fish, crustaceans, and fuelwood, and unsustained use can reduce mangrove resilience. Hence, a key adaptation response is improved management, education, and awareness-building as well as community involvement in mangrove area management. Reduction of non-climate stressors may enhance ecosystem productivity, which has been shown to cause elevation gain of tidal wetlands (Langley et al. 2009).

The adaptive capacity of a mangrove system can be increased by improvement in mangrove condition, allowing more vigorous growth and reproductive success. Where the condition of mangroves is already degraded, climate change is likely to make aspects of degradation worse. This positive feedback can be prevented by reducing existing human impacts on mangroves and by rehabilitating damaged areas.

Human pressure on mangroves (for example, through the gathering of food or fuelwood) has been reduced by many conservation practitioners through working with local communities and building their capacity to improve management and planning. Reduction of human impacts on mangroves is part of a "no regrets" strategy of improved management. In Cameroon, the local communities depend on mangrove wood as a fuel for cooking and smoking seafood and to provide poles for construction; and that wood is often gathered from unsuitable areas (Fig. 2). Mangrove wood gathering zones have now been designated, particularly excluding mangroves that are on or near the seaward edge or on the margins of creeks and waterways.

Fig. 2 Mangrove poles cut from a creek margin in Douala Estuary, Cameroon. At creek edges the mangrove are the least resilient owing to lower elevation, but unfortunately these areas are often the most accessible. (Photo: J. Ellison)

Fig. 3 Improved-efficiency wood smokehouses for cooking fish and shellfish in the Douala-Edea mangrove area, Cameroon. (Photo: J. Ellison)

Fuel wood extraction mangrove ecosystems in Cameroon's Douala-Edea Wildlife Reserve for commercial fish smoking using low fuel efficiency wood smoking systems is unsustainable (Feka et al. 2009). Improved smoking systems have been developed (Fig. 3) using about 50 % less wood and reducing health impacts on people. These smokehouses have been introduced to a number of mangrove-dependent communities, providing communal facilities for village use.

3.2 Improved Local Management

A key approach to the sustainable management and protection of wetlands such as mangroves is through the engagement of local communities, facilitated by their access to technical support and effective legislation (Ellison 2009). Awareness-building and education of community members improves the appreciation of mangrove values and improved sustainable use of mangrove resources, helps to reduce direct human impacts, and builds capacity for climate change adaptation. Community support for adaptation actions can be improved by education and capacity building, which are core tasks for many conservation institutions. Education and outreach programs regarding the value of mangroves allow change in attitudes and behaviour among individual members of the community (Gilman et al. 2008), bringing the overall community towards more informed decisions about mangrove resource use and management. Better management occurs in participatory community-based conservation areas where local committees close areas or restrict their use, based on the state of resources (Ellison 2009).

3.3 Improved Legislation

The engagement of local communities in the sustainable management of mangrove areas is absolutely indispensable, but this must also be supported by government legislation that protects mangrove ecosystems and allows management planning. Enforcement of existing legislation also needs to be effective. Legislative policies to protect mangroves are frequently weak or fragmented across a range of nonspecific laws that are administered by a range of governmental departments lacking the resources to implement them (Ellison 2009).

Such unspecific legislation for the protection or sustainable use of mangroves needs to be identified and improved, and conservation practitioners can assist in this process through ongoing stakeholder discussions, lobbying and advocacy. Management agencies can strengthen and further build capacity to enforce legislation and build staff understanding of wetland vulnerability and adaptation options. There is often a lack of understanding of the impacts of development on downstream mangroves.

Community and stakeholder involvement is an overarching objective in vulnerability assessment and adaptation planning (Sect. 2.2), and the sharing of results found of degraded mangrove areas, human impacts and local community feedback showing a need for changes to management approaches may help to support the case for legislative improvement. Such improvement can be facilitated by expert review of environmental legislation such as that carried out by the GEF International Waters Project in the Pacific Islands (Tavala and Hakwa 2004; Powell 2006; Evans 2006).

At the local level, policies for improved mangrove management can be successful through the participation of local communities in developing mangrove resource management committees (Ellison 2009). Improved management is helped by consultation and dialogue with key stakeholders such as local administration, local councils, and local government departments, especially forestry, environment, and fisheries.

3.4 Strategic Protected Areas

Protected areas provide refuges for wildlife and support centres of biodiversity. With climate change as a management concern it is important to designate areas that are likely to be resilient, also considering that mangroves are an ecosystem type that is under-represented in marine protected areas (Pomeroy et al. 2007). Mangrove settings that are strategic choices in consideration of future climate change are those that show lack of spatial change over recent decades (Ellison and Zouh 2012), have positive sediment supply, and high productivity and species diversity, as all of these factors enhance resilience.

Resilience building is one of the more risk-averse measures of the range of climate change adaptation options that conservation management may adopt (Heller and Zavaleta 2009). In anticipation of climate change related pressures, improved marine and land use planning that links adjacent habitat types that support each other into reserves is a precautionary measure to build resilience to climate change. For example, WWF has worked with communities in Tikina Wai, Fiji, since the late 1990s to establish three adjacent community mangrove reserves. These reserves are checked by community monitors who report to a marine resource committee with representatives from all towns in the sector. Permanent plots allow the basal area and biomass monitoring methodology recommended for evaluation of the success of protected area management (Pomeroy et al. 2007), and long-term monitoring of mangroves (Ellison et al. 2012). Community surveillance and monitoring enables feedback to the committee on any resource abuse or decline in fish or crab availability, and management decisions are made on this basis. Such community-managed mangrove protected areas increase resilience to climate change by reducing non-climate stressors and community surveillance of changes over time. Tikina Wai monitoring over 10 years has demonstrated the effectiveness of these protected areas, showing improvements in mangrove ecosystem health and productivity.

Climate change adaptation capacity of mangrove protected areas is increased if such areas have inland migration areas also reserved with low gradients to provide migration areas for up to 1 m and more of sea level rise. Protected areas that are larger may allow such migration as well as providing habitats for a variety of mangrove communities, which increases the adaptive capacity provided by biodiversity.

3.5 Rehabilitation of Degraded Mangroves

Rehabilitation of degraded mangrove systems is an effective strategy for building climate change resilience, particularly where sections of an otherwise healthy system are degraded. Degraded areas are more likely to show climate change impacts relative to mangroves that are healthy (McKee et al. 2007). Healthy mangroves promote higher levels of sediment accretion, while degraded mangrove areas are more likely to have reduced resilience to coastal erosion. Dense seedlings also enhance sediment accretion (Huxham et al. 2010; Kumara et al. 2010). Degraded mangrove locations within a particular forest area can be identified through forest survey assessment, spatial analysis evidence of forest decline, or compilation of local community knowledge.

There is a wealth of experience in mangrove reforestation, restoration, and replanting in many countries (Hong 1994; Chan 1996; Primavera and Esteban 2008; Biswas et al. 2009) that can be used to help enhance adaptive capacity. One successful example is a community forest at Yadfon in Thailand where, as part of a larger cooperative program to help fishing activities, a small committee promoted mangrove replanting (Quarto 1999). Fairly soon after these activities commenced villagers noticed increase in near-shore fish catches and greater abundance of rare species. This example shows the positive benefits of engagement of local community as well as stakeholders support.

A further successful example comes from the Upper Gulf of Thailand where, following coastal erosion, in 1987 the Thai government approved a national mangrove management plan including mangrove rehabilitation (Winterwerp et al. 2005). Following this, benefits observed included increased habitat for fish and crabs as well as migrating and local birds, and increased sedimentation. These projects demonstrate the "win-win" or "no regrets" outcomes of mangrove rehabilitation.

Although there is increasingly good Internet coverage of mangrove replanting activities, such as the Mangrove Action Project (2006) guide, there have been unsuccessful projects where most or all seedlings have died (reviewed by Lewis 2005). Successful replanting involves the flowchart of potential activities shown in Fig. 4.

3.5.1 Selection of "Climate-Smart" Mangrove Species

Understanding the ecology of local species is an important early step in successful mangrove restoration (Bosire et al. 2008), particularly in terms of choosing sites that have suitable hydrological regimes with respect to the frequency and duration of tidal flooding (Mangrove Action Project 2006; Bosire et al. 2008). Restoration also needs to consider changing future conditions of hydrology and mangrove habitats (Erwin 2009) in planning areas for replanting and selecting species. "Climate smart" species are those with the highest adaptive capacity, i.e. those that have more plastic ranges than others. These are the best to choose for rehabilitation.

The greatest climate change sensitivity of mangroves is to relative sea level rise, as the key stress to trees is increase of inundation periods. Sea level rise is projected to increase over the lifetime of mangrove trees. Although they can migrate, they will have to tolerate a continually rising sea level during their lifetime. Therefore, the most resilient species to changing sea level will be those with tolerance of a wider elevation bracket, which can be determined by survey of elevations at which different mangrove zones occur. In microtidal Cameroon's Douala Estuary, this was found to be *Rhizophora racemosa*, which occupies a c. 48 cm elevation bracket within a tidal range of about 1.2 m, while *Laguncularia* and *Avicennia* both have narrower elevation brackets, which would make these species less able to tolerate rising sea level (Ellison and Zouh 2012).

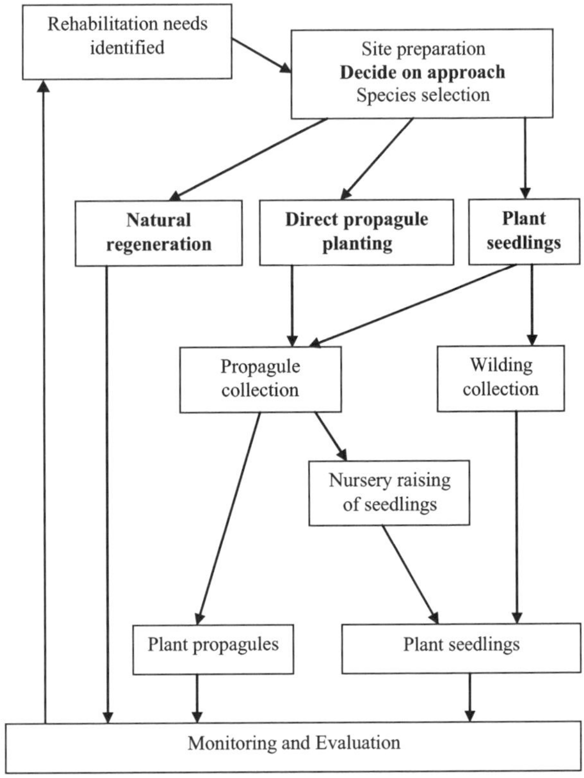

Fig. 4 Flowchart of approaches (*shown in bold*) and stages of mangrove rehabilitation

3.6 Promotion of Accretion in Mangroves

Sedimentation in mangroves allows the mangrove substrate to keep pace with sea level rise, which works as a natural adaptation process in mangrove systems and to reduce impacts of increased inundation stress. Sediment accretion allows the mangrove substrate to grow upwards. This can be facilitated by managers understanding the sedimentation processes and allowing accretion to occur.

Major external sources of sediment to the mangrove area include input from rivers in riverine settings, longshore transport, and gains from offshore which occur mostly during high magnitude storm/tsunami events. In such sudden extreme events excess sediment deposition may kill mangroves (Ellison 1998; Terrados et al. 1997). Sediment losses to the mangrove system include longshore transport down-coast, and erosion which can be enhanced by higher energy conditions such as boat wakes, which also tend to impact the vulnerable mangroves of seaward edge and creek margin locations. Accretion is also influenced by the internal supply and decomposition of organic matter and root mat growth (McKee et al. 2007).

Root mat growth has been shown to lower under dwarf or scrub mangroves and be higher under a dense, healthy mangrove forest (McKee 2011). Increased productivity of vegetation leads to marsh elevation gain (Langley et al. 2009). Dense seedlings also promote sedimentation including from root mat development, causing sediment surface elevation gain under densely replanted mangroves at both high and low tidal sites (Huxham et al. 2010). Seedling density enhances accretion rates in addition, by providing friction to tidal water movement to promote sediment flocculation and settling (Huxham et al. 2010; Kumara et al. 2010). Actions to enhance root mat productivity include reduction of non-climate stressors (Sect. 3.1) and replanting of degraded mangrove areas with dense seedlings (Sect. 3.5), which will enhance substrate accretion.

Reduced coastline sediment supply can result from increasing coastal human populations and associated development such as jetties (Appeaning Addo 2011). Foreshore developments can reduce longshore drift of sediment, thus reducing the supply of sediment into mangrove areas. These include coastal engineering structures such as groynes that tend to restrict sediment supply to down-drift coastal sections.

Fluvial dam construction can reduce riverine sediment supply and water discharge to coastal mangroves (Arthuron and Korateng 2006), leading to a sediment supply deficit. As mangrove resilience to rising sea level depends on sediment supply, these activities all contribute to the increased vulnerability. Adaptation actions that maintain and increase mangrove substrate accretion therefore should include river management agencies, infrastructure managers and coastal planners, to ensure designs that maintain sediment supply to the mangrove areas.

The management actions that can maintain and enhance mangrove sediment accretion are summarised in Fig. 5.

3.7 *Proactive Planning for Changed Conditions*

Active planning for conditions during rising sea level includes zoning inland for future mangrove migration areas as part of multi-sectoral regional coastal planning that integrates mangroves into an overall adaptation strategy. Reserve of potential migration areas inland of mangroves would improve future resilience of mangroves.

The largest impediment to migration of intertidal wetlands with sea level rise is barriers that may prevent wetland communities from accessing suitable adjacent areas (Lovelock and Ellison 2007; Gilman et al. 2008). Coastal lowlands behind mangroves are prime locations for transport corridors such as roads and railways, which tend to block access of tidal exchange and mangrove propagules to slightly higher land on the landward side.

Areas of greatest concern are mangrove areas that have a micro-tidal range, because of the amount of relocation of the mangroves onto a soil surface that was previously not mangrove habitat. A sea level rise of 0.18–0.59 m is projected to occur by 2099 (IPCC 2007), so considering mid-range sea-level rise of a 30 cm, in micro-tidal areas this would near totally relocate the intertidal zone upslope relative to a macro-tidal mangrove area. Micro-tidal areas of the Caribbean and Pacific feature low lying islands, where sediment and freshwater inputs from rivers and

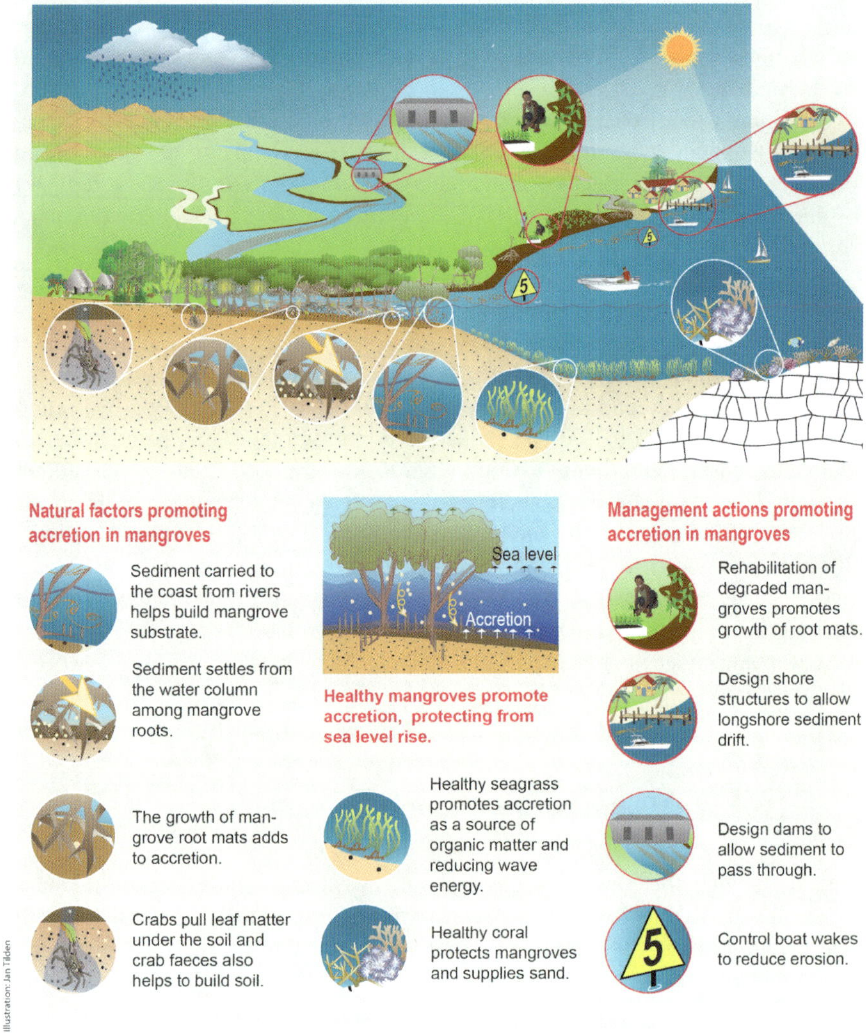

Fig. 5 Actions to enhance substrate accretion in mangroves (drawn by Jan Tilden)

groundwater are low relative to continental coasts, hence the capability of mangrove substrate to keep up will also be lower.

Planning for such future sea level rise should include the following considerations: elevation and gradient of land behind mangroves; barriers to migration, such as roads or railway tracks; background relative sea level trends of the area; sedimentation rates within mangroves and areas behind mangroves; and any development that may become problematic if inundated, such as rubbish dumps or local communities that may need relocation. The involvement of local communities in planning for changing conditions is exemplified by WWF work at Tikina Wai in Fiji.

Through a series of WWF facilitated community meetings in the villages of the district, the community systematically identified problems and their causes related to climate change vulnerability of their village and district (Table 3). Identification of these problems allowed formulation of solutions that would provide adaptation and guide community planning for changed conditions.

Identification of potential problems also facilitates communication to stakeholders and higher levels of government to allow sufficient lead time to develop socially acceptable, economically viable and environmentally sound management measures (Gilman et al. 2008). Rolling easements may allow legal options of eventual abandonment to be acceptable (Titus, 1991).

Such planning requires multi-sectoral collaboration, an enabling policy environment, and adaptive institutions at local and national levels (and maybe international). This planning can be instigated through the use of results of the vulnerability assessment process to point out needs to governmental and other stakeholders. Guidelines for larger-scale multi-sectoral regional coastal planning that integrates mangroves in an overall adaptation strategy are given by guides such as that provided by the U.S. Agency for International Development (2009).

3.8 Ongoing Monitoring and Evaluation

Science-based management is dependent on multi-scale understanding of biocomplex systems to allow for sound decisions (Day et al. 2008), and such understanding is provided through monitoring. Given the uncertainties about future sea level rise and climate change, as exemplified by the ranges and error margins of the projected changes (IPCC 2007), and the uncertainty of how increased CO_2 and changes in rainfall and sea level rise will combine to affect mangrove ecosystems and people, ongoing monitoring could be the most important adaptive management activity of all. Standardized methods between sites allow the separation of local influences on change from global influences as caused by sea level rise and global climate change (English et al. 1997; CARICOMP 2000; Ellison et al. 2012). The following parameters are best for ongoing monitoring of climate change impacts on mangroves: mangrove extent and condition; permanent plots showing community structure, biomass, recruitment and mortality; and sedimentation rates under mangroves.

Management of mangroves is best guided by such information about mangrove extent and condition. Ongoing repeat surveys will provide useful monitoring information on management success, needs, and climate change impacts. Ongoing monitoring also allows the evaluation of success of adaptation options once they are implemented, providing data on how the systems (both mangrove and local communities) respond. Community involvement with ongoing monitoring allows information on mangrove condition to directly inform local management decisions.

Table 3 WWF community consultation results from Tikina Wai, Fiji, and adaptation planning solutions

Location	Problem identified	Cause attributed	Solution, adaptation and community planning
Land	Freshwater shortage in dry season	Drought affects agricultural productivity and seasonality of traditional calendar of plants	Increase water storage capacity and improve district delivery of water
		Shift in the planting of traditional crops and increasing dependence on purchased food (with limited income source)	Increase understanding of alternative more climate smart crops
	Extreme rainfall events	Roads become impassable, breeding of mosquitoes and a rise in water-borne diseases (dengue, diarrhoea and skin diseases)	Increase school attendance flexibility Improve roads Develop better local income earning opportunities Improve community health education
Tidal	Sediment deposition in the inter-tidal areas	Logged pine forest areas associated with periods of heavy rain causes landslides and soil erosion	Improve catchment management, such as logging only in the dry season, riparian buffers
	Increasing shallowness of rivers and loss of wetlands near waterways Deeper areas in the tidal zone becoming shallow	Absent buffer zones between pine forests and the river exacerbate siltation within the river system	Improved understanding that sediment supply to the mangrove area is important for mangrove resilience to rising sea level
	Coastal flooding and erosion	Encroachment of the high tide mark inland as compared to the past	Improve survey points in the village to allow accurate comparison of land levels with sea levels Raise house bases
		Mangroves encroaching to previously exposed salt pans mean loss of a cultural heritage, the art of making traditional salt for which the district is renowned for	Gain funding for and build a more secure salt making facility on the highest section of the salt pan close to the village, also to facilitate tourism visits
	Excessive removal or cutting of mangroves from shoreline	Needs for wood. Inadequate surveillance and community education	Appoint Mangrove Monitors in each village for mangrove surveillance reporting to resource management committee, and require those who cut mangroves to replant mangroves Improve the traditional practise of bark harvesting so it does not damage tree health Rehabilitate and replant mangroves

Table 3 (continued)

Location	Problem identified	Cause attributed	Solution, adaptation and community planning
Coral reefs	Coral bleaching events observed	Correlation with ENSO events such as 2000	Develop partnership with a local dive shop for sea surface temperature monitoring on the barrier reef
	Increased crown of thorns during drought years	Unknown	Increase observation and communication among lagoon users to allow monitoring, reporting to resource management committee
	Fish spawning seasonality uncertain (compared to historical timelines)	Changed climate and coastal conditions	Ban commercial fishing in the marine protected areas Improve communication among fishing folks to pool community knowledge on fish spawning patterns

4 Conclusions

In coastal areas, three broad types of risk mitigation policies have emerged to address projected sea level rise: engineered coastal protection, accommodation in situ, and retreat or relocation inland (Alexander et al. 2012). Accommodation involves reduction of sensitivity and/or exposure to the effects of sea level rise, which for the existing large mangrove areas is the only feasible option. Planning for inland migration of mangroves in case such accommodation is not successful, especially if higher sea level rise projections occur in the future, is a pre-emptive back-up plan.

Three categories of adaptation management actions have been described here for mangrove systems. Actions that reduce existing threats include the improvement of local management, improving legislation that facilitates mangrove protection and sustained use, establishment of strategic protected areas and rehabilitation of degraded mangrove areas. These all would serve to increase mangrove resilience to impacts from climate change. Direct adaptation actions include the selection of "climate-smart" species in any rehabilitation, along with the creation and maintenance of protected areas, management actions to promote sediment accretion in mangroves, and proactive planning for changed conditions. Ongoing monitoring actions allow future planning as climate change increasingly occurs and adaptation options are implemented.

The overall framework for evaluation of adaptation strategies undertaken for coastal wetlands should be judged in how it has interconnected within its system as well as with other sectoral drivers (Burley et al. 2012). The success of adaptation outcomes may include its flexibility, inclusiveness, consistency and equitability (Burley et al. 2012), and evaluation strategies must be set in place as part of management. Stakeholder involvement in the adaptation planning process allows awareness of adjacent sectoral plans in coastal management, coastal policy, planning, economic development and transport and infrastructure, to allow compatibility and most efficient use of regional and national resources across these. Engagement and involvement of local communities around the mangroves, along with promoting their access to technical resources and advice, is most important in achieving climate change adaptation success.

Acknowledgments This research was funded by the UNEP Global Environmental Facility project "Coastal Resilience to Climate Change: Developing a Generalizable Method for Assessing Vulnerability and Adaptation of Mangroves" awarded to the World Wildlife Fund (WWF).

References

Adger WN, Arnell NW, Tompkins EL (2005) Successful adaptation to climate change across scales. Global Environ Change 15:77–86

Adger WN, Agrawala S, Mirza MMQ, Conde C, O'Brien K, Pulhin J, Pulwarty R, Smit B, Takahashi K (2007) Assessment of adaptation practices, options, constraints and capacity. In: Parry ML, Canziani OF, Palutikof JP, van der Linden PJ, Hanson CE (eds) Climate Change 2007: Impacts, Adaptation and Vulnerability. Contribution of Working Group II to the Fourth Assessment Report of the Intergovernmental Panel on Climate Change. Cambridge University Press, Cambridge, pp 717–743

Alexander KS, Ryan A, Measham TG (2012) Managed retreat of coastal communities: understanding responses to projected sea level rise. J Environ Plann Manag 55:409–433

Appeaning Addo K (2011) Changing morphology of Ghana's Accra coast. J Coast Conserv 15: 433–443

Arthuron R, Korateng K (2006) Coastal and marine environments, Sect. 2: Environmental state and trends: 20-year retrospective. In: United Nations Environment Programme (ed) Africa Environment Outlook 2–Our Environment, Our Wealth. UNEP, Nairobi, pp 155–195

Ball MC, Cocharane MJ, Rawson HM (1997) Growth and water use of the mangroves *Rhizophora apiculata* and *R. stylosa* in response to salinity and humidity under ambient and elevated concentrations of atmospheric CO_2. Plant Cell Environ 20:1158–1166

Biswas SR, Mallik AU, Choudhury JK, Nishat A (2009) A united framework for the restoration of Southeast Asian mangroves—bridging ecology, society, and economics. Wetlands Ecol Manage 17:365–383

Björk M, Short F, Mcleod E, Beer S (2008) Managing sea grasses for resilience to climate change. IUCN, Gland

Bosire JO, Dahdouh-Guebas F, Walton M, Crona BI, Lewis RR, Field C, Kairo JG, Koedam N (2008) Functionality of restored mangroves: a review. Aquat Bot 89:251–259

Burley JG, McAllister RRJ, Collins KA, Lovelock CE (2012) Integration, synthesis and climate change adaptation: a narrative based on coastal wetlands at the regional scale. Reg Environ Change 12:581–593

CARICOMP (Caribbean Coastal Marine Productivity) (2000) Manual of methods for mapping and monitoring of physical and biological parameters in the coastal zone of the Caribbean. Methods Manual Level 1. Kingston, Jamaica and St. Petersburg, FL, University of the West Indies and University of South Florida

Chan HT (1996) Mangrove reforestation in peninsular Malaysia. In: Field CD (ed) Restoration of mangrove ecosystems. International Society for Mangrove Ecosystems. Okinawa, Japan, pp 64–75

Day JW, Christian RR, Boesch DM, Yáez-Arancibia A, Morris J, Twilley RR, Naylor L, Schaffner L, Stevenson C (2008) Consequences of climate change on the ecogeomorphology of coastal wetlands. Estuar Coast 31:477–491

Duke NC, Meynecke JO, Dittmann S, Ellison AM, Anger K, Berger U, Cannicci S, Diele K, Ewel KC, Field CD, Koedam N, Lee SY, Marchand C, Nordhaus I, Dahdouh-Guebas F (2007) A world without mangroves. Science 317:41
Ellison JC (1998) Impacts of sediment burial on mangroves. Mar Poll Bull 37:420–426
Ellison JC (2009) Wetlands of the Pacific Island Region. Wetl Ecol Manag 17:169–206
Ellison JC (2012) Climate change vulnerability assessment and adaptation planning for mangrove systems. WWF US, Washington DC
Ellison JC, Zouh I (2012) Vulnerability to climate change of mangroves: Assessment from Cameroon, Central Africa. Biology 1:617–638. doi:10.3390/biology1030617
Ellison JC, Jungblut V, Anderson P, Slaven C (2012) Manual for mangrove monitoring in the Pacific Islands Region. SPREP, Apia
Ellison JC (2013) Vulnerability of mangroves to climate change. In: Ibrahim F-H, Latiff A, Hakeem KR, Ozturk M (eds) Mangrove ecosystem of Asia: status, challenges and management strategies. Springer, New York, Chapter 10
English S, Wilkinson C, Baker V (1997) Manual for survey of tropical marine resources, 2nd edn. Townsville, Australian Institute of Marine Science
Erwin KL (2009) Wetlands and global climate change: The role of wetland restoration in a changing world. Wetlands Ecol Manage 17:71–84
Evans N (2006) Natural resources and the environment in Fiji: a review of existing and proposed legislation. International Waters Project-Pacific Technical Report 21, SPREP, Apia
FAO (2003) Status and trends in mangrove area extent worldwide. In: Wilkie ML, Fortuna S (eds) Forest Resources Assessment Working Paper No. 63. Forest Resources Division, FAO, Rome
FAO (2007) The world's mangroves 1980–2005: A thematic study prepared in the framework of the Global Resources Assessment 2005. FAO Forestry Paper 153, Rome
Feka NZ, Chuyong GB, Ajonina GN (2009) Sustainable utilization of mangroves using improved fish-smoking systems: a management perspective from the Douala-Edea wildlife reserve, Cameroon. J Trop Conserv Sci 2:450–468
Gehrke PC, Sheaves MJ, Terry JP, Boseto DT, Ellison JC, Figa BS, Wani J (2011) Vulnerability of freshwater and estuarine fish habitats in the tropical Pacific to climate change. In: Bell JD, Johnson JE, Hobday AJ (eds) Vulnerability of fisheries and aquaculture in the Pacific to climate change. Secretariat of the Pacific Community, Noumea, pp 369–431
Gilman E, Ellison JC, Jungblut V, Van Lavieren H, Wilson L, Areki F, Brighouse G, Bungitak J, Dus E, Henry M, Kilman M, Matthews E, Sauni I, Teariki-Ruatu N, Tukia S, Yuknavage K (2006) Adapting to Pacific Island mangrove responses to sea level rise and climate change. Clim Res 32:161–176
Gilman E, Ellison JC, Coleman R (2007) Assessment of mangrove response to projected relative sea-level rise and recent historical reconstruction of shoreline position, American Samoa. Environ Monit Assess 124:105–130
Gilman E, Ellison JC, Duke NC, Field C, Fortuna S (2008) Threats to mangroves from climate change and adaptation options: a review. Aquat Bot 89:237–250
Giri C, Ochieng E, Tieszen LL, Zhu Z, Singh A, Loveland T, Masek J, Duke N (2011) Status and distribution of mangrove forests of the world using earth observation satellite data. Glob Ecol Biogeogr 20:154–159
Glick P, Stein BA (2010) Scanning the conservation horizon: a guide to climate change vulnerability assessment. National Wildlife Federation, Washington DC
Grimsditch GD, Salm RV (2006) Coral reef resilience and resistance to bleaching. IUCN, Gland
Gunderson LH (2000) Ecological resilience—In theory and application. Annu Rev Ecol Syst 31:425–439
Hansen LJ, Biringer JL, Hoffmann JR (2003) Buying time: a user's manual for building resistance and resilience to climate change in natural systems. World Wildlife Fund, Washington DC
Hansen J, Hoffman J, Drews C, Mielbrecht E (2009) Designing climate smart conservation: guidance and case studies. Conserv Biol 24:63–69
Heller NE, Zavaleta ES (2009) Biodiversity management in the face of climate change: a review of 22 years of recommendations. Biol Conserv 142:14–32

Hoegh-Guldberg O, Bruno JF (2010) The impact of climate change on the world's marine ecosystems. Science 328:1523–1528

Hong PN (1994) Reafforestation of mangrove forests in Vietnam. In: Proceedings of an International Timber Organisation Workshop: Development and Dissemination of Reafforestation Techniques of Mangrove Forests. Japanese Association for Mangroves, Okinawa, pp 141–165

Huxham M, Kumara MP, Jayatissa LP, Krauss KW, Kairo J, Langat J, Mencuccini M, Skov MW, Kirui B (2010) Intra- and interspecific facilitation in mangroves may increase resilience to climate change threats. Philos Trans R Soc Lond B Biol Sci 365:2127–2135

IPCC (Intergovernmental Panel on Climate Change) (2007) Climate change 2007: synthesis report. Contribution of Working Groups I, II, and III to the Fourth Assessment Report of the Intergovernmental Panel on Climate Change. Geneva, Intergovernmental Panel on Climate Change

Kelly PM, Adger WN (2000) Theory and practice in assessing vulnerability to climate change and facilitating adaptation. Climat Ch 47:325–352

Khan AS, Ramachandran A, Usha N, Arul Aram I, Selvam V (2012) Rising sea and threatened mangroves: a case study on stakeholder, engagement in climate change communication and non-formal education. Int J Sustain Dev World Ecol 19:330–338

Krauss KW, Cahoon DR, Allen JA, Ewel KC, Lynch JC, Cormier N (2010) Surface elevation change and susceptibility of different mangrove zones to sea-level rise on Pacific high islands of Micronesia. Ecosystems 13:129–143

Kumara MP, Jayatissa LP, Krauss KW, Phillips DH, Huxham M (2010) High mangrove density enhances surface accretion, surface elevation change, and tree survival in coastal areas susceptible to sea-level rise. Oecologia 164:545–553

Langley JA, McKee KL, Cahoon DR, Cherry JA, Megonigala JP (2009) Elevated CO_2 stimulates marsh elevation gain, counterbalancing sea-level rise. Proc Natl Acad Sci 106:6182–6186

Lewis RR (2005) Ecological engineering for successful management and restoration of mangrove forests. Ecol Eng 24:403–418

Lim B, Spanger-Siegfried E, Burton D, Malone E, Huq S (2004) Adaptation policy frameworks for climate change: developing strategies, policies and measures. University of Cambridge Press, Cambridge

Lovelock CE, Ellison JC (2007) Vulnerability of mangroves and tidal wetlands of the Great Barrier Reef to climate change. In: Johnson JE, Marshall PA (eds) Climate Change and the Great Barrier Reef: A Vulnerability Assessment. Great Barrier Reef Marine Park Authority and Australian Greenhouse Office, Townsville, pp 237–269

Mangrove Action Project (2006) Five steps to successful mangrove restoration. Mangrove Action Project, Yogyakarta, Indonesia

Marshall P, Schuttenberg J (2006) A reef manager's guide to coral bleaching. Great Barrier Reef Marine Park Authority. Townsville, Australia

McKee KL, Cahoon DR, Feller IC (2007) Caribbean mangroves adjust to rising sea levels through biotic controls on change in soil elevation. Glob Ecol Biogeogr 16:545–556

McKee KL (2011) Biophysical controls on accretion and elevation change in Caribbean mangrove ecosystems. Estuar Coast Shelf Sci 9:475–483

McLeod E, Salm RV (2006) Managing mangroves for resilience to climate change. IUCN, Gland

Mertz O, Halsnæs K, Olesen JE, Rasmussen K (2009) Adaptation to climate change in developing countries. Environ Manag 43:743–752

Nicholls RJ, Wong PP, Burkett VR, Codignotto JO, Hay JE, McLean RF, Ragoonaden S, Woodroffe CD (2007) Coastal systems and low-lying areas. In: Parry ML, Canziani OF, Palutikof JP, van der Linden PJ, Hanson CE (eds) Clim Change 2007: Impacts, Adaptation and Vulnerability. Contribution of Working Group II to the Fourth Assessment Report of the Intergovernmental Panel on Climate Change. Cambridge University Press, Cambridge, pp 315–356

Nitschke CR, Innes JL (2008) Integrating climate change into forest management in South-Central British Columbia: an assessment of landscape vulnerability and development of a climate-smart framework. For Ecol Manag 256:313–327

Polidoro BA, Carpenter KE, Collins L, Duke NC, Ellison AM, Ellison JC, Farnsworth EJ, Fernando ES, Kathiresan K, Koedam NE, Livingstone SR, Miyagi T, Moore GE, Nam VN, Primavera JH, Salmo III SG, Sanciangco JC, Sukardjo S, Wang Y, Yong JWH (2010) The loss of species: Mangrove extinction risk and geographic areas of global concern. PLoS ONE 5(4):e10095. doi:10.1371/journal.pone.0010095

Pomeroy RS, Parks JE, Watson LM (2007) How is your MPA doing? A guidebook of natural and social indicators of evaluating marine protected area management effectiveness. IUCN, Gland

Powell G (2006) Analysis of environment-related legislation in Tonga. International Waters Project-Pacific Technical Report 32, SPREP, Apia

Primavera JH, Esteban JMA (2008) A review of mangrove rehabilitation in the Philippines: successes, failures and future prospects. Wetlands Ecol Manage 16:345–358

Quarto A (1999) Local community involvement in mangrove rehabilitation: Thailand's Yadfon. In: Streever W (ed) An International Perspective on Wetland Rehabilitation. Kluwer Academic Publishers, Dordrecht, pp 139–142

Tavala YDT, Hakwa MT (2004) Review of environmental legislation and policies in Vanuatu. International Waters Project, SPREP, Apia

Terrados J, Thampanya U, Srichai N, Kheowvongsri P, Geertz-Hansen O, Boromthanarath S, Panapitukkul N, Duarte CM (1997) The effect of increased sediment accretion on the survival and growth of *Rhizophora apiculata* seedlings. Estuar Coast Shelf Sci 45:697–701

Titus JG (1991) Greenhouse effect and coastal wetland policy: how Americans could abandon an area the size of Massachusetts at minimum cost. Environ Manage 15:39–58

US Agency for International Development (2009) Adapting to coastal climate change: a guidebook for development planners. US Agency for International Development, Washington DC

Valiela I, Bowen JL, York JK (2001) Mangrove forests, one of the world's threatened major tropical environments. BioScience 51:807–815

Waycott M, McKenzie LJ, Mellors JE, Ellison JC, Sheaves MT, Collier C, Schwarz AM, Webb A, Johnson JE, Payri CE (2011) Vulnerability of mangroves, sea grasses and intertidal flats in the tropical Pacific to climate change. In: Bell JD, Johnson JE, Hobday AJ (eds) Vulnerability of fisheries and aquaculture in the Pacific to climate change. Secretariat of the Pacific Community, Noumea, pp 297–368

Winterwerp JC, Borst WG, de Vries MB (2005) Pilot study on the erosion and rehabilitation of a mangrove mud coast. J Coast Res 2:223–230

Zhao HX, Wu SH, Jiang LG (2007) Research advances in vulnerability assessment of natural ecosystem response to climate change. Ying Yong Sheng Tai Xue Bao 18:445–450

Managing Mangrove Habitat Expansion in New Zealand

Carolyn J. Lundquist, Donald J. Morrisey, Rebecca V. Gladstone-Gallagher and Andrew Swales

Contents

C. J. Lundquist (✉) · A. Swales
National Institute of Water & Atmospheric Research Ltd.,
PO Box 11115, Hamilton, New Zealand
e-mail: carolyn.lundquist@niwa.co.nz

D. J. Morrisey
National Institute of Water & Atmospheric Research Ltd.,
PO Box 893, Nelson, New Zealand

C. J. Lundquist
Institute of Marine Science, University of Auckland,
PO Box 349, Warkworth, New Zealand

R. V. Gladstone-Gallagher
Department of Biological Sciences, University of Waikato,
Private Bag 3105, Hamilton, New Zealand

I. Faridah-Hanum et al. (eds.), *Mangrove Ecosystems of Asia*,
DOI 10.1007/978-1-4614-8582-7_19,

Abstract While mangroves are indigenous to northern New Zealand and an integral part of functioning estuaries, rapid expansion of mangrove forests has occurred in recent decades, resulting in widespread support for estuarine restoration projects focusing on mangrove removals. Mangrove expansion is primarily associated with changes in land-use that increase terrestrial sediment erosion and deposition into coastal and estuarine environments. Objectives for mangrove removal in northern New Zealand often include a desire by local residents to restore open estuary sandflat conditions in areas that have been colonised by mangroves since the 1950s, and reinstate the navigational, recreational and amenity value of these areas. However, the likelihood of successful restoration is rarely considered in consent decisions, and minimal information is available on long-term trends in ecosystem health from areas where mangroves were cleared. Here, we discuss methods of mangrove removal, and recovery trajectories at numerous mangrove removal sites to identify physical and biological attributes of sites that are associated with limited (or fast) recovery, and minimal adverse impacts. We also discuss cost-effective management strategies to manage further spread of mangroves in New Zealand. Within a challenging and politically vibrant topic, we are informing the 'mangrove debate' with science to create better outcomes for estuarine health.

1 Introduction

In the past half century, mangroves have increased in extent in estuaries and tidal creeks throughout the upper half of the North Island of New Zealand (Morrisey et al. 2010). While mangroves are native and an integral part of functioning estuaries in northern New Zealand, their relative increase and association with fine sediments has resulted in an increasing number of consent applications for mangrove removals, with goals of estuarine areas returning to sandier, un-vegetated states (Green et al. 2003; Harty 2009). Most recently, draft guidelines in the 2013 Auckland Unitary Plan include provisions to allow the removal of mangrove areas that were established post-1996, subject to the methods of removal and disposal having only minor adverse effects in the coastal marine area (draft Auckland Unitary Plan available on: www.aucklandcouncil.govt.nz).

Legal and illegal removals have occurred to date in all four New Zealand regions where mangroves occur (Northland, Auckland, Waikato, and Bay of Plenty). Unfortunately, minimal information is available on long-term recovery (and likelihood of success) of the clearings. While increased sediment loading into estuaries is clearly linked to increasing mangroves, we lack understanding of most aspects of mangrove clearing, including the physical and ecological processes underlying differences in the timing and likelihood of success of mangrove removals. For example, we would predict that hydrodynamic differences (e.g., wind-wave exposure, tidal currents) between sites might influence the natural removal of sediments and organic material that have built up within the mangrove habitat. Rehabilitation might also depend on the influence of further terrestrial-based sediment loading and freshwater influx, local sediment characteristics, and colonization by organisms that will assist in the recovery process (e.g., via bioturbation). We also have a poor understanding of the potential barriers to restoration, such as sediments or organic material remaining from mangroves and how this might affect colonization and successional processes of benthic communities. Finally, we have limited knowledge of the long-term financial costs associated with improvements in catchment management that aim to reduce long-term sediment yields, as well as the maintenance costs associated with removal of seeds and seedlings in order to maintain an area cleared of mangroves.

Here, we review historical distributions of mangroves in New Zealand, and physical influences that are associated with the expansion of mangrove habitats. We discuss management actions taken in recent decades to manage the spread of mangrove habitats, and the success rates of different techniques aimed at reinstating the recreational, navigational, and amenity values that communities perceive are lost when mangrove expansion occurs in estuaries. We conclude with perspectives on the future of mangrove management in New Zealand, and the need for more holistic estuarine restoration, and likely changes to mangrove distributions due to changes in climate.

2 Distribution of New Zealand Mangroves

2.1 Taxonomy of New Zealand Mangroves

New Zealand contains a single species of mangrove, the grey mangrove or manawa (*Avicennia marina* subsp. *australasica*). This species has the largest geographical range of all mangroves, ranging from 25° N in Japan to 38° S in Australia. The subspecies *A. marina* subsp. *australasica* occurs in northern New Zealand, southeastern Australia, Lord Howe Island and New Caledonia (Morrisey et al. 2010). While mangroves are generally found in tropical locations where winter air temperatures do not drop below 20 °C, the occurrence of mangroves in southeastern Australia and New Zealand is postulated to be due to either transport during infrequent warm ocean currents, or as relict populations from historic periods with warmer climate (Duke et al. 1998).

2.2 Historical and Current Distribution

Mangroves have been present in New Zealand for approximately 19 million years, and specimens belonging to the genus *Avicennia* have been identified from silicified woods associated with lower Miocene rocks (Sutherland 2003). Additional evidence from pollen samples indicates the presence of the species *A. marina* in New Zealand for at least 11,000 years (Pocknall 1989). Today, mangroves occur only in the North Island of New Zealand, from 34°27′ S in the far north to 38°05′ S at Kawhia Harbour on the west coast and 38°03′ S at Ohiwa Harbour on the east coast.

2.3 Historical Loss of Mangrove Habitat

Human activity, associated with urbanisation and agricultural development following European colonization of New Zealand, generally resulted in loss of mangrove habitat due to construction of causeways, marinas, and other structures that restrict tidal flows and/or elevate water levels, as well as land reclamation of coastal areas for ports, landfills, airports, agriculture and stock grazing, and industrial and urban development (Dingwall 1984; Crisp et al. 1990). National and local legislation from the 1970s prohibited reclamation of coastal lands for agriculture and stock grazing, preventing further loss of mangrove habitat (Dingwall 1984). However, ongoing changes in catchment land-use to support agriculture, industrial and urban development have resulted in increased sediment rates that promote expansion of mangrove habitat (Green et al. 2003). In recent decades, many consent applications have been submitted by coastal communities and resource agencies both to carry out removal of mangroves and to prevent further spread of mangrove habitats (Morrisey et al. 2007).

2.4 Physical Influences on Mangrove Distribution in New Zealand

2.4.1 Geomorphic Distribution

New Zealand mangroves, like those of other temperate regions, occur mainly in sheltered coastal environments, rather than deltas and muddy open coasts as they do in the tropics. The largest forest areas occur in estuaries with large terrigenous sediment supplies, such as the Firth of Thames (1,100 ha) and Rangaunu Harbour (2,415 ha), or barrier-enclosed estuaries such as the Kaipara (6167 ha) and Tauranga (623 ha) Harbours (Hume et al. 2007). Mangroves in New Zealand also occur in drowned river valleys, tidal basins and lagoons.

2.4.2 Latitudinal Limits

Compared to most tropical species, members of the genus *Avicennia* are relatively tolerant of high variation in environmental factors such as water salinity, low air and water temperatures, frequency and severity of frosts, and day length (Chapman 1976; Duke 1990; Stuart et al. 2007; Krauss et al. 2008). While other factors also contribute, the distribution of *A. marina* in New Zealand is primarily constrained by its physiological limitations to low temperatures (Sakai and Wardle 1978; Walbert 2002; Beard 2006). Mangroves are vulnerable to abiotic stressors at low temperatures (e.g. tissue damage from freezing) (Duke 1990; Saenger and Snedaker 1993), and the frequency and duration of frost events are suggested as a primary constraint on mangrove expansion at the limits of its latitudinal distribution in New Zealand.

2.4.3 Tidal Elevation

A range of factors influence the upper elevation limit of temperate mangrove forests. In the estuaries of northern New Zealand, mangrove forests generally occur up to mean high water spring (MHWS) in tidal elevation (Chapman and Ronaldson 1958; Swales et al. 2007a), but may occur above MHWS in locations where episodic storm tides increase the hydroperiod in these forests (Swales et al. 2007a). Maximum tidal height is important in maintaining substratum porewater salinity, and preventing establishment of freshwater plants (Chapman 1976; Gillanders and Kingsford 2002; Mitsch et al. 2009). Tidal currents and variations in tidal range also control the maximum tidal height for dispersal of seeds in the intertidal zone (Saintilan and Williams 1999; Rogers et al. 2005).

Physiological constraints determine the lower elevation limits of New Zealand mangrove. The lower elevation limit is roughly at mean sea level (MSL), such that mangrove habitats are submerged for no more than ~ 6 h per tidal cycle (Hume 2003). Field surveys in 17 Auckland estuaries showed high variability in average lower elevation limits (− 0.05–0.76 m MSL) (Swales et al. 2009). Newly colonised seedlings showed a lower distribution (−0.41–0.21 m MSL, average −0.15 m MSL) (Swales et al. 2009), though the lower range of this distribution is likely associated with low survivorship as mangrove seedlings are intolerant of continuous submersion (Clarke and Hannon 1970). Prolonged submersion in excess of ~ 4 h is also associated with decreased root growth and seedling development (Curran et al. 1986; Hovenden et al. 1995).

In addition to physiological constraints, hydrodynamics also controls seedling establishment through transport of seeds with wind-waves and currents (de Lange and de Lange 1994). Disturbances to the substratum by wind-waves reduce substratum stability, and result in dislodgement of newly established seeds and seedlings (Clarke and Myerscough 1993; Green et al. 1997; Swales et al. 2007b; Swales et al. 2009). In the Firth of Thames, El Niño periods associated with a reduction in wind and wave energy are associated with periodic expansions of mangrove habitat, whereas La Niña periods with strong northeast winds are associated with higher wave energy and thus higher seedling mortality (Swales et al. 2007b; Lovelock et al. 2010). Prolonged calm conditions are likely to allow seedlings to develop larger and deeper root systems, increasing their capacity to withstand strong wind-waves (Balke et al. 2013).

2.5 *Recent Expansion of Mangrove Habitat*

2.5.1 Patterns and Extent of Expansion

Despite historical losses of mangrove habitat after the arrival of Europeans, increases in mangrove distribution have occurred in many estuaries in northern New Zealand (summarised in Morrisey et al. 2010). Rates of increase have been calculated from analyses of historical aerial photographs from the 1940 and 1950s, and show rates

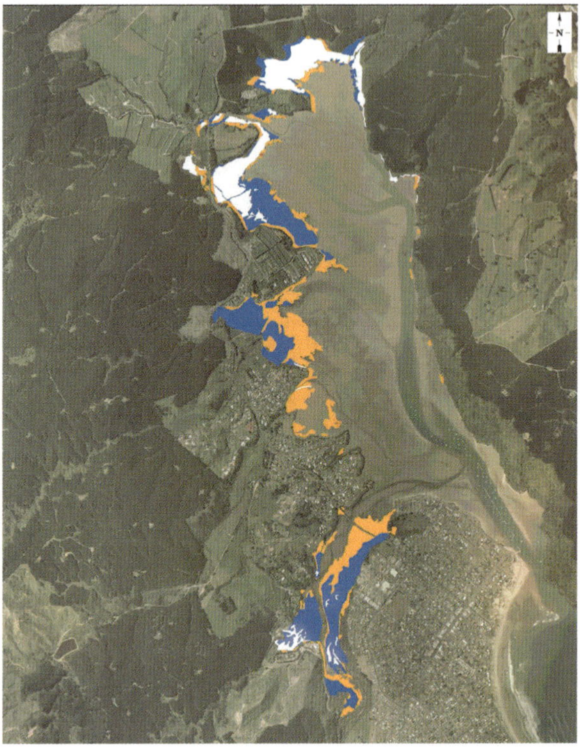

Fig. 1 Map demonstrating expansion of mangrove habitat based on aerial photographs for Whangamata Harbour, New Zealand. Mangrove extent is marked in *white*, *blue* and *yellow*, representing mangrove distributions in the Harbour in 1944, 1978 and 2002, respectively

varying between estuaries from < 1–20 % increase in areas covered by mangrove habitat per year (Morrisey et al. 2010, their Table 9). Mangrove colonization occurred in the early 20th century in many estuaries with large catchments that were subject to deforestation in the mid- to late 1800s (e.g., Swales et al. 1997; Swales et al. 2002). In other estuaries, mangrove expansion has occurred in more recent decades. In fact, expansion of mangrove forest in the Firth of Thames did not occur until the 1950s (Swales et al. 2007a; Lovelock et al. 2010), though mangroves were recorded in the delta of the Waihou River in 1769 by Captain James Cook (Beaglehole 1968). Other estuaries throughout northern New Zealand have also shown mangrove expansion in recent decades, with increases of 50 % and 75 % in Mangemangeroa and Waikopua Creek in the Whitford embayment, Auckland, between 1955 and 2000 (Nicholls and Ellis 2002); increases of 15 % in mangrove cover in Whangamata Harbour, Coromandel Peninsula, since the 1940s (Singleton 2007) (Fig. 1); and increases of > 50 % in mangrove cover in the sub-estuaries within Tauranga Harbour, Bay of Plenty, from 240 ha in 1943 to 545 ha in 1999 (Park 2004; Stokes et al. 2010).

2.5.2 Causes of Mangrove Expansion

Expansion of mangrove habitat in New Zealand has been attributed to estuary infilling and vertical accretion of tidal flats, increased nutrient inputs, climate warming, changes in relative sea level due to sedimentation and/or subsidence or a combination of all/some of these factors (Young and Harvey 1996; Schwarz 2002; Swales et al. 2007a; Morrisey et al. 2010). A regional study of Auckland estuaries confirmed that the largest increases in percent of estuary covered by mangrove habitat over the last 50–60 years have occurred in the smallest (i.e., $< 5\,\text{km}^2$) systems (Swales et al. 2009). One of the region's largest estuaries, the $65\,\text{km}^2$ Waitemata Harbour, showed a net 8 % loss of mangrove area due to reclamations associated with motorway construction, industrial development and refuse landfills in the 1950–1970s (Swales et al. 2009).

Erosion of catchment soils and coastal margins results in deposition and accumulation of terrigenous sediments in temperate mangrove systems (Ellison 2009), and human activities in catchments have increased sediment yields by an order of magnitude or more (Walling 1999; Nichol et al. 2000; Swales et al. 2007a). Catchments draining to estuaries with mangroves in northern New Zealand average $239\,\text{km}^2$ in area (range, $6–4194\,\text{km}^2$) (Hume et al. 2007). The small, steep basins that characterize New Zealand geomorphology have limited storage capacity for eroded sediments, and sediment delivery is generally high during episodic intense rainstorms that are typical in northern New Zealand (Griffiths and Glasby 1985; Milliman and Syvitski 1992).

At the peak of catchment land use changes following the arrival of Europeans (mid-late 1800s), catchment sediment yields were as much as several orders of magnitude higher than pre-European values of several t $\text{km}^{-2}\ \text{yr}^{-1}$ (Prosser et al. 2001). Reforestation, destocking and/or reduced sediment supply have decreased yields over the past century but they remain several times higher than pre-European values (Healy 2002; Swales et al. 2002). Current high sediment accumulation rates ($10–100\,\text{mm yr}^{-1}$) are associated with high rates of terrigenous sediment supply and proximity to catchment outlets (e.g., tidal creeks). Changes in mangrove distribution are closely linked to substratum elevation relative to sea level. On accreting tidal-flats, expansion of mangrove habitat occurs when the bed elevation has increased sufficiently for mangrove seedlings to colonise (Chapman and Ronaldson 1958; Swales et al. 2002; Swales et al. 2007a). Within the mangrove forest and along forest margins, drag-induced current dampening due to the presence of pneumatophores promotes sediment retention (Young and Harvey 1996; Furukawa et al. 1997).

3 Managing Mangrove Expansion

3.1 Methods of Mangrove Removal

Expansion of mangrove habitats in recent decades has resulted in increasing numbers of consent applications and proposed changes to regional council plans to facilitate removal of mangrove adults, and manage spread of mangrove habitats via seedling removal (Green et al. 2003; Harty 2009). Both legal and illegal mangrove removals have occurred in northern New Zealand in response to high rates of mangrove expansion in recent decades. Objectives for mangrove removal in northern New Zealand often include a desire by local residents to restore open estuary sandflat conditions in areas that have been colonised by mangroves since the 1950s, and reinstate the navigational, recreational and amenity value of these areas. Consents generally include provisions to maintain suitable habitat for wading and roosting birds, including both shorebirds that will benefit from newly unvegetated sandflats as forage areas for benthic infauna, and birds such as banded rail (*Gallirallus philippensis assimilis*) that utilise mangrove and saltmarsh vegetation as predator refugia (e.g., Resource Consent Application #65505 for Tauranga Harbour; Resource Consent Applications #122986 and #122987 for Whangamata Harbour).

A number of methods have been used to remove adult mangrove forests in New Zealand, ranging from manual labour using chainsaws or slash cutters, to mechanical removal utilising wide track machinery (Table 1). Disposal of vegetation has varied between removals, ranging from complete removal of above-ground vegetation from the coastal environment and disposal off-site, to burning on-site, to leaving either intact or mulched vegetative material on-site to decompose (Table 1).

3.2 General Trends in Recovery After Mangrove Management

Our understanding of recovery after mangrove removal in New Zealand estuaries is limited, as long-term trajectories in sediment characteristics and benthic community structure are poorly documented at most mangrove removal sites. A few areas have returned to sandier habitats in ~5 years since clearing (e.g., Patiki Bay, Whangamata (Coffey 2002), as well as manually cleared areas in Matua, Waikaraka and Waikareao estuaries, Tauranga Harbour (Wildlands Consultants 2003; C. Lundquist, pers. observation)). However, other sites (e.g., Moanaanuanu estuary in Whangamata Harbour; mechanically cleared areas in 11 sub-estuaries of Tauranga Harbour) have not yet returned to sandier habitats (Felsing 2006; Stokes 2009; Lundquist et al. 2012; Park 2012). Site-specific factors that are anticipated to affect the return to sandflats after mangrove removal include differences in hydrodynamics (wind-wave exposure, tidal currents), terrestrial-based sediment inputs, freshwater influx, and local sediment characteristics. Sites that have returned to sandflats are generally located in exposed areas with high fetch or near tidal creeks with strong tidal currents (C. Lundquist, pers. observation). In addition, the size of mangrove removal area potentially correlates with trends toward sandier sediment, with smaller removal areas (e.g., removal of less than 20 m in width) showing faster change in sediment erosion and root decomposition. However, for large removal areas, it is difficult to separate size and mechanical machinery impacts, as all large removals to date have been performed using mechanical methods, and were associated with slow recovery times (Lundquist et al. 2012). For example, quantitative surveys of changes in benthic community and sediments at sites with mechanical mangrove removals and *in situ* mulch deposition in Tauranga Harbour revealed significant impacts on benthic

Table 1 Selected examples of mangrove removals in New Zealand to illustrate the range of techniques used for extraction and disposal of mangrove vegetative material

Location	Date	Size of clearance	Tool	Disposal of above ground biomass	Disposal of below ground biomass
Whangamata Harbour (consented)[a]	2013 (in process)	22.91 ha	Mechanical removal by low psi tractor; manual removal in sensitive areas	Burned on site	Biomass associated with main stump removed; pneumatophore and fine root biomass left in situ
Whangamata Harbour, Patiki bay (consented)[b]	2000	0.48 ha	Manual removal using chain saws	Transported offsite for disposal	Four trials varied from full below ground biomass left in situ, to complete removal of below ground biomass
Whangamata Harbour, Moanaanuanu estuary (unconsented)[c]	2005	~ 2 ha	Manual removal using chain saws; further removal of a subset of the area using tractor	Burned on site	Left in situ to decompose except in small area
Mangawhai Estuary, Northland[d]	2004	0.26 ha	Manual removal using chain saws	Transported offsite for disposal	Left in situ to decompose
Pahurehure Inlet No. 2, Manukau Harbour[e]	2010	3 ha	Manual removal using chain saws	Transported off site by helicopter	Left in situ to decompose
Pahurehure Inlet No. 2, Manukau Harbour[f]	2012	24 ha	Mechanical removal using low psi tractor	Burn piles left on site	Left in situ to decompose
Pahurehure Inlet No. 1, Manukau Harbour[g]	2012	~ 0.2 ha	Manual removal (unconsented) using chain saws	Main branches removed from stump; all woody debris left in situ	Left in situ to decompose
Auckland airport site, Manukau Harbour[h]	2011	13 ha	Mechanical removal using low psi tractor	Mulchate left in situ	Left in situ to decompose

Table 1 (continued)

Location	Date	Size of clearance	Tool	Disposal of above ground biomass	Disposal of below ground biomass
Tauranga Harbour (11 sub-estuaries with resource consents to local community groups)[i]	2005–2010	< 0.5 ha per site per year, usually in strips of < 20 m in width	Manual removal by chainsaw or slash cutter (consented)	Either removed from coastal marine area, or burned on site	Left in situ to decompose
Tauranga Harbour (11 sub-estuaries, consented)[j]	2010–2011	~ 110 ha	Mechanical removal by low psi tractor	Mulchate left in situ	Left in situ to decompose

[a]Resource consent #122986 and #122987, Rivers and Catchment Services, Waikato Regional Council
[b]Resource consent #102475, Waikato Regional Council
[c]Unconsented mangrove removal (Felsing 2006)
[d]Environmental permit #CON20031099401, Northland Regional Council
[e]Permit #35053, Papakura District Council (trials)
[f]Permit #35053, Papakura District Council (consented removal)
[g]Unconsented mangrove removal (C. Lundquist, pers. obs.)
[h]Resource consent #38862, Auckland Council
[i]Resource consents #62776, 62054, 63941, and 64154 held by Estuary Care Groups, Bay of Plenty Regional Council
[j]Resource consent #65505 and #65693, Bay or Plenty Regional Council

communities, and few signs of recovery towards a typical sandflat (in terms of sediment characteristics or benthic community composition) over a 12-month period (Lundquist et al. 2012).

3.3 Decomposition of Remaining Mangrove Material

To date, most mangrove removal techniques have left below-ground root biomass *in situ* following both mechanical and manual tree removal (Table 1; Stokes 2009; Lundquist et al. 2012). In addition, methods such as mechanical mulching have left above-ground material on site, either in the form of stumps and/or mulchate (Table 1; Lundquist et al. 2012). Consequently, the recovery of these sites will depend, in part, on the decay rates of the remaining below- and/or above-ground biomass. A number of studies in New Zealand have measured decomposition of above- and below-ground mangrove material (e.g., Albright 1976; Woodroffe 1982; Oñate-Pacalioga 2005; Gladstone-Gallagher 2012; Gladstone-Gallagher et al., unpublished manuscript), and these studies are useful to provide estimates of recovery periods after mangrove removal based on decomposition rates of vegetative material. The recovery of mangrove removal sites is likely to be influenced by the type of material that is left *in situ* (i.e., below-ground vs. above- and below-ground biomass),

Table 2 Summary of t_{50} (days taken to decay by 50 % of original weight) values describing the decay of different mangrove materials, at three study sites in northern New Zealand. Footnotes indicate source of data

Mangrove material type	t_{50} (± SE; days taken to decay by 50 % of original weight)		
	Whangamata Harbour[a] (37°10′43.0″S, 175°51′36.9″E)	Pahurehure Inlet, Manukau Harbour[b] (37°02′663″S, 174°54′351″E)	Whangarei Harbour[b] (35°49′596″S, 174°25′696″E)
Surficial leaves	63 ± 3	36 ± 16	54 ± 4
Buried leaves	88 ± 6	71 ± 5	48 ± 14
Surficial wood	460 ± 28	788 ± 269	466 ± 39
Buried wood	613 ± 43	1223 ± 427	833 ± 177
Surficial pneumatophores	317 ± 30	428 ± 263	303 ± 100
Buried pneumatophores		530 ± 4	542 ± 169
Surficial fibrous roots		309 ± 157	192 ± 8
Buried fibrous roots		470 ± 81	5403 ± 3983

[a]Gladstone-Gallagher 2012; Gladstone-Gallagher et al. submitted manuscript
[b]Gladstone-Gallagher et al. unpublished manuscript

because the decay rates of different material types have been shown to differ significantly (Table 2; Gladstone-Gallagher 2012; Gladstone-Gallagher et al., unpublished manuscript). Leaf material appears to be of little concern to site recovery, as it decomposes rapidly within a period of months (Table 2; Albright 1976; Woodroffe 1982; Oñate-Pacalioga 2005; Gladstone-Gallagher 2012; Gladstone-Gallagher et al., unpublished manuscript). In contrast, *in situ* decay experiments have estimated that above-ground woody branch material (branch diameter 5–10 mm) will decay to 50 % of its original biomass after 2–5 years (Table 2; Gladstone-Gallagher 2012; Gladstone-Gallagher et al., unpublished manuscript), and larger woody debris and stumps left *in situ* will likely take much longer (based on wood disk decomposition rates in Florida, USA; Romero et al. 2005). Macroalgal blooms on mangrove mulchate create surface anoxia (Lundquist et al. 2012), and anoxic decay of woody material is 1.3–1.8 times longer than oxic decay on the sediment surface (Table 2; Gladstone-Gallagher 2012; Gladstone-Gallagher et al., unpublished manuscript). As a result, the breakdown and loss of this material will be prolonged in areas where significant macroalgal blooms occur. Observations of mangrove removals confirm these low rates of decay of woody debris, with mulchate showing minimal decomposition over a period of 2–3 years, and stumps still apparent at removal sites that are approaching a decade since mangrove removal (Fig. 2a; Stokes 2009; Lundquist et al. 2012; C. Lundquist, pers. observation).

Below-ground mangrove biomass consists of both fibrous root mass and pneumatophores. Most of the New Zealand mangrove removals have left below-ground material *in situ*, due to the logistical issues with removing it from the sediment (Table 1). Such below-ground biomass has been observed to remain for up to 8 years following clearances (R. Gladstone-Gallagher & C. Lundquist, pers. observation, Whangamata Harbour) and this is likely because decay is slow under anoxic conditions in the sediment (Table 2; Albright 1976; Gladstone-Gallagher 2012; Gladstone-Gallagher et al., unpublished manuscript). Pneumatophores and

Fig. 2a Photos of stumps and woody debris after mangrove clearing, Moanaanuanu Estuary, Whangamata Harbour, cleared in 2005; photo taken in August 2010. **b** Macroalgal bloom atop mangrove mulchate, Omokoroa Estuary, Tauranga Harbour, cleared in January/February 2010; photo taken in September 2010

fibrous roots can differ in their decay rates (Table 2), though one New Zealand study measured similar decay rates between the two root types (Albright 1976). *In situ* experiments have suggested that pneumatophore decay will take at least 2 years (Table 2; Albright 1976; Gladstone-Gallagher 2012; Gladstone-Gallagher et al., unpublished manuscript), while buried fibrous roots could remain for decades (Table 2; Gladstone-Gallagher et al. unpublished manuscript). However, fibrous roots that are exposed to the surface decompose faster than pneumatophores (Table 2). Consequently, sites with greater wind-wave exposure may result in faster fibrous root degradation, as surface root layers become exposed to oxic decay during sediment erosion (Gladstone-Gallagher et al., unpublished manuscript).

3.4 Physical Changes After Mangrove Removal

Mangrove management sites vary in both timing and success of return from mud to sandier sediments. Smaller removals have generally shown greater rates of change from muddy to sandier sediments (e.g., Waikaraka Estuary, Tauranga Harbour; Patiki Bay, Whangamata Harbour), with removals from highly exposed sites with sandier sediments showing fastest change (e.g., Coffey 2001; Wildlands Consultants 2003). In contrast, most sites with mechanical mulching showed slow or no change from mud to sandier sediments in the mulch zone over a 12-month period (Park 2012; Lundquist et al. 2012), though one estuary (Matua Estuary, Tauranga Harbour) with a small area cleared by *in situ* mulching (< 0.5 ha) changed from 70 to 35 % mud content over the course of 2 years (Park 2012). Sediments at many mangrove removal sites continue to be more similar to mangrove than near-by sandflat zones in a number of sediment metrics (mud content, organic content, chlorophyll content, and sediment cohesiveness), and high mud content remains years after removals (e.g., Moanaanuanu Estuary, Whangamata Harbour [Stokes 2009]; Waikaraka Estuary, Tauranga Harbour [Stokes et al. 2009]).

Some sites have also been associated with persistent anoxic sediments after mangroves are removed. For example, mulched sites in Tauranga Harbour showed persistent anoxic conditions within surface sediments, with oxic layer depths in the mulch zone generally reduced to < 2 mm for at least 24 months, except within 5–10 m of the seaward edge at some sites (Lundquist et al. 2012; C. Lundquist, pers. observation). In some locations, anoxic sediments have been exacerbated by the presence of macroalgal blooms on top of the mulch material (Fig. 2b).

Erosion rates of muddy sediments after mangrove removals have often been slow, and correlated with hydrodynamic exposure at a site. In Moanaanuanu Estuary (a sub-estuary in Whangamata Harbour), at a site experiencing restricted flow due to construction of a causeway near the mouth of the estuary, sediment erosion has been slow, with very slow winnowing of mud particles over time, and high remaining levels of root biomass (> 1 kg root biomass m^{-2}) (Stokes 2009). In contrast, reductions in sediment elevation of 9–38 mm y^{-1} were observed within 2 years of 0.96 ha of mangrove removal at Waikaraka Estuary, Tauranga Harbour, an exposed site with sandier sediments (Stokes et al. 2009). After mechanical mulching in 2010, perimeters of mulched areas measured in four estuaries in Tauranga Harbour showed very slow erosion of the mulch zone over 12 months (range 5–19 % decrease in total area across five estuaries), with mulch biomass relatively consistent throughout the mulch zone except for within 5–10 m of the seaward edge, and no apparent decrease in mulch biomass between 6 and 12-month samples (Lundquist et al. 2012).

3.5 Benthic Community Changes After Mangrove Removal

Benthic community recovery has been limited at the sites that have been monitored for benthic community changes after mangrove removal. Recovery after non-mechanical clearing of a small mangrove area (0.26 ha fringe) in Mangawhai Estuary was associated with rapid decreases in mud content and colonization by benthic macrofauna, though long-term trends were confounded by a large sediment deposition event associated with terrestrial development projects in the local catchment (Alfaro

2010). Low abundance of macrofauna was observed three years after removals at Moanaanuanu Estuary, with temporal trends indicating an increase over time, but the removal sites had not recovered within three years (Stokes 2009). At mulched sites in Tauranga Harbour, macrofaunal core samples indicate that, while some colonization was occurring, the resulting community was dominated by opportunistic species (oligochaetes, *Capitella* spp., and dipterid (fly) larvae), most of which are tolerant of anoxic conditions, and not representative of typical sandflat (or mangrove) communities (Lundquist et al. 2012). In contrast, observations of manual removal sites at Waikaraka and Matua estuaries, adjacent to mechanically cleared areas, suggest the potential for rapid change (within 5 years) to sandflat communities, and strongly contrast with observations of recovery trajectories with the mechanical mulching and *in situ* disposal method (Lundquist et al. 2012).

In comparing recovery from mangrove removal to other natural disturbance events in estuaries, such as catastrophic sediment deposition from large storm events, disturbance experiments in intertidal and shallow subtidal New Zealand estuaries suggest at least partial, if not full, recovery from disturbance after 6 months to 2 years, with recovery time dependent on depth of deposited sediment (Thrush et al. 2003, 2004, 2008). While some, though not all, non-mechanical mangrove removal trajectories have shown positive recovery trends over similar timelines, most recent mechanical removals (since 2010) have not yet trended toward sandflat habitats in either benthic community or sediment properties, and instead are displaying properties of high remaining vegetative biomass and depauperate opportunistic faunal communities.

3.6 Unanticipated Impacts of Mangrove Removal

Macro algal blooms have been recorded in monitoring surveys after mangrove removals in Tauranga Harbour (recorded in surveys between 2010 and 2012 at Waikaraka, Te Puna, Waikareao, Omokoroa, and Matua Estuaries), at Patiki Bay, Whangamata Harbour (Coffey 2002), and in Manukau Harbour (Pahurehure Inlet) (Lundquist et al. 2012; Coffey 2010). Macroalgal blooms on mangrove removal sites that left mulch *in situ* often reached 100 % cover in large (> 1 ha) patches at some locations, though seasonal and between-site variability in macroalgal abundance was apparent (Fig. 2b; Lundquist et al. 2012). Most macroalgal bloom species observed on removal sites were Ulvaceae, an algal family known to respond to increased nutrient concentrations with increasing growth rates (Heesch et al. 2007). *Hormosira banksii* is also occasionally abundant in intertidal estuaries associated with mangrove fringes (Morton and Miller 1973), but was not observed forming extensive macroalgal blooms on mangrove clearings (Lundquist et al. 2012). While *Ulva* sp. (sheet form, "sea lettuce") was common at the edge of many removal patches, and could potentially have washed in from outside the mulch patches, most macroalgal blooms consisted of other species which were clearly attached to pneumatophores or remaining mulch material, including *Ulva* sp. (tubular form) and filamentous species such as *Rhizoclonium* spp. (Cladophoraceae) and *Percursaria percursa* (Ulvaceae).

Prior to observations on mangrove mulch in Tauranga Harbour in 2011, *P. percursa* had not been documented on New Zealand's North Island, and its current distribution in New Zealand includes only one highly modified hypersaline lagoon (Pratt et al. 2013). Macroalgal blooms are not limited to sites with *in situ* mulching, and have occurred at sites that were not mulched (e.g., Patiki Bay, Whangamata Harbour; Pahurehure Inlet No. 2, Manukau Harbour) where below-ground biomass was left intact (Coffey 2002, 2010), though size and temporal extent of blooms were reduced compared to blooms on mulched sites that have recurred over at least three annual seasons (Lundquist et al. 2012).

The presence of large blooms of green macroalgae are generally associated with nutrient release, which suggests that leaving above- or below-ground biomass to decompose *in situ* may be contributing to these blooms. Laboratory experiments have demonstrated high rates of nutrient release associated with the deposition and decomposition of mangrove mulch, resulting in elevated concentrations of phosphorus and ammonia in the water column (Lundquist et al. 2012). Field sampling of pore water nutrients has also confirmed laboratory predictions of elevated concentrations of phosphorus and ammonia associated with mulch zones in Waikareao and Waikaraka estuaries in Tauranga Harbour (Lundquist et al. 2012).

Mangrove removals could also potentially cause decreased water column oxygen saturation levels. Decreases in oxygen concentrations below 80 % saturation were detected over both mulch zones and mangrove zones inshore of mulch zones, with levels as low as 50 % reached within a tidal cycle (Lundquist et al. 2012). The strongest decreases were observed in the latter half of the tidal cycle, when the water column over the mulch zone was no longer being refreshed with the influx of harbour water (Lundquist et al. 2012). Further comprehensive surveys are required to confirm what proportion of the tidal cycle is subject to decreased oxygen levels, and to what extent, and whether this results in changes in distribution of benthic and pelagic estuarine communities.

3.7 Management of Spread of Mangrove Propagules

Removal of adult mangroves is not in itself a solution to mangrove expansion, as colonization by mangrove seeds will continue to occur unless the full extent of mangroves is eradicated from New Zealand. Maintenance by regular seedling removal is expensive (Beca Carter 2007), and limited information exists on site-specific variability in colonization, survival, and growth of seedlings, from which to develop cost-effective strategies for seedling management.

Dispersal of seeds is dependent on tidal currents and wind-waves, and the buoyancy and longevity of seeds (Clarke 1993; de Lange and de Lange 1994). Propagules are obligate dispersers for a minimum of 5 days (based on Australian studies of *A. marina;* Clarke 1993) before seeds are capable of establishing roots in the sediment. However, propagules can remain viable for up to 5 months even if submerged (Clarke 1993). Regardless, observations suggest in both New Zealand and Australia that few propagules travel further than 10 km from their parent tree (Clarke 1993), and that dispersal is primarily confined to within the local estuary or tidal creek (de Lange and de Lange 1994). Assuming hypotheses of small dispersal distances are correct, it can be assumed that local clearing of adults should reduce costs of seedling management by reducing local seed productivity.

Cost-effective guidelines for seedling management are being developed, taking into account physical characteristics of sites that enhance seedling retention and survivorship, and natural seedling mortality rates due to wind-wave and storm exposure (C. Lundquist, unpublished data). High mortality rates of seedlings occur at exposed sites (e.g., Lovelock et al. 2010), suggesting that performing seedling maintenance in winter, after many seedlings have been dislodged by early winter storms, will minimise effort and cost (A. Swales, unpublished data). In contrast, sheltered sites may have lower mortality rates due to lower chance of dislodgement, requiring higher effort to remove seedlings. Methods of mangrove removal may also be correlated with likelihood of seedling recruitment at a cleared site, as biogenic structures like pneumatophores (if left *in situ*) may enhance drag and serve as effective hydrodynamic traps for propagules (Brinkman et al. 1997).

4 The Future of Mangroves in New Zealand

4.1 Future Expansion of Mangroves in New Zealand

Sedimentation rates are still high relative to pre-European periods, suggesting that rates of infilling of New Zealand estuaries will continue to exceed historical, natural rates (Swales et al. 2002). While removing mangroves and managing spread of mangrove seeds and seedlings to new habitats addresses a symptom of changing land-use, estuarine restoration should consider the causes of mangrove expansion through improved catchment management (Harty 2009). Estuarine management plans need to be holistic, focussing on issues beyond mangrove removal to include management of sedimentation, eutrophication and other sources of pollution, and invasive species (Borja et al. 2010). Recent mangrove resource consent applications (e.g., Tairua Harbour) are for estuarine restoration, with mangrove management being one of many goals to improve recreational and ecological health within the harbour (e.g., Tairua Harbour Estuarine Management Resource Consent, Waikato Regional Council). Mangrove removal consents should be informed by: the proposed extent of the clearing relative to historical distribution; the proposed methodology and likelihood of avoiding or mitigating adverse impacts on the removal area and on neighbouring habitats; and the ecosystem services provided by mangroves in a harbour that would be lost if mangroves were removed (e.g., principles proposed in Harty 2009). Consent applications to date have assumed that the mangrove removals will result in successful change to pristine ecosystems suitable for recreational, navigational, or amenity values. However, the limited monitoring that has occurred on mangrove removals suggests this is often not the case, particularly for large, mechanical mangrove removals, and that consent approval should also depend on recovery after mangrove removal and the potential for mangrove removals to result in an improvement in ecological health. In addition, costs of long-term maintenance to keep removal areas clear from colonization by seedlings should be included when determining costs of consents.

4.2 *Mangroves and Climate Change*

Climate change is anticipated to result in changes in mangrove distribution in New Zealand. Increased concentrations of carbon dioxide in the atmosphere as well as increased average temperatures may lead to increased rates of photosynthesis and growth, and potential for habitat expansion to higher latitudes (McLeod and Salm 2006; Lovelock and Ellison 2007). While reduced frequency of frost may increase suitability of higher latitudes for mangrove colonization, expansion is constrained by the limited number of estuaries south of the current mangrove latitudinal limit, and by limitations in seed dispersal potential based on current velocities and between-estuary transport distances (de Lange and de Lange 1994). Accelerated sea-level rise is anticipated to result in climate-change-related changes in mangrove distribution (Ellison 1994; Field 1995). Changing storm frequency with increasing numbers of episodic, intense rainfall events, and associated increases in sedimentation rates are another likely driver of future mangrove expansion (Lundquist et al. 2011).

The ability of mangroves to respond to sea-level rise is determined by the ability of species to colonise and extend shoreward, the availability of suitable substratum, and whether sediment accretion balances erosional processes (Done and Jones 2006). Shoreward migration of coastal habitats including mangroves due to sea-level rise may be restricted in some locations due to the presence of existing coastal structures (e.g., embankments, reclamations, and sea walls) (Ellison 2004; Gilman et al. 2008). Globally, mangrove response to recent sea-level rise has showed varied responses, with some locations where sediment accretion has kept up with sea-level rise (Alongi 2008). However, even with positive increments in surface accretion due to the sediment trapping function of mangrove ecosystems, many mangrove ecosystems show a relative loss of elevation due to sub-surface subsidence (Krauss et al. 2003; Cahoon et al. 2006; Gilman et al. 2008). In northern New Zealand, average relative sea-level rise of 1.4 mm yr^{-1} (Hannah 2004), combined with sedimentation accumulation rates of 1–5 mm yr^{-1} over the last 50–100 years (Swales et al. 2002; Ellis et al. 2004; Swales et al. 2009), suggest that increases in tidal-flat elevation have historically outpaced relative sea-level rise in northern New Zealand. Models of climate-change related impacts on mangrove distribution have been developed for Auckland east coast estuaries, based on historical trends, and Intergovernmental Panel on Climate Change (IPCC) mid-range and upper-range projections of sea-level rise for two future periods (2050s and 2090s), and using historical sedimentation rates of 3.8 mm y^{-1} (Swales et al. 2009; Lundquist et al. 2011). Scenarios based on current trends in sea-level rise predict an increase of 8 % in potential mangrove habitat by the 2050s and 14 % by the 2090s (Swales et al. 2009). However, scenarios with higher rates of sea-level rise (IPCC mid-range and upper-range projections) predict reductions in mangrove habitat of 10.2 % and 27 % by the 2090s for IPCC mid-range and upper-range projections, respectively. Mangroves are predicted to be displaced from intertidal flats in the main body of estuaries, but remain in tidal creek refuges (Lundquist et al. 2011).

4.3 Ecosystem Services Provided by Mangroves in New Zealand

4.3.1 Trophic Role

Temperate mangroves, such as those in New Zealand, comprise < 2 % of the world's mangrove areas, and consequently have received very little attention (Twilley et al. 1992; reviewed in Morrisey et al. 2010). Generally, temperate mangroves have smaller tree heights and are less productive than their tropical counterparts (Saenger & Snedaker 1993), as well as holding lower associated diversity of both marine and terrestrial species (Alfaro 2006; Morrisey et al. 2010). Their trophic role has been investigated, suggesting that productivity, primarily in terms of seasonal litterfall, is at the lower range of productivity estimates of tropical species (range of NZ mangrove productivity 1.3–8.10 t ha^{-1} yr^{-1}; tropical range 3.74–18.7 t ha^{-1} yr^{-1}; Morrisey et al. 2010; Gladstone-Gallagher 2012). In comparison to other temperate estuarine vegetation types, the range of productivity of New Zealand mangroves shows comparable productivity to other coastal vegetation including saltmarsh and seagrass meadows (Morrisey et al. 2007; Turner 2007). It is possible that trophic contributions by mangroves may be able to compensate for decreases in seagrass and saltmarsh where these have occurred (e.g., Park 2004; Gladstone-Gallagher et al., in press).

4.3.2 Capacity of Mangroves to Mitigate Coastal Hazards

A number of factors affect the ability of mangrove forests to mitigate storm erosion and flooding. These include forest width, degree of sediment compaction, tree density, and tree morphology (height, root structure, ratio of above- to below-ground biomass) (Alongi 2008). Variation in resistance to bed erosion among mangrove species largely reflects differences in root structure. For example, species like *Avicennia* spp. with predominantly shallow, horizontal root networks (< 0.5 m in depth) are more likely to become detached by erosion during storms (Othman 1994; Swales et al. 2007b). Loss of trees during storms (and tsunamis) is most likely on the seaward edge of mangroves, and wider mangrove stands are predicted to be more resilient to storm-related damage.

Vegetation-induced drag in mangrove forests attenuates the height of wind waves and swell (Brinkman et al. 1997), with a 50–70 % reduction in wave energy occurring within 20 m of the edge of a mixed *Avicennia* and *Rhizophora* forest (Phuoc and Massel 2006). Penetration of storm waves into shallow-water coastal and estuarine environments is enhanced by elevated sea levels (storm tides) related to meteorological factors and by spring tides, and mangroves may provide less effective protection in this situation (Massel et al. 1999; Alongi 2008).

5 Conclusion and Future Perspectives on Mangrove Management in New Zealand

Mangroves are an indigenous aspect of New Zealand estuaries, and before we can adopt effective estuarine management strategies, we must first address the causal factors around mangrove expansion (i.e., catchment management), as well as the services that mangroves provide to temperate estuaries. Accumulation of monitoring data from mangrove clearings has promoted recognition that mangrove removals are not guaranteed to result in the objectives of the mangrove removal consent, i.e., to 'turn the clock back' to pristine sandflat habitats. Councils are also becoming aware of the long-term costs of maintaining mangrove clearings, and require cost-effective guidelines for seedling removal strategies that minimise adverse impacts. Compiling comprehensive scientific evidence of success of different mangrove management methods will enable better guidance for community groups and councils on methods that are cost-effective, and that both minimise adverse impacts such as nutrient release and macroalgal blooms, and maximise the potential for sediment erosion and decomposition of mangrove vegetative biomass and a return of the area to a more desired state.

The relationship between people and mangroves in New Zealand is complex and driven by emotional responses to historical changes in estuarine habitats that are a symptom of catchment land-use practices. In managing mangrove expansion, a balance must be found between maintaining people's values and maintaining any valuable ecosystem services that mangroves provide. This requires careful and informed management of mangroves to ensure that firstly what people desire from mangrove removal is achieved and secondly that mangrove areas of ecological and functional importance are also maintained. In the absence of adequate information we run the risk of ad-hoc, unsuccessful removals doing more harm than good, and threatening ecosystem services and biodiversity in New Zealand estuaries.

Acknowledgments Support toward preparation of this manuscript was provided to CL by the New Zealand Ministry for Business, Innovation, and Employment Contract #CO1X1002, Rehabilitation of Aquatic Ecosystems. We thank Auckland Council for supporting the initial review of temperate mangroves in New Zealand. We thank Waikato Regional Council for providing map information for Whangamata Harbour.

References

Albright LJ (1976) *In situ* degradation of mangrove tissues (note). New Zeal J Mar Fresh Res 10(2):385–389

Alfaro AC (2006) Benthic macro-invertebrate community composition within a mangrove/seagrass estuary in northern New Zealand. Estuar Coast Shelf Sci 66:97–110

Alfaro AC (2010) Effects of mangrove removal on benthic communities and sediment characteristics at Mangawhai Harbour, Northern New Zealand. ICES J Mar Sci 67:1087–1104

Alongi DM (2008) Mangrove forests: resilience, protection from tsunamis, and responses to global climate change. Estuar Coast Shelf Sci 76:1–13

Balke T, Webb EL, van den Elzen E, Galli D, Herman PMJ, Bouma TJ (2013) Seedling establishment in a dynamic sedimentary environment: a conceptual framework using mangroves. J App Ecol 50:740–747. doi:10.1111/1365-2664.12067

Beaglehole JC (1968) The journals of Captain Cook on his voyages of discovery. The voyage of the Endeavour 1768–1771. Cambridge University Press, Cambridge

Beard CM (2006) Physiological constraints on the latitudinal distribution of the mangrove *Avicennia marina* (Forsk.) Vierh. subsp. *australasica* (Walp.) J. Everett (Avicenniaceae) in New Zealand. Unpublished PhD thesis, University of Waikato, p 229

Beca C (2007) Whangamata Mangrove removal options report. Environment Waikato Internal Series 2007/15, Hamilton, New Zealand, p 38

Borja A, Dauer DM, Elliott M, Simenstad CA (2010) Medium- and long-term recovery of estuarine and coastal ecosystems: patterns, rates and restoration effectiveness. Estuar Coast 33:1249–1260

Brinkman RM, Massel SR, Ridd PV, Furukawa K (1997) Surface wave attenuation in mangrove forests. Centre for Advanced Engineering, University of Canterbury, Christchurch

Cahoon DR, Hensel PF, Spencer T, Reed DJL, McKee K, Saintilan N (2006) Coastal wetland vulnerability to relative sea-level rise: wetland elevation trends and process controls. In: Verhoeven JTA, Beltman B, Bobbink R, Whigham DF (eds) Wetland and natural resource management. Springer-Verlag, Berlin

Chapman VJ (1976) Mangrove vegetation. J. Cramer Verlag, Germany, p 447

Chapman VJ, Ronaldson JW (1958) The mangrove and saltmarsh flats of the Auckland isthmus. New Zealand Department of Scientific and Industrial Research, Bulletin 125

Clarke PJ (1993) Dispersal of grey mangrove (*Avicennia marina*) propagules in southeastern Australia. Aquat Bot 45:195–204

Clarke LD, Hannon NJ (1970) The mangrove swamp and salt marsh communities of the Sydney district. III. Plant growth in relation to salinity and waterlogging. J Ecol 58:351–369

Clarke PJ, Myerscough PJ (1993) The intertidal distribution of the grey mangrove (*Avicennia marina*) in southeastern Australia; the effects of physical conditions, interspecific competition, and predation on establishment and survival. Aust J Ecol 18:307–315

Coffey BT (2001) Resource Consent 102475: Trial clearance of mangroves at Patiki Place Reserve, Whangamata Harbour. Final monitoring report—January 2001, plus appendices, p 9

Coffey BT (2002) Resource Consent 102475: trial clearance of mangroves Patiki Place Reserve, Whangamata Harbour. Environment Waikato, Hamilton, p 66

Coffey BT (2010) During clearance and immediate post-clearance ecological monitoring to meet condition 6 of Permit 35053 that authorises Papakura District Council to clear mangroves from Pahurehure Inlet No. 2. Report prepared for Papakura District Council by Brian T. Coffey and Associates, Ltd, p 55

Crisp P, Daniel L, Tortell P (1990) Mangroves in New Zealand—trees in the tide. GP Books, New Zealand, p 69

Curran M, Cole M, Allaway WG (1986) Root aeration and respiration in young mangrove plants (*Avicennia marina* (Forsk.) Vierh.). J Exp Bot 37:1225–1233

de Lange WP, de Lange PJ (1994) An appraisal of factors controlling the latitudinal distribution of mangrove (*Avicennia marina* var. *resinifera*) in New Zealand. J Coast Res 10(3):539–548

Dingwall PR (1984) Overcoming problems in the management of New Zealand mangrove forests. In: Teas HG (ed) Physiology and management of mangroves. Dr. W. Junk Publishers, The Hague, pp 97–106

Done T, Jones R (2006) Tropical coastal ecosystems and climate change prediction: global and local risks. In: Phinney JT, Hoegh-Guldberg O, Kleypas J, Skirving W, Strong A (eds) Coral reefs and climate change: science and management. Coastal and Estuarine Studies 61. American Geophysical Union, pp 5–32

Duke NC (1990) Phenological trends with latitude in the mangrove tree *Avicennia marina*. J Ecol 78(1):113–133

Duke NC, Ball MC, Ellison JC (1998) Factors influencing biodiversity and distributional gradients in mangroves. Glob Ecol Biogeog Lett 7(1):27–47

Ellis J, Nicholls P, Craggs R, Hofstra D, Hewitt J (2004) Effects of terrigenous sedimentation on mangrove physiology and associated macrobenthic communities. Mar Ecol-Progr Ser 270:71–82

Ellison JC (1994) Climate change and sea level rise impacts on mangrove ecosystems. IUCN Marine Conservation and Development Report, Gland, Switzerland, pp 11–30

Ellison J (2004) Vulnerability of Fiji's mangroves and associated coral reefs to climate change. World Wildlife Fund, University of Tasmania, Launceston

Ellison JC (2009) Geomorphology and sedimentology of mangroves. In: Perillo GME, Wolanski E, Cahoon DR, Brinson MM (eds) Coastal wetlands—an integrated ecosytem approach. Elsevier, Amsterdam, pp 565–591

Felsing M (2006) Benthic impacts of mangrove removal at Whangamata Harbour—analysis of Nov 05 samples, Internal Memorandum, Environment Waikato, Hamilton, New Zealand, p 10

Field CD (1995) Impact of expected climate change on mangroves. Hydrobiologia 295:75–81

Furukawa K, Wolanski E, Mueller H (1997) Currents and sediment transport in mangrove forests. Estuar Coast Shelf Sci 44(3):301–310

Gillanders BM, Kingsford MJ (2002) Impact of changes in flow of freshwater on estuarine and open coastal habitats and the associated organisms. Oceanogr Mar Biol 40:233–309

Gilman EL, Ellison J, Duke NC, Field C (2008) Threats to mangroves from climate change and adaptation options: a review. Aq Bot 89(2):237–250

Gladstone-Gallagher RV (2012) The production and decay of mangrove (*Avicennia marina* subsp. *australasica*) detritus and its effects on coastal benthic communities. MSc thesis, University of Waikato, New Zealand

Gladstone-Gallagher RV, Lundquist CJ, Carter KR, Pilditch CA (unpublished manuscript). Latitudinal comparisons of mangrove (*Avicennia marina* subsp. *australasica*) productivity and decomposition in two New Zealand estuaries. Prepared for Aq Bot

Gladstone-Gallagher RV, Lundquist CJ, Pilditch CA (in press). Mangrove (*Avicennia marina* subsp. *australasica*) litter production and decomposition in a temperate estuary. Submitted to New Zeal J Marine Fresh

Green MO, Black KP, Amos CL (1997) Control of estuarine sediment dynamics by interactions between currents and waves at several scales. Mar Geol 144:97–116

Green MO, Ellis J, Schwarz A-M, Green N, Lind D, Bluck B (2003) For and against mangrove control. NIWA Information Series, No 31, p 8

Griffiths GA, Glasby GP (1985) Input of river-derived sediment to the New Zealand continental shelf: 1. Mass. Estuar Coast Shelf Sci 21:773–787

Hannah J (2004) An updated analysis of long-term sea level change in New Zealand. Geophys Res Lett 31(3):L03307. doi:03310.01029/02003GL019166

Harty C (2009) Mangrove planning and managemenet in New Zealand and South East Australia—A reflection on approaches. Ocean Coast Manage 52:278–286

Healy T (2002) Muddy coasts of mid-latitude oceanic islands on an active plate margin—New Zealand. In: Healy T, Wang Y, Healy J-A (eds) Muddy coasts of the world: processes, deposits and function. Elsevier, Amsterdam, pp 347–374

Heesch S, Broom JE, Neill K, Farr TJ, Dalen J, Nelson WA (2007) Genetic diversity and possible origins of New Zealand populations of *Ulva*. Biosecurity New Zealand technical paper no. 01
Hovenden MJ, Curran M, Cole MA, Goulter PFE, Skelton NJ, Allaway WG (1995) Ventilation and respiration in roots of one-year-old seedlings of grey mangrove *Avicennia marina* (Forsk) Vierh. Hydrobiologia 295(1–3):23–29
Hume TM (2003) Estuaries and tidal inlets. In: Rouse HL, Goff JR, Nichol S (eds) The New Zealand coast: Te Tai o Aotearoa. Dunmore Press, Palmerstone North, pp 191–213
Hume TM, Snelder T, Weatherhead M, Liefting R (2007) A controlling factor approach to estuary classification. Ocean Coast Manage 50(11–12):905–929
Krauss KW, Allen JA, Cahoon DR (2003) Differential rates of vertical accretion and elevation changes among aerial root types in Micronesian mangrove forests. Estuar Coast Shelf Sci 54:251–259
Krauss KW, Lovelock CE, McKee KL, Lopez-Hoffman L, Ewe SML, Sousa WP (2008) Environmental drivers in mangrove establishment and early development: a review. Aq Bot 89(2):105–127
Lovelock CE, Ellison J (2007) Vulnerability of mangroves and tidal wetlands of the Great Barrier Reef to climate change. In: Johnson JE, Marshall PA (eds) Climate change and the Great Barrier Reef. Great Barrier Reef Marine Park Authority and Australian Greenhouse Office, Townsville, pp 237–270
Lovelock CE, Sorrell B, Hancock N, Hua Q, Swales A (2010) Mangrove forest and soil development on a rapidly accreting shore in New Zealand. Ecosystems 13:437–451. doi:10.007/s10021-010-9329-2
Lundquist CJ, Ramsay D, Bell R, Swales A, Kerr S (2011) Predicted impacts of climate change on New Zealand's biodiversity. Pac Conserv Biol 17:179–191
Lundquist C, Hailes S, Cartner K, Carter K, Gibbs M (2012) Physical and ecological impacts associated with mangrove removals using *in situ* mechanical mulching in Tauranga Harbour. NIWA technical report 137, p 103
Massel SR, Furukawa K, Brinkman RM (1999) Surface wave propagation in mangrove forests. Fluid Dyn Res 24(4):219–249
McLeod E, Salm R (2006) Managing mangroves for resilience to climate change. Gland, Switzerland, IUCN Global Marine Programme Resilience Science Working Paper Series no. 2, p 64
Milliman JD, Syvitski JPM (1992) Geomorphic/tectonic control of sediment discharge to the oceans: the importance of small mountainous rivers. J Geol 100:525–544
Mitsch WJ, Gosselink JG, Anderson CJ, Zhang L (2009) Wetland ecosystems. Wiley, Chichester
Morton JE, Miller MC (1973) The New Zealand sea shore. Collins, London
Morrisey D, Beard C, Morrison M, Craggs R, Lowe M (2007) The New Zealand mangrove: a review of the current state of knowledge. Auckland Regional Council Technical Report, no. 325, Auckland, New Zealand, p 156
Morrisey DJ, Swales A, Dittmann S, Morrison MA, Lovelock CE, Beard CM (2010) The ecology and management of temperate mangroves. Oceanogr Mar Biol 48:43–160
Nichol SL, Augustinus P, Gregory M, Creese R, Horrocks M (2000) Geomorphic and sedimentary evidence of human impact on the New Zealand coastal landscape. Phys Geogr 21:109–132
Nicholls P, Ellis J (2002) Fringing habitats in estuaries: the sediment-mangrove connection. Water & Atmosphere 10(4):24–25
Oñate-Pacalioga JA (2005) Leaf litter production, retention, and decomposition of *Avicennia marina* var. *australasica* at Whangateau Estuary, Northland, New Zealand. MSc thesis, University of Auckland, New Zealand
Othman MA (1994) Value of mangroves in coastal protection. Hydrobiologia 285(1–3):277–282
Park S (2004) Aspects of mangrove distribution and abundance in Tauranga Harbour. Bay of Plenty Regional Council Environmental Publication 2004/16, Hamilton, p 40

Park S (2012) Benthic macrofauna habitat and diversity associated with mangrove removal sites in Tauranga Harbour—year 2 of monitoring. Bay of Plenty Regional Council internal memorandum, File Ref: 4.00198, dated 9 March 2012

Phuoc VLH, Massel SR (2006) Experiments on wave motion and suspended sediment concentration at Nang Hai, CanGio mangrove forest, southern Vietnam. Oceanologia 48:23–40

Pocknall DT (1989) Late Eocene to early Miocene vegetation and climate history of New Zealand. J Roy Soc New Zeal 19:1–18

Pratt SJ, Lundquist CJ, Nelson W, Gemmill CEC (2013) A new record of *Percursaria percursa* (Ulvaceae, Ulvales) on the North Island, New Zealand. New Zeal J Bot 51:71–74

Prosser IP, Rutherford ID, Olley JM, Young WJ, Wallbrink PJ, Moran CJ (2001) Large-scale patterns of erosion and sediment transport in river networks. Mar Fresh Res 52:81–99

Rogers K, Saintilan N, Cahoon D (2005) Surface elevation dynamics in a regenerating mangrove forest at Homebush Bay, Australia. Wetlands Ecol Manage 13(5):587–598

Romero LM, Smith TJ III, Fourqurean JW (2005) Changes in mass and nutrient content of wood during decomposition in a south Florida mangrove forest. J Ecol 93:618–631

Saenger P, Snedaker SC (1993) Pan-tropical trends in mangrove above-ground biomass and litter fall. Oecologia 96:293–299

Saintilan N, Williams RJ (1999) Mangrove transgression into saltmarsh environments in south-east Australia. Global Ecol Biogeog 8(2):117–124

Sakai A, Wardle P (1978) Freezing resistance of New Zealand trees and shrubs. New Zeal J Ecology 1:51–61

Schwarz A-M (2002) The role of nutrients in contributing to mangrove expansion. NIWA Client report #HAM2002–051, Hamilton, New Zealand

Singleton P (2007) Draft Whangamata Harbour plan. Environment Waikato Internal Series 2007/14. p 84

Stokes D (2009) Assessment of physical changes after mangrove removals: Whangamata Harbour 2008. Environment Waikato Technical Report 2009/13. p 11

Stokes DJ, Healy TR, Cooke PJ (2009) Surface elevation changes and sediment characteristics of intertidal surfaces undergoing mangrove expansion and mangrove removal, Waikaraka Estuary, Tauranga Harbour, New Zealand. Int J Ecol Dev 12(W09):88–106

Stokes DJ, Healy TJ, Cooke PJ (2010) Expansion dynamics of mono-specific, temperate mangroves and sedimentation in two embayments of a barrier-enclosed lagoon, Tauranga Harbour, New Zealand. J Coast Res 26(1):113–122

Stuart S, Choat B, Martin K, Holbrook N, Ball M (2007) The role of freezing in setting the latitudinal limits of mangrove forests. New Phytol 173(3):576–583

Sutherland JI (2003) Miocene petrified wood and associated borings and termite faecal pellets from Hukatere Peninsula, Kaipara Harbour, North Auckland, New Zealand. J Roy Soc New Zeal 33:395–414

Swales A, Hume TM, Oldman JW, Green MO (1997) Holocene sedimentation and recent human impacts on infilling in a drowned valley estuary. 13th Australasian coastal and ocean engineering conference and the 6th Australasian ports and harbour conference. Christchurch, New Zealand

Swales A, Williamson RB, Van Dam LF, Stroud MJ, McGlone MS (2002) Reconstruction of urban stormwater contamination of an estuary using catchment history and sediment profile dating. Estuaries 25(1):43–56

Swales A, Bell RG, Ovenden R, Hart C, Horrocks M, Hermansphan N, Smith RK (2007a) Mangrove habitat expansion in the southern Firth of Thames: sedimentation processes and coastal-hazards mitigation. Environment Waikato technical report 2008/13, Hamilton, New Zealand

Swales A, Bentley SJ, Lovelock C, Bell RG (2007b) Sediment processes and mangrove-habitat expansion on a rapidly-prograding muddy coast, New Zealand. Coastal Sediments' 07—Proceedings of 6th International Symposium on Coastal Engineering and Science of Coastal Sediment Processes, Bournemouth University

Swales A, Gorman R, Oldman JW, Altenberger A, Hart C, Bell RG, Claydon L, Wadhwa S, Ovenden R (2009) Potential future changes in mangrove habitat in Auckland's east-coast estuaries. Auckland Council Technical Report 2009/079. Auckland, New Zealand

Thrush SF, Hewitt JE, Norkko A, Cummings VJ, Funnell GA (2003) Macrobenthic recovery processes following catastrophic sedimentation on estuarine sandflats. Ecol Appl 13:1433–1455

Thrush SF, Hewitt JE, Cummings VJ, Ellis JI, Hatton C, Lohrer A, Norkko A (2004) Muddy waters: elevating sediment input to coastal and estuarine habitats. Front Ecol Environ 2:299–306

Thrush SF, Halliday J, Hewitt JE, Lohrer AM (2008) Cumulative degradation in estuaries: The effects of habitat, loss fragmentation and community homogenization on resilience. Ecol Appl 18:12–21

Turner SJ (2007) Growth and productivity of intertidal *Zostera capricorni* in New Zealand estuaries. New Zeal J Mar Fresh Res 41:77–90

Twilley RR, Chen RH, Hargis T (1992) Carbon sinks in mangroves and their implications to carbon budget of tropical coastal ecosystems. Water Air Soil Poll 64:265–288

Walbert K (2002) Investigations in the New Zealand mangrove, *Avicennia marina* var. *australasica*. MSc thesis, University of Waikato

Walling DE (1999) Linking land use, erosion and sediment yields in river basins. Hydrobiologia 410:223–240

Wildlands Consultants (2003) Ecological restoration and enhancement of Waikaraka Estuary, Tauranga Harbour. Wildlands Consultants Ltd., Tauranga, p 12

Woodroffe CD (1982) Litter production and decomposition in the New Zealand mangrove, *Avicennia marina* var. *resinifera*. New Zeal J Mar Fresh Res 16:179–188

Young BM, Harvey LE (1996) A spatial analysis of the relationship between mangrove (*Avicennia marina* var. *australasica*) physiognomy and sediment accretion in the Hauraki Plains, New Zealand. Estuar Coast Shelf Sci 42:231–246

Mangroves are Wetlands, not Forests: Some Implications for Their Management

Brij Gopal

Contents

Abstract Mangroves are tree-dominated ecosystems occupying intertidal areas in the tropics and subtropics. The mangrove communities and ecosystem functions are governed primarily by the interaction between flows of freshwater, sediments and nutrients from the landward side, and tidal flooding from the sea. For about two centuries, mangroves have been exploited and managed as forests, with practically no concern for their hydrology. Studies during the past 50 years have recognised them as wetlands, characterised by hydrology as the major determinant of their structure and function. Mangroves are well recognised for their high biodiversity and contribution to coastal fisheries. Yet, current approaches to their management, resource use, conservation or restoration continue to ignore the role of the freshwater component of their hydrology. This paper stresses upon the need to manage mangroves as wetlands, and the crucial role of, and therefore the necessity to pay attention to freshwater flows, for sustaining their biodiversity and ecosystem services.

B. Gopal (✉)
Centre for Inland Waters in South Asia, 41-B Shiv Shakti Nagar,
Jagatpura Road, Rajasthan, 302017 Jaipur, India
e-mail: brij44@gmail.com

I. Faridah-Hanum et al. (eds.), *Mangrove Ecosystems of Asia*,
DOI 10.1007/978-1-4614-8582-7_20,

1 Mangroves: Forests or Wetlands?

Mangroves have been the subject of numerous studies worldwide and in recent years have attracted greater attention in the context of tsunamis and climate change. There is unanimity over the fact that the system includes trees and shrubs, and that they occur in intertidal areas in the tropics and subtropics. However, there continues to be confusion over their basic nature, or 'what are mangroves?' Whereas some prefer to restrict the use of the term only to woody plants (trees and bushes; FAO 1952), and call the ground ferns and other plants as 'mangrove associates' (Kathiresan and Bingham 2001), others apply it to the entire forest community, often described also as 'coastal woodland' or 'tidal forest'—terms which basically reflect their location (FAO 1994; Melana et al. 2000).

Mangroves however, differ considerably from other forests of the humid tropics in several of their ecosystem attributes. FAO (1994) clearly acknowledged some differences when stating that "mangroves depend on terrestrial and tidal waters for their nourishment, and silt deposits from upland erosion as substrate for support." It also identified the importance of the "substrate that is ever changing and dynamic over time." Studies during the 1960s and 1970s highlighted the distinctive nature of several tropical forests such as those of the Amazon floodplain, many riverine forests, Southeast Asian peat swamps and mangroves which were variously named and classified with other forest types. During the same period, a wide range of habitats, hitherto known as bogs, fens, marshes and swamps, were brought together under the term 'wetlands', which were characterised by prolonged or permanent flooding of their substrates, and a vegetation adapted to such flooding and associated depleted oxygen levels in the root zones (Cowardin et al. 1979). Further studies demonstrated the overriding role of flooding regimes in determining the community structure and ecosystem functions of wetlands (Gopal 1990; Gopal et al. 1990; Keddy 2010). Numerous studies during the past four decades have elaborated upon the wetland characteristics of mangroves, and these are summarised in several recent publications (Alongi 2009; Twilley and Day 2012). Like most other wetlands, mangroves also have, in general, a very low diversity of plants (often with only one or two tree species being dominant), but are quite rich in their faunal and microbial diversity (see Alongi 2009). The tree species present have several morphological, anatomical and physiological adaptations against hypoxic to anoxic conditions in the root zone, as well as to other stress factors (e.g., dynamic substrate and salinity). The primary production in mangroves is comparable to, or more than, that in humid tropical forests. Further, this organic matter produced contributes greatly to the faunal diversity and secondary production of the adjacent open waters (Alongi 2009). Interestingly, despite acknowledging the importance of hydrology to the mangrove community (Kjerfve 1990; Kjerfve et al. 1999), it has received little attention from mangrove managers. In the following pages, I discuss briefly the hydrology of mangroves, particularly those in Asia, and emphasise that the management of mangroves, whether for utilization, conservation or restoration, must be based on specific hydrological considerations, consistent with the objectives.

2 Hydrological Features of Mangroves

Hydrological processes of a region are influenced by climate and geomorphology and, at the same time in a feedback loop, regulate them to a great extent. The interplay of geomorphology, climate and hydrology becomes more complex in coastal environments because of the continuous movement of water from two opposite sides—tides from the sea and freshwater runoff or river discharge from the inland side. In coastal regions, diverse landforms are created by the interactions between rainfall, river discharge, tidal frequency and amplitude, wave power, and inputs of sediments derived from terrestrial erosion (Twilley et al. 1996). Also, coastal biota often act as 'engineers', to modify physiographic features. Thus, a wide range of hydrological regimes are obtained across a hierarchy of spatial and temporal scales, that in conjunction with other major variables such as nutrients, salinity and substrate characteristics, influence the mangrove community and functions.

However, until recently greater emphasis has been laid on the tidal component of the hydrological regimes. On the basis of 'local patterns of tides and terrestrial surface drainage' in North American tropics (where only four species of mangroves occur), Lugo and Snedaker (1974) differentiated five physiognomic types of mangroves. These are: (1) fringing mangroves: along the protected shorelines and islands, influenced by daily tides, (2) riverine mangroves: along rivers and creeks, flooded by daily tides and influenced by freshwater and nutrients from the rivers, (3) mangroves on small islands in shallow bays overwashed by high tides, (4) basin mangroves which are dwarf stands along drainage depressions, forming hammocks on elevated sites, and (5) scrub mangroves comprising of dwarf shrubs along flat coastal fringes. It is widely recognised that the species-rich mangroves of South and Southeast Asia differ greatly from the species-poor neotropical mangroves and this classification is not applicable to them (see Kjerfve 1990).

Thom (1982) and Galloway (1982) considered the role of the substrate and sedimentation, besides the tidal range, in recognising six broad categories of mangroves which covered most of those in Asia. These are:

1. Large deltaic systems (occurring in low tidal range, very fine allochtonous sediments, *e.g.* mangroves of Borneo, Sundarbans)
2. Tidal plains (where alluvial sediments are reworked by the tides, and there is the presence of large mudflats for the growth of mangroves)
3. Composite plains, under the influence of both tidal and alluvial conditions (*e.g.* lagoons formed behind wave-built barriers where mangroves grow)
4. Fringing barriers with lagoons (high wave energy conditions with autochthonous sediments of fine sand and mud, *e.g.* mangroves of the Philippines)
5. Drowned bedrock valleys (*e.g.* mangroves of Northern Vietnam, Eastern Malaysia or Andaman-Nicobar Islands)
6. Coral coasts (mangroves growing at the bottom of coral sand or on platform reefs, *e.g.* the mangroves of Indonesia and Singapore)

The role of river discharges was highlighted in the simpler classification scheme proposed by Woodroffe (1992) who grouped mangroves, on the basis of their hydrology, into three major types: (a) river-dominated, (b) tide-dominated and (c) interior mangroves (less influence of river/or tides). The river dominated mangroves are characterized by strong outwelling whereas the tide dominated mangroves experience bidirectional flux of water, energy and materials. Mangroves in the interior typically form sinks for sediment and nutrients.

Within tropical Asia, a large variability occurs within tidal regimes, river flow regimes and coastal geomorphology and hence, various mangrove sites experience a large variation in hydrological regimes. According to their frequency, tides are categorised as diurnal, semidiurnal (twice daily) and mixed. All three types of tides occur in the Indian Ocean although semidiurnal ones are the most widespread. Semidiurnal tides prevail on the coast along the Bay of Bengal whereas along the Arabian Sea, the tides are mixed. Tidal amplitudes are far more variable. Mauritius experiences a spring tidal range of only 0.5 m whereas the tides in the Arabian Sea reach their highest range at more than 11 m at Bhavanagar in the Gulf of Khambat and up to 7.8 m in the Gulf of Kachchh. Further northwest in Karachi (Pakistan), the tidal range is only 2.3 m. The tidal range is only 1 m at Cochin (southwest coast of India), 1.3 m at Chennai (Madras) on the eastern coast, and only 0.7 m at Colombo (Sri Lanka). Sagar Island, at the head of the Bay of Bengal, has a tidal range of 5.3 m and slightly higher tides are received at Diamond Harbor (5.78 m) and Yangon (Myanmar) (5.6 m). In the southeast Asian region tidal ranges vary similarly from > 1.5 m in Manila, about 2 m in Hong Kong and 2.5 m along Sulawesi coast to 5.8 m in New Guinea.

Freshwater flows into the mangroves come either from surface runoff from precipitation in nearby areas, or through rivers with large catchment areas. Most of the large rivers of South and Southeast Asia (Indus, Ganga, Brahmaputra, Irrawady, Mekong and Yangtse) originate in the Himalayan ranges and are fed by glaciers during the dry season and precipitation runoff during the monsoon season. Their low flows during the dry season and peak flows during the monsoons oscillate between extremes. The large spatial and temporal variability of monsoonal precipitation (which declines from east to west) results in further differences in freshwater flows in their deltas at different times of the year. Similar seasonal variations occur in river discharges in Indonesia, Malaysia and other countries.

Asian rivers, particularly those rising in the Himalayas, are also characterised by their high sediment loads, which have resulted in extensive delta and mangrove formations. The topography of the coastal belt also greatly influences the direction and rate of the flows of water into the system, and also the dispersion of sediments. In the lower part of the Ganga-Brahmaputra delta, an average gradient of only 5–10 cm per km results in meandering creeks and streams, temporary islands and pools, and the influence of flooding tides extends up to 50 km inland. A major part of the Sundarban thus oscillates seasonally between a river-dominated and tide-dominated system. Oag (1939) recognised three distinct seasonal phases in the tidal regime. During the period of the southwest monsoon, freshwater flows totally nullify the effect of the flood tides, leaving the ebb tides to strongly dominate the system. During the northeast monsoon (November to February), the effects of the flood tide are only slightly greater than that of the ebb tide. Later during the dry summer (May and June), before the southwest monsoon, the effect of the flood tides is much stronger than the ebb tides, and the estuary reaches maximum salinity (Chandra and Sagar 2003) .

Interactions between changing sediment loads, freshwater flows and tidal patterns and other factors cause local variations in land forms and hydrology (Dijksma et al. 2011). Sedimentation has resulted in the loss of connect between river Ganga and the major rivers in the eastern part of the Sundarban, the Saptamukhi, Thakuran, Matla, Gosaba and Harinbhanga. These rivers are now largely tidal in nature (Mitra et al. 2009). Similarly, during ebb tides the receding waters cause scouring of the top soil, creating innumerable tiny creeks originating from the centre of the moving islands. The receding waters carry huge volumes of silt deposited along the banks of rivers and creeks during high tides. This increases the height of the river banks, as compared to the interiors of the islands. Lands on the sea faces are both continually denuded by tidal waves, as well as built up by wave action depositing silt back onto the shores (Gopal and Chauhan 2006). Such a dynamic deltaic environment under the dominant influence of freshwater flows and sediments from one side and high tides from the other comprises a mosaic of hydrological conditions which strongly influence the characteristics of the mangroves.

It may be worth mentioning here that groundwater (sub-surface flows) also makes a significant contribution to mangrove hydrology in many areas (e.g. Wolanski and Gardiner 1981; Semeniuk 1983; Drexler and Ewel 2001; Drexler and DeCarlo 2002; Mazda and Ikeda 2006) but has received practically negligible attention.

3 Mangrove Management

Humans have used the mangroves in Asia for more than 1000 years. In the Ganga-Brahmaputra delta, besides using the trees for timber and fuel, mangrove vegetation was cleared to create rice-cum-fish farms. Historically, mangroves were common pool resources which were gradually taken over at different times by the rulers who increasingly controlled their management. During the 14th century in India, the clearing of trees and shrubs for cultivation of rice was actively promoted by the then Turk sultan rulers of Bengal, and followed by Moghul rulers until the area was taken over by the East India Company in the middle 18th century (see Eaton 1991) . Available evidence suggests that the Portuguese learned the traditional Indian technique of rice-fish farming in mangrove areas and during the 14th century and transferred this technology to Angola and Mozambique (Vannucci 1997; Kathiresan and Bingham 2001). In Indonesia and the Philippines the conversion of mangroves for rice and fish dates back to the early 15th century (Hora and Pillay 1962; Primavera 1995). Later during the British rule transformation of the Sundarban into agriculture fields was promoted actively and deliberately (Richards 1990; Richards and Flint 1991). The Sundarban mangrove was declared as a reserve forest by the British in the mid 1800s specifically for resource exploitation. The Commissioner for the Sundarban was charged with the task of "regulating and managing the waterlogged forests and swamps of the lower delta" and "to ensure that private landowners cleared, settled and reclaimed Sundarban forests and swamps for rice cultivation" (Richards 1990). The colonial forest department sought to preserve large areas of the remaining Sundarban tidal forest by giving them legal status as reserved or protected forests which were then intensively managed to provide a sustainable supply of timber and firewood. After all, "the ultimate goal of forest management, economic considerations aside, is to exploit to the fullest the natural energies and resources available for any given site so as to produce maximum carrying capacity for the production of the desired products" (FAO 1994). In several reports, FAO (1984, 1985, 1992) has focused primarily on promotion of utilization of mangroves and their afforestation.

Similar exploitation of mangroves for timber and fuel wood, and conversion to paddy fields and aquaculture farms has occurred throughout Asia over centuries. Expansion of shrimp cultivation and salt pans decimated the large Chokoria Sundarbans in the delta of the Matamuhury River (Bangladesh) to a small patch of few individuals of *Heritiera fomes* (Biswas and Choudhury 2007). Economic factors driven by developmental pressures play a decisive role in the formulation of management strategies, and most South and Southeast Asian countries are among the least developed. However, it must be noted that the recent exploitation and conversion to shrimp farms in Southeast Asia are driven by global demands (Gopal 2005).

3.1 Rehabilitation of Degraded Mangroves

After considerable areas of mangroves were lost, converted and degraded by exploitation and aquaculture (see Kairo et al. 2001; Dijksma et al. 2010), thereby affecting the availability of resources, people turned to rehabilitation through afforestation activities based on silvicultural practices. In Malaysia, the Matang Mangrove forest reserve of the state of Perak, has been systematically managed for fuel wood and poles since 1908 (Chong 2006) . The silviculture system which initially followed a variable rotation age of 20–40 years and maintained some seed trees in the logged-over areas for regeneration, has now been changed to 30 years rotation with clear felling without retaining seed plants. Also, the mangroves in the Klang Islands (Selangor) are managed solely for the production of poles, charcoal, woodchips and fishing stakes, and therefore, *Rhizophora apiculata* and *R. mucronata* are the preferred species for plantation (Chong 2006). Extensive plantation of *Sonneratia apetala* and *Avicennia officinalis* have been undertaken in Bangladesh since 1966 (Saenger and Siddiqi 1993), whereas *Rhizophora apiculata* has been planted over more 1300 km^2 in the Mekong delta in Vietnam (Blasco et al. 2001). In the Philippines, mangrove replantation started as community initiatives during the 1930s and government-sponsored projects were taken up in the 1970s which turned in the 1980s into large-scale international development assistance programs (Primavera and Esteban 2008). In Indonesia, the management of mangroves is regulated by the silvicultural practices in their harvesting and by leasing arrangements for allocating the mangroves (Kusmana 2012). In Java, mangrove rehabilitation by replanting abandoned shrimp ponds has been linked with poverty reduction and livelihood development (http://www.wetlands.org/?TabId=2291). In China, monocultures of *Kandelia obovata* , *Sonneratia caseolaris* and *Rhizophora stylosa* are promoted in mangrove reforestation despite known consequences for potential insect outbreaks (Chen et al. 2009). In India, mangrove plantation was started as 'restoration' activity in Tamil Nadu by the M.S. Swaminathan Research Foundation. After the Asian tsunami, mangrove plantations have been made on a large scale, especially in Gujarat which has the second largest area of mangroves in the country (Vishwanathan 2011).

These afforestations, most often creating monospecific stands of highly salt-tolerant species, cannot be considered as rehabilitation because they fail to restore (or even simulate) the high ecological values of the original forests (Sanyal 1998; De Leon and White 1999; Lewis 2005). According to Walters (2004), mangrove plantations are an efficient alternative to harvesting from unplanted, natural mangroves and their spread may reduce harvesting pressures on existing forests. However, mangrove plantations are very different in their structure and composition from natural forests which are gradually being replaced. Furthermore, plantations are not typically viewed by the planters for their environmental conservation value and are, hence, frequently cut and cleared to make space for alternative uses, especially fish farming and residential settlements.

3.2 *Management for Conservation*

The process of conservation of mangroves for biodiversity protection began only after India's independence and the partition of the Indian Sundarban between India and East Pakistan (now Bangladesh). In India, three wildlife sanctuaries were created (spread over three decades) within the Sundarban (Lothian Island, Sajnakhali Wildlife Sanctuary and Haliday Island). In 1973, an area of 2,585 km^2 was declared as a Tiger Reserve of which the core area of 1,330 km^2 was later designated as a National Park. These protected areas focused on characteristic wildlife such as the tiger, spotted deer, wild boar and rhesus macaque. After Bangladesh became a sovereign state, it created, in 1977, three wildlife sanctuaries on three disjunct deltaic islands in the Sundarban forest division of Khulna district. In 1987, the Sundarban National Park in India, and in 1997, parts of the Sundarban in Bangladesh, were inscribed on the World Heritage list (IUCN 1997). The entire Indian Sundarban area, including reclaimed lands, has also been designated as the Sundarban Biosphere Reserve of which the core zone comprises the national park and the Tiger reserve. Approaches to conservation, however, differ considerably between the two countries (for detailed discussion, see Seidensticker et al. 1991). India has designated Bhitarkanika mangroves on the eastern seacoast as a Ramsar site, while in Bangladesh most of its Sundarbans reserved forest has been designated as a Ramsar site. Some of the

Ramsar sites in Sri Lanka have small patches of mangroves within them whereas almost all of the mangroves in Pakistan are covered by the Indus delta Ramsar site.

China has an extensive network of wetland nature reserves of which several are important mangrove areas and have also been designated under the Ramsar convention. Most of the countries of Southeast Asia have established protected areas comprised of important mangroves. About 20 % of the total mangroves in Southeast Asia have thus been protected (Giesen et al. 2007), although the proportion of protected areas varies greatly among countries. Indonesia has the largest area of protected mangroves followed by Papua-New Guinea, but Cambodia has the largest proportion (49 %) of its mangroves within the protected area network. The proportion of mangroves protected in Indonesia, Papua-New Guinea and Thailand is about 27 %, 25 % and 10 %, respectively. It is interesting to note that Vietnam has promoted extensive regeneration of mangroves after their near total destruction during the war. These mangroves include the 42,630 ha Can Gio nature reserve in the Mekong delta that was declared a UNESCO biosphere reserve in the beginning of 2000. Within Southeast Asia, several important mangroves have also been designated as Ramsar sites.

3.3 Management Problems

The management of natural resources has moved over the decades from exploitation (sustained utilization) to conservation and rehabilitation (or restoration) of degraded ones. We often talk of sustainable management and ecosystem-based management which sustains the composition, structure, functioning and ecosystem services. It requires proper understanding of the ecological interactions and processes operating within an ecosystem, and also the setting up of explicit goals and policies (cf. Christensen et al.1996).

After the extensive loss and degradation of mangroves throughout Asia, some areas are now being protected, apparently for the conservation of biodiversity. Monocultures of mangrove species are also raised in degraded areas in recognition of their productive and protective function. However, these ‘management practices’ ignore the diversity of ecosystem services of mangroves and fail to address the negative changes in habitats due to various reasons, including invasive species, changing hydrology and salinity levels and deteriorating water quality. According to the FAO (1994), “Habitat protection is the ultimate goal of conservation, to which all other approaches are subsidiary.” For conservationists worldwide, mangroves present the great immediate challenge. Technically, mangroves are easier to manage compared to the species rich humid tropical forests. There are typically only a handful of mangrove species, many of which coppice or regenerate freely. However, whereas the terrestrial forester is concerned primarily with managing forests grown on stable and firm ground, in the tidal swamps he has to manage aquatic resources as well. It is the aquatic resources—the freshwater flows, tidal flows and the ecological processes in the aquatic environment—that are grossly ignored in mangrove management.

3.4 Tidal Regimes and Mangrove Species

Hydrologically, the tidal regimes are comprised of, besides the tidal amplitude, frequency, duration and timing, particularly if the tide is experienced in different parts of the intertidal zone. These components are then affected by the local variation in elevation profile due to sedimentation. Various species of mangroves respond differently to different tidal regimes. For example, in the Indian part of the Sundarban, a mangrove stand that experiences total diurnal inundation is dominated by *Avicennia marina* and *A. alba* while *Excoecaria agallocha, Ceriops dacandra* and *Acanthus ilicifolius* dominate at sites that are not completely inundated (Saha and Choudhury 1995). *Nypa fruticans* also seems to prefer sites with low level of tidal inundation (Siddiqi 1995). Van Loon et al. (2007) observed, in Vietnam, that in an area with an irregular tidal regime and/or an irregular elevation profile, the duration of inundation is more important, and the vegetation can be better characterised by the duration per inundation and per day. Experimental and field studies in China have shown that *Bruguiera gymnorrhiza* had lower tolerance to soil flooding than *Kandelia obovata* (Ye et al. 2003), while the optimal growth of the latter species was obtained at 2–4 h flooding per tidal cycle (Chen et al. 2004; 2005). *Sonneratia apetala* has greater tolerance to high tide (Chen et al. 2009).

3.5 Importance of Freshwater Flows

As mentioned earlier, the majority of the Asian mangroves lie in the deltas of major Himalayan rivers which carry enormous sediment loads to the oceans. The monsoonal climate with large spatial and temporal variability adds to the variability of freshwater flows to the mangroves. Thus, unlike other mangroves, the Asian mangroves are mostly river dominated. The greater combined freshwater flows from precipitation, surface runoff and river discharges are directly correlated with higher mangrove species richness, height, and productivity (Saenger and Snedaker 1993). Within India, there is a distinct and prominent correlation between freshwater discharge from the rivers and the mangrove species richness (see Selvam 2003). As the total annual discharge decreases in Krishna and Cauvery rivers, the mangrove species richness declines sharply (and *Avicennia marina* is becoming dominant), despite the fact that the annual rainfall remains similar or slightly higher than in the River Godavari's delta. The effect of higher rainfall seems to be nullified by the prolonged dry season. Similarly, absence of significant river discharges and low rainfall on the western coast of India are reflected in very low species richness, but heavy rainfall produces large freshwater flows in the Andaman-Nicobar Islands, resulting in species richness nearly as high as in the Sundarban. Similar relationships between river discharges and rainfall (duration and total amount) are evident in different islands of Indonesia, Malaysia and the Philippines. Although the importance of freshwater inflows to mangrove forests has been recognized for a long time (see Wolanski and Gardiner 1981), these very flows have been usually overlooked in Asian countries while the tidal hydrodynamics and influences are discussed in detail. According to Duke et al. (1998), although mangrove species differ in their tolerances across a wide range of salinities, none essentially requires saltwater to survive (see also Ball 2002).

The importance of freshwater flows, particularly to river dominated mangroves, is far more significant with respect to other ecosystem services. With reference to the protective function of mangroves against cyclonic storms and tsunamis, the height and density of trees would be an important factor. Many studies on the growth of mangrove species show that high salinities have negative effects on metabolism, growth, productivity and height (see Cintr'on et al. 1978; Naidoo 2010; Feller et al. 2010). In Vietnam, recently Loi (2008) has reported on the hydrology and its effects on the mangrove community structure and functions in the Can Gio biosphere reserve. It is well known that over the past several decades, the mangrove structure in Sundarban is becoming simpler and the average height of the trees is decreasing. As a long-term consequence of decline in freshwater flows and increase in salinity, *Heritiera*is being replaced by *Excoecaria* (Christensen 1984) and *Nypa fruticans* and *Phoenix paludosa*are declining rapidly. It is estimated that in the Bangladesh part of the Sundarban, 0.4 % of the forest area is replaced by dwarf species every year. This also causes a decline in the habitat for birds, monkeys and other tree-dwelling species.

The dependence of coastal fisheries on mangroves has been a major theme of discussion among mangrove researchers (see Baran 1999; Sukardjo 2004; Islam and Wahab 2005; Manson et al. 2005). It is also pointed out that river-dominated mangroves characterized by high nutrient influx and strong out-welling from mangroves, play a significant role in maintaining the fishery production of the adjacent coastal waters (Selvam 2003). Thus, reduction in freshwater flow affects the amount of nutrient exported to the coastal environment and thereby, the fishery production.

Recently, Ewel (2010) discussed the impacts of changes in freshwater inputs, such as those caused by water diversion upstream, to mangrove forests that often "lead at first, to subtle changes in function and eventually to dramatic changes in species composition." She points out that "these changes may not become apparent for years or even decades, but they may have important consequences for coastal food chains and for the socio-economic benefits they extend to indigenous people" (Ewel 2010). Further, impacts arise from the management of freshwaters upstream at different times of the year in relation to periods of rainfall, and consequently, the timing of freshwater flows into the mangroves. The impacts of freshwater flow diversion from river Ganga with reference to the Sundarbans have been a matter of intense discussion and dispute between India and Bangladesh. Unfortunately, neither have systematic studies of the ecosystem structure and functioning being undertaken, nor have the needs of freshwater flows into Sundarban been assessed to date. Similar changes in freshwater flow have also occurred in the Indus river, with consequence for the mangroves.

In the case of river-dominated mangroves of the Asian region, the impacts of human activities extend beyond the diversion of river flows. Various activities also impact the sediment load and nutrients in runoff. While a considerable amount of sediments are trapped behind dams, erosion is also accelerated by several land-based activities. Nutrients and pollutants invariably reach the mangroves, thereby affecting the ecosystem structure and function, even if these changes are not perceptible in the tree community.

Mangroves of some countries in Southeast Asia are not river dominated but are influenced most by freshwater runoff during the monsoonal rainfall as the dry periods are short.

The importance of the freshwater flows and the need to understand the freshwater requirement of the mangrove species used in plantation and rehabilitation also cannot be ignored. Selvam (2003) reports that the attempts to reintroduce *Sonneratia apetala*, *Xylocarpus granatum* and *Bruguiera gymnorhiza* in Pichavaram and Muthupet mangroves failed because of high soil salinity. He clearly emphasises that species with low tolerance to salinity cannot be reintroduced successfully without increasing the freshwater flow. Strangely, the view of the politicians and the water resource managers that 'not a drop of water should go waste to the sea', simply ignores the multiple ecosystem services of the mangroves that can be sustained only by freshwater flows, and are not restricted to wood production alone.

Another usually overlooked or underestimated contribution of freshwater flows is towards the mitigation of sea level rise impacts. The reduction in freshwater flows in the rivers because of withdrawal and diversion (as also due to altered precipitation regimes), is likely to have a synergistic effect on the decline of mangroves. Mangroves of low relief islands in carbonate settings that lack rivers are likely to be the most sensitive to sea-level rise, owing to their sediment-deficit environments. In the absence of sediment transport to river dominated mangroves, the balance between subsidence processes and accretion will be lost, aggravating the impacts of sea level rise. A combination of sustained erosion, subsidence and sea level rise implies that the lower areas and islands will continue to fall below sea level and will disappear with time.

4 Conclusion

The problems of mis-managing mangroves results from our failure to understand the true nature of the system, to recognise hydrology as the key driving variable, and above all to appreciate that Asian mangroves are very different from those of African tropics or the neo-tropics. Researchers, managers and policy makers must come out of the colonial mindset, and must not readily fall prey to outdated, inapplicable concepts and approaches usually developed outside the region. The geological, geomorphic, ecological, biological, and socio-cultural peculiarities of the Asian region, together with the diversity within it, cannot be and should not be ignored in any discussion on the human-nature relationship. Though mangrove ecosystems are receiving increasing attention, we are still far from understanding their dynamics. There has been little effort to synthesise relevant studies for a holistic management and implementation action plans.

The long term and sustainable conservation of mangroves for their multiple ecosystem services requires that they be treated as distinct from other forests. Unlike the evergreen or deciduous forests, mangroves experience a highly dynamic environment and are in a state of continued flux. They are governed by their specific flooding regimes which govern salinity gradients, nutrients and the supply of fine sediments. These primary drivers are directly controlled by the biophysical, climatic and anthropogenic processes in their watershed—the basins of the rivers whose delta they occupy. We should avoid transforming mangroves to plantations, and afforestation programmes that focus on particular species only, while ignoring ecosystem attributes, their relation to offshore impacts on fisheries and organic matter transport, etc., and instead focus on their management as integrated, holistic ecosystems with considerable biotic diversity and diverse ecosystem services.

References

Alongi DM (2009) Paradigm shifts in mangrove biology. In: Perillo GME, Wolanski E, Cahoon DR, Brinson MM (eds) Coastal Wetlands—an integrated ecosystem approach. Elsevier, New York, pp 615–640

Baran E (1999) A review of quantified relationships between mangroves and coastal resources. Mar Biol Cent Res Bull 62:57–64

Alongi DM, Brinkman R (2011) Hydrology and biogeochemistry of mangrove forests. In: Levia DF, Moses DC, Tanaka T (eds) Forest hydrology and biogeochemistry—synthesis of past research and future directions (Ecological Studies 216). Springer, Dordrecht, pp 203–219

Ball MC (2002) Interactive effects of salinity and irradiance on growth: implications for mangrove forest structure along salinity gradients. Trees 16:126–139

Biswas SR, Choudhury JK (2007) Forests and forest management practices in Bangladesh: the question of sustainability. Int Forest Rev 9(2):627–640

Blasco F, Aizpuru M, Gers C (2001) Depletion of the mangroves of continental Asia. Wetlands Ecol Manage 9:245–256

Chandra G, Sagar RL (2003) Fisheries in the Sundarbans: problems and prospects. http://ssrncom/abstract=2084014. Accessed 31 July 2013

Chen LZ, Wang WQ, Lin P (2004) Influence of waterlogging time on the growth of *Kandelia candel* seedlings. Acta Oceanol Sin 23:149–158

Chen LZ, Wang WQ, Lin P (2005) Photosynthetic and physiological responses of *Kandelia candel* L Durce seedlings to duration of tidal immersion in artificial seawater. Environ Exp Bot 54:256–266

Chen LZ, Wang W, Zhang Y, Lin G (2009) Recent progresses in mangrove conservation, restoration and research in China. J Pl Ecol 2:45–54

Chong VC (2006) Sustainable utilization and management of mangrove ecosystems of Malaysia. Aquat Ecosyst Health Manage 9(2):249–260

Christensen B (1984) Ecological aspects of the Sundarbans FAO: TCP/BGD/2309 (Mf) FAO, Rome, p 42

Christensen NL, Bartuska Ann M, Brown JH, Carpenter SR, D'Antonio C, Francis R, Franklin JF, MacMahon JA, Noss RF, Parsons DJ, Peterson CH, Turner MG, Woodmansee RG (1996) The report of the ecological society of America committee on the scientific basis for ecosystem management. Ecol Appl 6:665–691

Cintr'on G, Lugo AE, Pool DJ, Morris G (1978) Mangroves of arid environments in Puerto Rico and adjacent islands. Biotropica 10:110–121

Cowardin LM, Carter V, Golet PC, LaRoe ET (1979) Classification of wetlands and deepwater habitats of the United States. United States Fish and Wildlife Service, Washington, DC

DeLeon TOD, White AT (1999) Mangrove rehabilitation in the Philippines. In: Streever W (ed) An international perspective on wetland rehabilitation. Kluwer Academic Publishers, The Netherlands

Dijksma R, Van Loon AF, TeBrake B, Van Huijgevoort MHJ (2011) Mangrove rehabilitation under climate change: towards a hydrological tool for quantifying the effect of expected sea level rise on mangroves. (paper to Delta 2011 Conference). http://webdocs.alterra.wur.nl/internet/corporate/nieuws/abstract_Dijksma_delta2011.pdf. Accessed 31 July 2013

Dijksma R, Van Loon AF, Van Mensvoort MEF, Van Huijgevoort MHJ, TeBrake B (2010) An extended hydrological classification for mangrove rehabilitation projects; a case study in Vietnam. In: Suan-Pheng K, Ismail AM, Hoanh CT, Szuster BW (eds) Tropical deltas and coastal zones; food production, communities and environment at the land-water interface. CABI, Wallingford, pp. 384-397

Drexler JZ, DeCarlo E (2002) Source water partitioning as a means of characterizing hydrologic function in mangroves. Wetlands Ecol Manage 10:103–113

Drexler JZ, Ewel KC (2001) Effects of the 1997–1998 ENSO-related drought on hydrology and salinity in a Micronesian wetland complex. Estuaries 24:347–356

Duke NC, Ball MC, Ellison JC (1998) Factors influencing biodiversity and distributional gradients in mangroves. Global Ecol Biogeogr 7:27–47

Eaton R (1991) Human settlement and colonization in the Sundarbans, 1200–1750. In: Seidensticker J, Kurin R, Townsend AK (eds) The commons in South Asia: societal pressures and environmental integrity in the Sundarbans. The International Center, Smithsonian Institution, Washington, DC

Ewel KC (2010) Appreciating tropical coastal wetlands from a landscape perspective. Front Ecol Environ 8(1):20–26

FAO (1952) Tropical silviculture, vol 1. FAO, Rome

FAO (1984) Mangrove forests in Asia-Pacific region. Special study on forest management, afforestation and utilization of forest resources in the developing regions. GCP/RAS/106/JPN. FAO, Rome, p 38

FAO (1985) Mangrove management in Thailand, Malaysia and Indonesia. FAO environment paper 4, FAO, Rome

FAO (1992) Mangrove for charcoal. A vanishing woodfuel resource system (The case of Yeesarn, upper gulf of Thailand) FAO/GCP/RAS/131NET Field document No. 30, FAO, Bangkok.

FAO (1994) Mangrove forest management guidelines. Forestry Paper 117, FAO Forestry Department, Rome

Feller IC, Lovelock CE, Berger U, McKee KL, Joye SB, Ball MC (2010) Biocomplexity in mangrove ecosystems. Annu Rev Mar Sci 2:395–417

Galloway RW (1982) Distribution and physiographic communities of Australian mangroves. In: Clough BF (ed) Mangrove ecosystems in Australia. Australian National University Press, Canberra, pp 31–54

Giesen W, Wulffraat S, Zieren M, Scholten L (2007) Mangrove guidebook for Southeast Asia. FAO and wetlands international, FAO regional office for Asia and the Pacific, Bangkok, Thailand

Gopal B, Kvet J, Löffler H, Masing V, Patten BC (1990) Definition and classification. In: Patten BC, Jorgensen SE, Dumont HJ, Gopal B, Koryavov P, Kvet J, Löffler H, Sverizhev Y, Tundisi JG (eds) Wetlands and shallow continental water bodies. Natural and Human Relationships, vol 1. SPB Academic Publishing, The Hague, pp 9–16

Gopal B (with a contribution from Masing V) (1990) Biology and ecology. In: Patten BC, Jorgensen SE, Dumont HJ, Gopal B, Koryavov P, Kvet J, Löffler H, Sverizhev Y, Tundisi JG (eds) Wetlands and shallow continental water bodies. Natural and Human Relationships, vol 1. SPB Academic Publishing, The Hague, pp 91–239

Gopal B (2005) Does inland aquatic biodiversity have a future in Asian developing countries? Hydrobiologia 542:69–75

Gopal B, Chauhan M (2006) Biodiversity and its conservation in the Sundarban mangrove. Ecosystem Aquat Sci 68:338–354

Hora SL, Pillay TVR (1962) Handbook of fish culture in Indo-Pacific regions. FAO Biol Tech Pap 14:32

Islam MS, Wahab MA (2005) A review of the present status and management of mangrove wetland habitat resources in Bangladesh with emphasis on mangrove fisheries and aquaculture. Hydrobiologia 542:165–190

IUCN (1997) Sundarban wildlife sanctuaries (Bangladesh): World heritage nomination—IUCN technical evaluation, p 12. http://whc.unesco.org/archive/advisory_body_evaluation/798.pdf. Accessed 31 July 2013

Kairo JG, Dahdouh-Guebas F, Bosire JO, Koedam N (2001) Restoration and management of mangrove systems: a lesson for and from the East African region. S Afr J Bot Pretoria 67(3):383–389

Kathiresan K, Bingham BL (2001) Biology of mangroves and mangrove ecosystems. Adv Mar Biol 40:81–251

Keddy PA (2010) Wetland ecology: principles and conservation, 2nd edn. Cambridge University Press, Cambridge

Kjerfve B (1990) Manual for investigation of hydrological processes in mangrove ecosystems. UNESCO/UNDP. New Delhi, India, p 79

Kjerfve B, Drude de Lacerda L, Rezende CE, Coelho-Ovalle AR (1999) Hydrological and hydrogeochemical variations in mangrove ecosystems. In: Yáez–Arancibia A, Lara–Domínquez AL (eds) Ecosistemas de Manglar en América. Tropical. Instituto de Ecología A.C., México, UICN/ORMA, Costa Rica, NOAA/NMFS, Silver Spring, MD, USA, pp 71–82

Kusmana C (2012) Management of mangrove ecosystem in Indonesia. Paper presented in workshop on Mangrove re-plantation and coastal ecosystem rehabilitation, Faculty of forestry, Gadjah Mada University, Yogyakarta (7 February)

Lewis RR (2005) Ecological engineering for successful management and restoration of mangrove forests. Ecol Engg 24:403–418

Loi Tan L (2008) Effect of hydrology on mangrove ecosystems: effect of hydrology on the structure and function of mangrove ecosystems in the Can Gio mangrove biosphere reserve, Vietnam. PhD thesis, Louisiana State university, Baton Rouge, USA, pp 232

Lugo AE, Snedaker SC (1974) The ecology of mangroves. Annu Rev Ecol Syst 5:39–64

Manson FJ, Loneragan NR, Skilleter GA, Phinn SR (2005) An evaluation of the evidence for linkages between mangroves and fisheries: a synthesis of the literature and identification of research directions. Oceanography and Marine Biology: an Annual Review 43:485–515

Mazda Y, Ikeda Y (2006) Behavior of the groundwater in a riverine-type mangrove forest. Wetlands Ecol Manage 14:477–88

Melana DM, Atchue J III, Yao CE, Edwards R, Melana EE, Gonzales HI (2000) Mangrove management handbook. Department of Environment and Natural Resources, Manila, Philippines, pp 96

Mitra A, Banerjee K, Sengupta K, Gangopadhyay A (2009) Pulse of climate change in Indian Sundarbans: a myth or reality. Natl Acad Sci Lett 32(1 & 2):1–7

Naidoo G (2010) Ecophysiological differences between fringe and dwarf *Avicennia marina* mangroves. Trees 24:667–673

Oag TM (1939) Report on the River Hooghly and its headwaters. Commissioner for the Port of Calcutta. Calcutta, India

Primavera JH (1995) Mangroves and brackish water pond culture in the Philippines. Hydrobiologia 295:303–309

Primavera JH, Esteban JMA (2008) A review of mangrove rehabilitation in the Philippines: successes, failures and future prospects. Wetland Ecol Manage 16:345–358

Richards JF (1990) Agricultural impacts in tropical wetlands: rice paddies for mangroves in South and Southeast Asia. In: Williams M (ed) Wetlands: a threatened landscape. Basil Blackwell, Oxford, pp 217–233

Richards JF, Flint EP (1991) Long-term transformations in the Sundarbans wetlands forests of Bengal. In: Seidensticker J, Kurin R, Townsend AK (eds) The commons in South Asia: societal pressures and environmental integrity in the Sundarbans. The international center, Smithsonian Institution, Washington, DC

Saenger P, Siddiqi NA (1993) Land from the sea: The mangrove afforestation program of Bangladesh. Ocean Coastal Manage 20:23–39

Saenger P, Snedaker SC (1993) Pantropical trends in mangrove above-ground biomass and annual litterfall. Oecologia 96:293–299

Saha S, Choudhury A (1995) Vegetation analysis of restored and natural mangrove forest in Sagar Island, Sunderbans, east coast of India. Indian J Mar Sci 24:133–136

Siddiqi NA (1995) Site suitability for raising Nypa fruticans plantations in the Sunderban mangroves. J Trop For Sci 7(3):405–411

Sanyal P (1998) Rehabilitation of degraded mangrove forests of the Sundarbans of India. In: Program of the international workshop on rehabilitation of degraded coastal systems, Phuket Marine Biological Center, Phuket, p 25

Seidensticker J, Kurin R, Townsend AK (eds) (1991) The commons in South Asia: societal pressures and environmental integrity in the Sundarbans of Bangladesh. Proceedings from a workshop held November 20–21, 1987, Washington, DC, The International Center Smithsonian Institution, Washington, DC

Selvam V (2003) Environmental classification of mangrove wetlands of India. Curr Sci 84:757–765

Semeniuk V (1983) Mangrove distribution in northwestern Australia in relationship to regional and local freshwater seepage. Vegetatio 53:11–31

Sukardjo S (2004) Fisheries associated with mangrove ecosystem in Indonesia: a view from a mangrove ecologist. Biotropia 23:13–39

Thom BG (1982) Mangrove ecology—a geomorphological perspective. In: Mangrove ecosystem in Australia. National Mangrove Workshop, ANU Press, Canberra, pp 3–17

Twilley RR, Day JW Jr (2012) Mangrove wetlands. In: Day JW, Crump BC, Kemp WM, Yáez-Arancibia A (eds) Estuarine ecology. Wiley, New York, pp 165–202

Twilley RR, Snedaker SC, Yaez-Arancibia A, Medina E (1996) Biodiversity and ecosystem processes in tropical estuaries: perspectives from mangrove ecosystems. In: Mooney H, Cushman H, Medina E (eds) Biodiversity and ecosystem functions: a global perspective. Wiley, New York, pp 327–370

van Loon AF, Dijksma R, van Mensvoort MEF(2007) Hydrological classification in mangrove areas: a case study in Can Gio, Vietnam. Aquat Bot 87:80–82

Vannucci M (1997) Supporting appropriate mangrove management. International News Letter of Coastal Management-Intercoast Network, Special edition 1, pp 1–3

Vishwanathan PK (2011) Mangroves and their multifunctionality: an analysis of the impacts of community based mangrove restoration in Gujarat. Paper presented at Indian society for ecological economics sixth biennial conference “Nature, economy and society: understanding the linkages” Centre for Economic and Social Studies (CESS), Hyderabad, October 20–22, 2011 (http://www.ecoinsee.org/6bconf/Theme%20A/AB.4.4%20Vishwanathan.pdf)

Walters BB (2004) Local management of mangrove forests in the Philippines: successful conservation or efficient resource exploitation? Human Ecol 32:177–195

Wolanski E, Gardiner R (1981) Flushing of salt from mangrove swamps. Austral J Mar Freshwat Res 32:681–683

Woodroffe CD (1992) Mangrove sediments and geomorphology. In: Robertson AI, Alongi DM (eds) American Geophysical Union, Washington, DC, USA, pp 7–41

Ye Y, Tam NFY, Wong YS, et al (2003) Growth and physiological responses of two mangrove species (*Bruguiera gymnorrhiza* and *Kandelia candel*) to waterlogging. Environ Exp Bot 49:209–221

Index

I. Faridah-Hanum et al. (eds.), *Mangrove Ecosystems of Asia*,

DOI 10.1007/978-1-4614-8582-7,

Printed by Printforce, the Netherlands